Study Guide and Solutions Manual to Accompany
Organic Chemistry

A. DAVID BAKER
QUEENS COLLEGE

ROBERT ENGEL
QUEENS COLLEGE

West Publishing Company
St. Paul   New York   Los Angeles   San Francisco

WEST'S COMMITMENT TO THE ENVIRONMENT
In 1906, West Publishing Company began recycling materials left over from the production of books. This began a tradition of efficient and responsible use of resources. Today, up to 95% of our legal books and 70% of our college texts are printed on recycled, acid-free stock. West also recycles nearly 22 million pounds of scrap paper annually—the equivalent of 181,717 trees. Since the 1960s, West has devised ways to capture and recycle waste inks, solvents, oils, and vapors created in the printing process. We also recycle plastics of all kinds, wood, glass, corrugated cardboard, and batteries, and have eliminated the use of styrofoam book packaging. We at West are proud of the longevity and the scope of our commitment to our environment.

Production, Prepress, Printing and Binding by West Publishing Company.

COPYRIGHT © 1992 by   WEST PUBLISHING CO.
                     610 Opperman Drive
                     P.O. Box 64526
                     St. Paul, MN 55164–0526

All rights reserved
Printed in the United States of America
99 98 97 96 95 94 93 92   8 7 6 5 4 3 2 1 0

ISBN 0–314–00490–4

TABLE OF CONTENTS

Introduction	1
Chapter 1 - Bonding in Organic Compounds	4
Chapter 2 - Introduction to Organic Compounds: Functional Groups, Nomenclature and Representations of Structure	15
Chapter 3 - Intermolecular and Acid-Base Interactions	23
Chapter 4 - Alkanes and Cycloalkanes. I. An Introduction to Structure and Reactions	30
Chapter 5 - Methanol and Ethanol	38
Practice Examination One	45
Chapter 6 - Alcohols	47
Chapter 7 - Types of Bond Cleavage - Carbocations and Radicals as Reaction Intermediates	56
Chapter 8 - Stereochemical Principles	63
Chapter 9 - Carbon-Carbon Doubly Bonded Systems. I. Structure, Nomenclature, and Preparation	72
Chapter 10 - Carbon-Carbon Doubly Bonded Systems. II. Reactions of Alkenes	79
Practice Examination Two	91
Chapter 11 - Carbon-Halogen and Carbon-Metal Bonds: Two Extremes of Polarity	93
Chapter 12 - Substitution and Elimination Reactions of Haloalkanes	100
Chapter 13 - Alkanes and Cycloalkanes. II. Reaction Mechanisms and Conformational Analysis	111
Chapter 14 - Ethers and Epoxides	120
Chapter 15 - Alkadienes and Alkynes	129
Practice Examination Three	138
Chapter 16 - Molecular Orbital Concepts in Organic Chemistry: Conjugated π Systems and Aromaticity	140
Chapter 17 - Physical Methods of Structural Elucidation: Infrared and Nuclear Magnetic Resonance Spectrometry	149
Chapter 18 - Reactions of Benzene and Its Derivatives	157
Chapter 19 - Aldehydes and Ketones: Preparation, Properties and Nucleophilic Addition Reactions	173
Chapter 20 - Amines and Related Compounds	186

Practice Examination Four	197
Chapter 21 - Carboxylic Acids	199
Chapter 22 - Derivatives of Carboxylic Acids	210
Chapter 23 - Enamines, Enolates, and α,β-Unsaturated Carbonyl Compounds	221
Chapter 24 - Carbohydrates	236
Chapter 25 - Ultraviolet/Visible and Mass Spectrometry	250
Practice Examination Five	255
Chapter 26 - Amino Acids, Peptides, and Proteins	257
Chapter 27 - Heterocyclic Compounds	265
Chapter 28 - Nucleosides, Nucleotides, and Nucleic Acids	274
Chapter 29 - Synthetic Polymers	280
Chapter 30 - Molecular Orbitals in Concerted Reactions	285
Practice Examination Six	291
Answers to Practice Examinations	294

ACKNOWLEDGMENTS

We wish to recognize the encouraging efforts of our Acquisitions Editor, Ron Pullins, and our Developmental Editor, Denise Bayko. Also, we should note the patience of our colleagues and families who have put up with our calls of "I need a sabbatical!"

INTRODUCTION

This *Study Guide and Solutions Manual* is a supplement to the text *Organic Chemistry* by Baker and Engel. Its purpose is to help you to learn organic chemistry more effectively. A course in organic chemistry contains a great deal of material, as you have probably heard, and unless you study effectively from day one of the course you will soon be in trouble. Fortunately, however, organic chemistry is a very logical discipline. Once you understand the underlying principles you will find that you can make connections between seemingly unrelated topics. This will make the learning process much less formidable than you might think at the outset. If you find that you are not making connections between topics as the course progresses, something is wrong with your study habits. The best indication that you are making the right kind of progress is consistent success at problem solving. It is imperative that you know that you are on the right track. This manual and the text itself provide hints for approaching problems, and give other advice and study aids that should smooth your passage through the course. We wish you the greatest success, and trust that you will enjoy the intellectual challenges you will face in learning about a subject that is of such fundamental importance to the well-being of modern society.

At times in the beginnings of your study of organic chemistry you may feel that you are being asked to learn a large volume of material without being able to be particularly creative in the endeavour. In a sense, this is true, but only temporarily. In order to generate new and original ideas in any discipline you must first have a grasp of that which has gone before. While typical college students have significant experience in other disciplines to allow them to be creative immediately in those disciplines, this is generally not true for organic chemistry. You must first develop a base of knowledge of that which has already been done. However, with diligence in your efforts, you should begin to develop the background and approach to attack quickly new problems in new ways.

Before You Start

In the course ahead you will need to master many new concepts, much new vocabulary, and become proficient at depicting the structures of organic molecules, naming them correctly, and understanding and even predicting their reactions. You should be aware from the very beginning that this will require a serious commitment on your part. Even so, we hope that you will not view the study of organic chemistry as drudgery, but rather as fun in the solution of new types of problems, and that you will enjoy the sense of satisfaction that comes with the mastery of new areas of knowledge.

Your effort in organic chemistry will be divided between formal course work and individual study. The formal course work will consist of lectures, laboratory, and possibly a problem solving or recitation class. Your lecturer and/or laboratory instructor will also hold office hours so that you can ask questions about any items that are giving you difficulties. At home or in the library you will need to study your notes about the lecture, your course text, and this manual. Above all, you will need to solve problems. If you work diligently at all these activities you will meet the main ideas of organic chemistry in a number of different settings, and thus will be reinforcing constantly your knowledge. The following points may help you to get the maximum benefit from each aspect of the course.

The Lecture

A good lecturer will be able to focus your attention on the most fundamental aspects of the course in a clear and organized fashion. Therefore, the lectures constitute an extremely important part of the course. You should go to the lecture sessions alert and ready to think in depth about the material the lecturer is presenting. You should be prepared with your reading of the text prior to the lecture. If not, much of the lecture will make little sense and your attention will wander. A course in *Organic Chemistry* is quite unlike a course in *Introductory Chemistry*. In *Introductory Chemistry* the nature of the topics presented generally changes with little obvious connection between them; your understanding of a new topic may not be hampered by your lack of understanding of an earlier topic. We repeat, *a course in Organic Chemistry is not like a course in Introductory Chemistry*. Ideas and strategies, once introduced, are used again and again. Being properly prepared for lecture by reading the text, you should *not* resort to mindless writing of every word the lecturer says or writes. Rather, follow the trend of the lecture material, noting those points of particular emphasis or new material not present in the text. Make sure that your notes are legibile and that you have copied structures correctly. Your lecturer may use colored chalk or pens to emphasize certain aspects of the material presented; be prepared with colored pens or pencils to reproduce such art. Do not be reluctant to ask questions if you do not understand what a lecturer is saying.

It is a good idea to rewrite your lecture notes while the material is still fresh in your mind, generally within a day of lecture. Rewriting (*not* photocoping) helps you to learn the material and organize your thinking. Is anything puzzling to you? If so, check your text. Usually this will allow you answer your questions and modify your notes accordingly. If you still have difficulties with the lecture notes, discuss the puzzling aspects with your fellow students and instructors.

The Text

Although we tried to make the text readable and student-friendly, it is certainly not a book that can be read like a novel. Nor should you attempt to do so. You cannot expect to master the material in a chapter by reading it once from beginning to end. Even if you think you understand everything you read, you will probably not immediately notice all the connections between the different topics in a chapter. Therefore, when you read a chapter, stop at the end of each section to think about what you have read. After the lecture, compare what the text told you with what the lecturer said. Is it clear they were talking about the same thing? Think critically. If something seems strange, you need to be concerned about it. Don't postpone the worry...it will be too late! If necessary, ask your lecturer to clarify your difficulty. As lecturers, we can usually predict which students will do well in the course. Such students will ask questions of the type, "In the text it says this, but in lecture you seemed to say something else...", or, "Why does this reaction involve carbocations while this seemingly similar reaction does not?" These students have thought about the material in some depth to spot apparent contradictions, and are obviously using good study methods. On the other hand we also see students who attempt to memorize everything without understanding any of it. These students spend a great amount of time studying, but very inefficiently. Your accomplishments in organic chemistry will be severely restricted if all you can do is repeat that which you have memorized; our objective is not to produce human text books. Rather, accomplishment in organic chemistry requires application of your knowledge to problem solving, at first relatively simple problems, and ultimately to those problems that no one has ever solved before. Accomplishment in organic chemistry requires you to integrate the material and make sensible predictions based on what you know. Therefore, when you use the text, read a section or two and think about the material carefully. When you review your lecture notes, attempt to integrate them with your text material. Further, as you study the text, work the in-chapter problems.

When you reach the end of a chapter you will be ready to review what you have learned. This study guide will help you to focus on the most important topics from each chapter, and will give you review problems to practice. When you are reasonably confident in your ability to handle the material presented in a chapter, you are ready to attempt the end-of-chapter problems. These problems vary greatly in their difficulty. Some are simple exercises designed to reinforce the understanding of key concepts, reactions or mechanisms. Others require that you bring together several ideas from the same chapter or from earlier chapters. We will have more to say about problem solving in the next section.

A final word about study away from the classroom. *Get yourself a molecular model kit*. It is absolutely essential to have access to such a kit for understanding of some of the more subtle aspects of organic chemistry, expecially when the three-dimensional structure of a molecule plays a central role in the discussion.

Problem Solving

We will begin by discussing the *wrong* way to do problems. You look at a problem, make some progress with it, and then get stuck. After a few moments of thought, you check the answer in this study guide. Immediately you recognize that you had simply not recognized a key point that you felt you really knew and understood quite well. Perhaps this point came up in an unexpected way, or maybe a structure was drawn in an unfamiliar manner, and you set off down the wrong track. You think to yourself, "Sure....I really do undestand this problem. It's actually quite straightforward." You then go on to the next problem. This kind of approach and analysis can be dangerous as you can easily decieve yourself that you are doing fine when you are not. You should always attempt an entire set of problems before looking up any of the answers. If you get stuck on any of the problems, look first to the text material or your lecture notes for help. If after this you are still stuck or get a wrong answer, try to focus on what it is that you don't recognize as familiar in attacking the problem. Make a note of that which caused you the problem, even though you felt that you understood the material quite well.

Many students who have difficulty in organic chemistry do so from a lack of understanding of what is expected of them. Because the subject is very logical, they are lulled into a false sense of security. A student who performs poorly on a test often says, "I understood the lectures so well; everything seemed so logical, but when it came to the test nothing seemed familiar." Such students genuinely do feel that they follow the lecture and text explanations quite well. Usually they do poorly because they do not realize that the performance of organic chemistry requires application of what has been learned beyond mere memorization. If you want to avoid this difficulty, you must work many problems before coming to the examinations. Only then will you see that a cursory familiarity with the lecture and text material is not enough to guarantee success, and only then will you see what organic chemistry is about and the enjoyment (yes, *enjoyment*) that can be had in its practice.

Preparing for Examinations

Do not leave preparation for the last day or two! You will never be able to absorb and integrate all the relevant material. Instead, you should work on examination material from day one by making a serious attempt to master the material

During the build-up to the examination, you should give yourself plenty of practice in solving problems under examination conditions. If old examinations are available, work those problems. (Good instructors do not repeat problems from examination to examination, but the *style* of problem tells you what is expected of you in organic chemistry.) If none are available, make up your own examination by choosing a collection of problems from the course text or other texts (check your library).

If you have had good and consistent success in answering problems under examination conditions, you should be confident going into the examination.

The Laboratory

In the laboratory you will actually handle many of the organic compounds and perform the reactions you read about in the text and hear about in lecture. It is extremely important that you do *not* view each experiment as a separate entity to be completed following a cook-book style recipe. Look for connections between experiments. Many of the things you do in the work-up of a reaction and the isolation of a product are common to many experiments. Think about what you are doing and why you are doing it. After a while you should be able to develop your own procedures for performing reactions. This is what practicing scientists must do. There are no printed instructions for reactions that no one has done before. Your laboratory instructors may be graduate students. Perhaps you are more comfortable asking questions of them than of your lecturer. You should certainly regard your laboratory instructor as an important source of information about the course - both the lecture and laboratory portions.

How This Study Guide/Solutions Manual is Organized

The chapters parallel those in the Baker-Engel text. Each chapter begins with a summary of the main skills, concepts, and reactions tht appear in the chapter. Integrated into the summary are Practice Problems related to the important topics. In each chapter, the summary is followed by solutions to the problems within the Baker-Engel text. Finally, answers are given to the Practice Problems of the study guide.

CHAPTER 1
BONDING IN ORGANIC COMPOUNDS

Key Points

• Be able to designate the electronic configuration of an atom.
Practice Problem 1.1
Describe the electronic configuration of
    (a) a fluorine atom
    (b) a silicon atom.
How many unpaired electrons are present in each atom?

• Learn the standard valences of H, C, O, N, S, P, the halogens, B, and Al (these are the atoms that you will meet most commonly in organic chemistry) and use them, along with the noble gas rule where appropriate, to draw Lewis structures of compounds. In this context remember that boron and aluminum have insufficient electrons to achieve a noble gas configuration while remaining uncharged.
Practice Problem 1.2
Draw Lewis structures for each of the following: PF_3, $(CH_3)_3B$, ClNO, BrCN, HOCN.

• Be able to assign formal charges to atoms in Lewis structures.

Learning Hint
    Although you can always assign formal charges by following the systematic rules given in Chapter 1 of the text, the procedure is time-consuming. Try to recognize by inspection which formal charges to associate with particular atoms in certain types of structural environments. Here are some guidelines that cover the most frequently encountered situations:
    • If an atom forms its usual number of bonds (see Table 1.1 of text), it has zero formal charge.
    • If a nitrogen or oxygen atom forms one more bond than normal, its formal charge is 1+.
    • If a nitrogen or oxygen atom forms one less bond than normal, while still having a noble gas configuration, its charge is 1-.
    • If a carbon atom forms only three bonds, its formal charge is 1- if it has a noble gas configuration (it then has one unshared pair of electrons at the valence level), and 1+ if it has only a sextet of electrons in its valence level (there are no unshared valence level electron pairs).
    • A boron or aluminum atom with four bonds has a formal charge of 1-.

    For example, look at the carbon atom in cyanide ion, $(:C{\equiv}N:)^{1-}$. It forms one less bond than normal (three instead of the usual four). Since this carbon atom has eight valence level electrons about it, it does have a noble gas congifuration, and the formal charge is 1-. Likewise the carbon atom in carbon monoxide $(:C{\equiv}O:)$ also has a formal charge of 1-. On the other hand, the oxygen atom in carbon monoxide forms one more bond than normal (three instead of the usual two). Thus, it has a formal charge of 1+.

Practice Problem 1.3
What is the formal charge associated with each of the designated atoms, and what is the overall charge associated with each of the following species?

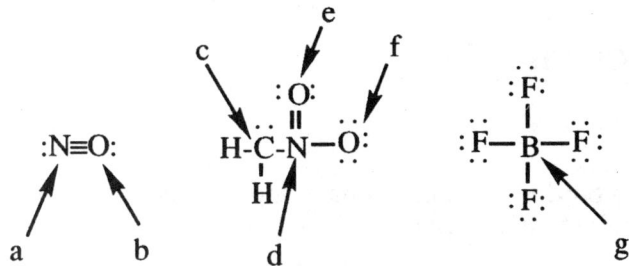

- Be able to draw the Lewis structure of an atom or ion in which the atoms do not form their usual numbers of bonds. Be sure to be able to indicate any formal charges.

Practice Problem 1.4
Draw a Lewis structure for ozone, O_3 (the three oxygen atoms are connected in the sequence O-O-O) and indicate any formal charges.

Practice Problem 1.5
Draw a Lewis structure for the methyldiazonium cation, $H_3CN_2^+$. Indicate formal charges. This ion contains a C-N-N sequence of atoms and all of the hydrogen atoms are bonded to carbon.

- Understand how to predict the geometric arrangement of bonds about an atom. (Hint: Draw the Lewis structure first, and then apply the VSEPR model.)

Practice Problem 1.6
Which molecule, H_2O or H_2Be, is bent? Explain your decision.

Practice Problem 1.7
What is the Cl-C-N bond angle in a molecule of cyanogen chloride (ClCN)?

Practice Problem 1.8
Predict the Cl-C-Cl bond angle in phosgene (Cl_2CO).

- Understand the hybridization model. Be able to predict the hybridization scheme associated with atoms in molecules.

Learning Hint
To decide on the hybridization of a particular atom in a molecule, first draw a Lewis structure if one is not given. Then, locate the atom in question and find the number of atoms bonded to it and the number of unshared valence level electron pairs it possesses. For example, the nitrogen atom of $H_3C-C\equiv N$: is bonded to one atom and has one unshared valence level electron pair. The total number of these two cases is two. This is the number of hybrid orbitals used by the atom. Since the number is two, the hybridization must be *sp* (there are two hybrid orbitals in the *sp* set). If the total were three, the hybridization would be *sp²*, and if the number were four, the hybridization would be *sp³*. Other types of hybridization are relatively rare in organic compounds.

Practice Problem 1.9
What type of hybridization is associated with the central atom in BeH_2 and in H_2O, and with the carbon atom in CO_2 and in H_2CO?

Practice Problem 1.10
Assign a hybridization type to each of the atoms *a* to *n* in the following structure.

:N≡C-C≡C-C(=O:)-Ö-CH₂-CH₂-CH=CH-CH₂-N̈=CH-CH₃

a b c d e f g   h   i   j   k   l m n

Practice Problem 1.11
Based on hybridization arguments, which compound would you expect to have the stronger (and shorter) N-H bond(s): ammonia, or methyleneimine (H₂C=NH)? Explain your decision.

Practice Problem 1.12
According to valence bond theory, the overlaps of which orbitals lead to the C-H bonds in acetylene (H-C≡C-H)? To the C-C sigma bond? To the C-C pi bonds?

- Understand how to draw and to interpret resonance structures.

Practice Problem 1.13
Draw two resonance structures for the nitrite ion, NO_2^- (the sequence of bonded atoms is ONO). What is the expected N-O bond order based on these structures?

Practice Problem 1.14
Draw a resonance structure for cyanate ion (CNO⁻) in which each atom has a noble gas configuration and a formal charge of 1- associated with carbon. Draw a second resonance structure in which all of the atoms still have a noble gas configuration, but in which the carbon atom has a formal charge of 2-.

Practice Problem 1.15
Which would you predict to contain the shorter N-O bond, nitrite ion or nitrous acid, HNO_2 (sequence of atoms is HONO)?

- Understand how to use curved arrows to indicate the shift of a pair of electrons.

Practice Problem 1.16
Draw the structures that result from shifting electrons according to the curved arrows, and indicate any formal charges in the resulting structures.

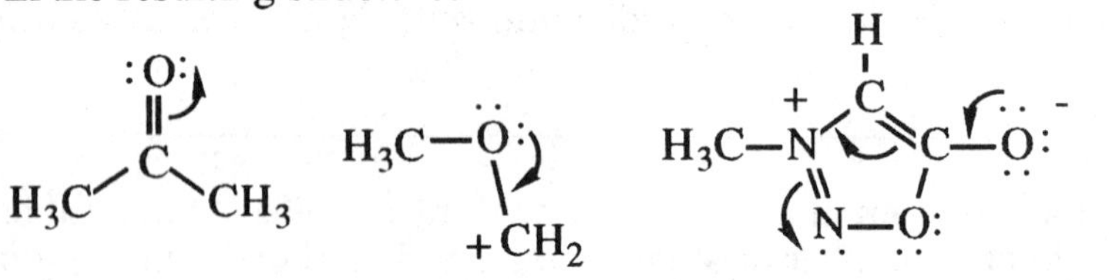

a   b   c

Practice Problem 1.17
Place curved arrows in the structures to the left so that the structures on the right are generated by the designated electron pair shifts.

a $\left[\begin{array}{c} \text{structure with } H_3C-N^+ \text{ ring, } C-\ddot{O}:^- , N-\ddot{O}: \end{array} \right. \longleftrightarrow \left. \begin{array}{c} \text{structure with } H_3C-N \text{ ring, } C-\ddot{O}:^- , N=O:^+ \end{array} \right]$

b $\left[\begin{array}{c} HC^+-CH \text{ ring with } CH_2, N-H \end{array} \longleftrightarrow \begin{array}{c} HC=CH \text{ ring with } CH_2, N^+-H \end{array} \right]$

- Understand the fundamental ideas of the molecular orbital model.

Practice Problem 1.18
Use the molecular orbital model to predict which of the following has or have a bond order > 0:
    $H_2, H_2^-, H_2^{2-}, He_2, He_2^+$

Practice Problem 1.19
Draw a valence shell molecular orbital energy level diagram for N_2 showing all orbitals and their occupancy by electrons. Use the molecular orbital model to predict the bond order in N_2, N_2^+, and N_2^-.

Solution of Text Problems

1.1 (a) two
     (b) one
     (c) one

1.2 (a) $:\ddot{Br}-C\equiv N:$ (b) $\ddot{O}=C=\ddot{O}$ (c) $:N\equiv N:$ (d) methanol structure (e) dimethyl ether structure

(f) $:\ddot{Br}-CH_2-CH_2-\ddot{O}-H$ (g) $:\ddot{F}-\ddot{F}:$ (h) formaldehyde structure

1.3 (a) $H_2C=CHBr$ (b) $:\ddot{Cl}-C(Cl)_2-C(Cl)_2-\ddot{Cl}:$ with H's (c) hydrazine H_2N-NH_2 (d) $Cl_2C=O$ (e) $:N\equiv C-C\equiv N:$

1.4 hydrogen sulfide ion: H, 0; S, -1. ammonium ion: N, +1; each H, 0. benzonitrile *N*-oxide: N, +1; O, -1. sulfuric acid: each H, 0; S, +2; two O of (OH), 0; remaining two O, each -1

1.5 (a) $H-\ddot{N}=\ddot{N}-H$ (b) ClO_4 with Cl 3+ and four $\ddot{O}:^-$ (c) $\ddot{O}=S^+-\ddot{O}:^-$ (d) $:\ddot{O}-\ddot{O}=\ddot{O}$ with - and + (e) $H-C\equiv\ddot{O}:^+$

1.6

8 Study Guide and Solutions Manual

1.6

There are 6 H-C-H bond angles, each 109°28'.

1.7 The planes are perpendicular to each other. That is, there is an angle of 90° between them. Similarly, the two rings of spiropentane are at a 90° angle relative to each other.

1.8

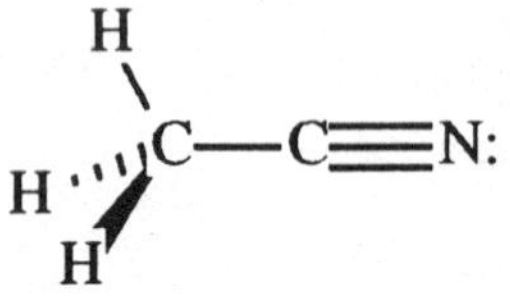

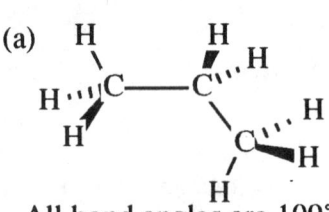

The three H-C-H bond angles are all 109°28', as are the three H-C-C bond angles. The C-C-N bond angle is 180°.

1.9

(a)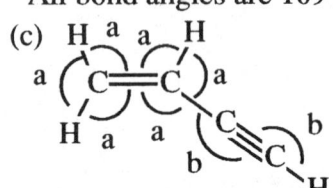

All bond angles are 109°28'.

(b)

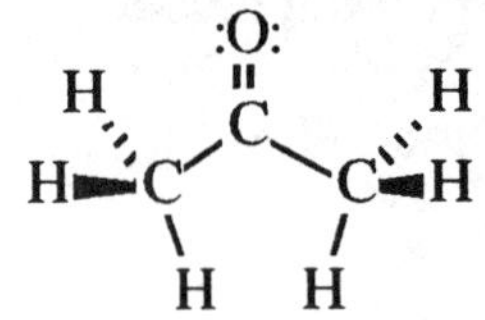

H   :O: H
 \\ || /
  C—C—C
 /   \\
H H   H H

The O-C-C bond angles and the C-C-C bond angle are 120°. All H-C-H and H-C-C bond angles are 109°28'.

(c)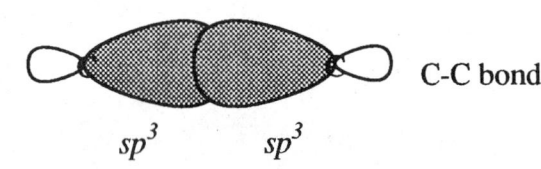

All bond angles indicated "a" are 120°, while all bond angles indicated "b" are 180°.

1.10

C-C bond sp^3 sp^3 All C-H bonds sp^3 $1s$

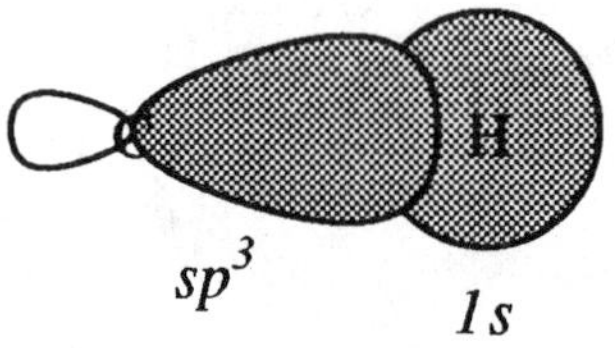

1.11 For each N-H bond the nitrogen uses an sp^3 hybrid orbital while the hydrogen uses a $1s$ orbital.

1.12 (a) The geometry about each carbon atom is trigonal planar, all six atoms lying in the same plane. All bond angles are 120°. All C-H bonds use sp^2 hybrid orbitals of carbon and $1s$ orbitals of hydrogen. For the two C-C bonds, one is a sigma bond using an sp^2 hybrid orbital from each carbon and the other is a pi bond using a p orbital from each carbon atom.

(b) The two C-H bonds are sigma bonds formed using sp^2 hybrid orbitals from carbon and $1s$ orbitals from hydrogen. Similarly, one of the C-N bonds is sigma using an sp^2 hybrid orbital from each of nitrogen and carbon. The N-H bond is a sigma bond using an sp^2 hybrid orbital from nitrogen and a $1s$ orbital from hydrogen. The second C-N bond is a pi bond using a p orbital from each of nitrogen and carbon. The entire molecule is planar with all bond angles being approximately 120°. An unshared valence electron pair is located on nitrogen in an sp^2 hybrid orbital.

(c) The two C-H bonds are sigma bonds formed using sp hybrid orbitals from carbon and $1s$ orbitals from hydrogen. One of the C-C bonds is also a sigma bond formed using an sp hybrid orbital from each of the carbon atoms. The remaining two C-C bonds are pi bonds formed using a pair of p orbitals from each carbon.

1.13 In sp^3 carbon the hybrid orbital is 25% s in character while in sp^2 it is 33 1/3% s in character and in sp it is 50% s in character. The C-H bond in the acetylene molecule are the shortest of any in the problem because the carbon in acetylene uses an sp hybrid orbital to bind to hydrogen whereas in the other molecules carbon uses an sp^2 hybrid orbital. The sp hybrid orbitals are shorter, held closer to the carbon nucleus, than are the sp^2 hybrid orbitals.

Chapter 1   9

1.14

(a) H₂C=N⁺=N⁻ ⟷ ⁻:CH₂—N⁺≡N:

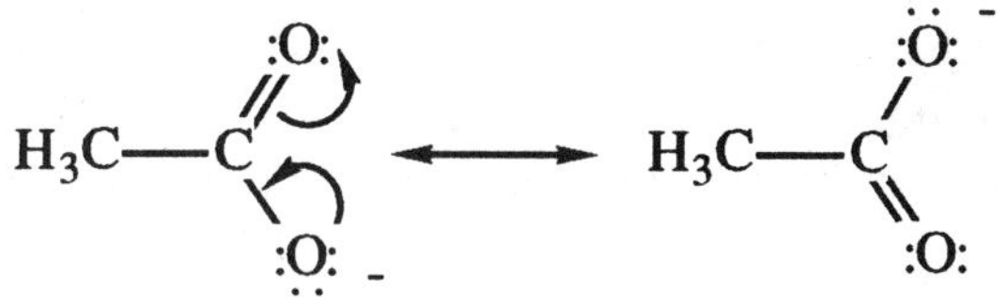

(b) HN=N⁺=N⁻ ⟷ ⁻:N(H)—N⁺≡N:

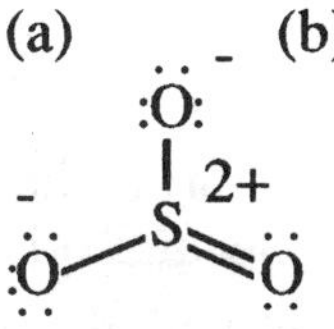

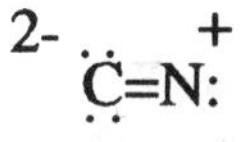

1.15

(a) ⁻:O—C(=O)—O:⁻ ⟷ O=C(—O:⁻)—O:⁻ ⟷ ⁻:O—C(—O:⁻)=O

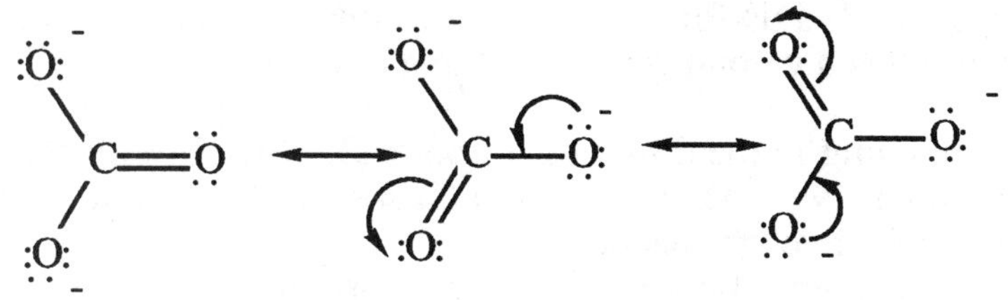

(b) ⁻N=C=O: ⟷ :N≡C—O:⁻

1.16

(a) ⁻:O—N=O: ⟷ :O=N—O:⁻

(b) H₃C—C(=O:)—O:⁻ ⟷ H₃C—C(—O:⁻)=O:

(c) ⁻N=C=S: ⟷ :N≡C—S:⁻

1.17

(a) ⁻:O—S(2+)(=O)—O:⁻ — yes

(b) C=N: with 2− and + — no: separation of unlike charges

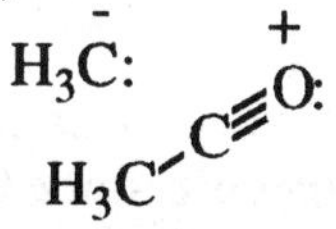

(c) :C=N:⁻ — no: carbon has only six electrons

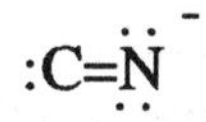

(d) H₂C⁻—N⁺≡N: — yes

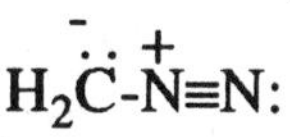

(e) H₃C—C(⁺)=O:⁻ with H₃C — no: separation of unlike charges

(f) H₃C—O⁺=NH₃ — no: five bonds to nitrogen

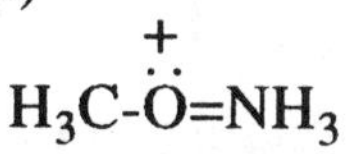

1.18

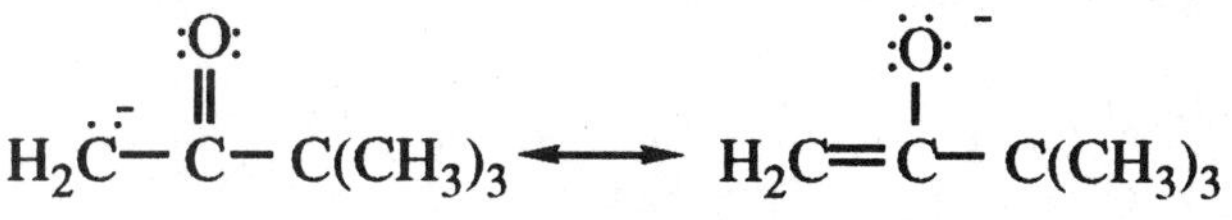

:O:
‖
H₂C⁻—C—C(CH₃)₃   ⟷   H₂C=C(—O:⁻)—C(CH₃)₃

resonance stabilized anion

H₃C—C(=O:)—C(CH₃)₂—CH₂⁻

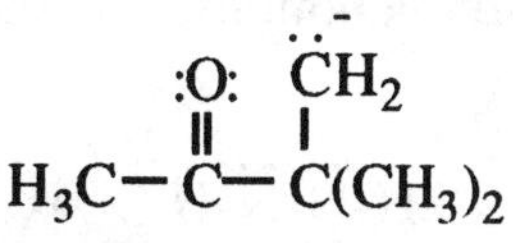

No other resonance structure can be drawn. The anion is not resonance stabilized.

We thus predict that the proton is more likely to be removed by base from methyl shown at the left side of the structure as it leads to the more stable anion.

1.19 Only a proton on carbon adjacent to the cyano group would be removed readily since its removal leads to a resonance stabilized anion whereas removal of any other proton does not.

1.20

[structure: resonance forms of benzene with curved arrows]

1.21 Carbon: sp^2 hybridized, 2 sp^2-sp^2 sigma bonds for each carbon to the adjacent carbons, 1 sp^2 sigma bond to hydrogen, an unhybridized p orbital of each carbon is involved in a p-p pi bond with adjacent carbons. The entire molecule would be planar.

1.22 Using the molecular orbital view of bonding in the dihydrogen molecule (Fig. 1.30, text), we place two electrons in the lowest available molecular orbital level no matter what their source. There is no difference in the H-H bond for dihydrogen molecules formed by the two approaches.

1.23 Using Fig. 13.2 of the text, we find the sigma bonding molecular orbital to be filled and the sigma antibonding molecular orbital to be half-filled (total of three electons). There remains a net 1/2 sigma bond stabilization between the two atoms. The remaining bonding energy is ΔE, or ~52 kcal/mole.

1.24 (a) no net bonding
(b) net 1/2 sigma bond
(c) net 1 sigma bond
(d) net 1/2 sigma bond

1.25 In removing an electron from dinitrogen to give N_2^+, the electron comes from a bonding molecular orbital. Thus the bonding is predicted by the molecular orbital model to be weaker in the cation than in the neutral dinitrogen molecule. However, with the removal of an electron from the difluorine molecule, the electron removed comes from an antibonding molecular orbital. Thus the bonding in the cation is stronger than in the neutral difluorine molecule.

1.26

[MO energy diagram with labels σ_2^*, π_3^*, π_4^*, σ_1, π_1, π_2] The bond order is two.

1.27 The bonding in the dioxygen cation is 1 sigma bond and 1 1/2 pi bond. The bonding in the cation is stronger than in the neutral molecule since the electron removed came from an antibonding molecular orbital.

1.28 Each carbon atom uses an sp^3 hybrid orbital to generate a sigma bonding molecular orbital and a sigma antibonding molecular orbital. With the ground state ethane molecule, the bondng molecular orbital is populated by two electrons. This description differs from the valence bond description only in the existence of an empty antibonding molecular orbital.

1.29 Each of carbon and nitrogen uses an sp hybrid orbital to generate a sigma bonding/antibonding pair of molecular orbitals, as well as two p orbitals from each to generate two pairs of pi bonding/antibonding molecular orbitals. In acetonitrile, the sigma bonding and the two pi bonding molecular orbitals are filled with electrons.

1.30
(a) H—Ö—Ö—H
(b) H-N-N-H with H H
(c) H₂C-S-H with H
(d) H-C-N-H structure with CH₃

1.31
(a) protonated structure with H-N⁺-C
(b) carbonate structure
(c) H-C-C-O⁻ structure
(d) :F:⁻

1.32 (a) 1+ formal charge on the oxygen atom, all other atoms 0 formal charge, overall 1+ on ion
(b) 1- formal charge on carbon atom with unshared valence level electron pair, all other atoms 0 formal charge, overall 1- on ion
(c) 1+ formal charge on carbon with only three bonds to it, all other atoms 0 formal charge, overall 1+ on ion

1.33
methane	sp^3	tetrahedral
dichloromethane	sp^3	tetrahedral
methylamine	sp^3	tetrahedral
bromomethane	sp^3	tetrahedral
methanol	sp^3	tetrahedral
acetonitrile	sp^3	tetrahedral
	sp	linear
acetylene	each sp	linear
carbon dioxide	sp	linear
ethylene	each sp^2	trigonal planar
formaldehyde	sp^2	trigonal planar
methyleneimine	sp^2	trigonal planar

1.34 (a) sp^3 and sp^2 (b) sp^2 and sp (c) three are sp^3 and two are sp^2 (d) two are sp^3 and one is sp
1.35 (a) 4 (b) 2 (c) 2
1.36 and 1.37

(a) 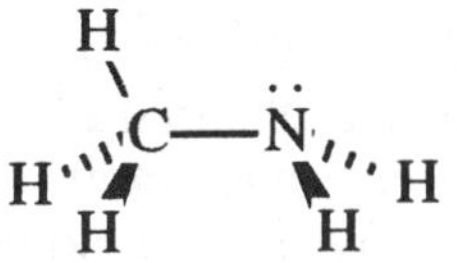 [structure of methylamine] (b) 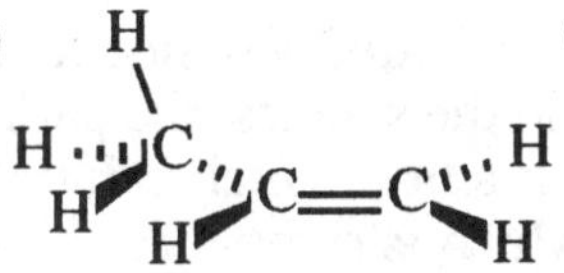 [structure of ethylene] (c) 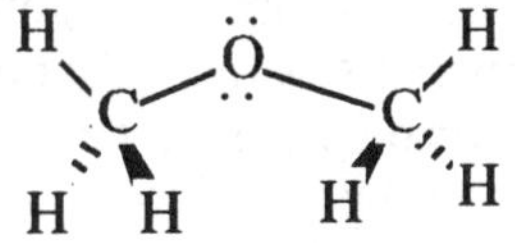 [structure of dimethyl ether]

(d) 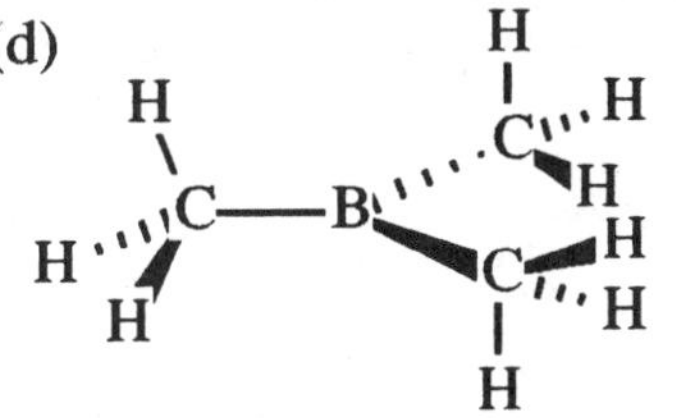 [structure with C—B and ring] (e)  $:N\equiv O:^+$

1.38 With the oxygen atom using sp^3 hybrid orbitals to form its bonds and hold its unshared valence electron pairs, we would anticipate the H-O-H bond angle to be 109°28'. However, the electron-electron repulsions between the two orbitals with the unshared pairs are greater than those between the two bonding pairs. This results in the bonds being pushed closer together than in a perfectly tetrahedral system, and a smaller bond angle.

1.39

(a)  $:N\equiv\overset{+}{N}-\overset{..}{\underset{..}{O}}:^- \longleftrightarrow {}^-:\overset{..}{N}=N=\overset{+}{\overset{..}{O}}$

(b)  $H_2C=\overset{+}{N}=\overset{..}{\underset{..}{N}}{}^- \longleftrightarrow {}^-H_2C-\overset{+}{N}\equiv N:$

(c)  $\overset{-}{H_2C}-CH=CH-\overset{\overset{:O:}{\|}}{C}-CH_3 \longleftrightarrow H_2C=CH-CH=\overset{\overset{:\overset{..}{O}:^-}{|}}{C}-CH_3$

(d) 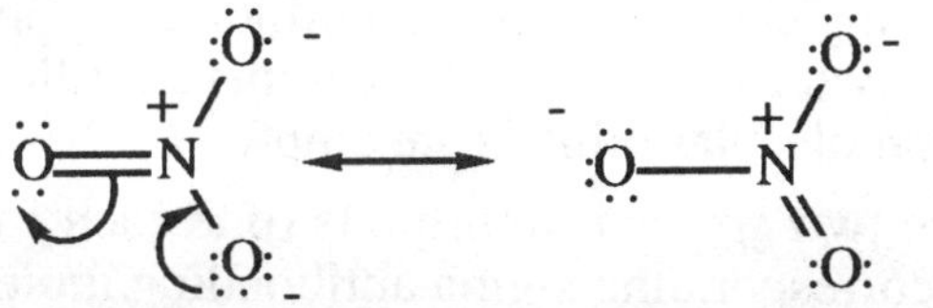 $\overset{..}{\underset{..}{O}}=\overset{+}{\underset{\underset{..}{\underset{:O:^-}{|}}}{N}}\overset{:\overset{..}{O}:^-}{} \longleftrightarrow {}^-:\overset{..}{\underset{..}{O}}-\overset{+}{\underset{\underset{..}{\underset{:O:}{\|}}}{N}}\overset{:\overset{..}{O}:^-}{}$

1.40 For the product ion structure shown to the left, resonance stabilization is present (delocalized from carbon to oxygen). For the product ion structure shown to the right, no such electron delocalization is possible.
1.41 (a) 10

(b)

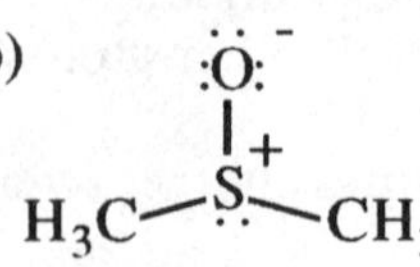

$$H_3C-\overset{\overset{:\overset{-}{O}:}{|}}{\underset{..}{S}}{}^{+}-CH_3$$

1.42 Structures a) and e) are acceptable as resonance structures. Structures b) and f) involve separation of opposite charges, an energy increasing factor. Structure c) has fewer bonds owing to electrons being unpaired. Structure d) has a "long bond", necessarily weaker than an ordinary one and necessitating a shift in the positions of the atoms.

1.43 The nitrite ion is bent with the nitrogen being sp^2 hybridized. The nitronium ion is linear with an sp hybridized nitrogen.

1.44 Drawing the Lewis structure of the azide ion we see that the central nitrogen is necessarily sp hybridized making the ion linear. We can not write any other resonance structure for the azide ion. Hydrazoic acid has a nitrogen-hydrogen bond present.

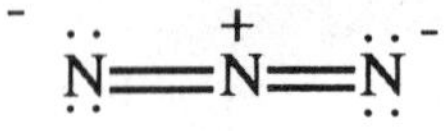

$$\overset{-}{\underset{..}{N}}=\overset{+}{N}=\overset{-}{\underset{..}{N}} \qquad \underset{H}{\overset{}{\searrow}}\overset{}{\underset{..}{N}}=\overset{+}{N}=\overset{-}{\underset{..}{N}} \longleftrightarrow \underset{H}{\overset{}{\searrow}}\overset{-}{\underset{..}{N}}-\overset{+}{N}\equiv N:$$

azide ion 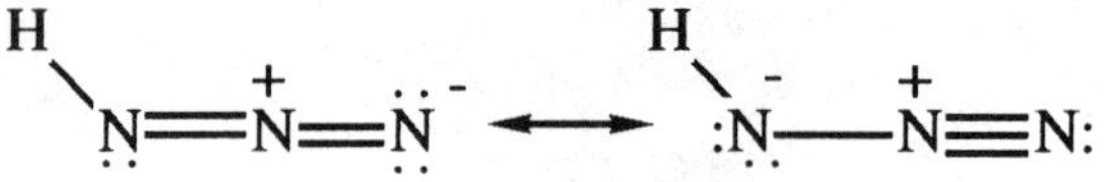   hydrazoic acid

1.45

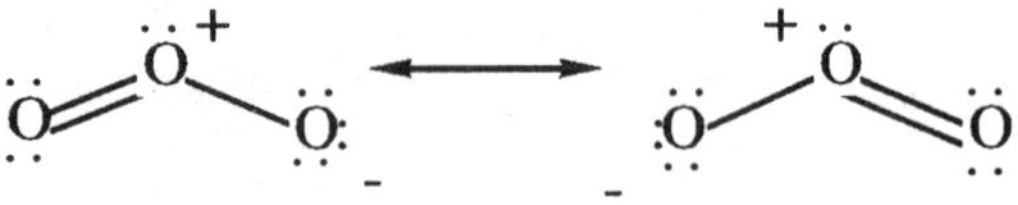

$$\overset{..}{\underset{..}{O}}=\overset{+}{\underset{..}{O}}\diagdown \overset{..}{\underset{-}{\underset{..}{O}}}: \quad \longleftrightarrow \quad :\overset{..}{\underset{\underset{-}{..}}{O}}\diagdown \overset{+\,..}{\underset{..}{O}}=\overset{..}{\underset{..}{O}}$$

The central oxygen is sp^2 hybridized, making the bond angle 120°.

1.46 One carbon has only three bonds to it, and no unshared valence level electron pairs. Thus, that carbon bears a positive charge. Because this is the only charge in the species, the entire structure is that of a cation.

1.47 Removal of a proton from carbon 1 generates an anion which can not be resonance stabilized. However, removal of a proton from carbon 3 generates an anion which is resonance stabilized. With the resonance stabilized anion each carbon-carbon bond has the same amount of sigma and pi character, whereas with the anion generated by loss of a proton from carbon 1, one carbon-carbon linkage is clearly a double bond while the other is only a single bond.

1.48

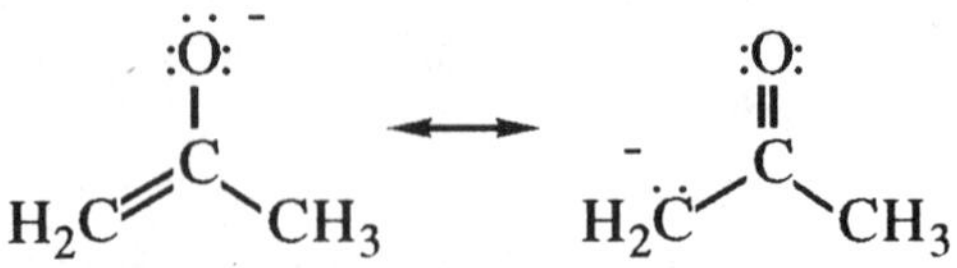

$$H_2C\overset{\overset{:\overset{-}{O}:}{|}}{\underset{}{C}}CH_3 \quad \longleftrightarrow \quad \overset{-}{H_2C}\overset{\overset{:O:}{||}}{\underset{}{C}}CH_3$$

1.49

$$:\overset{..}{\underset{..}{F}}-\overset{\overset{:\overset{..}{F}:}{\diagup}}{\underset{\diagdown \overset{..}{\underset{..}{F}:}}{B}} \quad \longleftrightarrow \quad \overset{+}{F}=\overset{\overset{:\overset{..}{F}:}{\underset{-}{\diagup}}}{\underset{\diagdown \overset{..}{\underset{..}{F}:}}{B}}$$

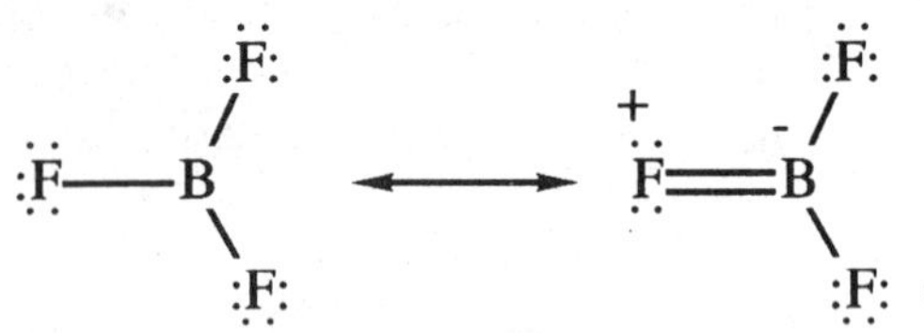

The observation of the shorter bond length for B-F in the neutral molecule than in the tetrafluoborate anion suggests that the resonance structure shown to the right above is of significance. Double bonds between a given pair of atoms are shorter than single bonds between those same two atoms owing to a change in hybridization required for double bond formation. We would expect all of the bonds in boron trifluoride to be of equal length. Further resonance structures may be written incorporating the double bond character between boron and each of the three fluorine atoms.

1.50 (a) 1 sigma and 1 pi (b) 1 sigma and 2 pi (c) 1 sigma and 1 1/2 pi (d) 1 sigma and 1 1/2 pi

1.51 (a) A sigma bonding molecular orbital formed from two sp hybrid orbitals (one from each carbon) is filled to give a sigma bond between the two atoms. The corresponding sigma antibonding orbital is empty. Further, two pi bonding molecular orbitals formed from four p orbitals (two from each carbon atom) are filled giving two pi bonds for the system. The corresponding two pi antibonding molecular orbitals are empty.

(b) A sigma bonding molecular orbital formed from two sp^2 hybrid orbitals (one each from carbon and oxygen) is filled giving a sigma bond for the system. The corresponding sigma antibonding molecular orbital is empty. Further, a pi bonding molecular orbital formed from two p orbitals (one each from carbon and oxygen) is filled giving a pi bond between the atoms. The corresponding pi antibonding molecular orbital is empty.

(c)   A   sigma   bonding   molecular   orbital   formed   from   an

sp^3 hybrid orbital of carbon and a p orbital from iodine is filled giving the sigma bond for the system. The corresponding sigma antibonding molecular orbital is empty.

1.52   Consideration of the *1s* orbitals and the molecular orbitals which would be generated by their combination does not change the bonding picture for the dinitrogen molecule. Both the resultant sigma bonding and antibonding molecular orbitals would be filled providing zero bonding or antibonding effect.

Solution of Study Guide Practice Problems

1.1

(a) F: 1s ↑↓, 2s ↑↓, 2p ↑↓ ↑↓ ↑

(b) Si: 1s ↑↓, 2s ↑↓, 2p ↑↓ ↑↓ ↑↓, 3s ↑↓, 3p ↑ ↑ _

1.2

:F—P̈—F: H₃C—B—CH₃   :C̈l—N̈=Ö   :B̈r—C≡N: H—Ö—C≡N:
   | |
  :F: CH₃

1.3   a: 0, b: 1+, c: 1-, d: 1+, e: 0, f: 1-, g: 1-

1.4

[Ö=Ö—Ö:]⁻
    +

1.5

         +
H₃C—N≡N:

1.6   The water molecule (H₂O) is bent while H₂Be is linear. The water molecule has two unshared valence level electron pairs associated with oxygen in addition to the two hydrogens. The VSEPR approach thus leads us to conclude that the molecule will be bent to minimize electron-electron interactions. However, the H₂Be has no unshared valence level electron pairs at beryllium. Thus, electron-electron interaction minimization is attained with the hydrogens as far apart as possible, in a linear arrangement about beryllium.

1.7   180°

1.8   ~120°

1.9   For H₂Be we would conclude that the beryllium hybridization is sp whereas it is sp^3 for the oxygen in water. In carbon dioxide the carbon hybridization is sp whereas the carbon hybridization in formaldehyde is sp^2.

1.10   a, b, c and d are sp hybridized

e, i, j, l and m are sp^2 hybridized

f, g, h, k and n are sp^3 hybridized

1.11   We would expect the N-H bond in methyleneimine to be stronger and shorter than the N-H bond in ammonia. Methyleneimine uses sp^2 hybrid orbitals to form sigma bonds to nitrogen whereas ammonia uses sp^3 hybrid orbitals for such bonds. Bonds using orbitals having a greater percentage s character are shorter and stronger than those with a smaller percentage s character.

1.12   The C-H bonds of acetylene are formed by overlap of carbon sp hybrid orbitals with hydrogen s orbitals whereas the carbon-carbon sigma bond is formed by the overlap of two sp hybrid orbitals and the carbon-carbon pi bonds result from overlap of two carbon p orbitals.

1.13

[Ö=N̈—Ö:⁻ ⟷ :⁻Ö—N̈=Ö]   The expected bond order of the O-N bonds is 1.5.

14 Study Guide and Solutions Manual

1.14

$$[:C\equiv\overset{+}{N}-\overset{-}{\ddot{O}}: \longleftrightarrow \overset{2-}{\ddot{C}}=\overset{+}{N}=\ddot{O}]$$

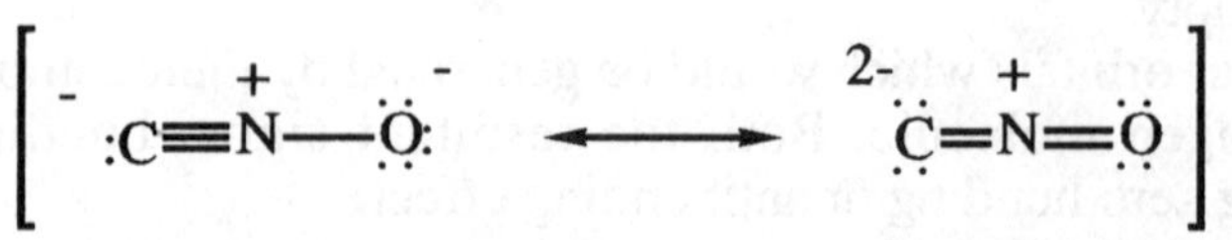

1.15 Nitrous acid should have one N-O bond that is shorter than either of those in nitrite ion. There is only one Lewis structure for nitrous acid, and it contains an N=O bond.

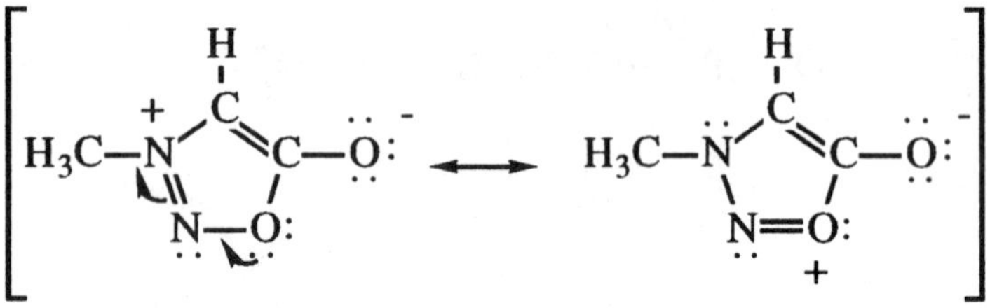

1.16

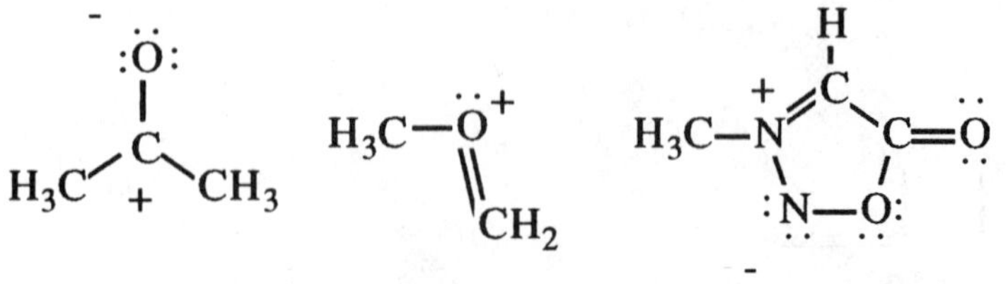

a, b, c (structures)

1.17

a, b (structures)

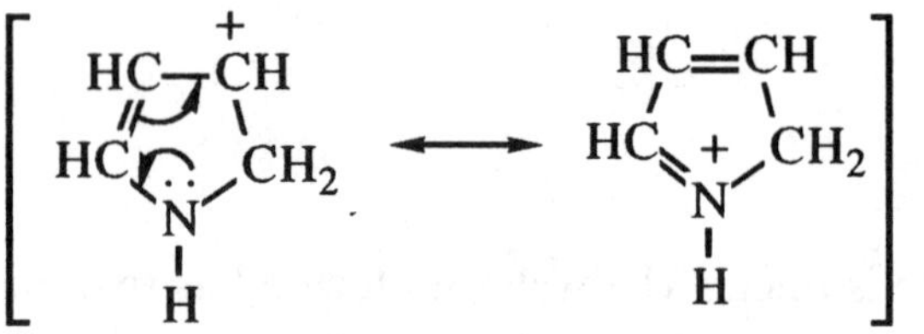

1.18 The H_2, H_2^- and He_2^+ species are predicted to have bond order > 0 using the molecular orbital approach.

1.19

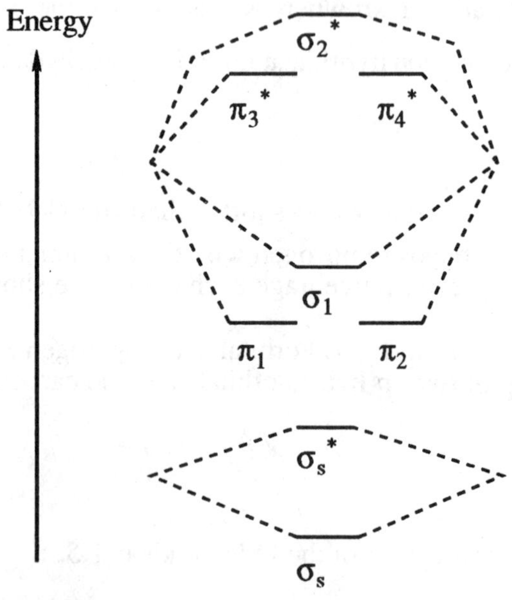

Electrons in each level in each species

Molecular orbital	N_2	N_2^+	N_2^-
σ_2^*	0	0	0
π_3^*, π_4^*	0	0	1
π_1, π_2	2	1	2
σ_1	4	4	4
σ_s^*	2	2	2
σ_s	2	2	2

The total bond order in N_2 is thus 3, whereas in both N_2^+ and N_2^- it is 2 1/2.

CHAPTER 2
INTRODUCTION TO ORGANIC COMPOUNDS; FUNCTIONAL GROUPS, NOMENCLATURE AND REPRESENTATIONS OF STRUCTURE

Key Points

• Be familiar with the functional groups present in the main classes of organic compounds - alkenes, alkynes, haloalkanes, carboxylic acids, aldehydes, ketones, amines, and ethers.

Practice Problem 2.1
Draw structures matching the given descriptions:
(a) an alcohol containing two carbon atoms
(b) an ether containing three carbon atoms
(c) a symmetric alkyne containing four carbon atoms
(d) a carboxylic acid containing three carbon atoms
(e) two different ketones each containing five carbon atoms
(f) the two lowest molecular weight aldehydes
(g) three ethers each containing four carbon atoms
(h) two different amines of molecular weight 45

• Know the different types of representations of structure. Remember that in skeletal type "stick" structures, carbon atoms are not specifically shown - they are understood to be at the ends of all lines and at the junctures of lines. Also, hydrogen atoms are not specifically shown if they are bonded to carbon, but are shown if they are bonded to any other atom.

Practice Problem 2.2
Which of the structural representations *B* through *D* represent the same molecule as structure *A*?

CH₃CH₂CH(CH₃)CH₂CH(CH₃)₂

B   *C*   *D*

Practice Problem 2.3
Draw skeletal (stick-type) representations of all alcohols and all ethers of the formula $C_4H_{10}O$.

• Know how to name alkyl groups. Learn how to assign a systematic name (see Table 2.3 of text, and accompanying discussion), and also know the common names isopropyl, isobutyl, *sec*-butyl, *tert*-butyl, isopentyl (also known as isoamyl), and neopentyl. Know how to name the following substituent groups: -F, -Cl, -Br, -I, -CN, -NO₂, -OCH₃, -NH₂, and -D (Table 2.4 in text).

Practice Problem 2.4
Without looking at Table 2.3, give the common names corresponding to the following alkyl groups: 2-methylpropyl, 3-methylbutyl, 1,1-dimethylethyl.

Practice Problem 2.5
Assign a systematic name to the alkyl group depicted below:

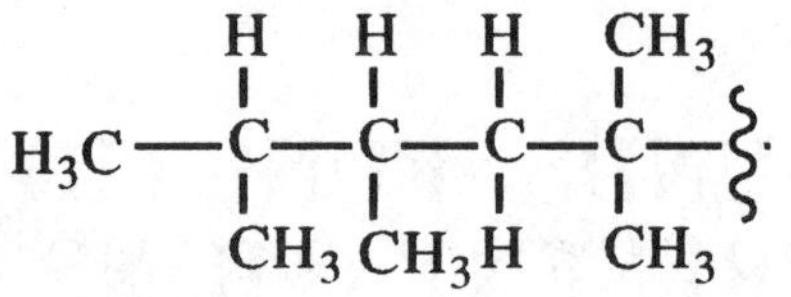

$$H_3C-\underset{\underset{CH_3}{|}}{\overset{\overset{H}{|}}{C}}-\underset{\underset{CH_3}{|}}{\overset{\overset{H}{|}}{C}}-\underset{\underset{H}{|}}{\overset{\overset{H}{|}}{C}}-\underset{\underset{CH_3}{|}}{\overset{\overset{CH_3}{|}}{C}}-\xi$$

• Know how to name branched and substituted alkanes. Carefully study and learn the rules given in section 2.2 of the text, "The Naming of Branched Alkanes."
Practice Problem 2.6
Why is it *not* correct to name the compound shown below as 2-ethylbutane? What is the correct name?

$$H_3C-\underset{\underset{CH_2-CH_3}{|}}{CH}-CH_2-CH_3$$

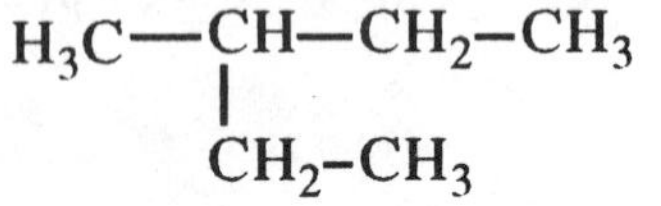

Practice Problem 2.7
Tell what is wrong with the name "3,4-dimethylpentane." What is the correct name of the compound whose structure would be indicated by that incorrect name?
Practice Problem 2.8
Assign a systematic name to the following molecule:

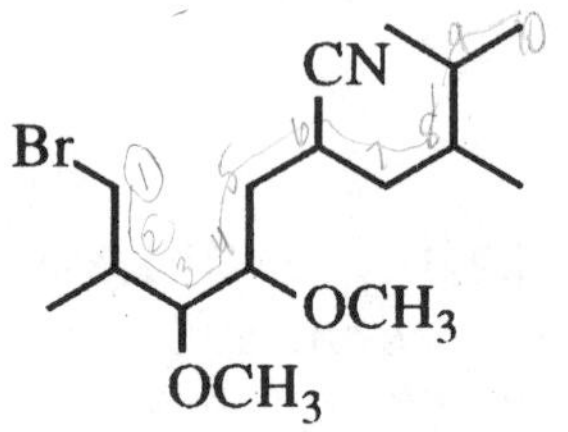

• Be familiar with the use of neo- and iso- in naming alkanes.
Practice Problem 2.9
What are the systematic names for neopentane and isohexane?

• Learn how to classify carbon atoms and hydrogen atoms as primary, secondary, *etc.* (study section 2.3 in the text). Also, know the primary, secondary, tertiary classification for haloalkanes and alcohols.
Practice Problem 2.10
How many carbon atoms and hydrogen atoms of each type (1°, 2°, 3°, 4°) are present in 3,3-dimethylpentane?
Practice Problem 2.11
Give systematic IUPAC names to all possible tertiary haloalkanes of formula $C_6H_{13}Br$.

• Learn the conventions used in the different types of projection drawings - wedge, sawhorse, Newman and Fischer - and understand what is meant by the terms conformation, staggered conformation and eclipsed conformation.
Practice Problem 2.12
Draw a wedge projection drawing of dibromomethane so that the two bromine atoms are in the plane of the paper. Also draw a Fischer projection representing a view of this molecule in which both bromine atoms are above the plane of the paper toward the viewer.
Practice Problem 2.13
Draw a Newman projection of (a) a staggered, and (b) an eclipsed conformation of 2-methylpropane, as seen by sighting along the C1-C2 bond.
Practice Problem 2.14
One of the conformations of a molecule is represented by the Newman projection shown below. Give a systematic name for the molecule.

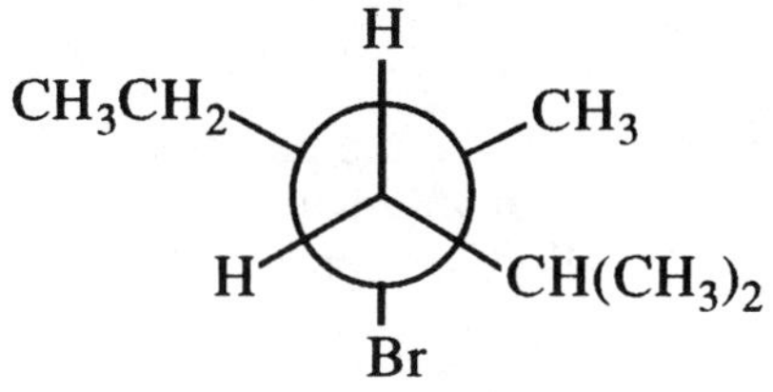

Solution of Text Problems

2.1

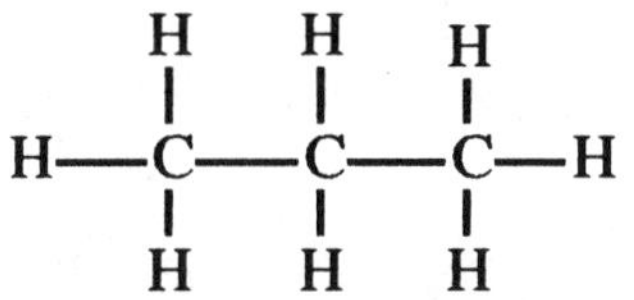

C—H sp^3 $1s$

C—C sp^3 sp^3

Both types are sigma bonds.

2.2 An alkane has the general formula C_nH_{2n+2}. The molecular weight of an alkane is thus: $12n + 2n + 2$. Here, $12n + 2n + 2 = 198$, and $n = 14$. Thus the formula is $C_{14}H_{30}$.

2.3 (a) alcohol; carboxylic acid
(b) amine; carboxylic acid
(c) alkene; ether
(d) alkene; alkene; alcohol
(e) alcohol; alcohol; alkene; carboxylic acid; ketone

2.4 (a) C_7H_{16}
(b) $C_6H_{14}O$
(c) $C_6H_{14}O$
(d) $C_8H_{16}O$

2.5 (a) Structures (i) and (ii) represent the same molecule, but structure (iii) represents a different molecule.
(b) Structures (i) and (iii) represent the same molecule, but structure (ii) represents a different molecule.

2.6

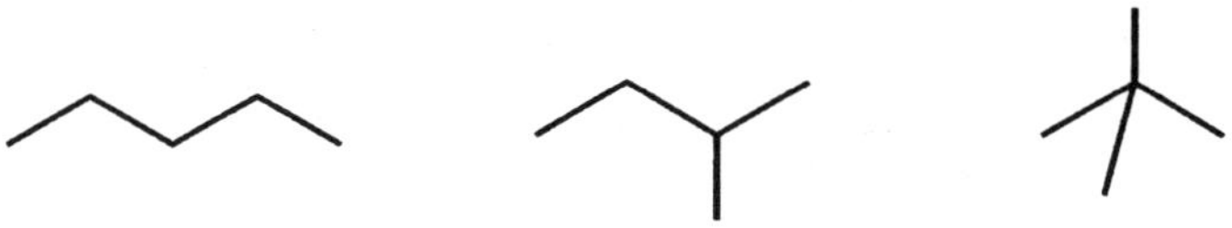

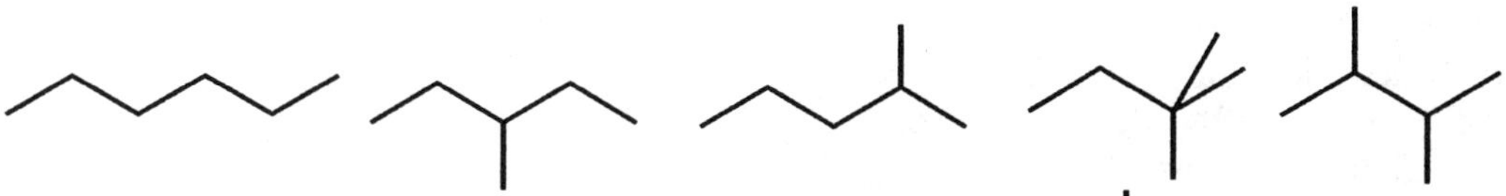

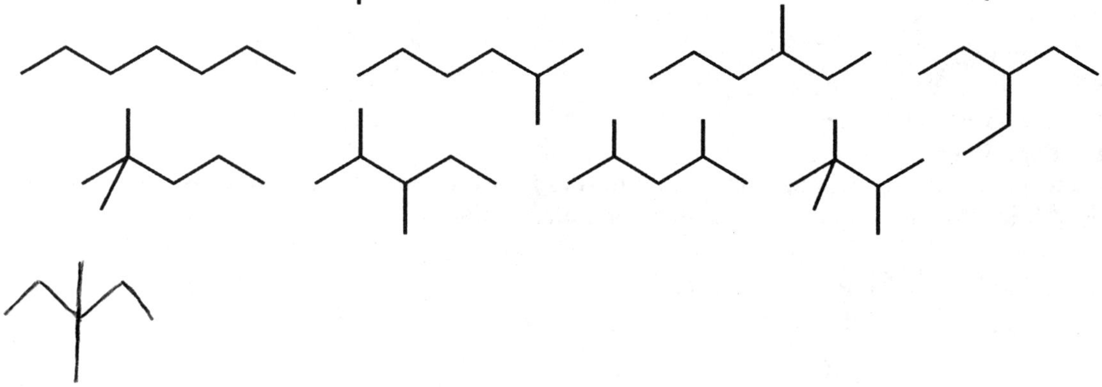

2.7

(a)

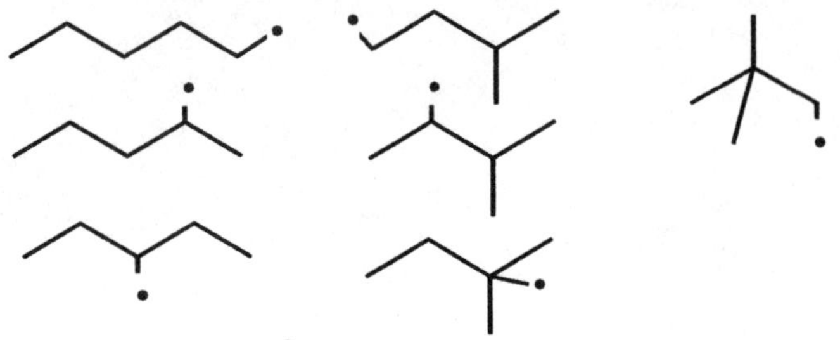

—• indicate the points of open valence of the alkyl groups

(b)

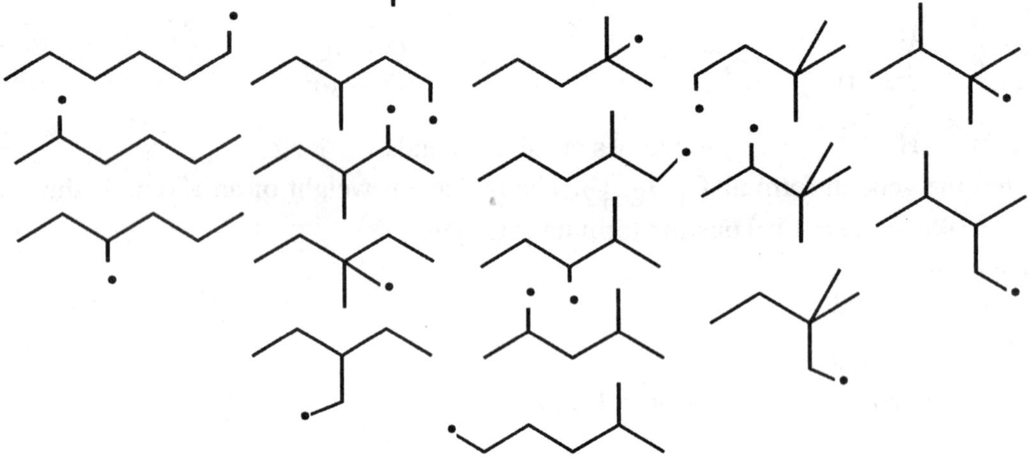

2.8

(a) 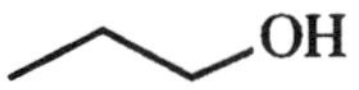 *n*-propyl alcohol 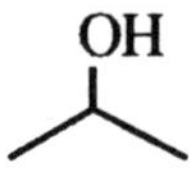 isopropyl alcohol

(b)  *n*-butyl alcohol 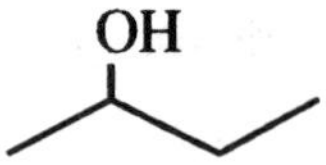 *sec*-butyl alcohol

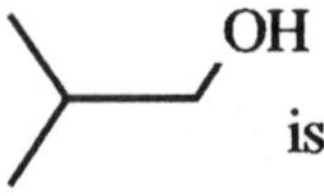 isobutyl alcohol 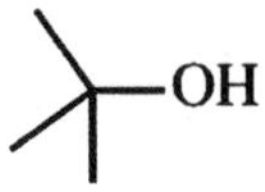 *tert*-butyl alcohol

2.9 (a) 2,2,4-trimethylpentane
(b) 4-ethyl-3-methylheptane
(c) 4-ethyl-5-(1-methylethyl)decane
(d) 3-ethyl-2-methylhexane

2.10 (a) 2,3-dimethylbutyl-
(b) 1,1,2-trimethylbutyl-

2.11 (a) 3,3-dichloro-2-methylpentane
(b) 1-cyano-3-ethylpentane
(c) 1,2-diaminoethane

2.12 The central carbon of neopentane is a quaternary carbon atom; the remaining four carbons are primary carbon atoms. All hydrogen atoms (12) of neopentane are primary.

2.13

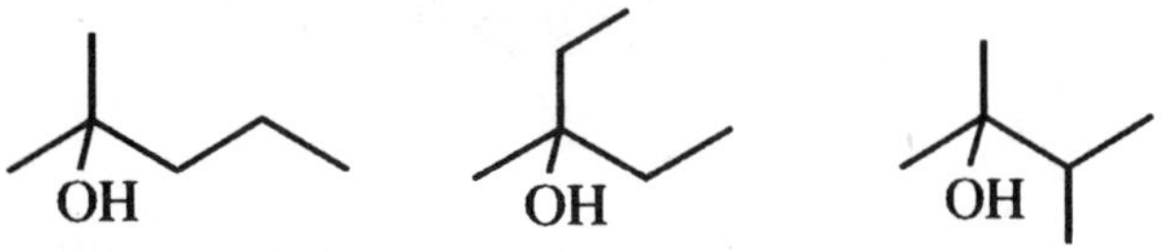

2.14

(a), (b), 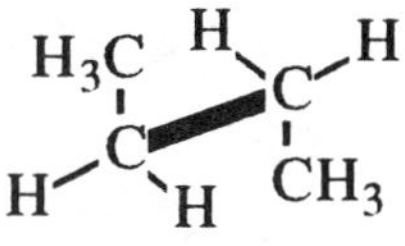 (c) [Newman-like sawhorse projections of substituted ethanes]

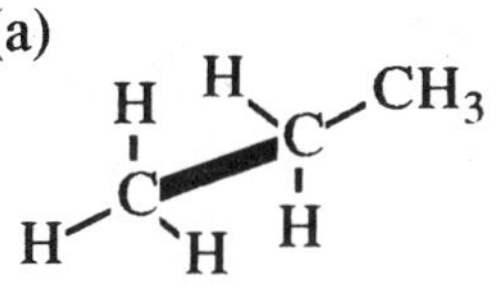

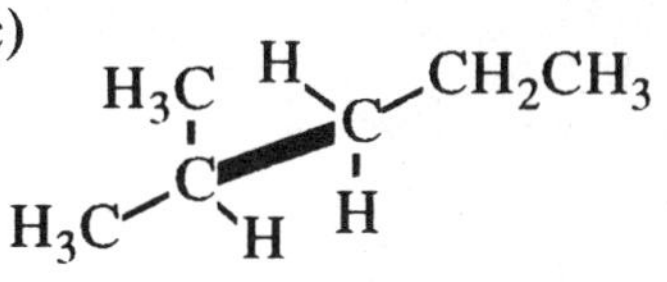

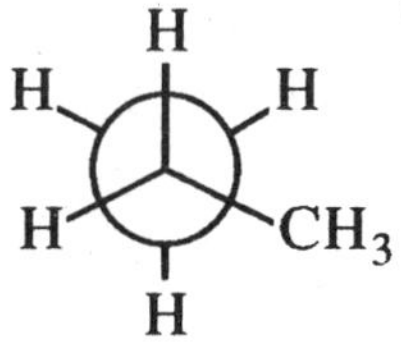

2.15

(a), (b), 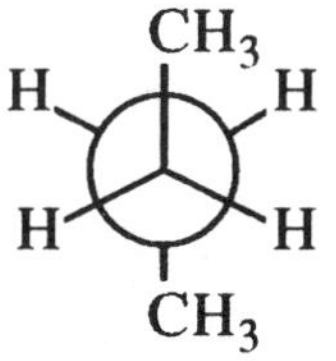 (c) [Newman projections]

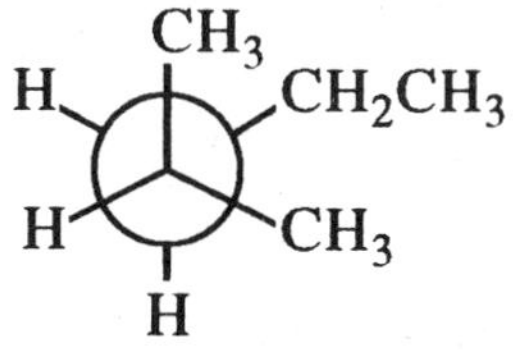

2.16

[Six Newman projections]

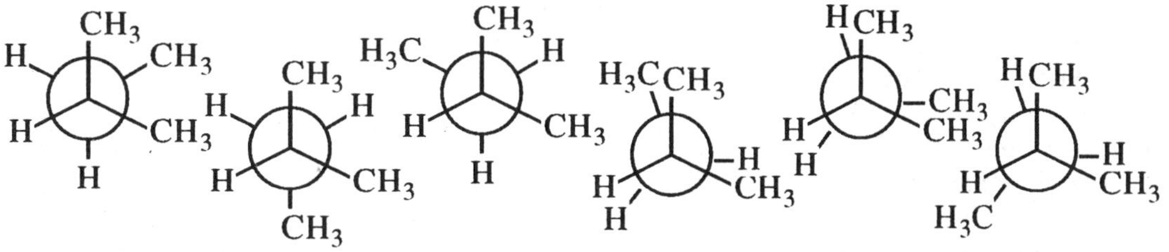

2.17 The two models are not identical, but are non-superimposable mirror images of each other. To make them identical it is necessary to switch two substituents on either one of the models. Thus there are two isomeric forms of bromochlorofluoroiodomethane. Isomers of this type will be considered in detail in Chapter 8.

2.18 The two models are identical. No matter how we construct them or exchange substituents, the two models are always completely superimposable.

2.19

(a) [Three alkyne structures]

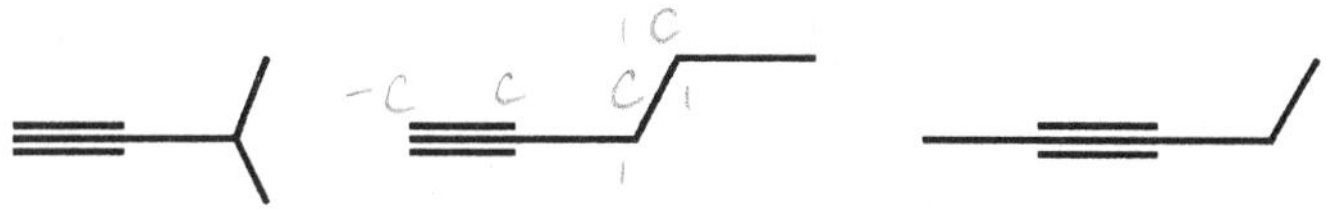

(b) [Four carboxylic acid structures]

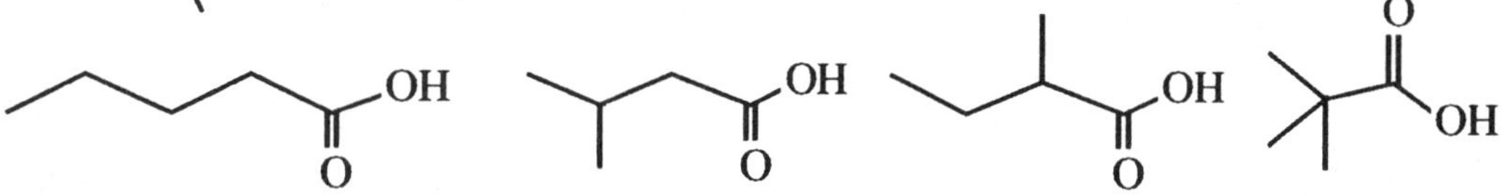

check (c) [Structures of isomeric alcohols]

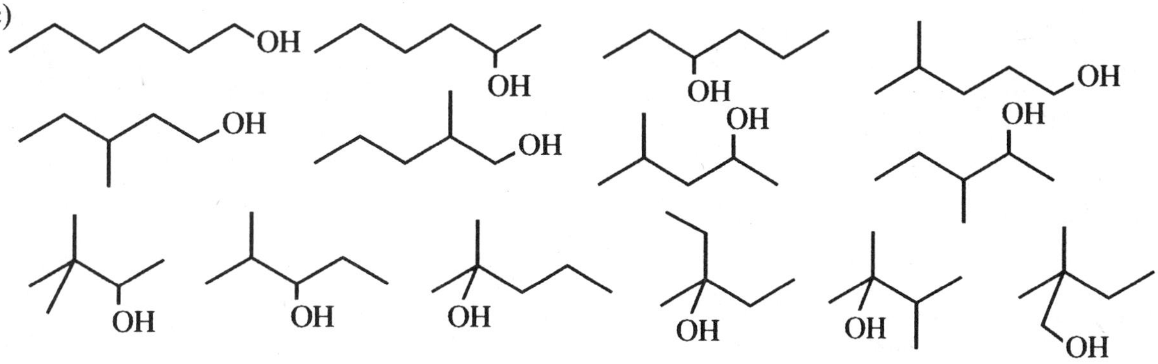

(d) [Aldehyde and ketone structures]

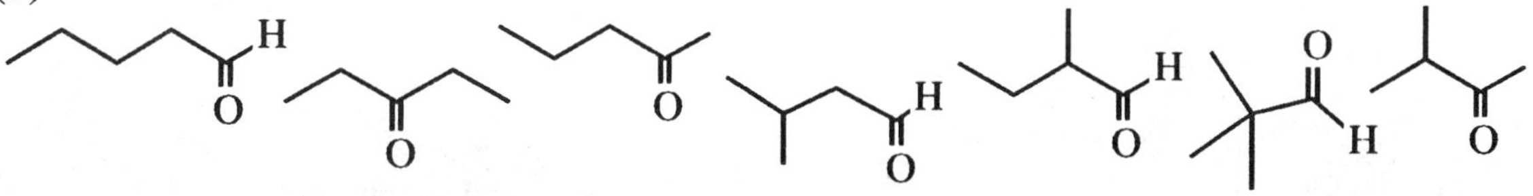

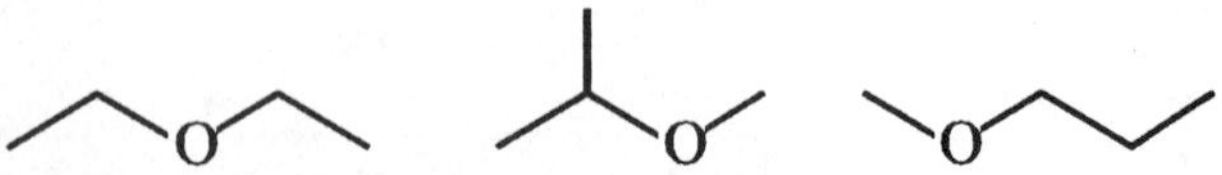

2.20 (a) 2,2-dimethylbutane
(b) 3-ethyl-4,4-dimethylhexane
(c) 3,3-diethyl-4-methyl-5-propyloctane
(d) 4,4-dipropylheptane
(e) 4-ethyl-3-methylheptane
(f) 4-isopropyloctane

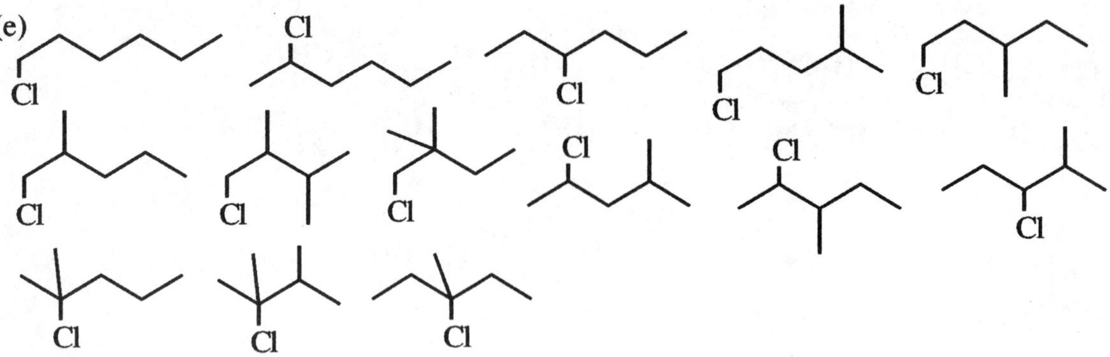

2.21 (a) 4 primary; 1 secondary; 1 quaternary
(b) 5 primary; 3 secondary; 1 quaternary; 1 tert.
(c) 6 primary; 7 secondary; 2 tertiary; 1 quaternary
(d) 4 primary; 8 secondary; 1 quaternary
(e) 4 primary; 4 secondary; 2 tertiary
(f) 4 primary; 5 secondary; 2 tertiary

g) 4P, 4S, 2T
h) 4P, 2S, 3T

2.22 (a) 2,2-dimethylpropane
(b) 2,2,3,3-tetramethylbutane, 2,3,4 trimethylpentane
(c) 2,2,4-trimethylpentane
(d) 2,3,4-trimethylpentane

2.23 (a) 1-chloro-2,2-dimethylpropane
(b) 1-bromo-4-methylpentane
(c) 2-methylpentane
(d) 2-bromobutane
(e) 2-bromo-2-methylpropane

2.24
(a)  3,3-dimethylhexane
(b)  3-ethyl-3-methylhexane
(c) 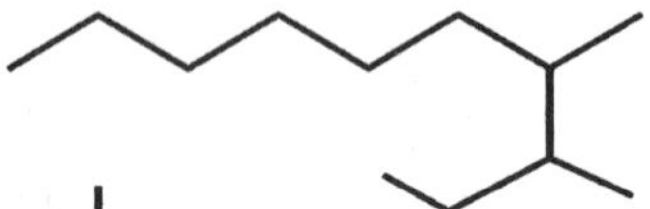 3,4-dimethyldecane
(d) 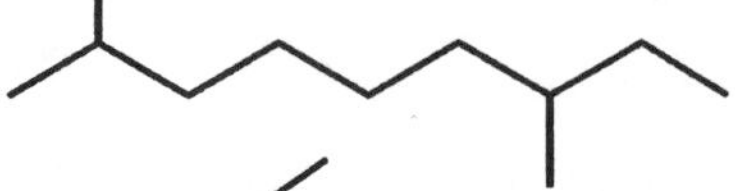 2,7-dimethylnonane
(e) 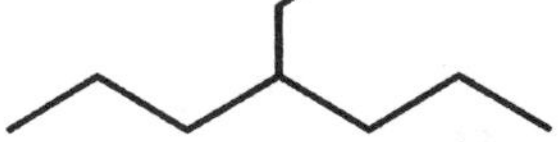 4-ethylheptane
(f) 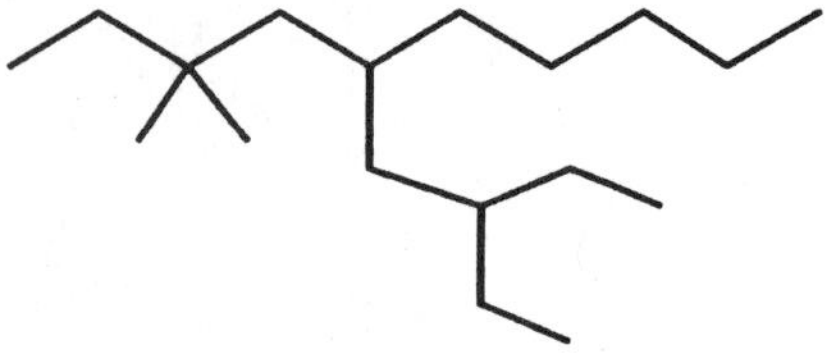 5-(2-ethylbutyl)-3,3-dimethyldecane

Chapter 2   21

2.25 (a) C$_8$H$_{18}$ 3-methylheptane
(b) C$_7$H$_{15}$Br   2-bromo-3,4-dimethylpentane
(c) C$_{15}$H$_{32}$ 2,7-dimethyl-5-(1-methylpropyl)nonane
(d) C$_{13}$H$_{28}$ 3-methyl-4-propylnonane   4-methyl-5-propylnonane
(e) C$_{16}$H$_{34}$ 4,5-diisopropyldecane

2.26 (a) (b) (c) (d) (e) (f)

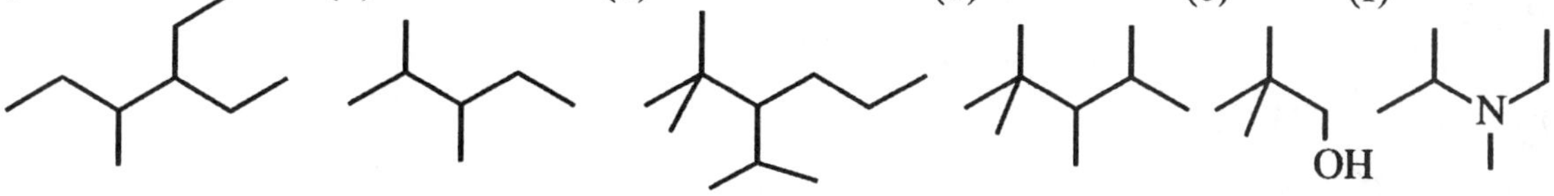

2.27 (a) (b) (c) (d)

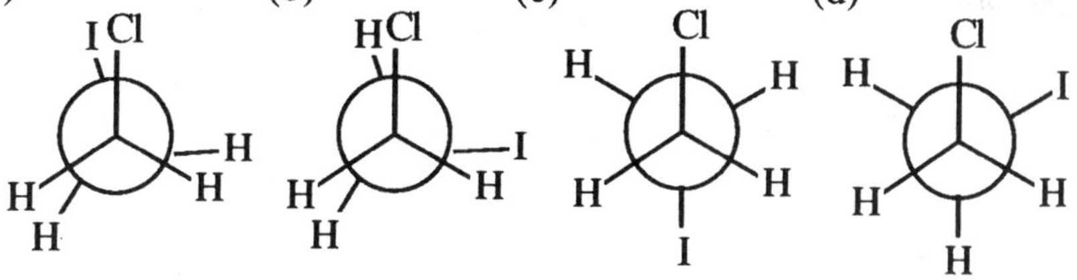

2.28

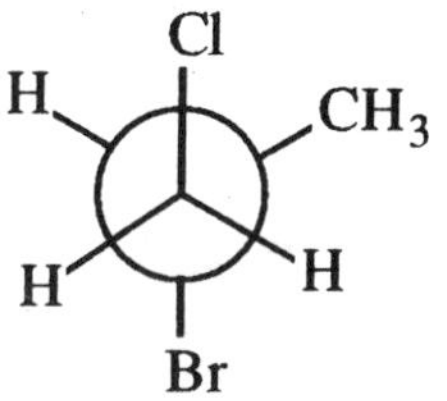

2.29 (a) 2-methylpentane
(b) 2,3-dibromo-3,6-dimethylheptane
(c) 2-bromo-2-methylpropane

2.30 (a) (b) (c)

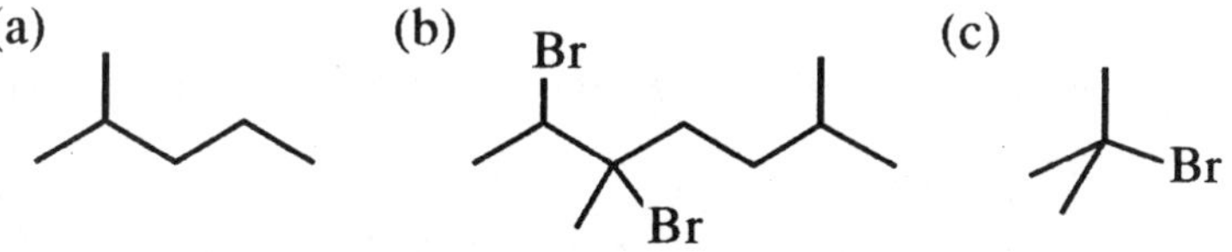

2.31 The two pertinent forms, as viewed along the C-2 to C-3 bond are: greatest, C-1 and C-4 methyl groups are staggered; least, C-1 and C-4 methyl groups are eclipsed. The ratio of greatest to least is ~3/2.

22 Study Guide and Solutions Manual

Solution of Study Guide Practice Problems

2.1
(a) CH$_3$CH$_2$OH
(b) CH$_3$OCH$_2$CH$_3$
(c) CH$_3$-C≡C-CH$_3$
(d) CH$_3$CH$_2$CO$_2$H
(e) CH$_3$-C-CH$_2$CH$_2$CH$_3$ and CH$_3$CH$_2$-C-CH$_2$CH$_3$
        ‖   ‖
        O   O
(f) H$_2$C=O and CH$_3$CHO
(g) CH$_3$CH$_2$OCH$_2$CH$_3$, CH$_3$OCH$_2$CH$_2$CH$_3$ and CH$_3$OCH(CH$_3$)$_2$
(h) (CH$_3$)$_2$NH and CH$_3$CH$_2$NH$_2$

2.2 All three (*B-D*) represent the same molecule as does *A*.

2.3

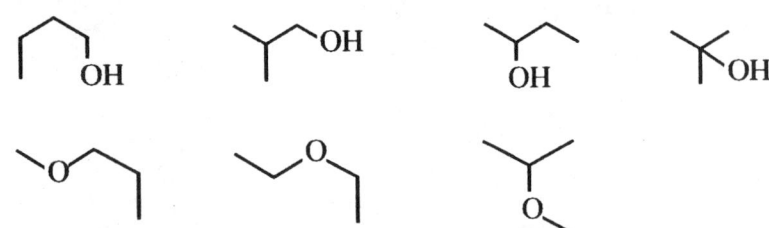

2.4 The corresponding common names are: isobutyl; isopentyl; *tert*-butyl.
2.5 The systematic name is: 1,1,3,4-tetramethylpentyl.
2.6 The longest continuous chain of carbon atoms has five carbon atoms. It should be named as a substituted pentane, specifically: 3-methylpentane.
2.7 We number the substituent positions beginning at the substituent carbon atom closest to an end of the longest continuous carbon chain. Thus, the correct name for the compound indicated is: 2,3-dimethylpentane.
2.8 The systematic name is: 1-bromo-6-cyano-3,4-dimethoxy-2,8,9-trimethyldecane.
2.9 The systematic names are: 2,2-dimethylpropane and 2-methylpentane.
2.10 For carbon atoms, there are: four primary, two secondary, and one quaternary. For hydrogen atoms, there are: twelve primary and four secondary.
2.11 There are only two possible compounds: 3-bromo-3-methylpentane, and 2-bromo-2-methylpentane.
2.12

2.13

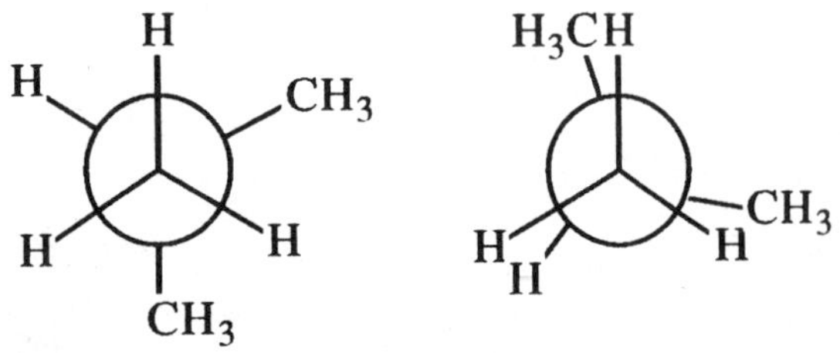

2.14 The IUPAC name is: 3-bromo-3,5-dimethylhexane.

CHAPTER 3
INTERMOLECULAR AND ACID-BASE INTERACTIONS

Key Points

• Understand the relationship between electronegativity and polar covalent bonds.
Review - Electronegativity is a measure of the relative ability of an atom in a bond to attract electron density. Atoms with high electronegativities attract electrons more effectively than atoms with low electronegativities. Covalent bonds are polar when the two atoms forming the bond have different electronegativities. If the difference in electronegativities is greater than ~1.7, the bond is usually regarded as being predominantly ionic rather than as polar covalent.

• Understand what is meant by a bond dipole.
Review - The more electronegative partner in a polar covalent bond has a partial negative charge, sometimes written as δ-, and the less electronegative partner has a partial positive charge, sometimes written as δ+. The polarity of the bond can also be indicated by an arrow (→) whose head points toward the more electronegative atom. A polar bond has an associated bond dipole, μ, defined by $\mu = (r)(q)$ where q is the absolute value of the charge on either of the atoms in the polar bond, and r is the bond length.

Practice Problem 3.1
Indicate using the arrow symbol the direction of the bond dipole (if any) in I_2, HCl, and ICl.

Practice Problem 3.2
Compare two molecules HX and HY. The H-X bond length is greater than the H-Y bond length, but the H-X bond dipole is smaller than the H-Y bond dipole. Which atom do you conclude is more electronegative, X or Y?

• Understand the relationship between the dipoles associated with individual bonds and the overall dipole moment of a molecule.
Review - The central point is that the net dipole moment of a molecule containing more than one bond is given by the vector sum of all of the individual bond dipole moments. Sometimes molecular dipole moments give an indication of structure. Important examples are found with molecules that are expected to have polar bonds, but are found to have zero molecular dipole moment. Such molecules can be inferred to have a symmetric structure so that the bond dipoles cancel. Another way to look at this is that the centers of positive and negative charge coincide. Molecules with no molecular dipole are classified as nonpolar, even though their individual bonds are polar.

Practice Problem 3.3
Both SO_2 and BF_3 are planar molecules, but only one has a molecular dipole moment. Use Lewis structures and VSEPR to predict which of the two has a non-zero molecular dipole moment.

Practice Problem 3.4
Predict which of the following molecules has the larger dipole moment: H_2O, H_2S, CF_4, CO_2.

Practice Problem 3.5
Which of the following molecules do you predict to be polar: PF_3, CCl_4, Cl_2O, BeH_2, CH_2Br_2?

• Learn about dipole-dipole interactions.
Review - Molecules that are polar associate by interaction of the positive end of one molecule with the negative end of another. These interactions are called dipole-dipole interactions, and constitute one type of intermolecular attractive force.

Practice Problem 3.6
Compare carbon monoxide (CO) and nitrogen (N_2). Both molecules have a molecular weight of 28. Suppose we take separate samples of carbon monoxide and nitrogen gas, under identical pressure conditions, and

progressively cool them. Which do you predict will turn into a liquid first (*i.e.* at a higher temperature)? Explain your choice.

• Learn about van der Waals forces
Review - These forces constitute another type of intermolecular attraction that can occur in nonpolar as well as in polar molecules. Oscillation of the electron charge cloud in a molecule can result in the generation of a temporary dipole, and this can induce another temporary dipole in a neighboring molecule. The van der Waals force results from the interaction of these temporary dipoles. The magnitude of the van der Waals attractive force between molecules increases with molecular weight. For low molecular weight substances the magnitude is very small, and in general, van der Waals forces are the weakest of intermolecular forces. The magnitude of the van der Waals force is also affected by the shape of the molecules involved. This can be important in comparing isomers of organic compounds. In general, the isomer with the largest surface area will have the largest van der Waals intermolecular attractive forces.

Practice Problem 3.7
Which of the CX_4 molecules (X = halogen) do you expect to have the highest boiling temperature? Explain your choice.

Practice Problem 3.8
Give systematic names to the C_9H_{20} isomers that you expect to have: (a) the lowest, and (b) the highest boiling temperature. Explain your choices.

• Be able to recognize the structural characteristics that are conducive to hydrogen bonding.
Review - For hydrogen bopnding to occur between two molecules one must be a hydrogen bond donor and one must be a hydrogen bond acceptor. The hydrogen bond donor must have a hydrogen atom bonded to a small, highly electronegative atom such as F, O, or N, and the hydrogen bond acceptor must also contain F, O, or N, but not necessarily bonded to hydrogen. In general, hydrogen bonds are the strongest of the intermolecular forces.

Practice Problem 3.9
Can hydrogen bonding occur between: (a) two molecules of methanol, CH_3OH, (b) two molecules of acetone, $(CH_3)_2C=O$, (c) a molecule of acetone and a molecule of methanol?

Practice Problem 3.10
Of the three main types of intermolecular forces, which will be present in each of the following (more than one possible for each substance): helium, pentane, 3-chloropentane, hydroxylamine (H_2NOH), formaldehyde ($H_2C=O$).

• Learn how to classify solvents as polar, apolar, protic, aprotic, and understand what is meant by dielectric constant.

Practice Problem 3.11
The dielectric constant of formic acid is 58, while that of acetic acid is 12.4. Which is the more polar solvent?

Practice Problem 3.12
Classify the following solvents as protic or aprotic: methanol, acetone (structure given in Practice Problem 3.9), liquid ammonia, diethyl ether (structure given in Table 3.2 of text), hexane, dichloromethane.

Practice Problem 3.13
In which solvent would you expect heptane to be least soluble: 1-propanol, 1-octanol, diethyl ether, dichloromethane?

• Learn how to classify acids and bases (Brønsted and Lewis). Also know some basic trends in acidity (across a periodic row, down a periodic group), and understand the use of K_a and pK_a. Remember that all acid-base equilibria lie to the side containing the weaker base and the weaker acid. Also remember the generalization that the stronger the acid, the weaker is its conjugate base, and that the stronger the base, the weaker is its conjugate acid.

Practice Problem 3.14
Which is the stronger acid, one with $pK_a = 5.0$ or one with $pK_a = 7.0$? Which acid has the stronger conjugate base?

Practice Problem 3.15
Predict the strongest acid of the group: CH_3OH, CH_3SH, CH_3SeH.

Chapter 3 25

Practice Problem 3.16
Which solution do you predict to have the lower pH: 1×10^{-3} M HF or 1×10^{-3} M HI?

Practice Problem 3.17
In its reaction with HCl, NH_3 acts in which of the following roles (more than one is possible): Lewis acid, Lewis base, Brønsted acid, Brønsted base?

Practice Problem 3.18
The equilibrium constants for all of the following reactions are > 1. Make a list of all of the substances that are acting as Brønsted acids in the forward or reverse reactions, placing them in decreasing order of acid strength. Use only the information given here.

(a) $HCO_3^- + HO^- \rightleftharpoons H_2O + CO_3^{2-}$
(b) $HOAc + HS^- \rightleftharpoons H_2S + OAc^-$ (HOAc = acetic acid)
(c) $H_2S + CO_3^{2-} \rightleftharpoons HCO_3^- + HS^-$
(d) $HSO_4^- + OAc^- \rightleftharpoons HOAc + SO_4^{2-}$

• Understand what is meant by frontier molecular orbitals.

Practice Problem 3.19
What type of orbital would be the HOMO in the reaction of water with hydrogen bromide to give a solution containing hydronium ions and bromide ions? Which orbital would be the LUMO?

Solution of Text Problems

3.1   Extended in three dimensions, each chloride ion is surrounded by six nearest neighbor sodium ions. The distance of closest approach of two sodium ions (center-to-center) is 3.76 angstroms. Extending the lattice in three dimensions each sodium ion is surrounded by twelve nearest neighbor sodium ions.

3.2

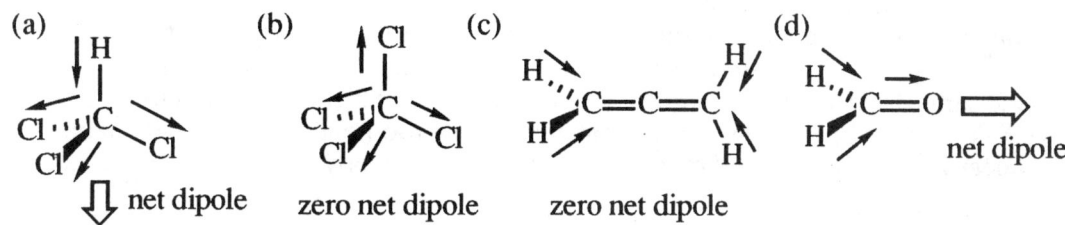

3.3   (a) propanone
       (b) dimethyl ether
       (c) fluoroethane

3.4 $F^- < Cl^- < Br^- < I^-$

3.5

highest   ⇐ increasing boiling point   lowest

3.6   neopentane < pentane < diethyl ether < 2-methyl-2-propanol < 1-butanol

3.7
(a) electron deficient $H_3C—I$ electron rich
(b) electron deficient $(CH_3)_3C—O—H$ electron rich
(c) electron deficient $(CH_3)_2C=O$ electron rich
(d) electron deficient $(CH_3)_2S—O$ electron rich

26 Study Guide and Solutions Manual

3.8 (a) net dipole (b) zero net dipole (c) net dipole (d) small net dipole

3.9 (a) heptane > hexane > pentane
(b) pentane > isopentane > neopentane
(c) octane > 2-methylheptane > 2,2,3,3-tetramethylbutane > diethyl ether
(d) tetrabromomethane > tetrachloromethane
(e) nonane > 3,3-diethylpentane

3.10 [structures in order of decreasing property]

3.11 (a) 1-butnaol
(b) methylpropylamine
(c) 2-butanone butanol
(d) 1-butanol

3.12 b, c, e, f, g, h

NO 3.13 Trichloromethane has the higher dipole moment. The vector sum of the individual bond dipoles is greater with trichloromethane than with dichloromethane.

3.14 In nitrogen trifluoride the vector sum of the three N-F bond dipole moments is in the direction opposite that of the moment from the nitrogen unshared valence level electron pair. With ammonia, however, the bond dipole moments are in the same direction as the moment from the unshared valence level electron pair.

3.15 If the molecule X-Y-X were linear, it would exhibit zero dipole moment. If it were bent, a non-zero dipole moment would result unless the individual bond moments exactly cancelled any moments due to unshared valence level electron pairs.

The ozone molecule is bent. It has a net dipole moment with the central oxygen being the positive end of the dipole. The O-O-O bond angle is ~120°.

3.16 (a) conjugate acid, H_2CO_3; conjugate base, CO_3^{2-}

(b) conjugate acid H_2SO_4; conjugate base, SO_4^{2-}

3.17 (a) base, triethylamine; acid, ethanol; conjugate acid, triethylammonium ion; conjugate base, ethoxide ion
(b) base, ethanol; acid, ethanol; conjugate acid, ethyloxonium ion; conjugate base, ethoxide ion
(c) base, water; acid, ethanol; conjugate acid, hydronium ion; conjugate base, ethoxide ion

3.18 H_3PO_4 > HCN > H_2O

3.19 (a) ethanol
(b) ethanol
(c) CH_3CH_2SH

3.20

HÖ:⁻ + HÖCH₂CH₃ ⇌ H₂Ö + :ÖCH₂CH₃⁻
                            conjugate   conjugate
                              acid   base

3.21 $CH_3O^- < CH_3S^- < CH_3Se^-$

3.22 $CH_3Se^- < CH_3S^- < CH_3O^-$

3.23

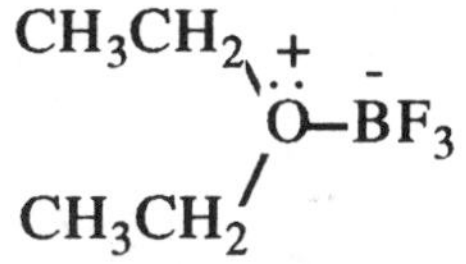

$$CH_3CH_2 \overset{+}{\underset{CH_3CH_2}{\diagdown}} \ddot{O} - \bar{B}F_3$$

3.24 $K_a \times K_b = K_w = 10^{-14}$ $1.8 \times 10^{-5} \times K_a = 10^{-14}$ $K_a = 5.6 \times 10^{-10}$

3.25 $K_a \times K_b = K_w = 10^{-14}$ $2.5 \times 10^{-11} \times K_b = 10^{-14}$ $K_a = 4 \times 10^{-4}$

3.26 (a) HA is the stronger acid
(b) 100
(c) B⁻ is the stronger base
(d) No, the equilibrium would lie to the opposite side (HB) because HA is a stronger acid than is HB.

3.27 (a) electron deficient

$$CH_3CH_2CH_2CH_2 \longrightarrow Br$$

electron rich

(b) electron deficient

$$CH_3CH_2S - H$$

electron rich

(c) electron deficient

$$(CH_3)_2C = NCH_3$$

electron rich

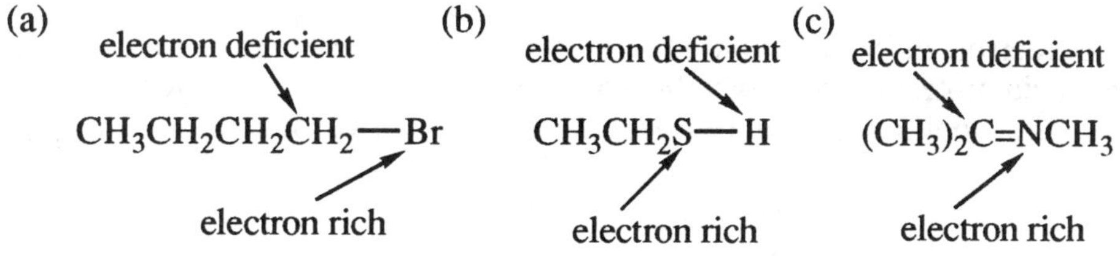

3.28

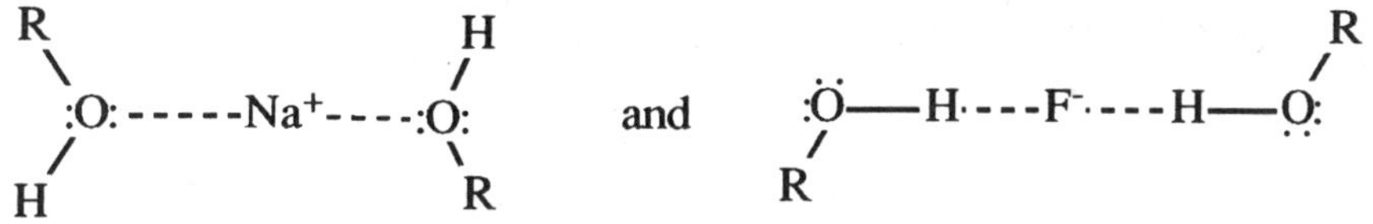

$$\underset{H}{\overset{R}{\diagdown}} \ddot{O}: \text{-----} Na^+ \text{----} :\ddot{O}\underset{R}{\overset{H}{\diagup}} \quad \text{and} \quad \underset{R}{\overset{R}{\diagup}}\ddot{O} - H \cdots F^- \cdots H - \ddot{O}\underset{}{\overset{R}{\diagdown}}$$

In NaI and NaF there are present ionic interactions attracting oppositely charged ions and repelling the like charged ions. We find that NaI is more soluble in water than is NaF owing to the weaker ionic interactions holding the lattice of the NaI crystal together as compared to the NaF crystal.

3.29 (a) HOMO is σ, LUMO is σ*; (b) HOMO is π_1 (or π_2), LUMO is π_3* (or π_4*); (c) HOMO is π_1, LUMO is π_2*.

3.30 We would add water to the hexane-ethanol mixture. Only the ethanol would dissolve to any measurable extent in the water leaving the hexane (with some ethanol still in it) as a separate layer. Several successive treatments of the mixture with pure water would remove virtually all of the ethanol from the hexane.

3.31 The conformation of glyoxal shown to the right is the more stable. In this conformation the C=O dipoles are in opposite directions, whereas in the conformation shown to the left the C=O dipoles are aligned with each other. In an aligned conformation there would result a repulsive interaction of the negative ends of the dipoles which could be relieved by rotating to the conformation shown to the right. The net dipole moment of the glyoxal molecule to be expected having the bond dipoles in opposite directions is zero.

3.32

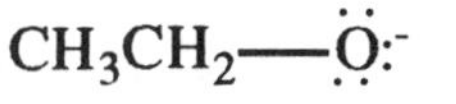

$$CH_3CH_2 - \ddot{O}:^-$$

longer C-O bond

$$CH_3 - \overset{\overset{\displaystyle :O:}{\|}}{\underset{:\ddot{O}:^-}{C}}$$

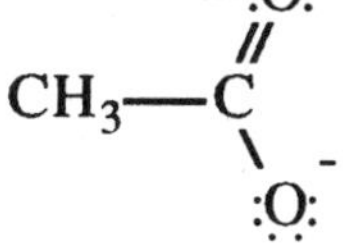

more stable, weaker base

3.33 The preparation of acetic acid would be the successful one.

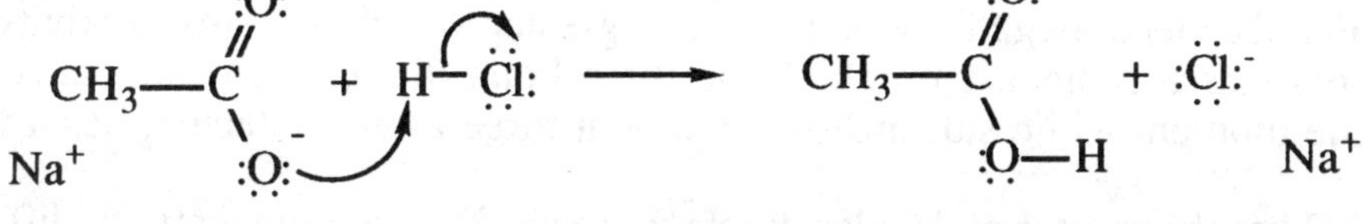

$$\underset{Na^+}{} CH_3 - \overset{\overset{\displaystyle :O:}{\|}}{\underset{:\ddot{O}:^-}{C}} + H - \ddot{C}\ddot{l}: \longrightarrow CH_3 - \overset{\overset{\displaystyle :O:}{\|}}{\underset{:\ddot{O} - H}{C}} + :\ddot{C}\ddot{l}:^- \quad Na^+$$

28 Study Guide and Solutions Manual

3.34

$$H_2\ddot{N}:^- \; + \; H-\ddot{O}-H \;\rightleftharpoons\; H_2\ddot{N}-H \; + \; {}^-{:}\ddot{O}-H$$
K⁺   (dissolves   K⁺
                   in water)

Amide ion is a much stronger base than is hydroxide ion. Fluoride ion is a weaker base than is hydroxide ion, and would not react appreciably with water to form HF.

3.35 In water HCl dissociates to form hydronium ion and chloride ion. The hydronium ion, the conjugate acid of water and the strongest acid which can exist to an appreciable extent in aqueous solution, is a weaker acid than is HCl. In hexane solution HCl remains undissociated and exhibits its full Brønsted acidity.

3.36 The packing of the two types of molecules in the liquid phase is tighter than in the two separate pure materials.

3.37 The surface area of the hydrophobic portion is minimized in *tert*-butyl alcohol, which is nearly spherical in shape.

3.38 In formate anion there is electron delocalization equally between the two oxygen atoms (resonance stabilization) whereas in the formic acid molecule the two carbon-oxygen bonds are not equivalent: one is a double bond to carbon while the other is a single bond to carbon (with a second bond to hydrogen).

3.39

(a) [Resonance structures of the anion of 2,4-pentanedione (acetylacetonate), showing three contributors with negative charge on central C, on one O, and on the other O]

(b) [Resonance structures showing two contributors of the anion derived from deprotonation at the CH₂ end of 2,4-pentanedione]

Removal of a proton from the central carbon generates an anion which has a greater degree of stabilization by electron delocalization (resonance).

3.40 Sulfuric acid is the stronger of the two acids. If it were not, we could not cause the equilibrium to favor the formation of HCl.

3.41

$$Na^+ \; H{:}^- \; + \; H-\ddot{O}H \;\longrightarrow\; H-H \; + \; {:}\ddot{O}H^- \; + \; Na^+$$

Lewis   Lewis
base   acid

The NaH and H-O-H are also acting as Brønsted bases in this reaction.

3.42 We would need to consider the LUMO as the σ* molecular orbital of the H-O bond, and the HOMO as the filled *1s* orbital of the hydride ion of sodium hydride.

Solution of Study Guide Practice Problems

3.1

I-I no bond dipole; H→Cl; I→Cl

3.2 Although the H-Y bond is shorter than the H-X bond, the H-Y dipole moment is still greater than the H-X dipole moment. Thus we conclude that the electronegativity of Y must be greater than the electronegativity of X.

3.3 Boron trifluoride is a trigonal planar molecule in which the three bond dipole moments cancel. Boron trifluoride has zero molecular dipole moment while sulfur dioxide, a bent three-atom molecule, has a non-zero molecular dipole moment.

3.4 Of the four possibilities, H_2O has the greater molecular dipole moment. The electronegativity difference in H_2S is less than in H_2O, and the other possibilities involve internal cancellation of the individual bond dipole moments.

3.5　The ones predicted to be polar are: PF_3, Cl_2O, and CH_2Br_2.

3.6　Carbon monoxide will have the higher boiling temperature. Carbon monoxide has a permanent dipole moment which provides an attractive intermolecular force for a collection of CO molecules whereas the nitrogen molecule has no permanent dipole.

3.7　We anticipate CI_4 to have the highest boiling temperature of the tetrahalomethanes. As the molecular weight is the greatest, the magnitude of the van der Waals forces (attraction) will be greatest.

3.8　We would predict 3,3-diethylpentane to have the lowest boiling temperature, as it has the smallest surface area, and nonane (*n*-nonane) to have the highest boiling temperature, as it has the largest surface area of the C_9H_{20} isomers.

3.9　(a) yes
　　　(b) no
　　　(c) yes

3.10　helium - van der Waals forces
　　　pentane - van der Waals forces
　　　3-chloropentane - dipole-dipole interactions and van der Waals forces
　　　hydroxylamine - hydrogen bonding, dipole-dipole interactions, and van der Waals forces
　　　formaldehyde - dipole-dipole interactions and van der Waals forces

3.11　Formic acid is a more polar solvent than is acetic acid.

3.12　methanol - protic
　　　acetone - aprotic
　　　liquid ammonia - protic
　　　diethyl ether - aprotic
　　　hexane - aprotic
　　　dichloromethane - aprotic

3.13　We would expect heptane to be least soluble in the most polar of the given solvents, that is it is least soluble in the 1-propanol.

3.14　The stronger acid is the one with the pK_a = 5.0. The acid with the pK_a = 7.0 has the stronger conjugate base.

3.15　The strongest acid of the three possibilities is CH_3SeH.

3.16　We predict that the HI solution will have the lower pH as it is more completely dissociated than is HF.

3.17　Ammonia is acting as both a Brønsted base and as a Lewis base in reaction with HCl.

3.18　The order of acid strength is: $HSO_4^- > HOAc > H_2S > HCO_3^- > H_2O$.

3.19　A non-bonded (filled) orbital on oxygen serves as the HOMO in this reaction while the LUMO would be the σ^* orbital of the HBr molecule.

CHAPTER 4
ALKANES AND CYCLOALKANES. I. AN INTRODUCTION TO STRUCTURE AND REACTIONS

Key Points

* Know how to compute the number of rings and/or π bonds in a hydrocarbon from its molecular formula.
Review - First, note the number of carbon atoms indicated by the formula. Multiply this number by two, and then add two. This gives the number of hydrogen atoms that would be present if there were no rings and no π bonds present. For every ring or π bond that is present, the number of hydrogen atoms is decreased by two. For a double bond the number is decreased by two and for a triple bond the number is decreased by four.

Practice Problem 4.1
How many rings or π bonds are present in a compound of formula $C_{20}H_{28}$?

Practice Problem 4.2
What is the formula of a hydrocarbon that contains eighteen carbon atoms, six of which are joined in a cyclohexane ring? Assume that no double or triple bonds are present.

Practice Problem 4.3
What is the formula of a compound containing 25 carbon atoms in which are located two five membered rings as well as two double bonds and one triple bond?

* Review the naming and representation of structure of branched alkanes.

Practice Problem 4.4
Give a systematic name to the compound $(CH_3)_2CHCH_2CH(CH_3)CH_2CH_2CH(CH_3)_2$.

Practice Problem 4.5
Draw a skeletal (stick type) structure for 3-ethyl-2,4-dimethylpentane.

Practice Problem 4.6
Name the compound represented below.

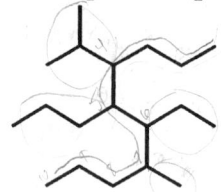

* Learn the nomenclature for cycloalkanes. When you are working with cyclic compounds, be aware that you may need to specify the compound as *cis-* or *trans-*.

Practice Problem 4.7
Name the following compounds.

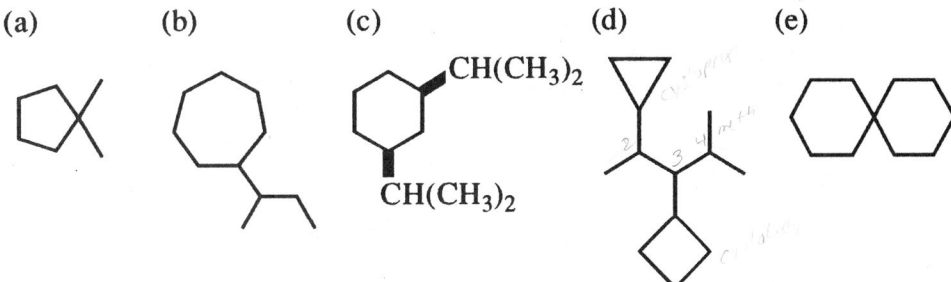

- Understand what is meant by the heat of combustion of a reaction. Understand the connection between the heat of combustion of isomers and the relative stabilities of the isomers. Understand how to use the heat of combustion of cycloalkanes to infer information about ring strain.

Practice Problem 4.8
Isomer A is more stable than isomer B. Which do you expect to have the more exothermic heat of combustion? For which combustion reaction will ΔH be more negative?

Practice Problem 4.9
Which compound, propane or hexane, do your predict will have the most exothermic heat of combustion (kcal/mole)? Explain your decision.

Practice Problem 4.10
Which is larger: (2 x the heat of combustion of cyclopropane) or (the heat of combustion of cyclohexane)? Explain your decision.

- Understand how to deduce a molecular formula from combustion analysis data.

Practice Problem 4.11
Squalene is a compound that can be extracted from shark oil. A 16.0 mg sample was subjected to combustion analysis, producing 51.2 mg of carbon dioxide and 17.5 mg of water. The molecular weight of squalene has been measured to be ~410. What is the molecular formula of the compound? No rings and no triple bonds are present in the squalene molecule; how many double bonds must be present?

- Remember the conditions for reaction of an alkane or cycloalkane with a halogen, and know the nature of the reaction and the nature of the products formed. Remember what is meant by the term *homolytic bond cleavage*.

Practice Problem 4.12
When cyclohexane and bromine are mixed at room temperature, no reaction occurs, but when light is directed onto the reaction mixture a gas is evolved and a bromine containing organic compound is formed. Why does the reaction occur when light is directed onto the reaction mixture? What is the gas that is evolved and what is the name, structure and formula of the organic bromine compound produced?

Practice Problem 4.13
Consider the reactions of (a) pentane and (b) spiro[3.3]heptane with bromine in the presence of light. In each case, how many different substitution products containing a single bromine atom could form?

- Know the different types of strain that can exist in organic molecules - angle strain, torsional strain, and steric strain. Be able to use and to draw diagrams showing potential energy as a function of rotation about the carbon-carbon bonds of simple alkanes.

Practice Problem 4.14
What type of strain makes an eclipsed conformation less stable than a staggered conformation of ethane?

Practice Problem 4.15
Although we often speak of van der Waals interactions as being attractive forces, under certain circumstances they can produce repulsive forces. Under what circumstances can this occur and what type of strain results from such repulsive van der Waals interactions?

Practice Problem 4.16
Draw Newman projections to show (a) the most stable, and (b) the least stable conformations of hexane, as viewed along the C3-C4 bond.

Solution of Text Problems
4.1 (a) C_nH_{2n+2} (b) C_nH_{2n} (c) C_nH_{2n-2}

32 Study Guide and Solutions Manual

4.2 cyclobutane and methylcyclopropane
4.3 A saturated *open-chain* alkane of ten carbons would have twenty-two hydrogens. This compound is deficient by six hydrogens from the open-chain norm, indicating that there are three rings present.
4.4 (a) 2,5-dimethyloctane
(b) 2,2,6,6-tetramethylheptane
(c) 2-methyl-3-(1-methylethyl)hexane
4.5 3 isopropyl 2-methylhexane

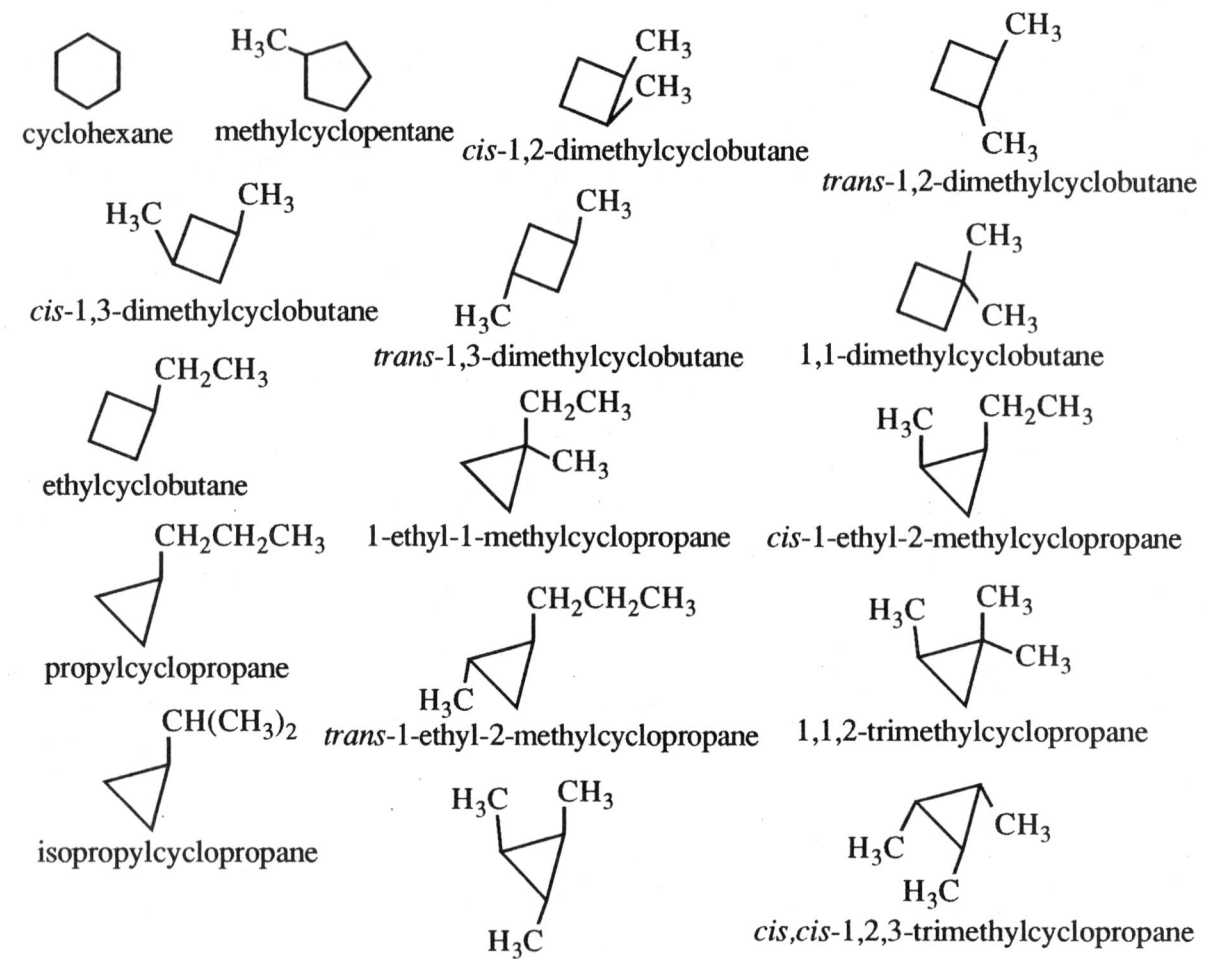

e) trans 1-3 dethylcyclooctane

4.7 (a) *cis*-1-ethyl-2-methylcyclohexane
(b) 1,1,2-trimethylcycloheptane
(c) *trans*-1,3-cyclooctane

4.8 (a) $C_6H_{14} + 19/2\, O_2 \rightarrow 6\, CO_2 + 7\, H_2O$

(b) $C_6H_{12} + 9\, O_2 \rightarrow 6\, CO_2 + 6\, H_2O$

(c) $C_7H_{14} + 21/2\, O_2 \rightarrow 7\, CO_2 + 7\, H_2O$

4.9 Carbon dioxide, of molecular weight 44.009, is 27.29% carbon. Thus, 9.42 mg of carbon was present in the starting sample.
Water, of molecular weight 18.015, is 11.19% hydrogen. Thus, 1.98 mg of hydrogen came from the starting sample. The total of carbon and hydrogen correlates with the weight of the starting sample, 11.40 mg.
Relative number of C atoms = 9.42/12.011 = 0.78
Relative number of H atoms = 1.98/1.008 = 1.96

(Number of C atoms)/(Number of H atoms) = 0.78/1.96 = 1/2.51, or $CH_{2.51}$, which has a formula weight of 14.54. With a molecular weight of 58, (58)/(14.54) = 3.99, the molecular formula should be $(CH_{2.51})_4$ or C_4H_{10}, as is the formula for butane.

4.10 The portion of the sample which is carbon is 7.28 mg. The portion of the sample which is hydrogen is 1.22 mg. This correlates with the weight of the original sample, 8.50 mg.
(Number of C atoms)/(Number of H atoms) = 0.606/1.210 or 1/2. Thus, the empirical formula is CH_2. *The formula shows that this cmpd could be a cycloalkane, but not an alkane.* With a molecular weight of 70, (70)/(14) = 5, for a molecular formula of $(CH_2)_5$ which is C_5H_{10}.

4.11
(a) One product is formed.

1-chloro-2,2,3,3-tetramethylbutane

(b) Three products are formed.

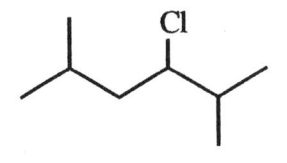

1-chloro-2,5-dimethylhexane   2-chloro-2,5-dimethylhexane   3-chloro-2,5-dimethylhexane

4.12 (a) 2,4-dimethylpentane
(b) pentane
(c) 3-methylpentane
(d) 4-(1-methylethyl)octane *or 4 isopropyloctane*
(e) spiro[4.4]decane *nonane*
(f) spiro[5.2]octane *2.5*
(g) *trans*-1,2-di-(1-methylethyl)cyclopentane *or trans 1,2 diisopropylcyclopentane*
(h) 2-cyclobutyl-1-cyclohexylpropane
(i) 3-ethyl-2,6,6-trimethyloctane
(j) *cis*-1,3-di-(1-methylpropyl)cyclobutane

4.13 (a) 2,5-dimethylhexane
(b) 3,4-dimethylbutane *hexane*
(c) 2,2,3,3-tetramethylbutane

4.14
(a) $(CH_3)_4C$ neopentane *1,1 dimethylcyclopropane*
(b) $(CH_3)_2CHCH_2CH_3$ isopentane
(c) $(CH_3)_4C$ neopentane
(d) $(CH_3)_4C$ neopentane
(e) □ cyclobutane

4.15

34 Study Guide and Solutions Manual

4.16

(a) cyclopropyl-CH(CH₃)₂

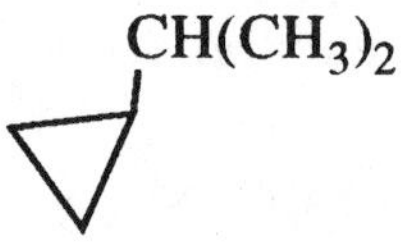

(b) cyclopropyl-CH₂CH(CH₃)₂

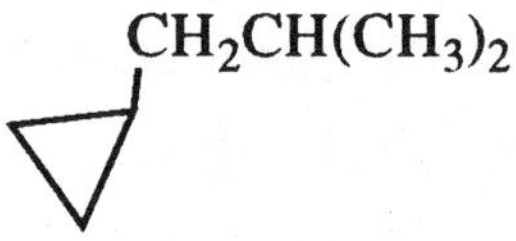

(c) cyclopropyl-CH₂CH(CH₃)₂ ; cyclopropyl with CH(CH₃)₂ and CH₃ ; cyclopropyl with CH(CH₃)₂ and CH₃ ; cyclopropyl with CH(CH₃)₂ and H₃C ; cyclobutyl-CH(CH₃)₂

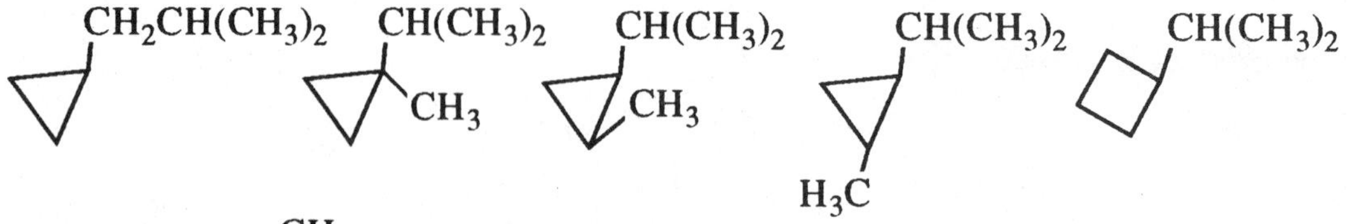

(d) cyclobutane structures with various methyl/ethyl substituents as shown

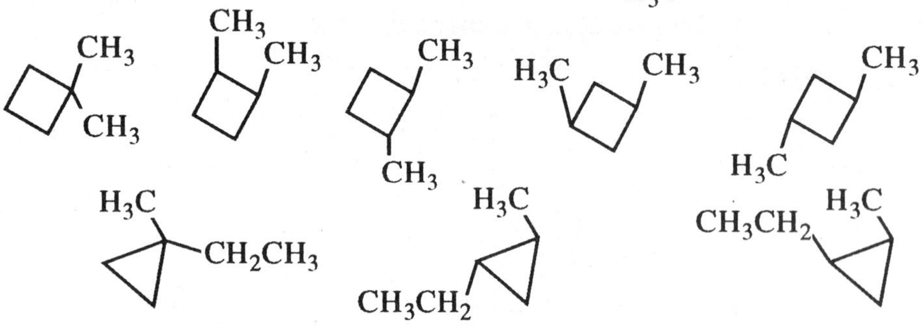

4.17
(a) nonane - has the greater surface area for a given volume
(b) hexane - has the lesser degree of branching, and thereby is of higher energy and liberates more energy upon combustion
(c) methylcyclopropane - has the greater energy because of ring strain
(d) cyclopentane - C_5H_{10} vs. C_5H_8
(e) cyclobutane - butane has two methyl groups
(f) *cis*-1,2-dimethylcyclopropane - methyl groups in a *cis* relationship raise the energy of the compound

4.18

(a) Newman projection with CH₂CH₂CH₃ (front top), H, H, and CH₂CH₂CH₃ (back bottom), H, H

(b) Newman projection with CH₂CH₂CH₃ (front top), H, H, and CH₂CH₂CH₃ (back top-right), H, H

(c) Newman projection with CH₂CH₂CH₃ (front top), H, H, and CH₂CH₂CH₃ (back top-left), H, H

(d) Newman projection with CH₂CH₂CH₃ (front top), H, H, and CH₂CH₂CH₃ (back bottom), H, H

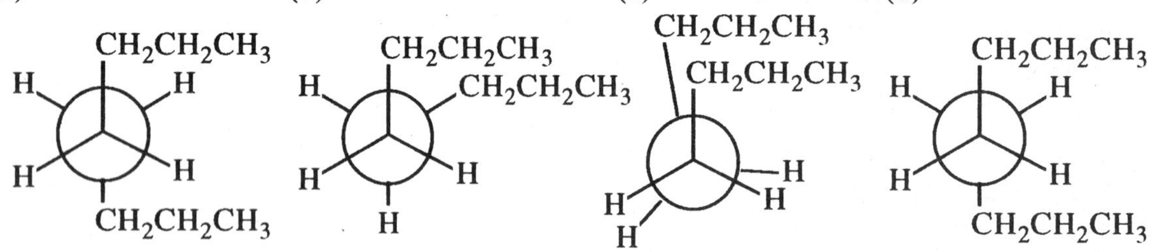

4.19

[structures:]
1,1,2-trimethylcyclobutane

1,1,3-trimethylcyclobutane

cis,cis-1,2,3-trimethylcyclobutane

cis,trans-1,2,3-trimethylcyclobutane

trans,trans-1,2,3-trimethylcyclobutane

4.20
1-chlorohexane; 2-chlorohexane; 3-chlorohexane

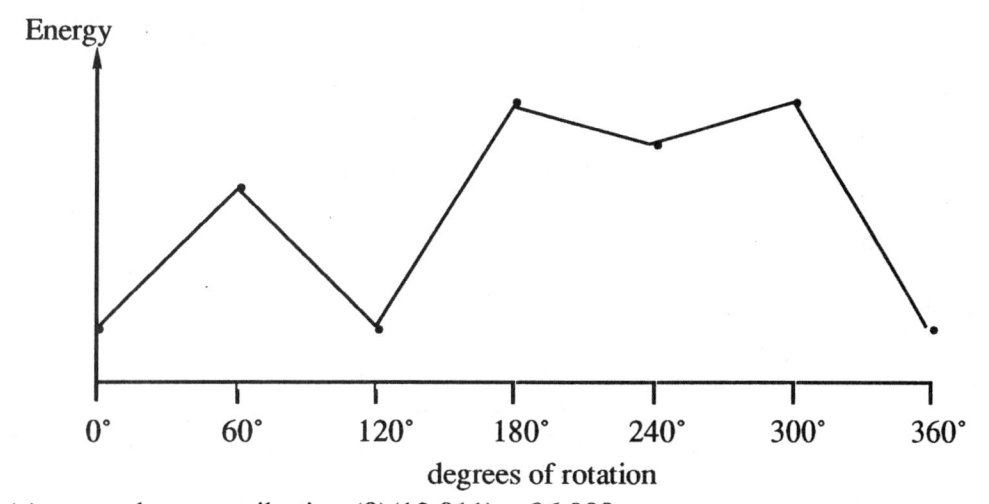

4.21

[Newman projections at 0° (360°), 60°, 120°, 180°, 240°, 300°]

[Energy diagram vs. degrees of rotation from 0° to 360°]

4.22
(a) carbon contribution (8)(12.011) = 96.088
hydrogen contribution (16)(1.008) = 16.128
molecular weight = 112.216
% carbon = (96.088)/(112.216) x (100) = 85.63%
% hydrogen = (16.128)/112.216 x (100) = 14.37%

(b) carbon contribution (5)(12.011) = 60.055

hydrogen contribution (8)(1.008) = 8.064
molecular weight = 68.119
% carbon = (60.055)/(68.119) x (100) = 88.16%
% hydrogen = (8.064)/(68.119) x (100) = 11.84%

4.23 From Problem 4.9 we know that % carbon in carbon dioxide is 27.29% and the % hydrogen in water is 11.19%. Thus:
(31.18)(0.2729) = 8.51 mg carbon in sample
(8.49)(0.1119) = 1.00 mg hydrogen in sample
(8.51)/(12.011) = 0.709 relative number of carbon atoms
(1.00)/(1.008) = 0.992 relative number of hydrogen atoms
(0.709)/(0.992) leads to an empirical formula of C_5H_7

This is obviously not the empirical formula as it can not represent a neutral, stable molecule; it would require an odd valence for carbon.

4.24 No. Both have the same *empirical formula* (CH_2) and would thereby produce the same combustion analysis.

4.25 It is not described by a single energy, but rather a range of energies. Rotations about other carbon-carbon bonds result in a raising or lowering of the total energy for the conformation eclipsed about the central carbon-carbon bond.

Solution of Study Guide Practice Problems

4.1 There are a total of 7 rings and/or π bonds.
4.2 The formula is $C_{18}H_{36}$.
4.3 The formula is $C_{25}H_{40}$.
4.4 The name is 2,4,7-trimethyloctane.
4.5

4.6 The name of the compound is 5-ethyl-7-isopropyl-4-methyl-6-propyldecane.
4.7 (a) 1,1-dimethylcyclopentane
     (b) 2-cycloheptylbutane
     (c) *cis*-1,3-diisopropylcyclohexane
     (d) 3-cyclobutyl-2-cyclopropyl-4-methylpentane
     (e) spiro[5.5]undecane

4.8 Isomer B will have the more exothermic heat of combustion, and it follows that it has the more negative ΔH for the combustion reaction.

4.9 Hexane will have the more exothermic heat of combustion because it is larger and there are more carbons and hydrogen sites to be oxidized.

4.10 Twice the heat of combustion of cyclopropane is larger than the heat of combustion of cyclohexane. The cyclohexane molecule contains the same number of carbon and hydrogen atoms as two molecules of cyclopropane, but there is no strain in the cyclohexane ring system.

4.11 We calculate the molecular formula $C_{30}H_{50}$ for squalene. Squalene thereby contains six double bonds.

4.12 Reaction occurs only when light is directed onto the reaction mixture because the formation of free radicals is required for the reaction to occur. Light causes the splitting of bromine molecules into bromine atoms (free radicals) allowing reaction to occur. The gas evolved is hydrogen bromide. Bromocyclohexane is formed of formula $C_6H_{11}Br$ and structure:

4.13 For pentane three different products containing a single bromine atom could be isolated. With spiro[3.3]heptane only two different products containing a single bromine atom could be isolated.

4.14 Torsional strain makes the eclipsed conformation less stable than the staggered conformation.

4.15 Steric strain occurs when two groups within a single molecule approach closer than their van der Waals radii would normally allow.

4.16

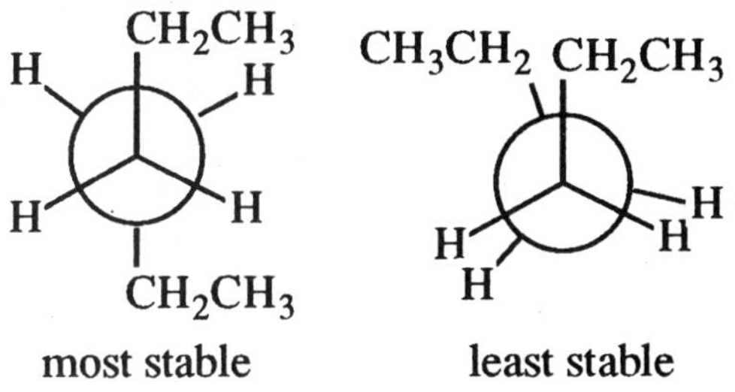

most stable   least stable

CHAPTER 5
METHANOL AND ETHANOL

Key Points

• Understand the role hydrogen bonding plays in influencing the physical and chemical properties of alcohols.

Practice Problem 5.1
Explain why methane (molecular weight 16) has a lower boiling temperature than ethane (molecular weight 30), while water (molecular weight 18) has a higher boiling temperature than methanol (molecular weight 32).

Practice Problem 5.2
Explain why water has a higher boiling temperature than does ethanol, but hydrogen sulfide (H_2S) has a lower boiling temperature than ethanethiol (CH_3CH_2SH).

Practice Problem 5.3
Explain why 1,2-ethanediol ($HOCH_2CH_2OH$) has a significantly higher boiling temperature (198°C) than that of either 1-propanol (97.4°C) or 2-propanol (82.4°C) even though all three substances have approximately the same molecular weight and all three can participate in hydrogen bonding.

Practice Problem 5.4
The boiling temperature of 1,5-pentanediol (260°C) is significantly higher than those of isomeric pentanediols. For example, 1,2-pentanediol and 2,3-pentanediol have boiling temperatures of 212°C and 188°C respectively. Propose a likely explanation for this observation.

• Learn the reactions that convert alcohols to alkoxides (either reaction with an active metal such as Na or K, or with a strong base). Understand that the base used to convert an alcohol to an alkoxide should be appreciably stronger than the alkoxide itself, or the position of equilibrium will be unfavorable.

Practice Problem 5.5
When sodium metal is added to ethanol, a gas is evolved. What is this gas?

Learning Hint

Upcoming sections of the text will be making extensive use of the curved arrow formalism introduced in Chapter 1 to depict electron shifts in chemical reactions. You should begin practicing with the curved arrow formalism now. Remember that each arrow shows the shift of a pair of electrons. The head of the arrow shows the destination of the electron pair. For example, consider the reaction of amide ion with methanol. This reaction is depicted in Eqn. 5.6 of the text and is reproduced below for convenience.

$$CH_3\ddot{O}-H + \dot{N}H_2^- \rightleftharpoons CH_3\ddot{O}:^- + :NH_3$$

By comparing the products with the reactants we can see which bonds are broken and which are formed in the reaction, and thus decide which electrons have been shifted. Here we see that the O-H bond of the starting methanol is broken and a new N-H bond is created. Also, we see that the nitrogen atom begins with two unshared valence level electron pairs but ends up with only one in the product, and that the oxygen atom begins with two unshared valence level electron pairs and ends up with three in the product. Putting everything together we can postulate that one of the unshared valence level electron pairs on nitrogen is used to form a bond to the hydroxylic group hydrogen atom of methanol, and that the electron pair constituting the O-H bonds of methanol remains on the oxygen atom as a new unshared valence level electron pair. The curved arrow representation for this reaction is shown below.

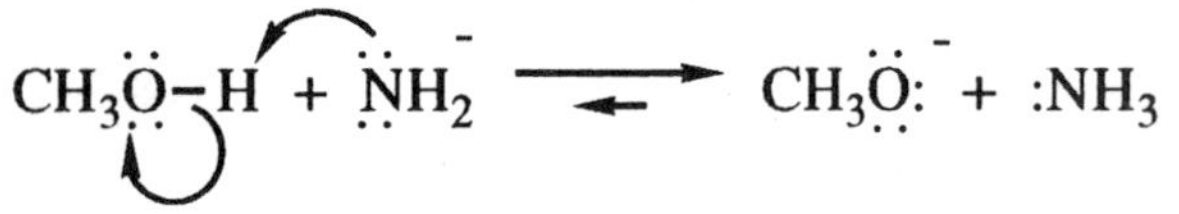

$$CH_3\ddot{O}-H + \dot{N}H_2^- \rightleftharpoons CH_3\ddot{O}:^- + :NH_3$$

A typical error that beginning students make is to concentrate on the movement of atoms rather than on electrons. This can lead to an erroneous use of curved arrows. For example, in the reaction shown above some students incorrectly focus on the transfer of a hydrogen atom from methanol to amide ion and write the curved arrow representation shown below.

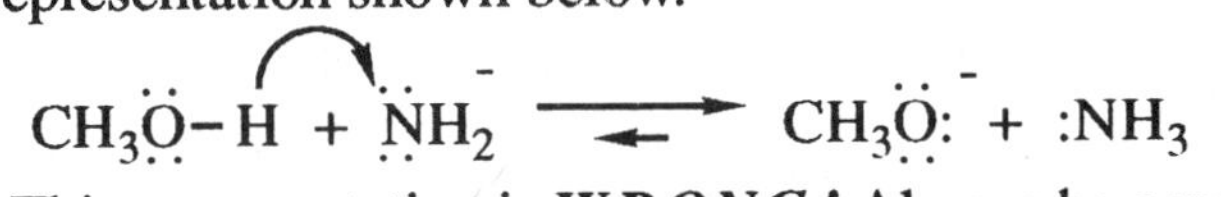

$$CH_3\ddot{O}-H + \ddot{N}H_2^- \rightleftarrows CH_3\ddot{O}:^- + :NH_3$$

This representation is **WRONG!** Always be careful to place the curved arrows appropriately.

Practice Problem 5.6
Use the curved arrow formalism to depict the reaction between methoxide ion and water to produce methanol and hydroxide ion.

Practice Problem 5.7
Use the curved arrow formalism to depict the reaction of hydrogen chloride and water to produce hydronium ion and chloride ion.

Practice Problem 5.8
Predict the products of a reaction in which electrons are shifted as indicated by the curved arrows in the reactants shown below.

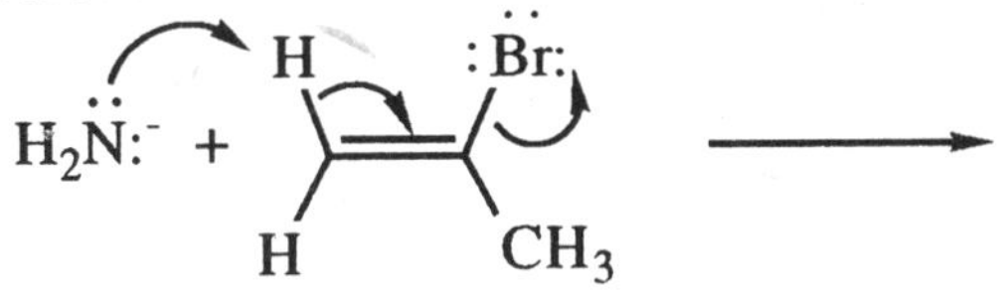

$$H_2\ddot{N}:^- + \text{(alkene with H, H, :Br:, CH}_3\text{)} \longrightarrow$$

- Learn how to recognize if a particular conversion is an oxidation, a reduction, or neither (section 5.7 of text).

Practice Problem 5.9
Are the conversions shown below oxidations, reductions, or neither?
- (a) ethanol converted to ethane
- (b) ethanol converted to ethanal (acetaldehyde; CH_3CHO)
- (c) methanol converted to methyloxonium ion
- (d) acetic acid (CH_3COOH) converted to ethanal

- Study all aspects of the reaction between ethanol and HBr as presented in Chapter 5 of the text. Methanol reacts in the same manner as does ethanol, and HCl and HI react similarly to HBr.

Practice Problem 5.10
Write out the two-step mechanism for the reaction of methanol with HI to produce iodomethane and water. Show the curved arrow formalism for both steps.

Practice Problem 5.11
Which of the descriptions given below best fits a nucleophile?
- (a) Lewis base
- (b) Lewis acid
- (c) neither Lewis acid nor Lewis base
- (d) could be either

Practice Problem 5.12
What orbital is the HOMO of the bromide ion? Is the energy of this orbital higher, lower, or the same as the energy of the HOMO of the chloride ion?

Solution of Text Problems

5.1 weight carbon in sample = (13.96)(0.2729) = 3.81 mg
weight hydrogen in sample = (11.43)(0.1119) = 1.28 mg
Therefore, there must be (10.16) - (5.09) = 5.07 mg oxygen in the sample.
relative number of carbon atoms = (3.81)/(12.011) = 0.317
relative number of hydrogen atoms = (1.28)/(1.008) = 1.270
relative number of oxygen atoms = (5.07)/(15.999) = 0.317
Dividing through by the smallest number, we arrive at an empirical formula of CH_4O. Since the molecular weight is found to be 32, the empirical formula is also the molecular formula.

5.2 weight carbon in sample = (6.00)(0.2729) = 1.64 mg
weight hydrogen in sample = (1.63)(0.1119) = 0.18 mg
Therefore, there must be (4.00) - (1.82) = 2.18 mg oxygen in the sample.

relative number of carbon atoms = (1.64)/(12.011) = 0.136
relative number of hydrogen atoms = (0.18)/(1.008) = 0.178
relative number of oxygen atoms = (2.18)/(15.999) = 0.136
Dividing through by the smallest number, we arrive at an empirical formula of $C_3H_4O_3$. Since the molecular weight is 128, the molecular formula must be twice the empirical formula, or $C_6H_8O_6$.

5.3 relative number of carbon atoms = (60.02)/(12.011) = 5.00
relative number of hydrogen atoms = (13.33)/(1.008) = 13.22
relative number of oxygen atoms = (26.65)/(15.999) = 1.66
Dividing through by the smallest number we arrive at an empirical formula of C_3H_8O. Since the molecular weight is ~60, the empirical formula is the molecular formula. Since the molecule is saturated and does not contain a hydroxyl group, it must be an ether. The structure must therefore be: $CH_3OCH_2CH_3$.

5.4
(a) Li–CH₃ + H–Ö–H ⟶ H₃C–H + Li⁺ + HÖ:⁻
(b) Li–CH₃ + H–Ö–CH₃ ⟶ H₃C–H + Li⁺ + H₃CÖ:⁻
(c) Li–CH₃ + H–Ö–CH₂CH₃ ⟶ H₃C–H + Li⁺ + CH₃CH₂Ö:⁻

5.5
$$CH_3CH_2\text{-}\overset{\overset{O}{\|}}{C}\text{-}H \qquad CH_3CH_2\text{-}\overset{\overset{O}{\|}}{C}\text{-}OH$$

5.6
$$CH_3\text{-}\overset{\overset{O}{\|}}{C}\text{-}\overset{\overset{O}{\|}}{C}\text{-}OH$$

5.7
CH₃CH₂–Ö–H + H–Br: ⟶ CH₃CH₂–Ö⁺(H)(H) + :Br:⁻

:Br:⁻ + CH₃CH₂–Ö⁺(H)(H) ⟶ CH₃CH₂–Br: + H–Ö–H

5.8 (a) C, 83.24%; H, 16.76%
(b) C, 83.90%; H, 16.10%
(c) C, 29.84%; H, 6.26%; O, 19.87%
(d) C, 40.02%; H, 6.71%; O, 53.28%

5.9 (a) $C_4H_{10}O_2$
(b) $C_6H_{10}N_4O_4$
(c) $C_{14}H_{12}O_4$
(d) $C_6H_9N_3O_2$

5.10 C_3H_6O

5.11 C_5H_8

5.12 There is one phosphorus atom per molecule of protein.

5.13 There are four atoms of iron per molecule of protein.

5.14 $C_8H_8O_2$

5.15 $C_5H_8NO_4Na$

5.16 $C_4H_8O_2$

5.17 Ethanol is capable of participating in extended intermolecular hydrogen bonding whereas pure dimethyl ether has no such capability. The presence of this hydrogen bonding makes it much more difficult to cause individual ethanol molecules to go into the gas phase from the liquid phase than is the case with dimethyl ether.

5.18   With two hydroxyl groups the hydrogen bonding capabilities of ethylene glycol in the liquid phase are much greater than those for butanol.

5.19   *t*-Butyl alcohol is nearly spherical in shape, minimizing the surface area of the non-polar regions of the molecule. The linear 1-butanol maximizes the non-polar surface area. The lower the surface area of the non-polar regions, the fewer unfavorable interactions with the surrounding water will occur.

5.20   (a) ethanol
      (b) ethanol

5.21
   (a) $CH_3CH_2OH + Na \rightarrow CH_3CH_2O^- \, Na^+ + 1/2 \, H_2$

   (b) $CH_3OH + NaNH_2 \rightarrow CH_3O^- \, Na^+ + NH_3$

   (c) $CH_3CH_2OH + HBr \rightarrow CH_3CH_2\overset{+}{O}H_2 + Br^-$

5.22
   (a) H—Ö—H

   (b) CH₃—Ö—H

   (c) CH₃CH₂—$\overset{+}{\text{Ö}}$H₂

   (d) H—C̈l:

5.23
   (a) CH₃C(=O)—Ö:⁻

   (b) :B̈r:⁻

   (c) CH₃—Ö:⁻

   (d) H₂Ö

   (e) CH₃—ÖH

5.24
   (a) $CH_3\text{-}\ddot{O}\text{-}H + H\text{-}O\text{-}S(=O)_2\text{-}OH \rightarrow CH_3\overset{+}{O}H_2 + {}^-\!:\ddot{O}\text{-}S(=O)_2\text{-}OH$

   (b) $:\ddot{B}r:^- + CH_3\text{-}\overset{+}{O}H_2 \rightarrow :\ddot{B}r\text{-}CH_3 + H_2\ddot{O}$

   (c) $CH_3CH_2\text{-}\ddot{O}\text{-}H + {}^-\!\ddot{N}H_2 \rightarrow CH_3CH_2\text{-}\ddot{O}:^- + \ddot{N}H_3$

5.25   (a) reduction
      (b) oxidation
      (c) reduction
      (d) oxidation
      (e) oxidation

5.26   Carbon dioxide, O=C=O, is the oxidation product in which carbon has a +4 oxidation number. In formaldehyde, H₂C=O, carbon has an oxidation number of zero.

5.27

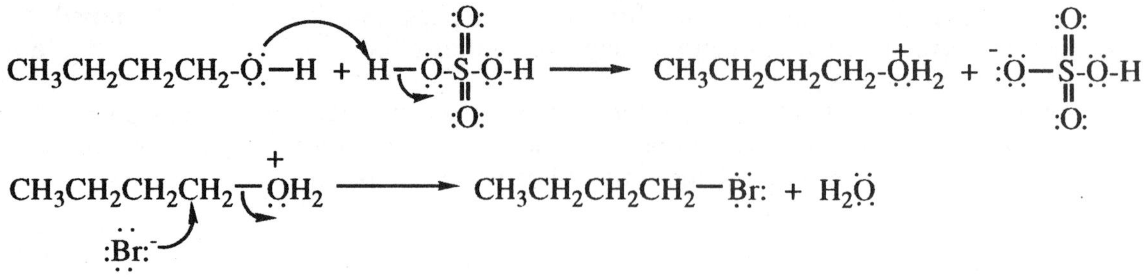

CH₃CH₂CH₂CH₂-Ö-H + H-Ö-S(=O)(=O)-Ö-H ⟶ CH₃CH₂CH₂CH₂-ÖH₂⁺ + ⁻:Ö-S(=O)(=O)-Ö-H

CH₃CH₂CH₂CH₂-⁺ÖH₂ ⟶ CH₃CH₂CH₂CH₂-Br: + H₂Ö
        :Br:⁻

5.28

(a) H-C≡N: ⁻:C≡N:

(b) 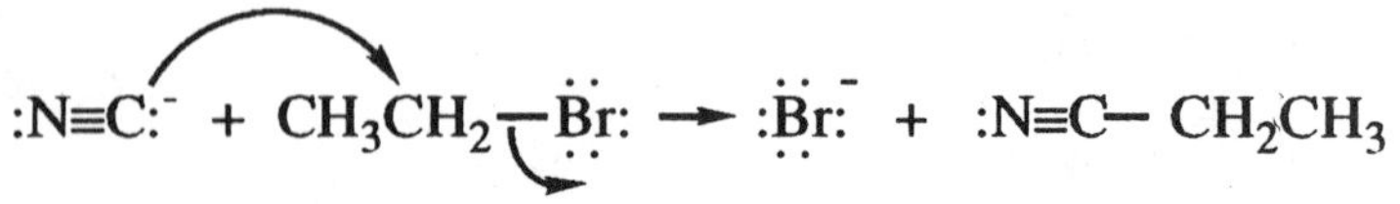 :N≡C:⁻ + CH₃CH₂-Br: ⟶ :Br:⁻ + :N≡C-CH₂CH₃

5.29   CH₃CH₂OCH₂CH₃

5.30   The oxidation number of the hydronium ion is -2, and the charge is 1+. There is no general relationship between formal charge and oxidation number.

5.31   The ethanol is first protonated at oxygen, followed by attack by water at the oxonium ion carbon site to displace the ¹⁸O in the form of H₂¹⁸O and leave a protonated ethanol molecule with ¹⁶O. This oxonium ion then loses a proton to another water molecule to give ethanol with ¹⁶O present.

5.32   Addition of ethanol to D₂O results in equilibrium formation of the oxonium ion [CH₃CH₂OHD]⁺. The H on this oxonium ion is then lost to either a D₂O molecule or a DO⁻ giving the deuterated ethanol.

5.33

CH₃CH₂-Ö-D + Na· ⟶ CH₃CH₂-Ö:⁻ Na⁺ + 1/2 D₂ gas

5.34

(a) 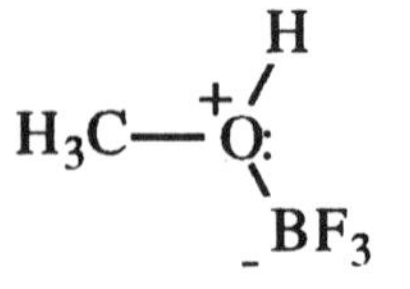 H₃C—⁺Ö(H)—⁻BF₃

(b) CH₄ + Li⁺ :C≡CH⁻

(c) CH₃CH₂Ö:⁻ Li⁺ + HC≡CH

(d) CH₃⁺ÖH₂ + H₂PO₄⁻

5.35

CH₃CH₂-Ö-H + H-Ö-S(=O)(=O)-ÖH ⟶ CH₃CH₂-⁺ÖH₂ + ⁻:Ö-S(=O)(=O)-ÖH

CH₃CH₂-⁺ÖH₂ + CH₃CH₂-ÖH ⟶ (CH₃CH₂)₂Ö⁺-H + H₂Ö

(CH₃CH₂)₂Ö⁺-H + ⁻:Ö-S(=O)(=O)-ÖH ⟶ (CH₃CH₂)₂Ö: + HÖ-S(=O)(=O)-ÖH

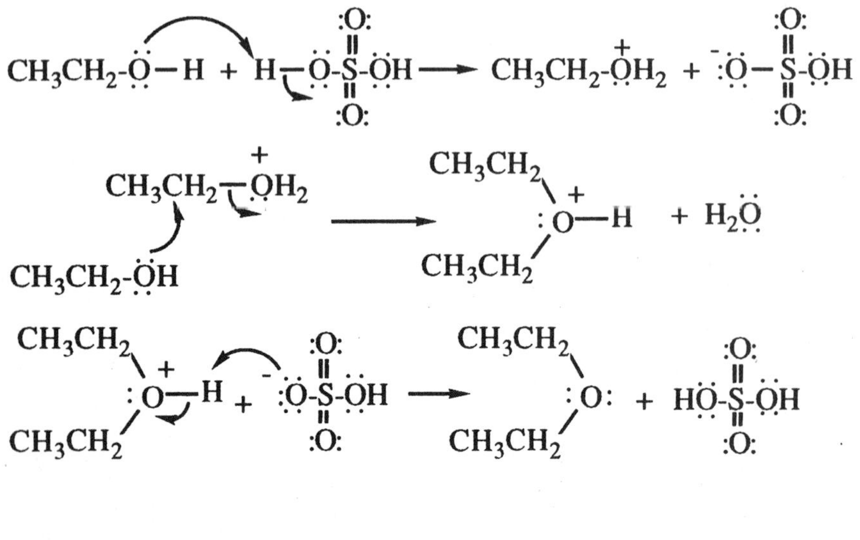

5.36

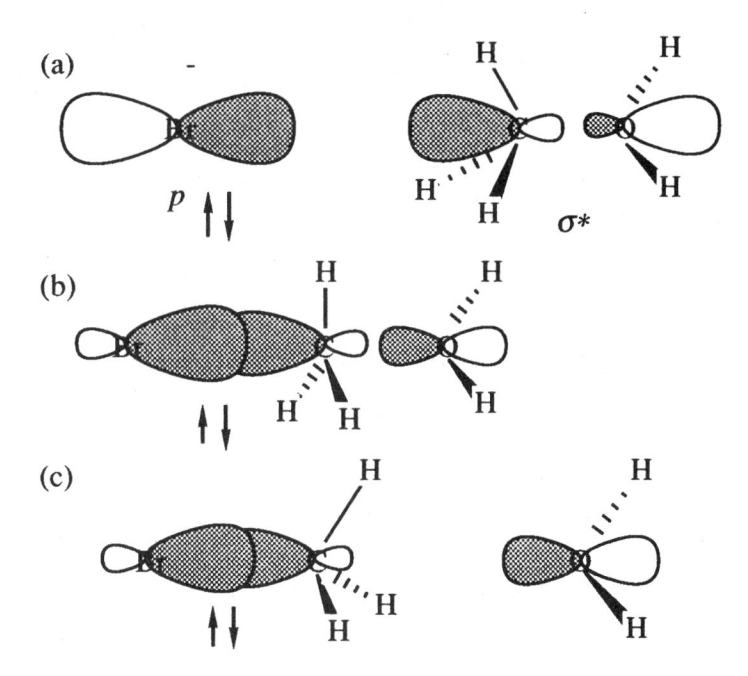

(a) [orbital diagram with p and σ*]

(b) [orbital diagram]

(c) [orbital diagram]

5.37 The difference in the reaction of the methyloxonium ion (Problem 5.36) and the ethyloxonium ion is in the groups attached to carbon. With the ethyloxonium ion one of the hydrogens has been replaced by a methyl group. This replacement makes the approach by bromide ion to the back side of the carbon more difficult and slows the reaction.

Solution of Study Guide Practice Problems

5.1 With methane and ethane the intermolecular associations are the result only of van der Waals forces. Since those forces are greater with the larger ethane than with the smaller methane, ethane has the higher boiling temperature. Compare now water and methanol: hydrogen bonding is the strongest type of intermolecular attractive force, and this effect is greater with water than with methanol since water has two hydrogens on each oxygen available for hydrogen bonding whereas methanol has only one.

5.2 The boiling temperature of water is higher than that of ethanol because of the greater ability of water to participate in hydrogen bonding (two available hydrogens on oxygen for water as compared to one for ethanol). However, although hydrogen sulfide and ethanethiol formally have the same relationship as do water and ethanol, the sulfur sites in each do not serve as hydrogen bonding acceptors, and no hydrogen bonding occurs. The boiling temperatures of hydrogen sulfide and ethanethiol are determined by dipole-dipole and van der Waals forces, these being more significant for ethanethiol than for hydrogen sulfide.

5.3 For 1,2-ethanediol the presence of two hydroxyl groups allows a greater number of hydrogen bonding interactions for each molecule than can occur with either of the propanols, each of which contains only one hydroxyl group.

5.4 The two hydroxyl groups of 1,5-pentanediol are sufficiently distant from each other that they participate almost exclusively in *inter*molecular hydrogen bonding, interactions which increase the boiling temperature of the compound. With 1,2-pentanediol and 2,3-pentanediol, however, the hydroxyl groups are located on adjacent carbon atoms and are capable of participating in *intra*molecular hydrogen bonding, an interaction which does not hold together separate molecules and thereby does not increase the boiling temperature.

5.5 The evolved gas is hydrogen (H_2).

5.6.

$$CH_3\ddot{O}:^- + H-\ddot{O}-H \longrightarrow CH_3\ddot{O}-H + {:}\ddot{O}-H$$

5.7

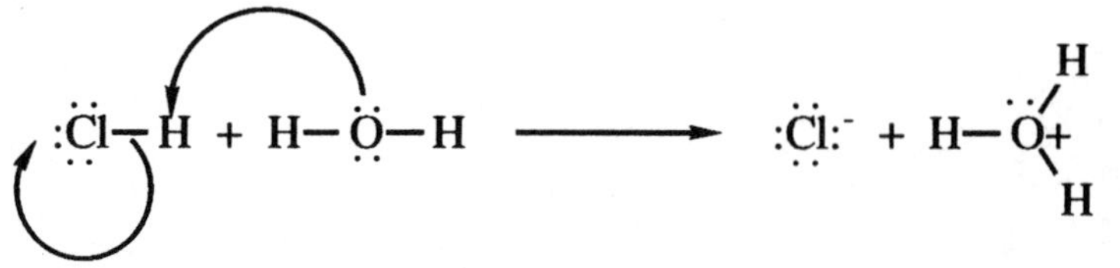

5.8 The products are: ammonia, propyne (CH_3-C≡C-H), and bromide ion.

5.9 (a) reduction
(b) oxidation
(c) ~~oxidation~~ neither, if H is added as a proton, H^+, reaction is acid-base, not redox
(d) reduction

5.10

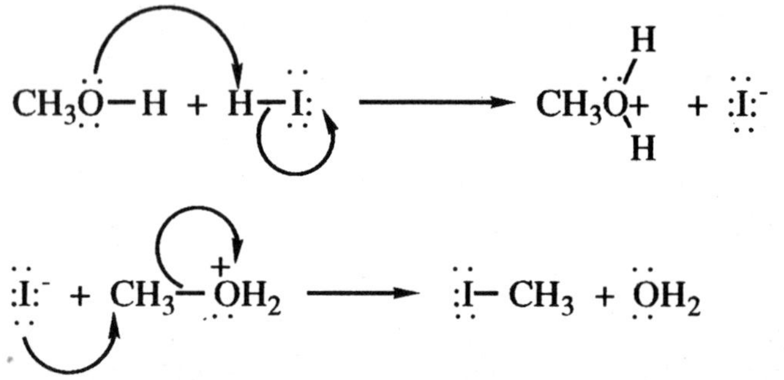

5.11 The best description of a nucleophile is as a Lewis base.

5.12 The HOMO of the bromide ion is a 4*p* orbital. This is of higher energy than the HOMO of a chloride ion, which is a 3*p* orbital.

PRACTICE EXAMINATION ONE

For most effective use of these practice examinations you should adhere to the standard examination procedures. That is, a time limit of 90 minutes should be set for completion of the entire practice examination, answers should be written out completely as they would be when presented for independent grading, no text or supplemental materials should be consulted during the testing period, and you should not check your answers until you have worked out the complete examination and the time limit has been reached.

1. Give systematic IUPAC names for each of the substances represented by the following five structures (4 points for each name):

(a) $(CH_3)_2CHCH_2CH(CH_3)_2$

(b) structure with NO_2 group

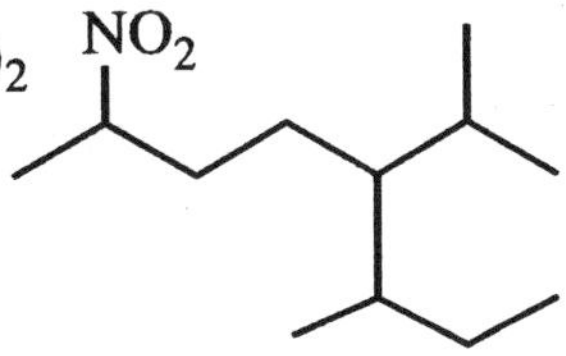

(c) Newman projection with CH_2Br, H, H_3C, $CH(CH_3)_2$, CH_3, H

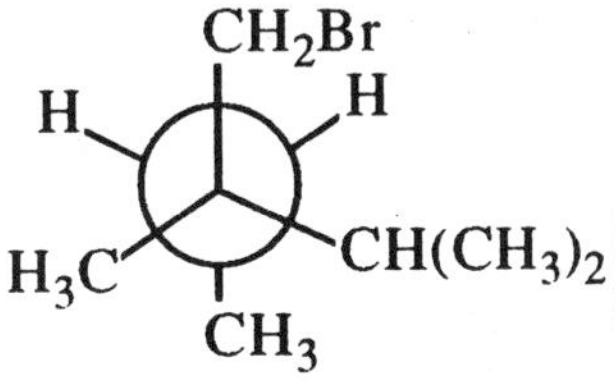

(d) spiro bicyclic structure

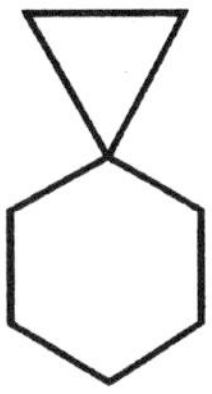

(e) cyclohexane with $CH_2CH(CH_3)_2$, H, H, $CH_2CH(CH_3)_2$

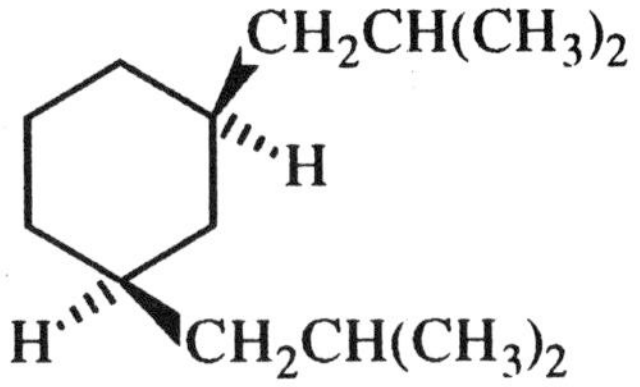

2. Draw structures to match the following five descriptions (4 points for each structure):
   (a) a Newman projection for 1,2-dibromopropane in which the two bromine atoms are eclipsed.
   (b) a wedge representation of *sec*-butyl alcohol in which a hydroxyl group is directed toward the viewer, and an ethyl group is directed away from the viewer.
   (c) a Fischer projection for 2-bromobutane in which a bromine atom and a methyl group are pointed away from the viewer.
   (d) a compound of formula C_5H_{10} which contains only single bonds and has one quaternary carbon atom.
   (e) a compound containing only carbon, hydrogen, and oxygen which analyzes for 6.67% hydrogen and 40.00% carbon and whose molecular formula is identical to its empirical formula.

3. Consider the S_2O molecule in which the sequence of bonded atoms is S-S-O, and provide answers for each of the following (5 points for each):
   (a) draw a Lewis structure in which the oxygen atom has a formal charge of 1-. Show all unshared valence level electron pairs and formal charges.
   (b) use the curved arrow formalism to generate a second resonance structure (in which all atoms obey the noble gas rule) from the one noted in part (a) of this problem, again showing all valence level unshared electron pairs and formal charges.
   (c) predict whether the S_2O molecule is linear or bent, and explain your choice.

4. Predict the products to be formed in the reaction occurring with the electron pair shifts as depicted below (5 points).

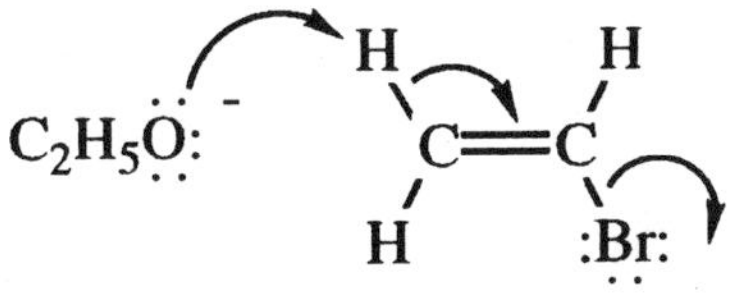

$C_2H_5\ddot{O}:^-$ with H-C=C-H, :Br: showing curved arrows

5. For each of the following choose the molecule (or molecules) that exhibit the indicated characteristic and briefly justify your choice (2 points for each):
   (a) is a secondary alcohol: isopropyl alcohol or isobutyl alcohol.
   (b) has a dipole moment: H_2S or H_2Be.
   (c) has the higher boiling temperature: 2,2-dimethylpropane or pentane.

(d) is a protic solvent: ethanol or water.
(e) is a stronger acid: water or hydrogen sulfide.
(f) has the more exothermic heat of combustion per -CH$_2$- group: cyclopropane or cyclohexane.
(g) contains carbon in a higher oxidation state: acetaldehyde (CH$_3$CHO) or acetic acid (CH$_3$COOH).
(h) always behaves as a Lewis base: a nucleophile or a Brønsted base.
(i) is formed when methyllithium (a source of the base H$_3$C:$^-$) is added to water: methane, methanol, lithium hydride, or lithium hydroxide.
(j) the gas produced when sodium metal reacts with methanol: methane, hydrogen, diethyl ether, or ethane.

6. Consider the reaction of ethanol with HCl and answer the following questions (4 points for each):
   (a) what is the observed organic product, and what are the nucleophile and the leaving group involved in its formation?
   (b) write equations for the two mechanistic steps, using the curved arrow formalism to indicate electron shifts.
   (c) explain the function zinc chloride serves (when added to the reaction mixture) in speeding the reaction.

7. Consider the reaction of ethanol with potassium amide (KNH$_2$) and answer the following questions (4 points for each):
   (a) what are the observed products?
   (b) write an equation for the process, using the curved arrow formalism to depict the electron shifts.

CHAPTER 6
ALCOHOLS

Key Points

• Make sure that you know how to classify alcohols as primary, secondary and tertiary, and learn the various types of nomenclature used for alcohols.
Practice Problem 6.1
Give a systematic name for each secondary alcohol of formula $C_5H_{12}O$, and also name these alcohols as carbinols.
Practice Problem 6.2
Give a systematic IUPAC name for *tert*-pentyl alcohol.
Practice Problem 6.3
Draw skeletal (stick-type) structures for (a) 1-methylcyclohexanol; (b) 2-bromo-1-propanol; (c) 4,5,5-trimethyl-1-3 heptanol.
Practice Problem 6.4
What is wrong with the name 3-methyl-2-propyl-1-butanol? What is the correct IUPAC name for the compound which was intended by this incorrect name?

• Learn the trends in acidities of alcohols, and understand the factors influencing acidity.
Practice Problem 6.5
Which alcohol in each pair do you expect to be the stronger acid in aqueous solution?
    (a) *sec*-butyl alcohol or isobutyl alcohol
    (b) 2-chloroethanol or ethanol
    (c) ethanol or 1-butanol

• Become familiar with the idea that an alcohol can act as a Lewis base because the oxygen has unshared valence electron pairs that can be used to form a new bond to another atom. In any reaction where such a bond is formed the alcohol is acting as a Lewis base. It is acting as a *Brønsted base* if the new bond is to a hydrogen atom, or as a *nucleophile* if the new bond is to any atom other than hydrogen. If an originally attached group is displaced from the site of nucleophilic attack by the oxygen atom, we refer to the reaction as being a *nucleophilic substitution reaction*.
Practice Problem 6.6
Write an equation for a reaction between methanol and nitric acid in which methanol acts as a Brønsted base and nitric acid acts as a Brønsted acid. Use the curved arrow formalism to depict the electron-pair shifts that are occurring.
Practice Problem 6.7
Write an equation for a nucleophilic substitution reaction between methanol and phosphorus tribromide in which methanol acts as the nucleophile and a bromide ion is displaced. Used the curved arrow formalism to depict the process.

• Learn that the reaction of alcohols with PX_3 (X = Cl, Br, or I) is a good method for preparing haloalkanes. Understand the nature of the mechanism. Learn the correlation between leaving group ability and basicity.
Practice Problem 6.8
Write out a two-step mechanism for the reaction of methanol with PI_3 to produce iodomethane and $HOPI_2$. Use the curved arrow formalism for depicting both steps.
Practice Problem 6.9
One of the following reactions occurs readily in the manner indicated by the curved arrow depiction, and one of the reactions does not. Which occurs readily? Explain your choice and show the products of that reaction.

(a) 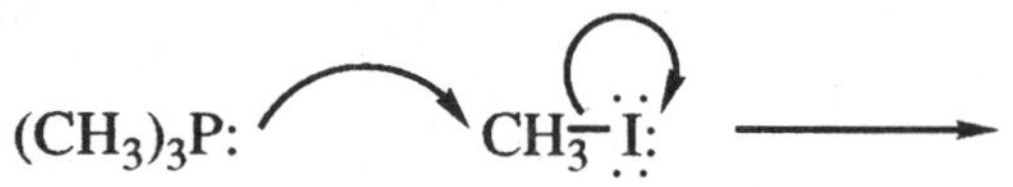 (CH₃)₃P: ⌒→ CH₃—I:

(b) 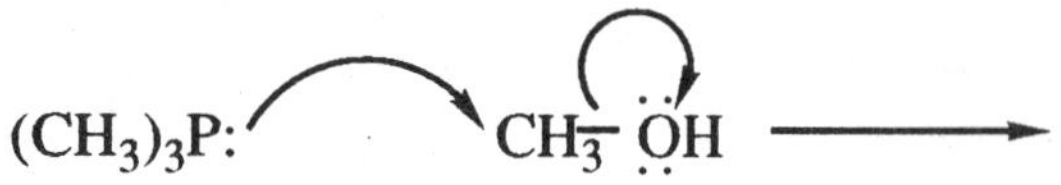 (CH₃)₃P: ⌒→ CH₃—OH

Practice Problem 6.10
Consider the conversion shown below.

$$3 H_2O + PCl_3 \longrightarrow P(OH)_3 + 3 HCl$$

Write a mechanism for this reaction, based on your knowledge of the mechanism of the reaction of alcohols with PCl₃. Used the curved arrow formalism to depict each step.

* Be able to recognize that an ester can be viewed as being derived from inorganic or organic acids containing the -OH group. One (or more) -OH groups of the acid is replaced by an -OR group(s) in the ester derivative. To name an ester, we first name the alkyl group or groups associated with the -OR portion. The rest of the name is derived from the name of the parent acid just as it would be in the name of a salt. For example, esters (and salts) of sulfuric acid are sulfates, esters (and salts) of acetic acid are acetates, *etc.* See Fig. 6.8 in the text for further examples.

Practice Problem 6.11
Ethyl nitrate forms when ethanol and concentrated nitric acid are mixed. It is an extremely dangerous compound, decomposing violently. (Be careful never to mix ethanol and concentrated nitric acid in the laboratory!) What is the structure of ethyl nitrate?

Practice Problem 6.12
Draw the structure of diisopropyl carbonate.

* Understand that the reaction of an acid halide with an alcohol can lead either to a haloalkane or to an ester (see Fig. 6.9 in the text and read the accompanying discussion).

Practice Problem 6.13
A substance of formula $C_{12}H_{27}PO_3$ can be isolated from the reaction mixture a short time after isobutyl alcohol has been allowed to stand in contact with PBr₃ at a low temperature. What is the structure of this substance?

Practice Problem 6.14
If the reactants of previous problem (6.1̷3̷) are allowed to react for a longer period of time, the product described above reacts further with the HBr that is also formed in this reaction to give isobutyl bromide. Using the curved arrow formalism, suggest a mechanism for this reaction to form isobutyl bromide.

Solution of Text Problems
6.1 (a) 2°, (b) 1°, (c) 3°, (d) 2°, (e) 2°, (f) 3°

6.2

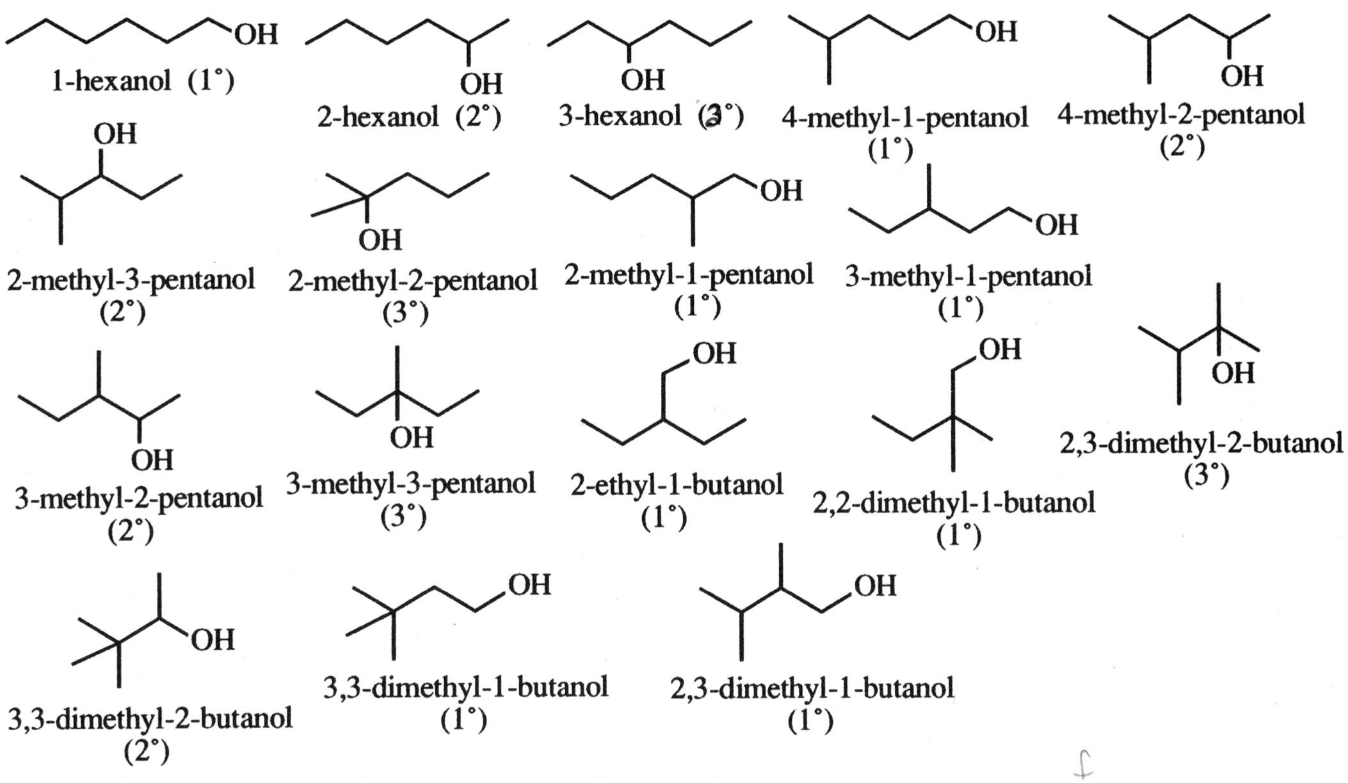

1-hexanol (1°); 2-hexanol (2°); 3-hexanol (3°); 4-methyl-1-pentanol (1°); 4-methyl-2-pentanol (2°); 2-methyl-3-pentanol (2°); 2-methyl-2-pentanol (3°); 2-methyl-1-pentanol (1°); 3-methyl-1-pentanol (1°); 2,3-dimethyl-2-butanol (3°); 3-methyl-2-pentanol (2°); 3-methyl-3-pentanol (3°); 2-ethyl-1-butanol (1°); 2,2-dimethyl-1-butanol (1°); 3,3-dimethyl-2-butanol (2°); 3,3-dimethyl-1-butanol (1°); 2,3-dimethyl-1-butanol (1°)

6.3 methanol; 1-butanol; 2-propanol; 2-methyl-2-propanol; 2-methyl-1-propanol; 2-butanol; 2-methyl-2-butanol; 3-methyl-1-butanol

6.4 $(CH_3)_3C\ddot{O}:^-$ The *tert*-butoxide ion is the conjugate base of the weakest acid of the three parent alcohols.

6.5 The enolate anion has charge delocalization stabilization over oxygen and carbon (resonance delocalization) while in the ethoxide ion the charge is localized on a single oxygen. Thus, ethoxide is the stronger base.

6.6 Chloroalkanes and fluoroalkanes are the only ones of those listed that would undergo nucleophilic substitution with hydroxide ion, and fluoroalkanes would be significantly less reactive than chloroalkanes.

6.7

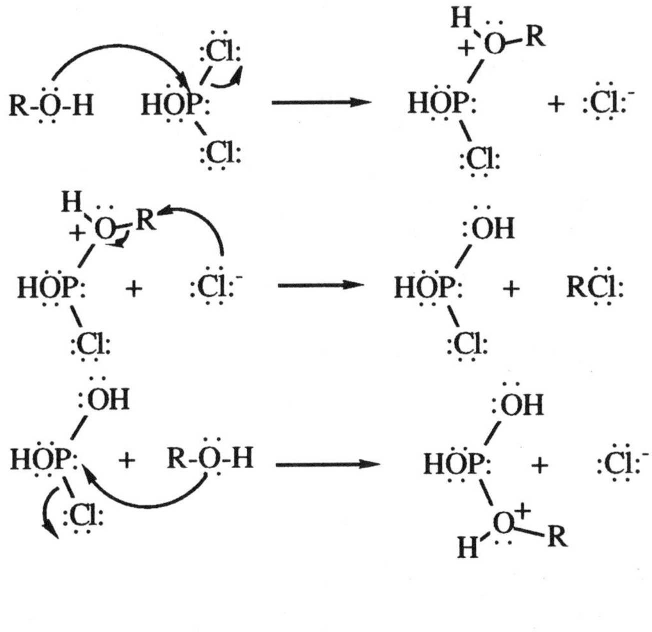

[Mechanism diagram showing three steps:

Step 1: R-Ö-H + HÖP: with :Cl-Cl: → H-Ö+-R bonded to HÖP: with :Cl: + :Cl:⁻

Step 2: H-Ö+-R bonded to HÖP: with :Cl: + :Cl:⁻ → HÖP: with :OH and :Cl: + RCl:

Step 3: HÖP: with :OH and :Cl: + R-Ö-H → HÖP: with :OH and :Cl:⁻ + H-Ö+-R]

50 Study Guide and Solutions Manual

[Mechanism scheme showing HOP(OH)(OH)-O(H)(+)-R attacked by Cl⁻ → (HO)₃P: + RCl]

6.8 The products are phosphoric acid and 1-chlorobutane.

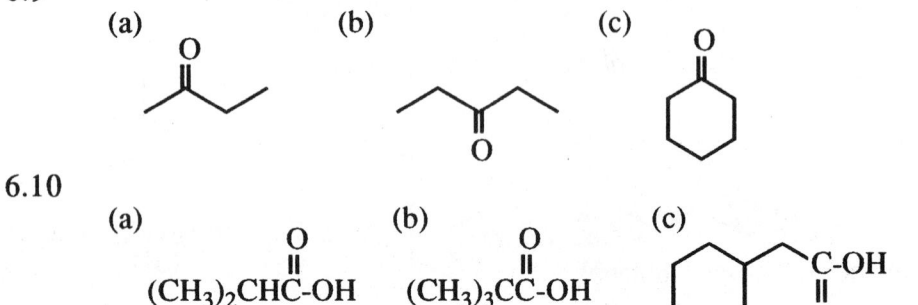

6.9 (a) [structure: butan-2-one] (b) [structure: pentan-3-one] (c) [structure: cyclohexanone]

6.10 (a) (CH₃)₂CHC(O)-OH (b) (CH₃)₃CC(O)-OH (c) [cyclohexyl-CH₂-C(=O)-OH]

6.11 Substance A must be a primary alcohol in order to lead to B as described. There are two primary alcohols that fit the given formula, and consequently two corresponding acids.
  A CH₃CH₂CH₂CH₂OH or (CH₃)₂CHCH₂OH
  B CH₃CH₂CH₂CO₂H or (CH₃)₂CHCO₂H

6.12 (a) 1-pentanol; (b) 2,3-dimethyl-1-pentanol; (c) 2-methyl-2-hexanol; (d) 2,4-dimethyl-3-hexanol; (e) 5-ethyl-2,2-dimethyl-3-heptanol; (f) 3,3-diethyl-1-pentanol; (g) 3-propyl-2-hexanol; (h) 6-ethyl-3-methyl-5-nonanol; (i) 3,4-dimethyl-3-hexanol; (j) 4-propyl-4-heptanol; (k) 2,4-pentanediol; (l) 2-ethyl-3-methyl-1-pentanol; (m) 3-methyl-2-butanol; (n) cyclopentanol; (o) 4-ethyl-3-heptanol; (p) 5,5-dimethylcyclooctanol

6.13 (a) primary; (b) primary; (c) tertiary; (d) secondary; (e) secondary; (f) primary; (g) secondary; (h) secondary; (i) tertiary; (j) tertiary; (k) both are secondary; (l) primary; (m) secondary; (n) secondary; (o) secondary; (p) secondary

6.14 [structures (a) through (n) of various alcohols, and (k) K⁺ ⁻OCH₂CH₃ (l) Na⁺ ⁻OCH(CH₃)₂ (m) (CH₃)₃CO⁻ K⁺]

6.15 (a) secondary; (b) secondary; (c) primary; (d) tertiary; (e) tertiary; (f) secondary; (g) tertiary; (h) primary; (i) tertiary; (j) tertiary

Chapter 6   51

6.16

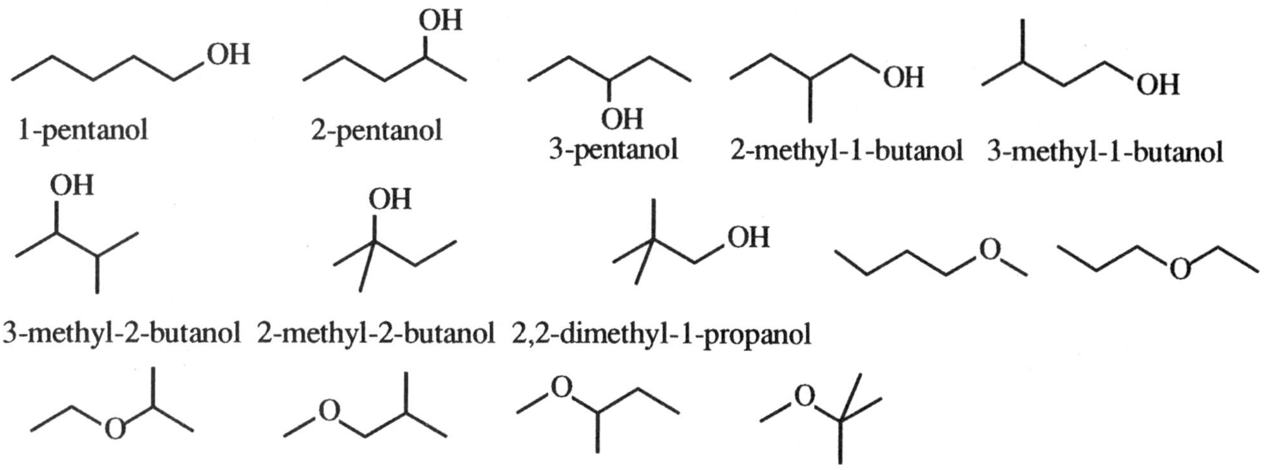

[Structures shown:]
1-pentanol, 2-pentanol, 3-pentanol, 2-methyl-1-butanol, 3-methyl-1-butanol

3-methyl-2-butanol, 2-methyl-2-butanol, 2,2-dimethyl-1-propanol, [two ether structures]

[four ether structures]

6.17

(a) CH$_3$O–S(=O)$_2$–OCH$_3$ (b) CH$_3$O–N$^+$(=O)O$^-$ (c) H–C(=O)–OCH$_3$

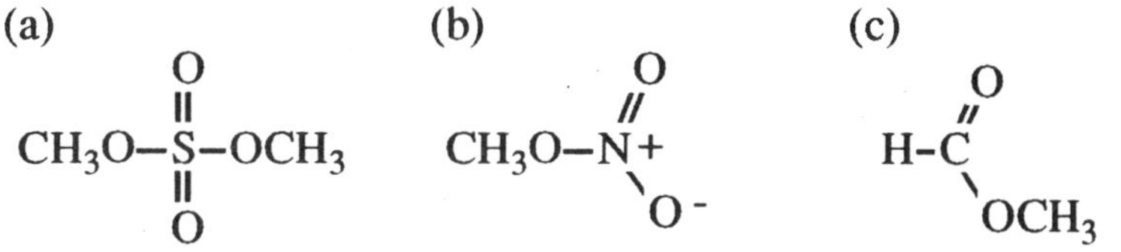

6.18

(a) CH$_3$CH$_2$CH(Cl)CH$_2$CH$_3$; (b) CH$_3$CH$_2$C(O)CH$_2$CH$_3$; (c) (CH$_3$)$_2$CHCH$_2$CHO;
(d) no reaction; (e) CH$_3$CH$_2$CH$_2$CH$_2$C(Br)(CH$_3$)$_2$; (f) (CH$_3$)$_2$CHCH(CH$_3$)CH$_2$Cl;
(g) CH$_3$C(O)CH(CH$_3$)CH(CH$_3$)CHO; (h) (CH$_3$)$_3$CCH$_2$CH$_2$CO$_2$Na; (i) (CH$_3$CH$_2$CH$_2$O)$_3$PO;
(j) CH$_3$CH$_2$CH$_2$CH$_2$C(O)CH$_3$; (k) CH$_3$CH$_2$CH(CH$_3$)CH$_2$CO$_2$H; (l) HO$_2$CCH$_2$CH$_2$CH$_2$CH$_2$CO$_2$H
          ĊH$_2$ C(O)CH$_3$

6.19

(a) 3-hexanol   CH$_3$CH$_2$CH(OH)CH$_2$CH$_2$CH$_3$
(b) 3-methyl-2-butanol   (CH$_3$)$_2$CHCH(OH)CH$_3$
(c) 2,2-dimethyl-1-propanol   (CH$_3$)$_3$CCH$_2$OH
(d) 4-methyl-1-hexanol   CH$_3$CH$_2$CH(CH$_3$)CH$_2$CH$_2$CH$_2$OH
(e) 3-pentanol   CH$_3$CH$_2$CH(OH)CH$_2$CH$_3$
(f) 4,4-dimethyl-2-pentanol   (CH$_3$)$_3$CCH$_2$CH(OH)CH$_3$
(g) 4-ethyl-1-heptanol   CH$_3$CH$_2$CH$_2$CH(CH$_2$CH$_3$)CH$_2$CH$_2$CH$_2$OH
(h) 3-methyl-1-pentanol   CH$_3$CH$_2$CH(CH$_3$)CH$_2$CH$_2$OH
(i) 5-ethyl-4,4-diethyl-2-heptanol   (CH$_3$CH$_2$)$_2$CHC(CH$_3$)$_2$CH$_2$CH(OH)CH$_3$
(j) 4-methyl-2,5-heptanediol   CH$_3$CH$_2$CH(OH)CH(CH$_3$)CH$_2$CH(OH)CH$_3$

6.20

(a) 4-methyl-3-hexanol   CH$_3$CH$_2$CH(CH$_3$)CH(OH)CH$_2$CH$_3$
(b) 5,5-dimethyl-2-hexanol   (CH$_3$)$_3$CCH$_2$CH$_2$CH(OH)CH$_3$
(c) 3,4-dimethyl-1-pentanol   (CH$_3$)$_2$CHCH(CH$_3$)CH$_2$CH$_2$OH
(d) 2-ethyl-2-methyl-1-butanol   (CH$_3$CH$_2$)$_2$C(CH$_3$)CH$_2$OH
(e) 4-ethyl-5-methyl-1,6-octanediol CH$_3$CH$_2$CH(OH)CH(CH$_3$)CH(CH$_2$CH$_3$)CH$_2$CH$_2$CH$_2$OH
(f) 3-methyl-3-pentanol   (CH$_3$CH$_2$)$_2$C(OH)CH$_3$
(g) 1-methylcyclobutanol   (CH$_2$)$_3$C(OH)CH$_3$
(h) ~~4-ethyl-5-nonanol~~   CH$_3$CH$_2$CH$_2$CH$_2$CH(OH)CH(CH$_2$CH$_3$)CH$_2$CH$_2$CH$_3$
      2-ethyl cyclohexanol

6.21

(a) CH$_3$CH$_2$CH(CH$_3$)CH(CH$_3$)CO$_2$H
(b) CH$_3$CH$_2$C(OH)(CH$_3$)$_2$
(c) CH$_3$CH$_2$CH(CH$_3$)C(O)CH(CH$_3$)$_2$
(d) CH$_3$CH$_2$C(CH$_3$)$_2$CH$_2$CH$_2$CH$_2$C(CH$_3$)$_2$CO$_2$H
(e) (CH$_3$CH$_2$)$_2$CHCH$_2$CH$_2$CO$_2$H

Any carboxylic acid will actually form as its (water

soluble) potassium salt under the alkaline reaction conditions. Aqueous acid must be added in the workup of the reaction to convert such salts to their parent acids.

6.22
(a) $CH_3CH_2CH(CH_3)CH(CH_3)CHO$
(b) $CH_3CH_2C(OH)(CH_3)_2$
(c) $CH_3CH_2CH(CH_3)C(O)CH(CH_3)_2$
(d) $CH_3CH_2C(CH_3)_2CH_2CH_2CH_2C(CH_3)_2CHO$
(e) $(CH_3CH_2)_2CHCH_2CH_2CHO$

6.23
(a) $CH_3CH_2CH(CH_3)CH(CH_3)CH_2Cl$
(b) $CH_3CH_2C(Cl)(CH_3)_2$
(c) $CH_3CH_2CH(CH_3)CH(Cl)CH(CH_3)_2$
(d) $CH_3CH_2C(CH_3)_2CH_2CH_2CH_2C(CH_3)CH_2Cl$
(e) $(CH_3CH_2)_2CHCH_2CH_2CH_2Cl$

6.24
$CH_3CH_2CH_2CH_2C(OH)(CH_3)_2 < CH_3CH_2CH(OH)CH_3 < CH_3OH < Cl_2CHCH_2OH$

The electronegativity effect of the two chlorines in 2,2-dichloroethanol renders it more acidic than methanol. The remaining order follows from the standard relationship (in solution) of tertiary alcohols being less acidic than secondary alcohols, which are less acidic than primary alcohols or methanol.

6.25 The conjugate bases are phenoxide and ethoxide ions, respectively. For the phenoxide ion the charge is greatly delocalized around the ring as can be shown with contributing resonance structures. There is no such delocalization with ethoxide ion. The charge with the ethoxide ion is localized on the oxygen atom. Thus, the phenoxide ion is relatively stabilized, and, in equilibrium with its parent acid, forms to a greater extent than does ethoxide ion. Phenol is thus more acidic than is ethanol.

6.26

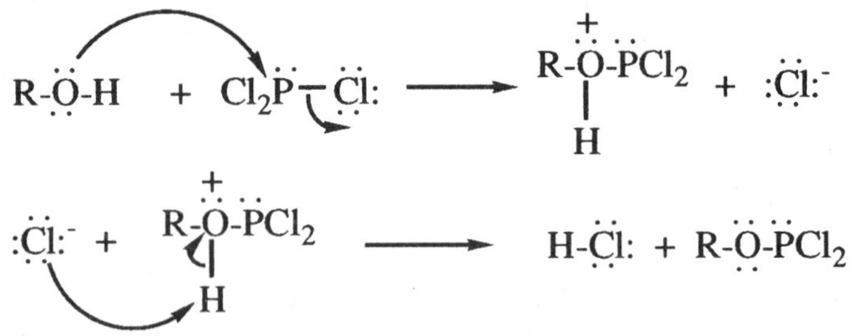

$$R\text{-}\ddot{O}\text{-}H + Cl_2P\text{-}\ddot{C}l: \longrightarrow R\text{-}\overset{+}{\underset{H}{\ddot{O}}}\text{-}PCl_2 + :\ddot{C}l:^-$$

$$:\ddot{C}l:^- + R\text{-}\overset{+}{\underset{H}{\ddot{O}}}\text{-}PCl_2 \longrightarrow H\text{-}\ddot{C}l: + R\text{-}\ddot{O}\text{-}PCl_2$$

Reaction of the remaining two chlorine substituents on phosphorus with two equivalents of alcohol occurs in a completely analogous manner as shown above for reaction of the one chlorine substituent.

6.27 (a) 2-Propanol reacts with aqueous basic permanganate whereas *tert*-butyl alcohol does not react. Thus, addition of 2-propanol to a dilute aqueous permanganate solution would remove its purple color, but addition of *tert*-butyl alcohol would not do so.

(b) Cyclohexanol would react with dilute aqueous potassium permanganate, serving to decolorize the solution, whereas 1-methylcyclohexanol would not react with this reagent.

(c) As in part (a), propanal will be oxidized by permanganate but *tert*-butyl alcohol will not be so oxidized.

6.28 $(CH_3)_2C(OH)CH_2CH_3$

6.29 Treatment of the primary alcohol 1-undecanol with chromic anhydride (CrO_3) in pyridine will give the undecanal directly.

6.30 The reaction of one of the chlorines attached to phosphorus and its replacement by a hydroxyl group is illustrated in the two steps shown below. The replacement of the remaining two chlorines occurs in a completely analogous manner. Note that this process is completely analogous to the mechanism illustrated for the reaction of an alcohol, ROH, with phosphorus trichloride shown in problem 6.26.

$$H\text{-}\ddot{O}\text{-}H \;+\; Cl_2P\text{-}\ddot{C}l\text{:} \longrightarrow H\text{-}\overset{+}{\underset{H}{\ddot{O}}}\text{-}PCl_2 \;+\; \text{:}\ddot{C}l\text{:}^-$$

$$\text{:}\ddot{C}l\text{:}^- \;+\; H\text{-}\overset{+}{\underset{H}{\ddot{O}}}\text{-}PCl_2 \longrightarrow H\text{-}\ddot{C}l\text{:} \;+\; H\text{-}\ddot{O}\text{-}PCl_2$$

6.31
    (a) $CH_3OH + NaOH \rightleftharpoons CH_3O^- Na^+ + H_2O$

    (b) $CH_3CH_2CH_2CH_2CH_2OH + HCl \longrightarrow CH_3CH_2CH_2CH_2CH_2Cl + H_2O$

    (c) no reaction

    (d) $CH_3CH_2OH + (CH_3)_3B \longrightarrow CH_3CH_2\overset{+}{\underset{H}{O}}-\bar{B}(CH_3)_3$

    (e) $CH_3CH_2CH_2CH_2OH + Na \longrightarrow CH_3CH_2CH_2CH_2O^- Na^+ + 1/2\, H_2$

    (f) no reaction

6.32 (a) treatment with chromic anhydride in pyridine
    (b) 1. treatment with potassium permanganate in aqueous base; 2. treatment with aqueous acid
    (c) treatment with phosphorus trichloride
    (d) treatment with phosphorus tribromide

6.33 From 1-heptanol there is formed $CH_3CH_2CH_2CH_2CH_2CH_2CO_2^-K^+$ which is soluble in water. From 3-heptanol there is formed $CH_3CH_2C(O)CH_2CH_2CH_2CH_3$ which is insoluble in water. The potassium salt of heptanoic acid is soluble since it has an ionic region which interacts with water in an attractive manner, while 3-heptanone has no functional group which interacts so strongly with water.

6.34 Treat the mixture with aqueous basic potassium permanganate. The unwanted 2-methyl-1-octanol will be oxidized to a carboxylate salt and remain soluble in the aqueous medium. The 2-methyl-2-octanol, being a tertiary alcohol, will not be oxidized and can be extracted from the reaction mixture with a solvent such as hexane.

6.35 (a) chromic anhydride, pyridine
    (b) phosphorus oxychloride in hexane solution at 0°
    (c) thionyl chloride
    (d) sodium dichromate, sulfuric acid
    (e) sodium dichromate, sulfuric acid

6.36 (a) chloride ion
    (b) chloride ion
    (c) formate anion
    (d) hydroxide ion
    (e) formate anion
    (f) acetate anion

6.37 D $HOCH_2CH_2CH_2CH_2OH$ E $HO_2CCH_2CH_2CO_2H$

6.38 F $CH_3CH(OH)CH_2CH_2OH$ G $CH_3CH(OH)CH(OH)CH_3$ H $CH_3C(O)CH_2CO_2H$
    I $CH_3C(O)C(O)CH_3$

6.39 We could not make the distinction. Each oxidation would lead to a product of formula $C_4H_6O_2$.

6.40 The molecular weight of 1-hexanol is 102. Thereby, 20 g of 1-hexanol is 0.196 mole. The minimum amount of PBr_3 required is 1/3 this number of moles, or 0.0653 mole. Since 270.2 is the molecular weight of PBr_3, the minimum weight of PBr_3 required is 17.7 g.

6.41 J $HOCH_2CH_2CHO$ K $BrCH_2CH_2CHO$ L $HO_2CCH_2CO_2H$

6.42 Acetic acid is a stronger acid than is methanol. The equilibrium lies to the left (methanol) side rather than to the acetic acid side. A stronger acid (such as sulfuric acid) is required to convert the sodium acetate to acetic acid.

6.43 The preparation of Q and R has the higher actiation energy.

Solution of Study Guide Practice Problems

6.1 There are three secondary alcohols of the given formula. These are: 2-pentanol, 3-pentanol, and 3-methyl-2-butanol. Carbinol names are methyl propyl carbinol, diethyl carbinol, isopropyl methyl carbinol.

6.2 The IUPAC name is 2-methyl-2-butanol.

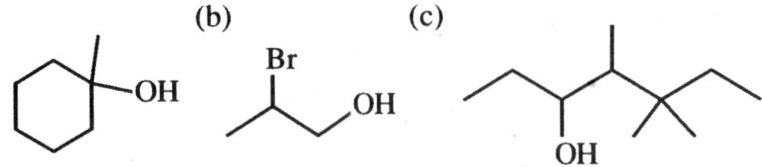

6.3
(a) 1-methylcyclohexanol structure with OH
(b) 2-bromo-1-propanol structure (Br on C2, OH on C1)
(c) 3,3-dimethyl-4-hexanol type structure with OH

6.4 The longest continuous carbon chain bearing the hydroxyl group is five carbons, not four carbons as the given name would imply. Considering the structure indicated by the improper name, the proper name would be: 2-isopropyl-1-pentanol.

6.5
(a) isobutyl alcohol
(b) 2-chloroethanol
(c) ethanol

6.6

$$CH_3-\ddot{O}H + H-\ddot{O}NO_2 \longrightarrow CH_3-\overset{+}{\underset{H}{O}}-H + {}^-\!:\!\ddot{O}-NO_2$$

6.7

$$CH_3-\ddot{O}H + PBr_3 \longrightarrow CH_3-\overset{+}{\underset{PBr_2}{O}}-H + :\!\ddot{B}r\!:^-$$

6.8

$$CH_3-\ddot{O}H + PI_3 \longrightarrow CH_3-\overset{+}{\underset{PI_2}{O}}-H + :\!\ddot{I}\!:^-$$

$$:\!\ddot{I}\!:^- + CH_3-\overset{+}{\underset{PI_2}{O}}-H \longrightarrow \ddot{I}-CH_3 + H\ddot{O}-PI_2$$

6.9 Reaction (a) occurs readily to produce tetramethylphosphonium iodide, $(CH_3)_4P^+\ I^-$. The displacement of an iodide ion, a good leaving group, the conjugate base of a strong acid, occurs readily whereas the displacement of a hydroxide ion, a poor leaving group, the conjugate base of a weak acid, does not occur at all easily.

6.10

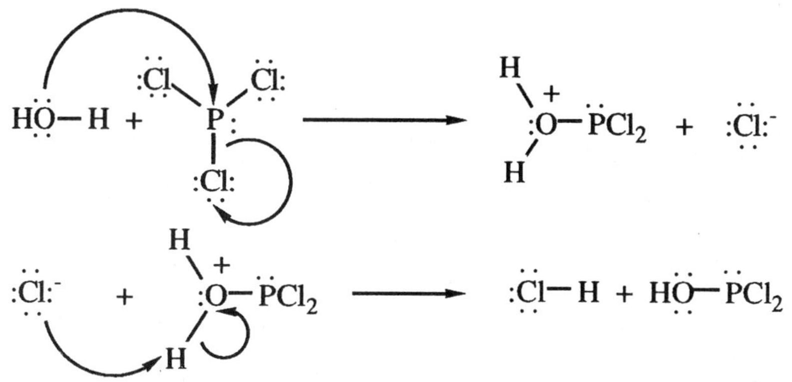

[Mechanism diagram showing:]

HÖ—H + PCl₃ (with Cl substituents shown) → H₂O⁺—PCl₂ + :Cl:⁻

:Cl:⁻ + H₂O⁺—PCl₂ → :Cl—H + HÖ—PCl₂

The reaction continues with the attaack of a second water molecule on the intermediate HOPCl₂. The mechanism of this attack is completely analogous to the first step shown above. That is, a water molecule uses an unshared valence level electron pair on oxygen to displace a chloride ion from phosphorus. This is followed by removal of a hydrogen from oxygen of the intermediate oxonium ion in a manner completely analogous to the second step shown above. Finally, a third water molecule attacks to displace the last chloride ion from phosphorus with subsequent hydrogen ion removal and formation of :P(OH)₃ and the third molecule of HCl.

6.11

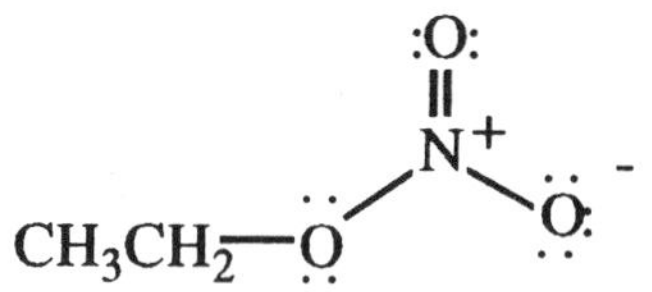

$$CH_3CH_2-\ddot{O}-\overset{\overset{:O:}{\|}}{\underset{\ddot{O}:^-}{N^+}}$$

6.12

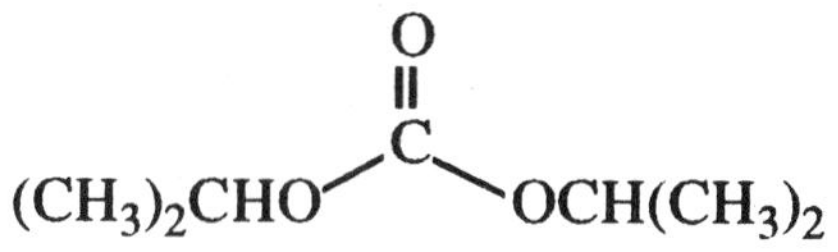

$$(CH_3)_2CHO-\overset{\overset{O}{\|}}{C}-OCH(CH_3)_2$$

6.13

:ÖCH₂CH(CH₃)₂
  /
:P—ÖCH₂CH(CH₃)₂
  \
:ÖCH₂CH(CH₃)₂

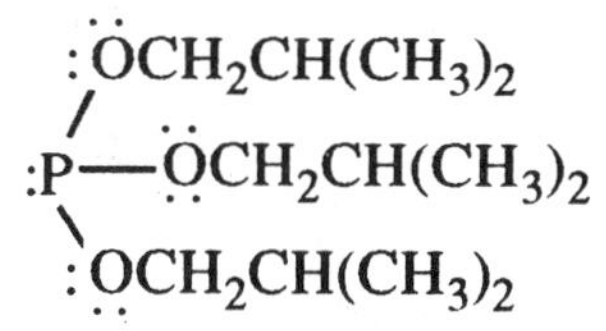

6.14 The ester of phosphorous acid (shown above; triisobutyl phosphite) undergoes reaction with HBr involving first the protonation of an oxygen to generate an oxonium ion followed by bromide ion attack on carbon to generate the isobutyl bromide and a partial ester of phosphorous acid, as shown below. The process repeats until three molecules of isobutyl bromide have been formed along with the phosphorous acid, :P(OH)₃.

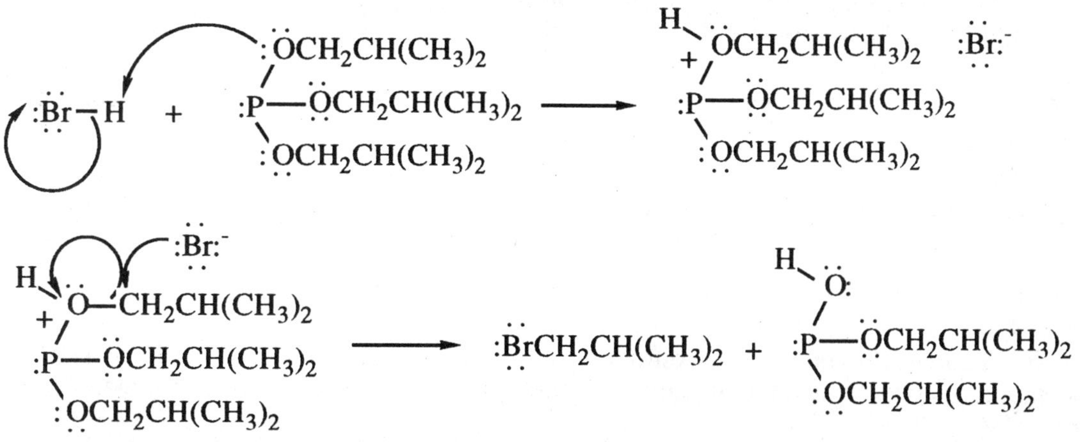

[Mechanism diagram:]

:Br—H + :P(OCH₂CH(CH₃)₂)₃ → [H—Ö⁺(CH₂CH(CH₃)₂)—P(OCH₂CH(CH₃)₂)₂] :Br:⁻

:Br:⁻ + H—Ö⁺—CH₂CH(CH₃)₂ on P(OCH₂CH(CH₃)₂)₂ → :BrCH₂CH(CH₃)₂ + H—Ö—P(OCH₂CH(CH₃)₂)₂

CHAPTER 7
TYPES OF BOND CLEAVAGE - CARBOCATIONS AND RADICALS AS REACTION INTERMEDIATES

Key Points

• Understand the difference between homolytic and heterolytic bond cleavage. Learn how to use curved arrows and fish-hook arrows to indicate the various types of bond cleavage.

Practice Problem 7.1
Use either curved or fish-hook arrows to denote the indicated bond cleavages, and show the structures of the products for each of the following:
      (a) homolytic cleavage of the O-O bond in dimethyl peroxide, CH_3-O-O-CH_3.
      (b) heterolytic cleavage of the C-Cl bond in *sec*-butyl chloride to produce a carbocation.

• Know how to use bond dissociation energies.

Practice Practice 7.2
The O-O bond of dimethyl peroxide (see the previous problem) is much weaker than the C-O bonds of the same compound. Given this, which of the following statements is/are true?
      (a) More energy is released when the O-O bond breaks than when a C-O bond breaks.
      (b) Less energy is released when the O-O bond breaks than when a C-O bond breaks.
      (c) More energy is required to break the O-O bond than a C-O bond.
      (d) Less energy is required to break the O-O bond than a C-O bond.

Practice Problem 7.3
All of the halogen molecules (X_2) react with methane as follows:

      $CH_4 + X_2 \longrightarrow CH_3X + HX$

Use the bond dissociation energies in Table 7.1 to determine which halogen reacts most exothermically with methane.

Practice Problem 7.4
Consider the reaction:

      $(CH_3)_3CH \longrightarrow (CH_3)_3C^+ + H:^-$

Using the bond dissociation energies from Table 7.1 and the data given below, calculate ΔH for this reaction.
      Ionization energy of a *tert*-butyl radical = 171 kcal/mole
      Electron affinity of a hydrogen atom = -17.2 kcal/mole

• Learn the order of stability of carbocations and radicals: 3° > 2° > 1° > methyl. (There are certain exceptions which will be discussed later.)

Practice Problem 7.5
What is the most stable carbocation of formula $C_5H_{11}^+$?

• Understand that some alcohols react with acidic reagents to form carbocations.

Practice Problem 7.6
Which of the following alcohols do you predict would most likely react with a given acid to form a carbocation?
      (a) cyclopentanol; (b) 1-methylcyclopentanol; (c) 2-methylcyclopentanol; (d) 3-methylcyclopentanol

Practice Problem 7.7
Both *tert*-butyl alcohol and 1-butanol react with HBr to give haloalkanes. However, the mechanisms for the two reactions are different. In what way do these mechanisms differ?

Chapter 7   57

• Add to your list of important reactions the conversion of alcohols to alkenes by treatment with concentrated sulfuric acid, and learn the mechanism of this reaction type, known as dehydration.
Practice Problem 7.8
Show a mechanism using the curved arrow formalism for the conversion of cyclohexanol to cyclohexene upon heating with concentrated sulfuric acid.
Practice Problem 7.9
Which alcohol, 1-pentanol or 2-pentanol, would you expect to undergo dehydration more readily? Explain your choice.

• Be aware that the skeletal rearrangements of carbocations (but not radicals) are common, and be able to predict the rearranged skeleton.
Practice Problem 7.10
When neopentyl alcohol is heated with sulfuric acid, a skeletal rearrangement occurs leading to a mixture of two alkenes. Draw structures for the rearranged carbocation which is formed as an intermediate in this reaction and the two alkene products.
Practice Problem 7.11
Is the rearrangement shown below one that would be expected to occur? Briefly justify your answer.

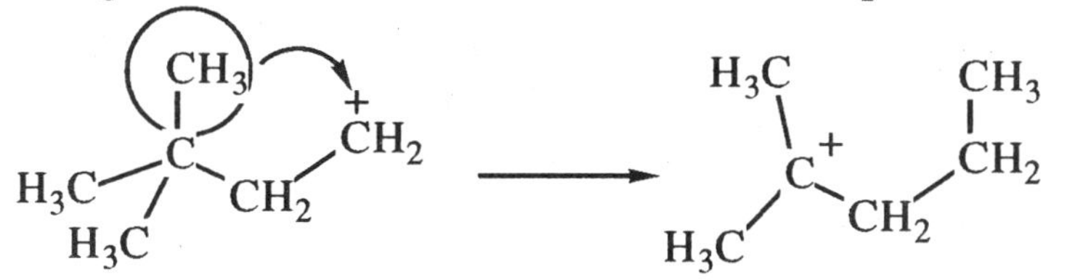

Solution of Text Problems

7.1   The ΔH for the reaction is the difference between the energies of the bonds broken and the bonds formed:
       ΔH = 104 - 119 = -15 kcal/mole
7.2   A C-H bond is broken and an H-Cl bond is formed
       ΔH = 104 - 103 = 1 kcal/mole
7.3   From the hydroperoxide there are formed:
       R-CH-Ö-CH$_2$R   and ·ÖH
          |
          :Ö: and from the organic peroxide there are formed:
          |
       two RCH$_2$-Ö-CH-Ö·
                      |
                      R
7.4   For the process forming H$^+$ and F$^-$ from H-F,
       ΔH = 136 + 314 - 79.3 = 370.7 kcal/mole
       For the process forming F$^+$ and H$^-$ from H-F,
       ΔH = 136 + 401 - 17.2 = 519.8 kcal/mole

7.5
   (a) CH$_3$CH$_2$CH$_2$CH$_2$$\overset{+}{C}H_2$ (CH$_3$)$_2$CHCH$_2$$\overset{+}{C}H_2$ CH$_3$$\overset{+}{C}HCH_2CH_3$ (CH$_3$)$_3$C$\overset{+}{C}$H$_2$
                                                                                                              H$_2$C+
   (b) CH$_3$CH$_2$CH$_2$$\overset{+}{C}HCH_3$ CH$_3$CH$_2$$\overset{+}{C}HCH_2CH_3$ (CH$_3$)$_2$CH$\overset{+}{C}$HCH$_3$
   (c) (CH$_3$)$_2$$\overset{+}{C}CH_2CH_3$
7.6   The structure is (CH$_3$CH$_2$CH$_2$)$_2$O. This ether forms by attack of a molecule of 1-propanol on the

58 Study Guide and Solutions Manual

oxonium ion derived by protonation of 1-propanol. A molecule of water is displaced and the reaction is completed by the loss of a proton from the protonated ether.

7.7 Different alkenes can be formed from the same intermediate carbocation by the removal of hydrogen from different carbons adjacent to the carbocation site. The two alkenes are:

$$\begin{array}{cc} H_3CCH_3 & H_3CCH_3 \\ C{=}C & CH{-}C \\ H_3CCH_3 & H_3CCH_2 \end{array}$$

7.8 There are several alkenes which can form in this reaction. 1-Pentene is not the only alkene that can form in this reaction. The *cis*- and *trans*-2-pentenes can also form, and in fact do so to a great extent.

7.9 The initially formed carbocation is:

$$CH_3\overset{+}{C}HCHCH_2CH_3 \quad \text{which undergoes a hydride shift to generate} \quad CH_3CH_2\overset{+}{C}CH_2CH_3$$
$$\phantom{CH_3\overset{+}{C}HC}\underset{CH_3}{|} \underset{CH_3}{|}$$

and a final product of $CH_3CH_2\underset{\underset{CH_3}{|}}{\overset{Cl}{\underset{|}{C}}}CH_2CH_3$

The hydride shift leads to a more stable tertiary carbocation whereas a methide shift would lead to another secondary carbocation.

7.10
(a)

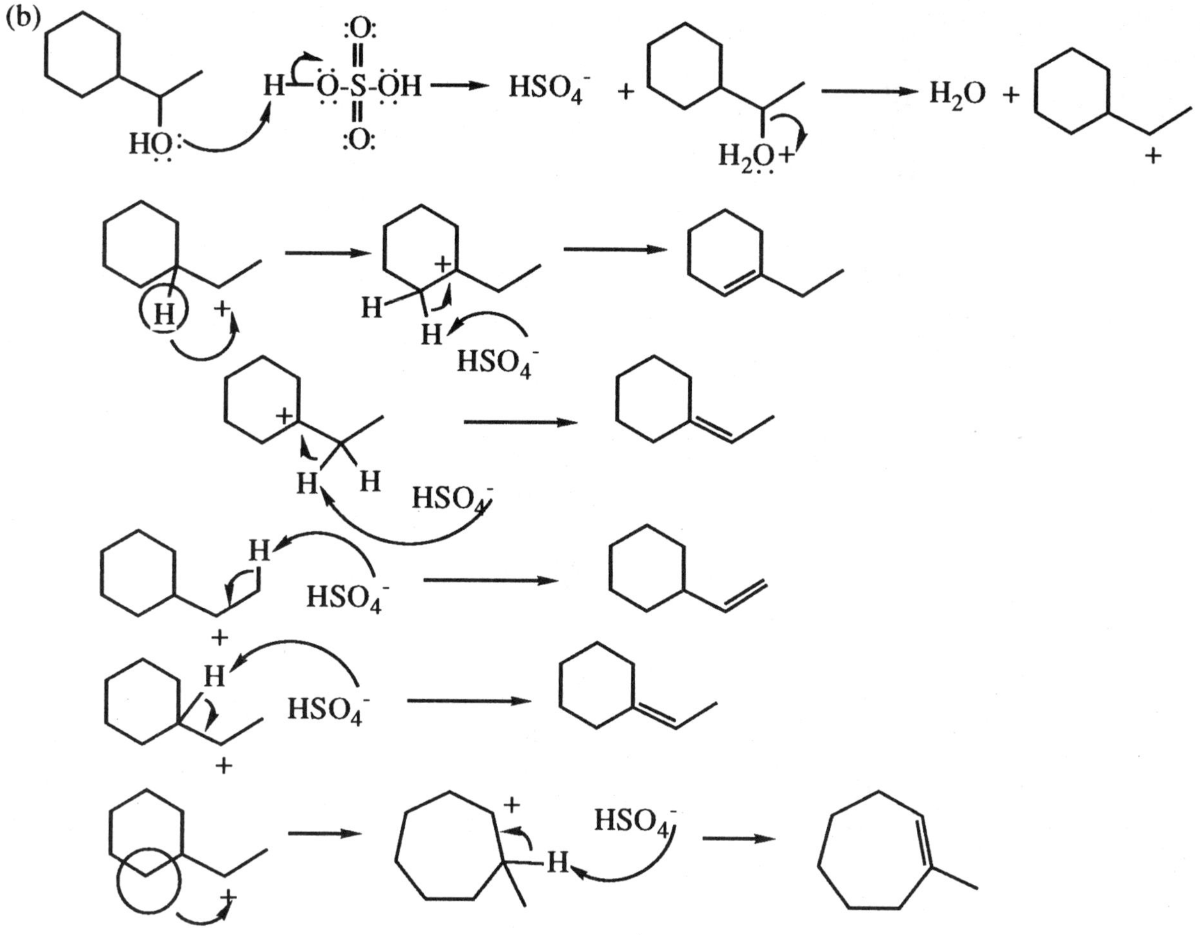

In the last of the reactions shown above the migrating alkyl group is part of a ring. When this type of migration occurs the ring size of the product is different from that of the starting material (see section 13.10 of the text for a more complete discussion).

7.11

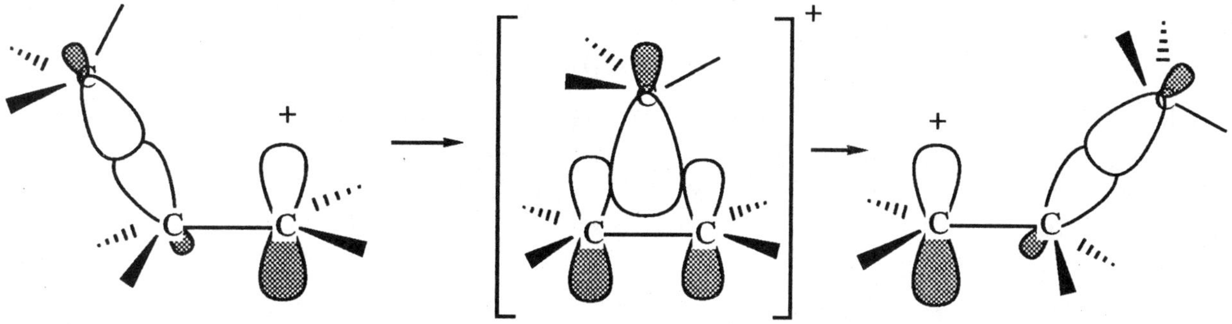

7.12 (a) 104 + 46 - 87.5 - 70 = -7.5 kcal/mole
(b) 98 + 58 - 81.5 - 103 = -28.5 kcal/mole
(c) 104 + 36 - 2(71) = -2 kcal/mole
(d) 92 + 103 - 78 - 119 = -2 kcal/mole

7.13 There are two types of bonds in ethane: the C-H bond with a strength of 98 kcal/mole and the C-C bond with a strength of 82 kcal/mole. The C-C bond is more easily broken and will form two methyl radicals.

60 Study Guide and Solutions Manual

7.14

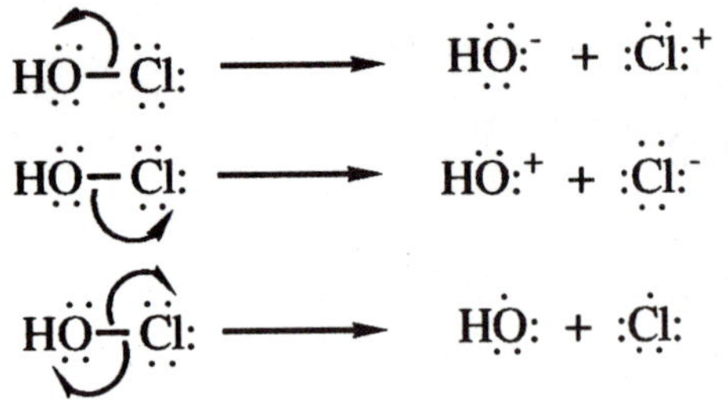

$HO-Cl: \longrightarrow HO:^- + :Cl:^+$

$HO-Cl: \longrightarrow HO:^+ + :Cl:^-$

$HO-Cl: \longrightarrow HO\cdot + \cdot Cl:$

7.15

[Structures showing carbocation rearrangements labeled a, b, c, d, e]

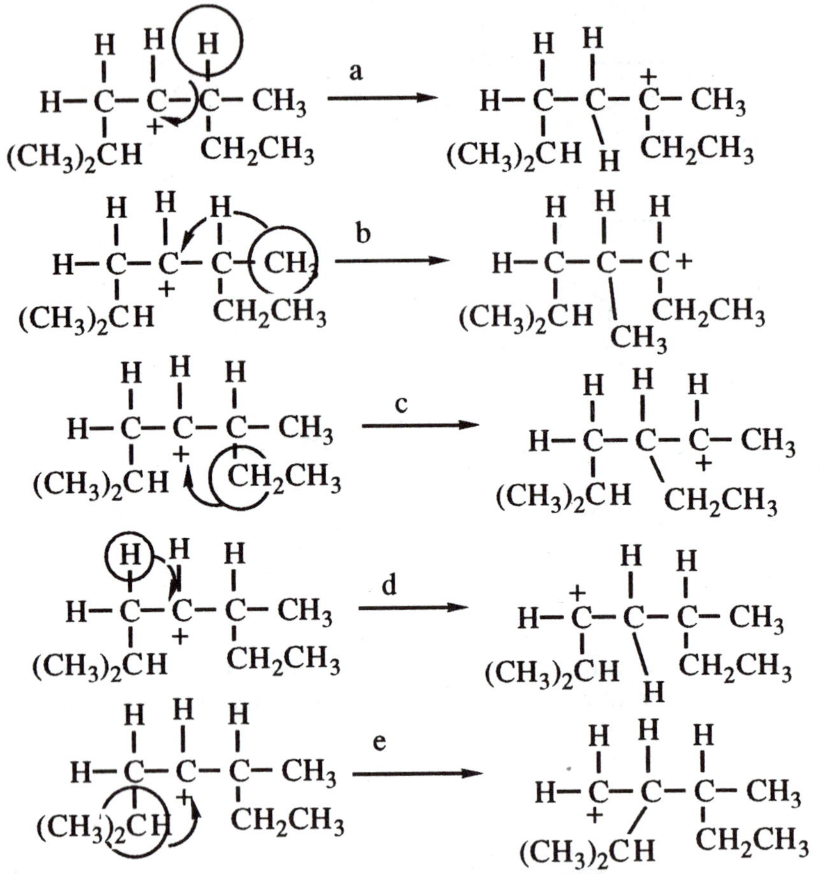

The carbocation formed by rearrangement (a) would be the dominant species among the possibilities as it is the only one that is a tertiary carbocation.

7.16 (a) $CH_3CH_2CH_2CH_2I$; (b) $CH_3CH=C(CH_3)_2$; (c) $(CH_3)_2C=C(CH_3)_2$;
(d) $CH_3CH_2CH_2(Cl)9CH_3)CH_2CH_3$; (e) $(CH_3CH_2)_2CHCH_2CH_2CH_2Cl$;
(f) $(CH_3CH_2)_2CHCH_2CH_2CH_2Br$ $CH_3CH_2CH_2C(Cl)(CH_3)CH_2CH_3$

7.17 (a) $CH_3CH_2CH(CH_3)CH(CH_3)CH_2Br$ 1-bromo-2,3-dimethylpentane
  (b) $CH_3CH_2C(Br)(CH_3)_2$ 2-bromo-2-methylbutane
  (c) $CH_3CH_2CH(CH_3)CH_2C(Br)(CH_3)_2$ 2-bromo-2,4-dimethylhexane   See sheet
  (d) $(CH_3)_3CCH_2CH_2CH(CH_3)C(Br)(CH_3)_2$ 2-bromo-2,3,6,6-tetramethylheptane
  (e) $(CH_3CH_2)_2CHCH_2CH_2CH_2Br$ 1-bromo-4-ethylhexane
  (f) $CH_3CH_2C(Br)(CH_3)_2$ 2-bromo-2-methylbutane

7.18 Both resultant carbocations have charge stabilization by resonance delocalization.

$CH_2=CH-\overset{+}{C}H_2 \longleftrightarrow \overset{+}{C}H_2-CH=CH_2$ $CH_3\overset{..}{O}-\overset{+}{C}H_2 \longleftrightarrow CH_3\overset{+}{O}=CH_2$

7.19 The rearranged tertiary carbocation shown below forms after protonation, loss of water, and a methide shift. This carbocation then leads to products by the routes shown.

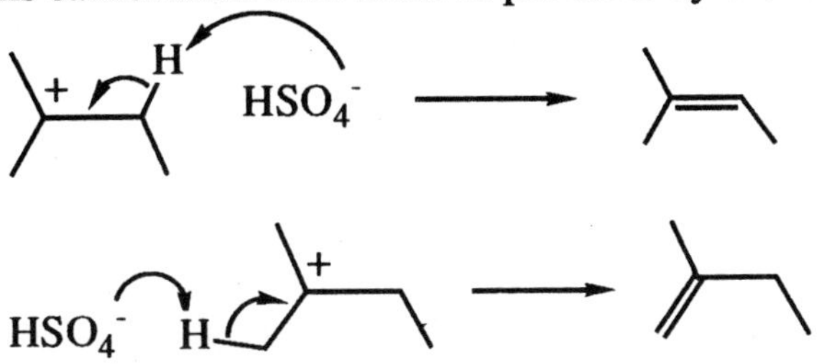

The product in the first of these two routes is formed in the greater amount.

7.20 (a) The 1-butyl carbocation would form.
(b) It would rearrange to the 2-butyl carbocation.
(c) 2-bromobutane
(d) The 1-butyl carbocation does not form. Rather, bromide ion displaces water from the oxonium ion.
(e) After rearrangement occurs to generate the 2-butyl carbocation, a proton is removed from the sp^3 internal carbon to generate the alkene product.

7.21 Once the oxonium ion is formed, it undergoes internal reaction as shown below.

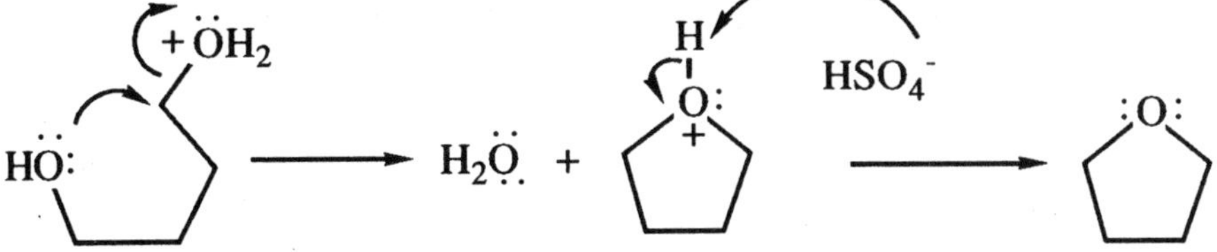

7.22 Once the *tert*-butyl carbocation has formed, the reaction progresses as shown below.

$$(CH_3)_3\overset{+}{C} + {}^{18}\ddot{O}H_2 \longrightarrow (CH_3)_3C{-}^{18}\overset{+}{O}H_2 \rightleftharpoons (CH_3)_3C{-}^{18}\ddot{O}H + H^+$$

7.23

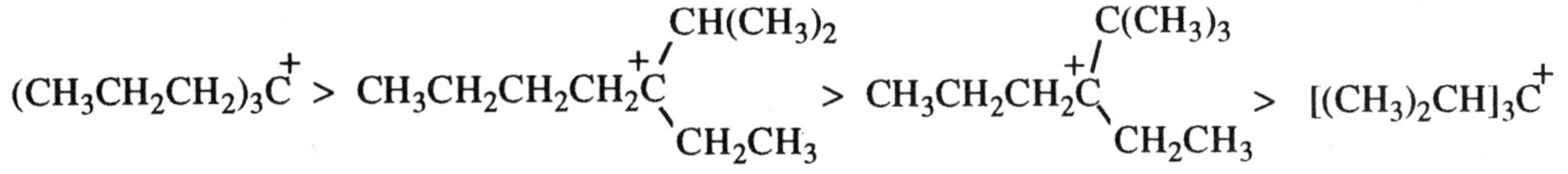

$$(CH_3CH_2CH_2)_3\overset{+}{C} > CH_3CH_2CH_2CH_2\overset{+}{\underset{CH_2CH_3}{C}}{-}CH(CH_3)_2 > CH_3CH_2CH_2\overset{+}{\underset{CH_2CH_3}{C}}{-}C(CH_3)_3 > [(CH_3)_2CH]_3\overset{+}{C}$$

The greater the number of hydrogens attached to carbon atoms adjacent to the carbocation site, the greater is the stabilization. Electrons of a C-H bond are more readily delocalized into an empty *p*-orbital than are electrons of a C-C bond.

7.24 After protonation of the sulfur of 1-butanethiol, the reaction proceeds as shown below.

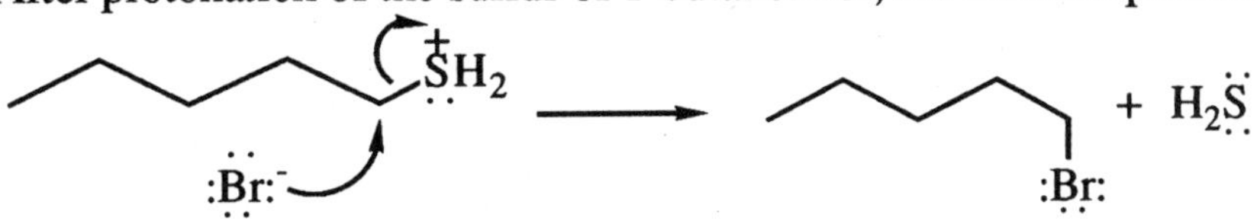

This reaction would proceed faster than the corresponding reaction with 1-butanol since H_2S is a better leaving group than is H_2O.

Solution of Study Guide Practice Problems

7.1

(a) $CH_3{-}\ddot{O}{-}\ddot{O}{-}CH_3 \longrightarrow 2\, CH_3{-}\ddot{O}\cdot$

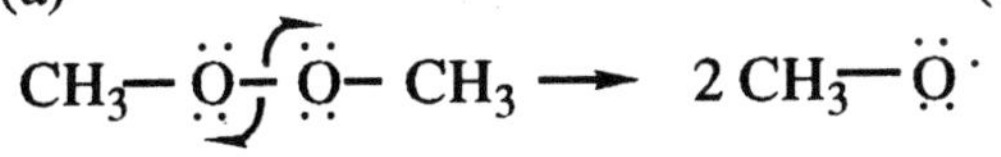

(b) $CH_3CH_2CHCH_3 \longrightarrow CH_3CH_2\overset{+}{C}HCH_3 + :\ddot{Cl}:^-$
         |
         :Cl:

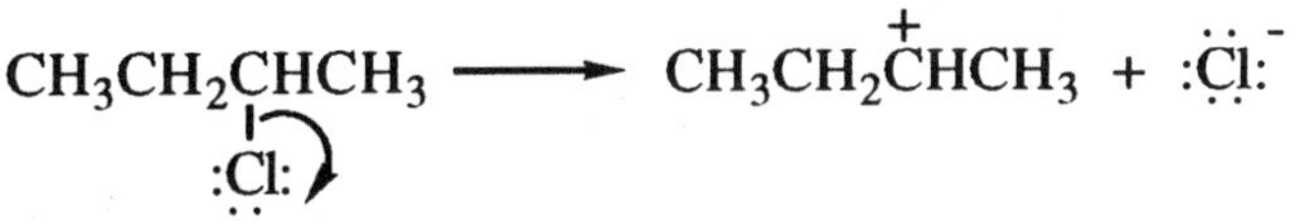

7.2 Less energy is required to break the O-O bond than a C-O bond, (d).
7.3 Of the halogens, fluorine will react most exothermically with methane.
7.4 $\Delta H = 91 + 171 - 17.2 = 244.8$ kcal/mole

7.5

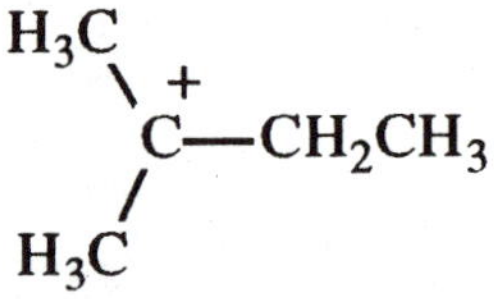

$$\text{H}_3\text{C} \underset{\text{H}_3\text{C}}{\overset{+}{\text{C}}}\text{—CH}_2\text{CH}_3$$

7.6   1-methylcyclopentanol

7.7   The difference between the two mechanisms is that with 1-butanol a molecule of water is displaced from the protonated alcohol by an incoming bromide ion, whereas with *tert*-butyl alcohol the protonated alcohol dissociates to a carbocation and a water molecule, the carbocation combining with a bromide ion.

7.8

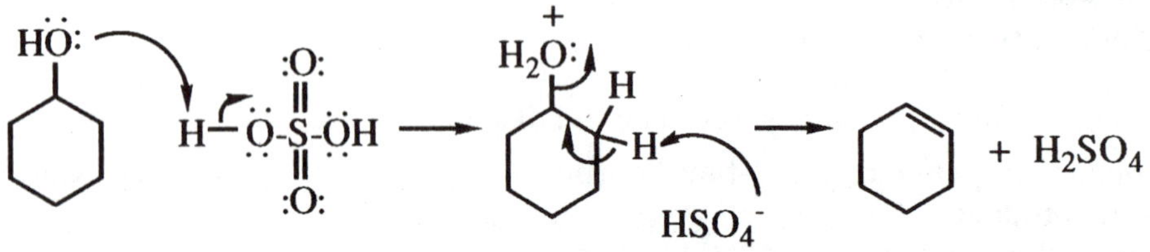

[Mechanism: cyclohexanol + H₂SO₄ → protonated cyclohexanol (with HSO₄⁻) → cyclohexene + H₂O → cyclohexene + H₂SO₄]

7.9   The 2-pentanol would be anticipated to undergo dehydration more readily as it has a greater availability of hydrogens on adjacent carbon atoms (relative to the protonated hydroxyl group) which can be removed for alkene formation.

7.10

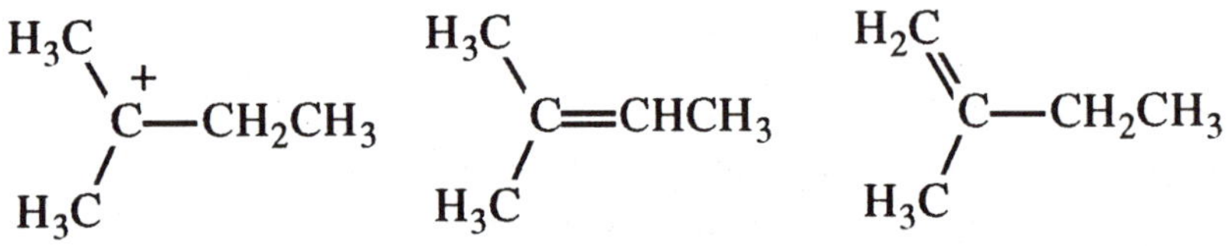

$$\text{H}_3\text{C} \underset{\text{H}_3\text{C}}{\overset{+}{\text{C}}}\text{—CH}_2\text{CH}_3 \qquad \text{H}_3\text{C} \underset{\text{H}_3\text{C}}{\text{C}}\text{=CHCH}_3 \qquad \text{H}_2\text{C} \underset{\text{H}_3\text{C}}{\overset{\|}{\text{C}}}\text{—CH}_2\text{CH}_3$$

7.11   We would not expect this rearrangement to occur. The methide group is not transferred between adjacent carbon sites, but across an intervening methylene group. The distance is too great in this situation for the proper alignment of orbitals needed for an alkide shift to occur.

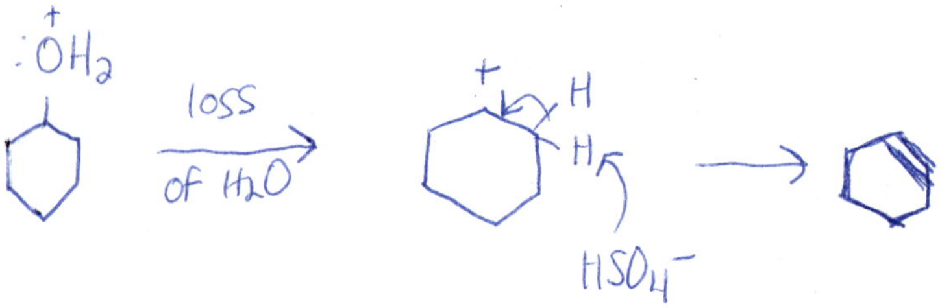

CHAPTER 8
STEREOCHEMICAL PRINCIPLES

Key Points

• Understand the relationship between isomer number and structure

Practice Problem 8.1
How many isomers of formula CH_2Br_2 would you expect if the bonding arrangement about the central carbon atom were planar or pyramidal?

• Be able to recognize stereogenic carbon atoms in open-chain and cyclic structures.

Practice Problem 8.2
Draw structures for all chiral isomers of formula $C_5H_{12}O$.

Practice Problem 8.3
How many stereogenic carbon atoms are present in each molecule of the compounds shown below?

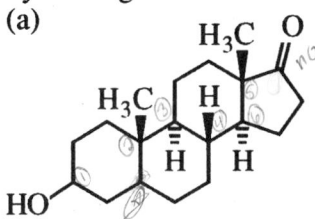

(a) androsterone; a male sex hormone

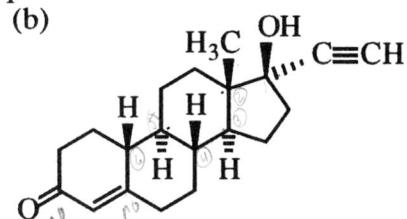

(b) norethidrone; active ingredient of oral contraceptives

Practice Problem 8.4
Which of the following are chiral?
   (a) 1-ethyl-1-methylcyclopropane
   (b) 3-methylcyclopentene
   (c) 4-methylcyclopentene
   (d) 2-chloropentane
   (e) 3-chloropentane

• Understand the relationship between chirality and optical activity

Practice Problem 8.5
A reaction is performed that leads to an organic product having a single stereogenic carbon atom. A chemist carefully isolates the product and analyzes in a polarimeter. No optical activity is observed. What conclusions can you draw regarding the nature of the product, the starting material, and the reaction itself?

64 Study Guide and Solutions Manual

Practice Problem 8.6
The pure (+) enantiomer of a certain substance has an optical rotation of +80°. A chemist measures the rotation of a sample containing both (+) and (−) enantiomers, finding a rotation for the sample of −40°. What are the percentages of (+) and (−) enantiomers in the mixture? What is the optical purity of the mixture?

- Know how to work with stereochemical projection drawings.

Practice Problem 8.7
Complete the structures B-G such that they all represent the enantiomer of structure A.

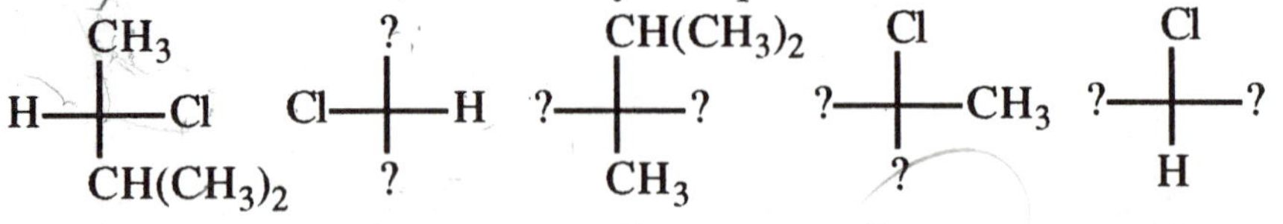

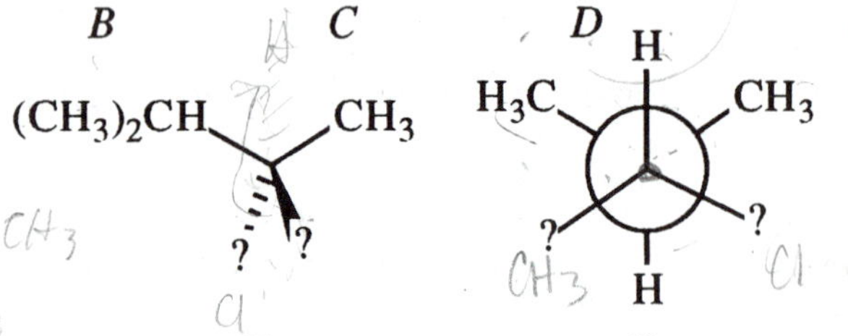

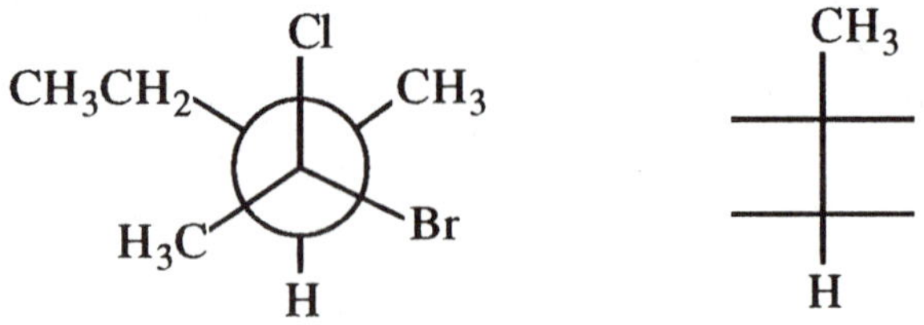

- Know how to assign R and S descriptors to stereogenic centers.

Practice Problem 8.9
Give complete names, including appropriate stereochemical descriptors, for the compounds shown below.

(a)   (b)   (c)   (d)

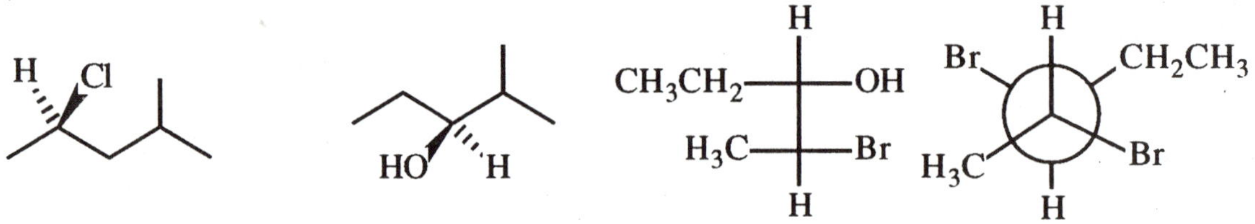

Practice Problem 8.8
Complete the Fischer projection on the right so that it represents the same stereoisomer as the Newman projection shown on the left.

- Be able to recognize the stereochemical relationship between a pair of diagrams.

Practice Problem 8.10
Compare each of the structures B-D with the reference structure A. Tell whether each structure is a different view of A itself or of a different conformation of A, or a view of the enantiomer of A, or a view of a diastereoisomer of A.

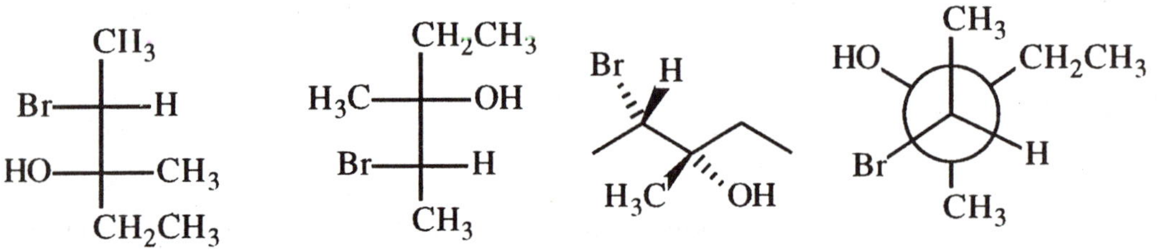

Practice Problem 8.11
Compare structures *B-D* with that of reference structure *A*. Tell whether each structure *B-F* is that of an enantiomer of *A*, or a diastereoisomer of *A*, or is identical to *A* itself or a different conformation of *A*.

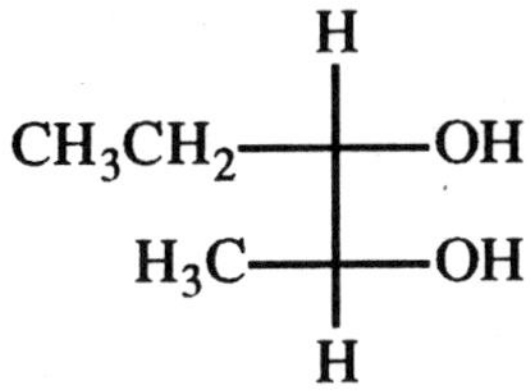

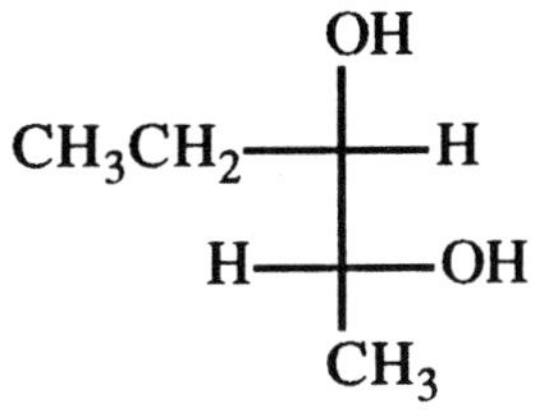

Solution of Text Problems

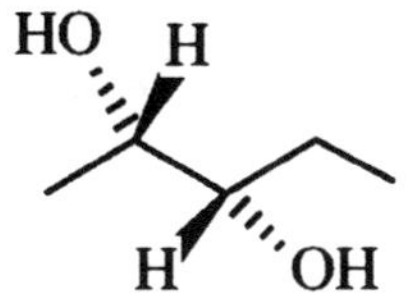

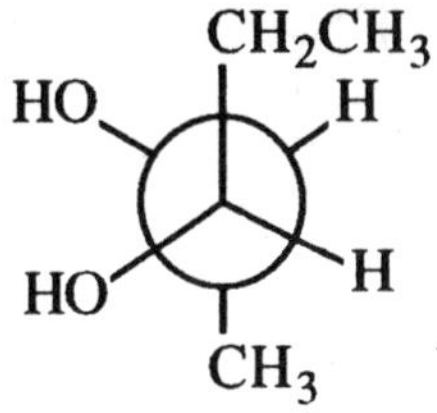

8.1

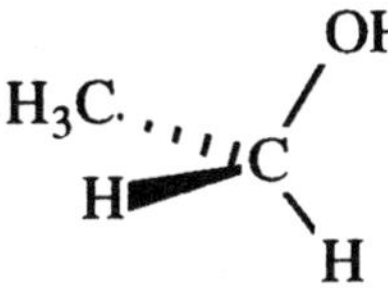

Only one ethanol structure can be made.

8.2

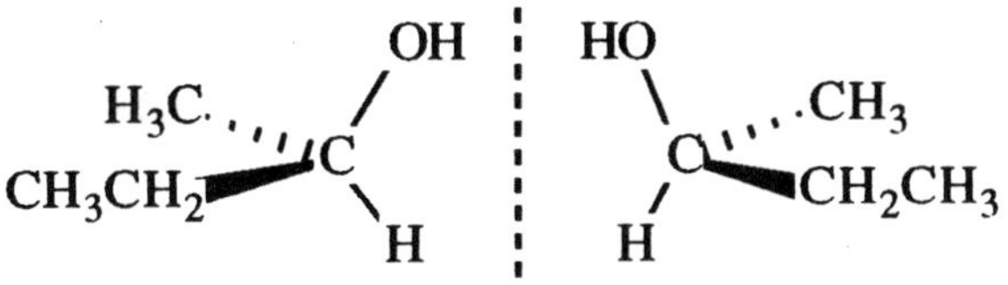

The two models are superimposable only after an odd number of group exchanges have been made.

8.3
(a)

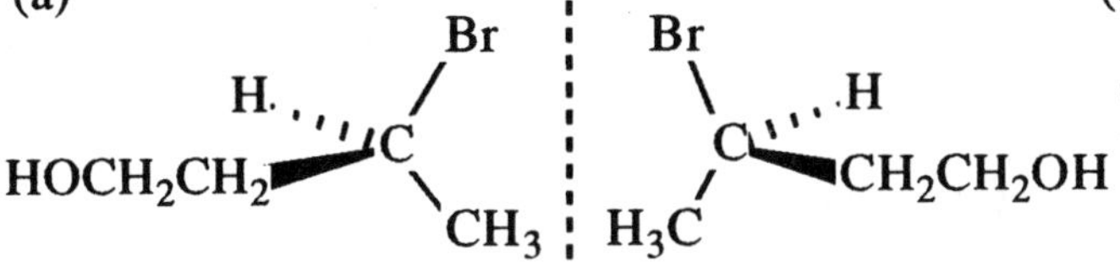

(b)

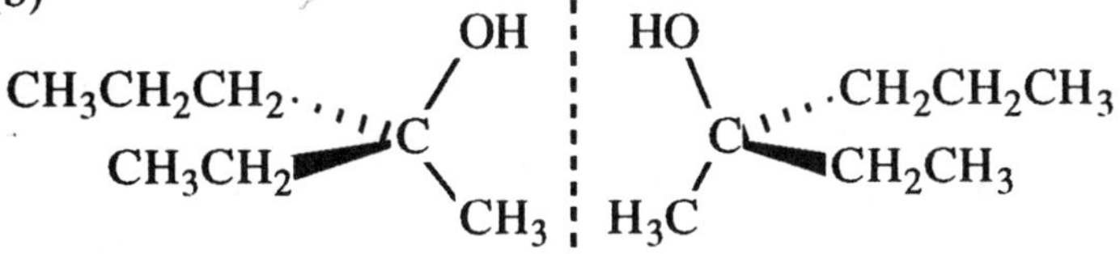

(c)

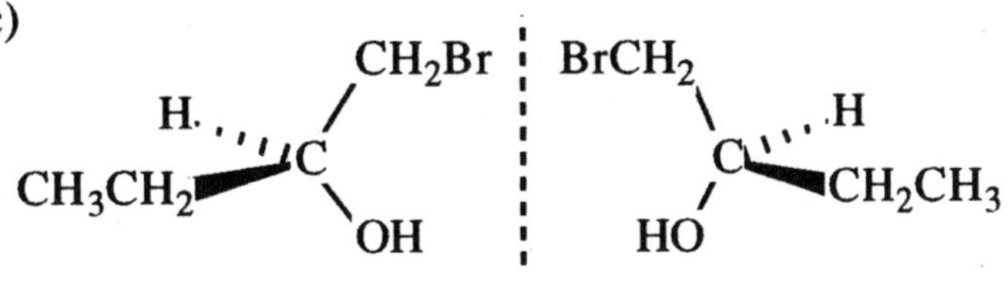

8.4 (a) -15.0°; (b) -40.0°; (c) +8.8°
8.5 (a) 26.4%; (b) 15.4%; (c) 89.6%
8.6 (a) identical; (b) identical; (c) identical; (d) enantiomers

8.7
(a)

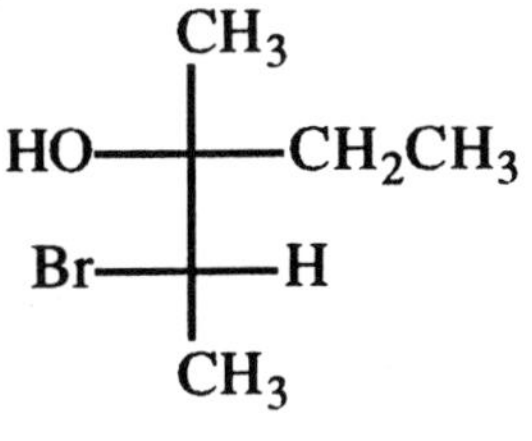

(b)

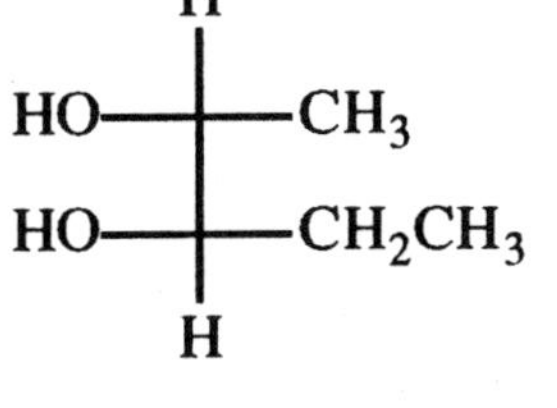

(c)

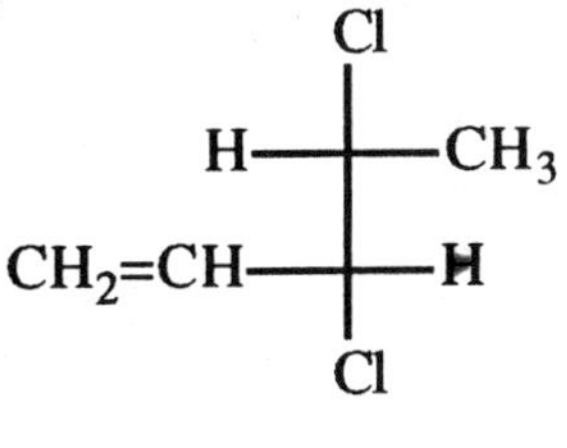

66 Study Guide and Solutions Manual

8.8, 8.9

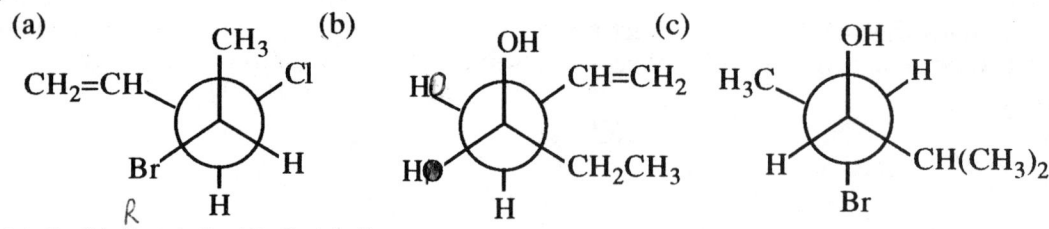

(a) [Newman projection: CH₂=CH, CH₃, Cl, Br, H, H] R

(b) [Newman projection: HO, OH, CH=CH₂, CH₂CH₃, HO, H]

(c) [Newman projection: H₃C, OH, H, CH(CH₃)₂, H, Br]

8.10 (a) S; (b) S; (c) R; (d) S; (e) S

8.11 To arrive at the designation (2R,3S)-2,3-butanediol we need only to begin numbering at the opposite end of the carbon chain.

8.12 Each of the stereogenic sites (three are present) has the R configuration.

8.13 There are twelve different valid Fischer projections for (R)-2-butanol, as shown below. (Me = methyl; Et = ethyl)

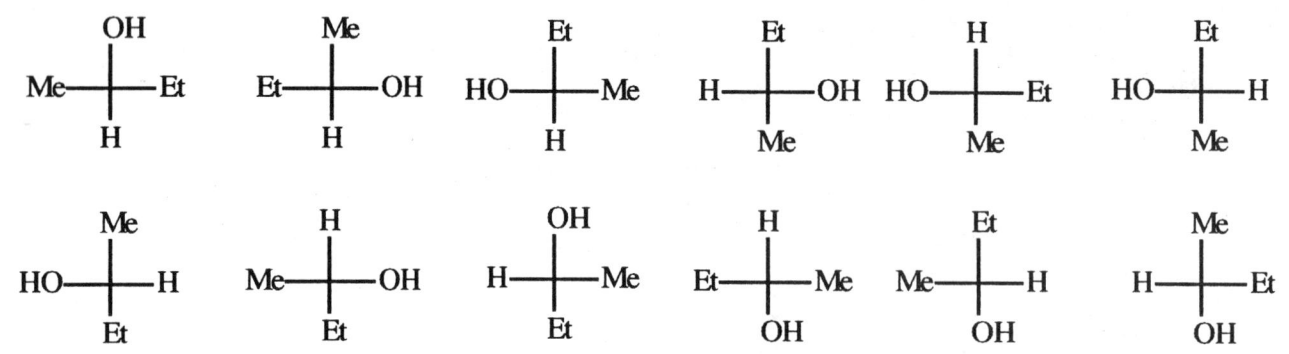

[12 Fischer projections shown]

OH   Me   Et   Et   H   Et
| | | | | |
Me—+—Et   Et—+—OH   HO—+—Me   H—+—OH   HO—+—Et   HO—+—H
| | | | | |
H   H   H   Me   Me   Me

Me   H   OH   H   Et   Me
| | | | | |
HO—+—H   Me—+—OH   H—+—Me   Et—+—Me   Me—+—H   H—+—Et
| | | | | |
Et   Et   Et   OH   OH   OH

If we flip each projection through 180° we find that we have produced one of the other twelve valid Fischer projections for (R)-2-butanol. Vertical bonds remain pointing behind the plane of the paper, and horizontal bonds remain pointing up from the plane of the paper. Moreover, there have been two exchanges in positions of substituents about the stereogenic center, thus regenerating the original configuration. However, if we rotate any of these projections through 90° either clockwise or counterclockwise, we no longer have a valid Fischer projection. The vertical bonds now point up from the plane of the paper and the horizontal bonds point back behind the plane of the paper, a situation which does not represent a Fischer projection.

8.14 (a) enantiomers; (b) conformational enantiomers; (c) conformational enantiomers

8.15 Stereogenic centers are indicated by an asterisk (*).

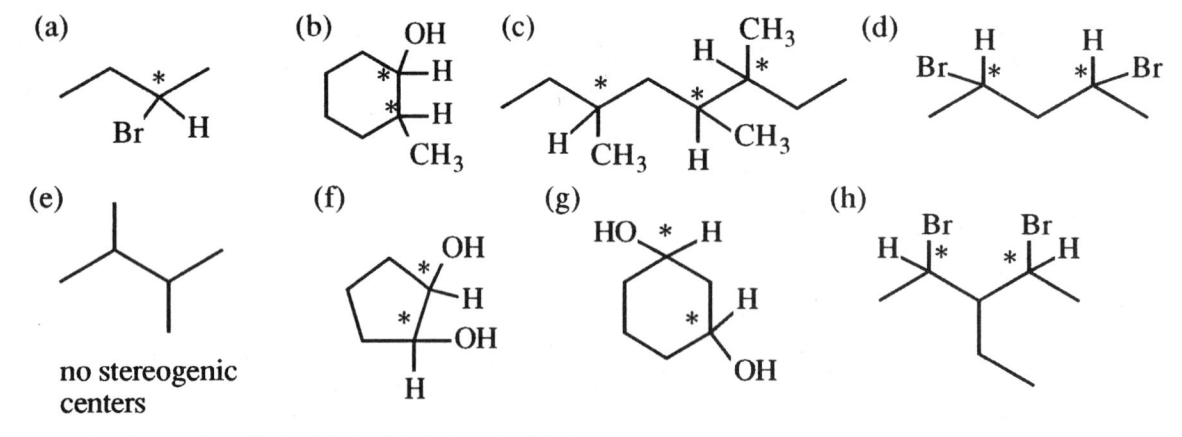

(a) [structure with Br, H, *]

(b) [cyclohexane with OH, H, H, CH₃, two *]

(c) [structure with CH₃, H, CH₃, H, two *]

(d) [structure with Br, H, H, Br, two *]

(e) [structure: no stereogenic centers]

(f) [cyclopentane with OH, H, OH, H, two *]

(g) [cyclohexane with HO, H, H, OH, two *]

(h) [structure with Br, H, H, Br, two *]

8.16 (a) 2; (b) 4; (c) 8; (d) 4; (e) 1; (f) 4; (g) 4; (h) 4

8.17-8.19

(d) meso (first structure)

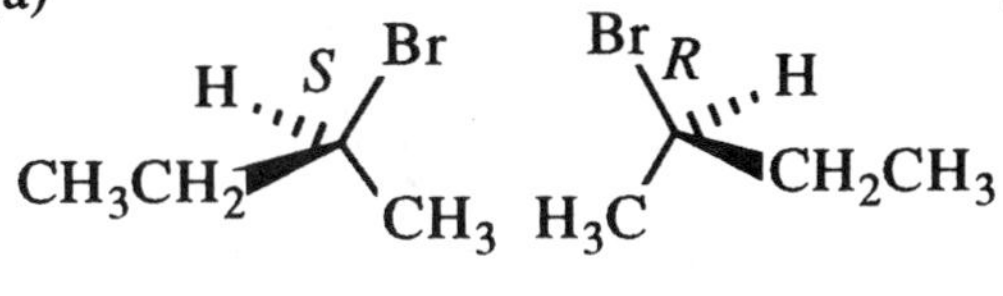

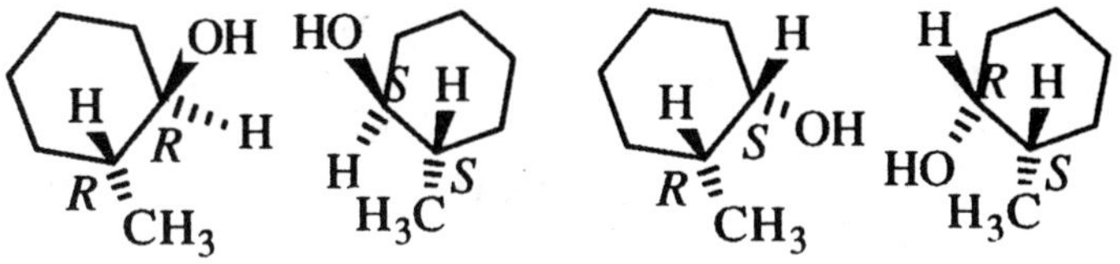

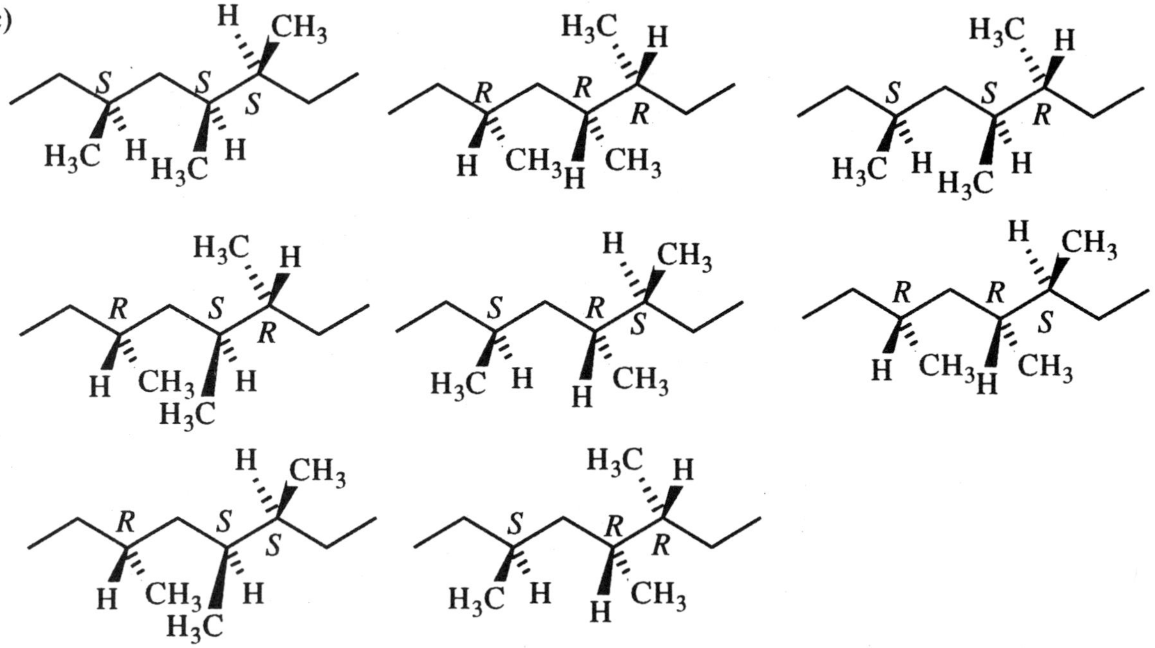

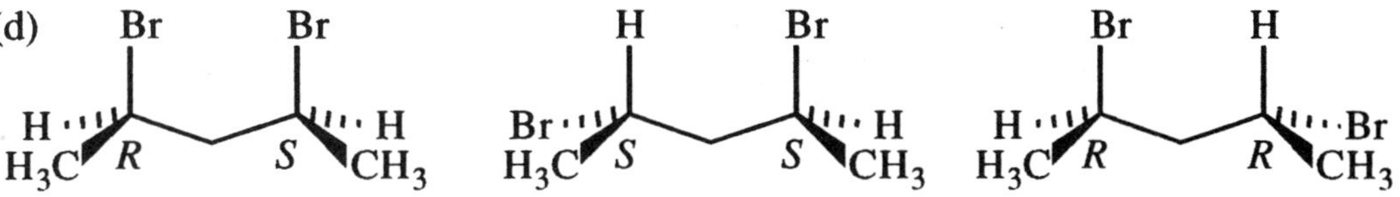

(e) only one structure

(f) meso (first structure)

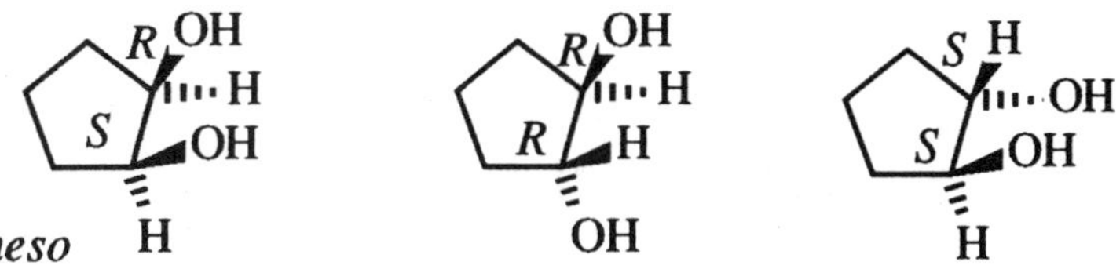

(g) meso (first structure)

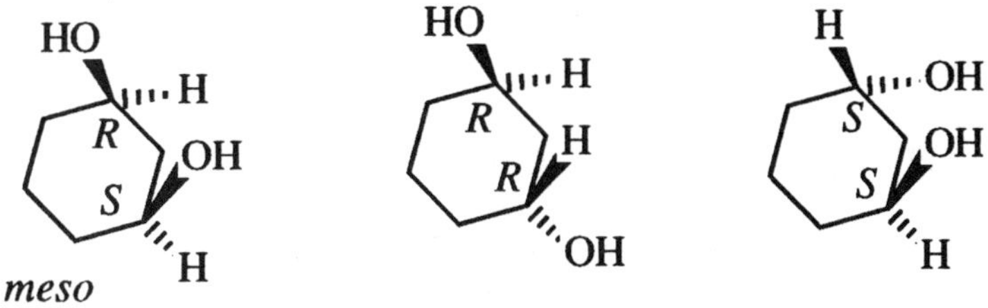

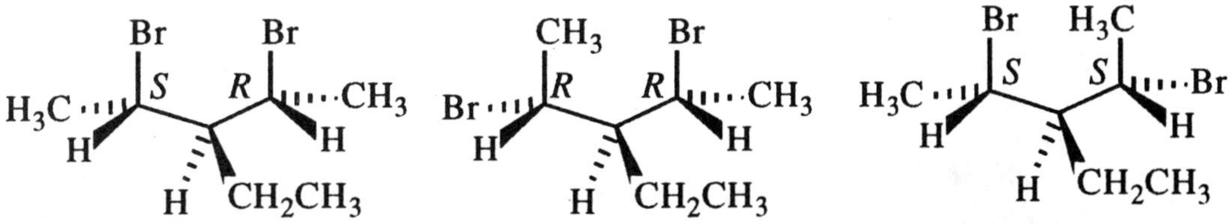

68 Study Guide and Solutions Manual

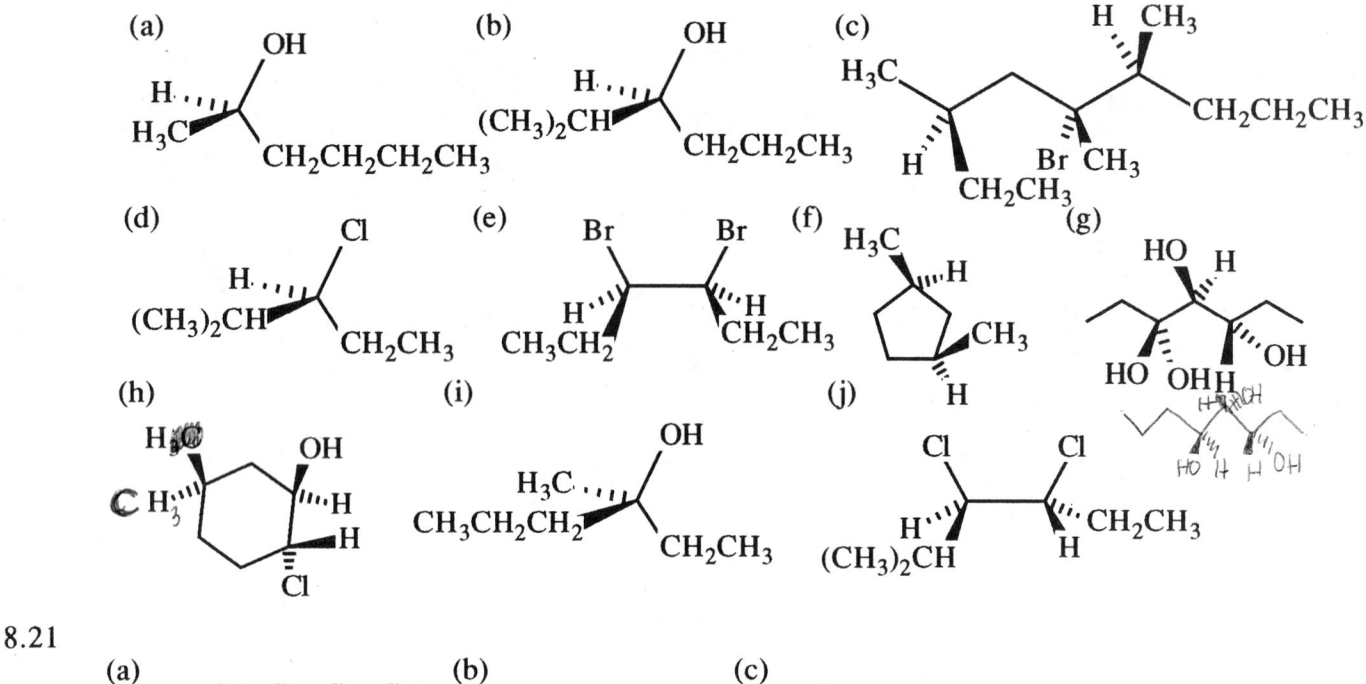

8.20

(a), (b), (c), (d), (e), (f), (g), (h), (i), (j) [structural formulas]

8.21

(a), (b), (c), (d), (e), (f), (g), (h), (i), (j) [Fischer projection structures]

8.22 75.5% optical purity
8.23 20.5% optical purity
8.24 (a) False - Chiral compounds having only one stereogenic center and no olefinic linkages can not have diastereoisomers.
   (b) False - Compounds that are *meso* contain stereogenic centers but are not optically active.
   (c) False - There is no *a priori* relationship of absolute configuration and sense of optical rotation at a particular wavelength.
   (d) False - The relative priorities of substituents about a stereogenic center can be changed without breaking any bonds to the stereogenic center.
8.25 (a) 0; (b) 1; (c) 4; (d) 1; (e) 9
8.26 (a) R; (b) S; (c) R; (d) S
8.27 (a) R; (b) (3R,4S); (c) (2R,3R)
8.28 (a) (3S,4R); (b) S; (c) (2S,3S)

8.29

(a) Fischer projection: Cl (top), H₃C—|—H, CH₃CH₂—|—H, Br (bottom)

(b) Fischer projection: Cl (top), H₃C—|—H, H—|—CH₃, Br (bottom)

8.30

(a) Newman projection with front CH₂CH₃, OH, CN (front carbon with CH₃ below); H, H on back

(b) Newman projection with front CH₃, CH=CH₂, Br; H, H, Br on back

8.31

$$\text{BrCH}_2\!-\!\overset{\overset{\displaystyle CH_3}{|}}{\underset{\underset{\displaystyle H}{|}}{C}}\!-\!\text{CH}_2\text{Cl}$$

(S)-1-bromo-3-chloro-2-methylpropane

8.32

$$\text{BrCH}_2\text{CH}_2\!-\!\overset{\overset{\displaystyle CH_3}{|}}{\underset{\underset{\displaystyle H}{|}}{C}}\!-\!\text{CH}_2\text{CH}_2\text{OH} \xrightarrow{\text{PBr}_3} \text{BrCH}_2\text{CH}_2\overset{\overset{\displaystyle CH_3}{|}}{\text{CH}}\text{CH}_2\text{CH}_2\text{Br}$$

8.33-8.34

(a) Eight Fischer projections of structures with CH₂OH and CH₃ termini with three stereocenters bearing OH/H.

(b) Four wedge-dash structures with Br, H on two carbons bearing H₃C and CH₂CH₃ groups.

(c) Three wedge-dash structures with Br, H on two carbons bearing H₃C and CH₃ groups.

meso - only three actually exist of four theoretically possible structures

8.35 Measure the rotation of the sample diluted by a known amount, such as by 50%, and compare the resultant rotation with the original sample. If the rotation of the original sample were really +5.0°, the diluted sample should have a rotation of +2.5°. If the rotation of the original sample were really +365°, the rotation of the diluted sample would be +182.5°.

8.36 We calculate the optical purity of the sample to be 64.4%. This means that there is a 64.4% excess of the dextrorotatory form; the remainder of the sample is an equal mixture of the two forms. Thus the mixture is 82.2%

dextrorotatory and 17.8% levorotatory.

8.37 (A) diastereoisomers; (b) identical; (c) identical; (d) enantiomers; (e) diastereoisomers

8.38 As there are four stereogenic centers in the general structure, there are theoretically sixteen stereoisomers. However, due to the symmetry of the system, fewer than sixteen actually exist. There are two *meso* structures for this formula, as shown below.

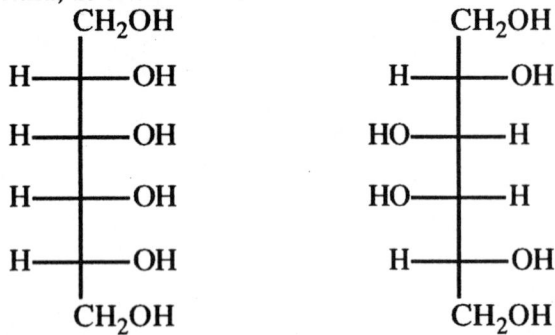

8.39 There are eight optically active stereoisomers of the general structure given, as shown below. This is an unusual system in which only ten of the theoretically possible stereoisomers actually exist, due to the particular symmetry of the structure.

8.40 (a) R; (b) R; (c) R

8.41 There are three stereogenic centers in the molecule.

8.42 It remains R. No bonds have been broken to the stereogenic center and there has been no change in the relative priorities of the attached groups.

8.43 (2S,3R)-2,3-dibromopentane

8.44 (a) identical; (b) diastereoisomer; (c) enantiomer; (d) identical; (e) identical

8.45 (a) non-chiral; (b) chiral; (c) chiral

8.46

8.47 (a) -N=O
(b) -P(O)(OH)$_2$

(c) -C(CH$_3$)$_2$CH(CH$_3$)$_2$
(d) -CH$_2$CH=CH-

Solution of Study Guide Practice Problems

8.1 We would expect two isomers of the formula CH$_2$Br$_2$ if the bonding about carbon were square planar or pyramidal.

8.2 There are four sets of enantiomers of the given formula, one of each set being shown below.

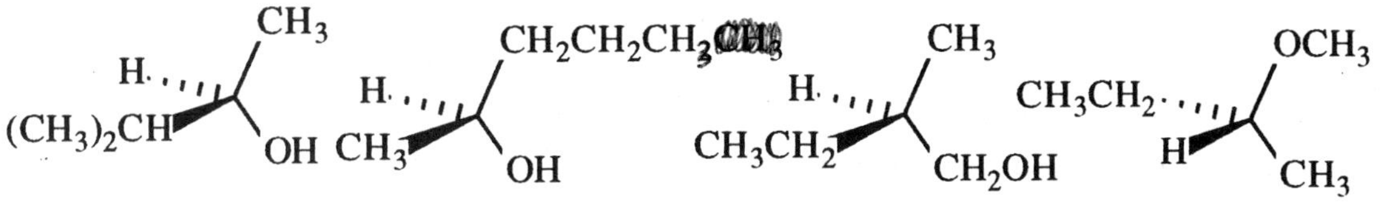

8.3 (a) 7; (b) 6

8.4 (a) not chiral; (b) chiral; (c) not chiral; (d) chiral; (e) not chiral

8.5 The lack of optical activity in the product could be the result of one (or more) of several factors. The product could be racemic, in which case we could have begun with a racemic starting material, or an achiral starting material. Also, the reaction could have involved the reaction of an optically active starting material which underwent a racemization during the reaction or proceeded through an achiral intermediate. Less likely, but still a possibility, is that the product is a single enantiomer, but with a negligibly small rotation at the wavelength of light used for the measurement of optical activity.

8.6 The optical purity is 50%. The mixture contains 75% of the (-) enantiomer and 25% of the (+) enantiomer.

8.7

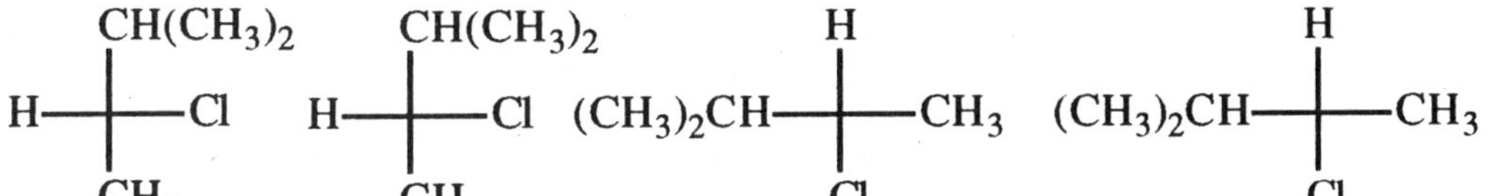

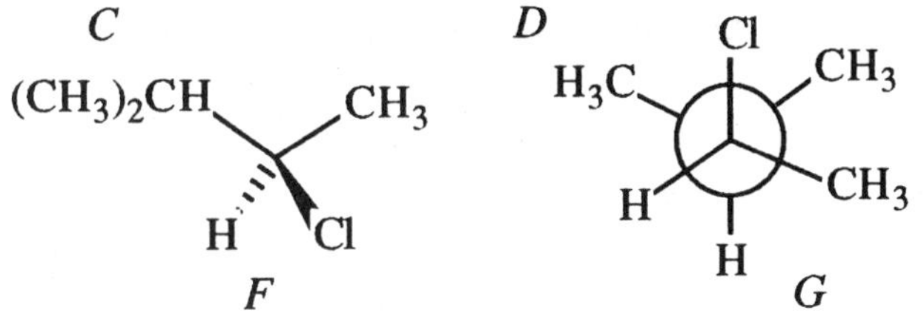

8.8

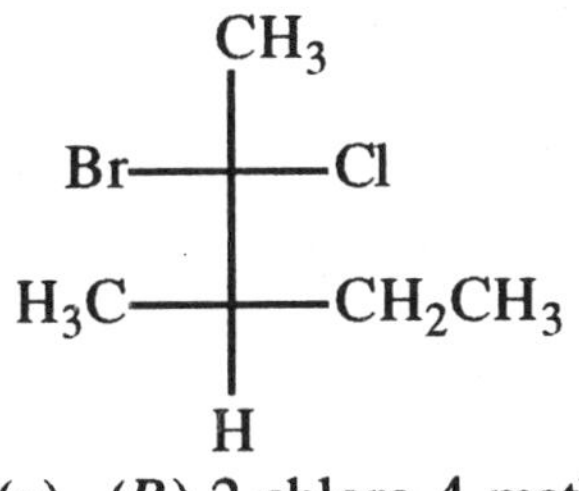

8.9
(a) (R)-2-chloro-4-methylpentane
(b) (S)-2-methyl-3-pentanol
(c) (R)-2-bromo-3-pentanol (2R-3S)
(d) (2R,3S)-2,3-dibromopentane

8.10 Structures B and C are diastereoisomers of A, while D is identical to A.

8.11 Structure B is an enantiomer of A, while C and D are diastereoisomers of A.

CHAPTER 9
CARBON-CARBON DOUBLY BONDED SYSTEMS I. STRUCTURE, NOMENCLATURE, AND PREPARATION

Key Points

• Understand the σ–π model for the carbon-carbon double bond, and related aspects of bonding.
Practice Problem 9.1
Which bond, σ or π, matches each of the following descriptions?
  (a) is stronger
  (b) is cylindrically symmetrical
  (c) has associated with it a higher energy filled molecular orbital
  (d) releases the most energy when formed
Practice Problem 9.2
Which type of C-H bond is weakest in a molecule of propene? Explain your choice.

• Know the following points of nomenclature relating to compounds with carbon-carbon double bonds:
  • basic nomenclature for the alkene linkage
  • nomenclature of compounds containing the alkene linkage as well as other functional groups (Table 9.2)
  • the *cis/trans* and *E/Z* descriptor systems
  • the classification of alkenes as mono-, di-, tri-, or tetrasubstituted
Practice Problem 9.3
Name the following compounds using the appropriate *E/Z* descriptor when appropriate.

(a) BrCH$_2$CH$_2$CH=CH$_2$
(b), (c), 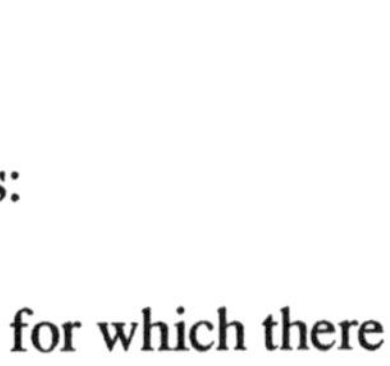 (d), 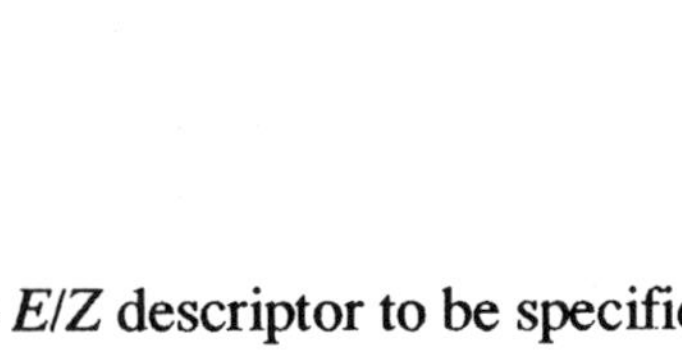 (e) [structures shown]

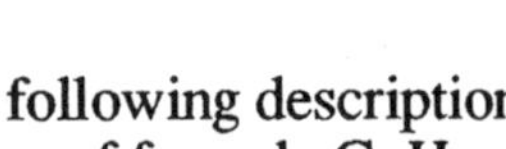

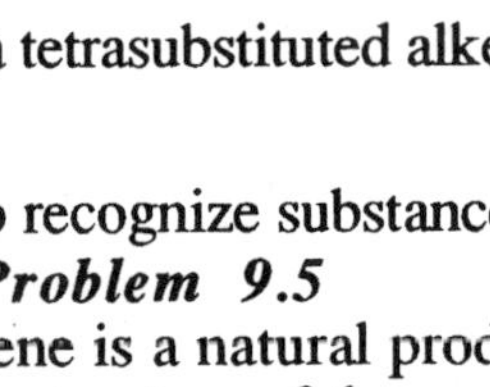

Practice Problem 9.4
Suggest structures matching the following descriptions:
  (a) trisubstituted Z alkenes of formula C$_7$H$_{14}$
  (b) a tetrasubstituted alkene of formula C$_8$H$_{16}$ for which there is no *E/Z* descriptor to be specified

• Be able to recognize substances composed of isoprene units.
Practice Problem 9.5
Caryophyllene is a natural product obtained from clove oil. Given that it is known to be composed of isoprene units, which structure of those shown below could *not* be that for caryophyllene?

A   B

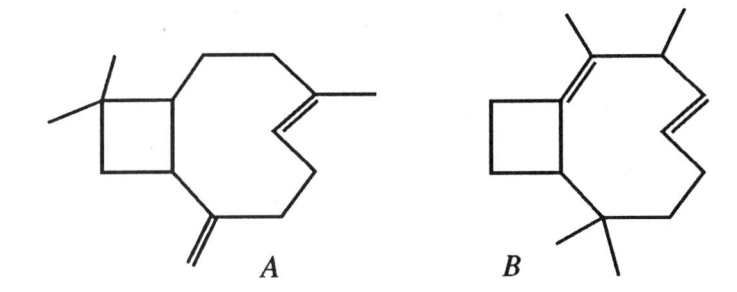

- Know the reagents used and the mechanisms involved in the syntheses of alkenes by:
  - dehydration of alcohols (review Chapter 7)
  - dehydrohalogenation of haloalkanes

Practice Problem 9.6
Which preparative method for alkenes, dehydrohalogenation or dehydration, occurs by a mechanism involving more than one step?

Practice Problem 9.7
Name all bromoalkanes of formula C_4H_9Br that will yield a single alkene on treatment with a base.

Practice Problem 9.8
Suppose you wish to convert a sample of 1-hexanol to 1-hexene which is free of isomeric alkenes. How should you proceed?

Solution of Text Problems

9.1   (a) 2-ethyl-1-hexene
      (b) 2,3,4-trimethyl-2-hexene
      (c) 3-ethyl-4-(1-methylbutyl)-3-nonene
      (d) 2,3-dimethylcyclohexene
      (e) 4-ethylcyclohexene

9.2   (a) 3-bromo-2-methyl-1-propene
      (b) 4-bromo-1-butene
      (c) 5-chloro-1,3-cyclohexadiene
      (d) 2-bromo-4-methyl-4-penten-1-ol
      (e) 1-bromomethylcyclohexene
      (f) 2,3-dimethyl-2-pentene

9.3

cis-1,2-dibromoethene   trans-1,2-dibromoethene   The cis-1,2-dibromoethene has the higher dipole moment, and thereby is expected to have the higher boiling temperature of the two isomers.

9.4

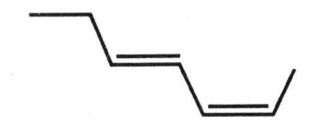

cis,cis-2,4-heptadiene   cis,trans-2,4-heptadiene

9.5   (a) (E)-3-ethyl-2,6-dimethyl-4-propyl-3-heptene
      (b) (Z)-1-bromo-2-chloro-1-butene
      (c) (Z)-cyclohexene
      (d) (E)-1-bromocyclohexene
      (e) (Z)-1-bromocycloheptene

9.6

(a) [structure with Br] (b) [diene structure] (c) [cyclohexadiene structure]

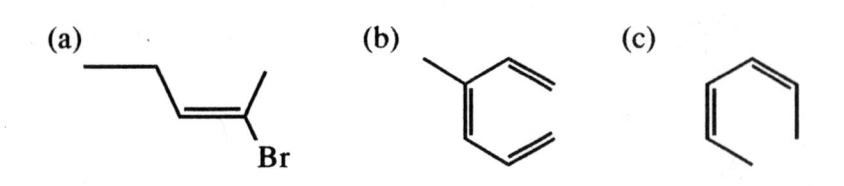

9.7

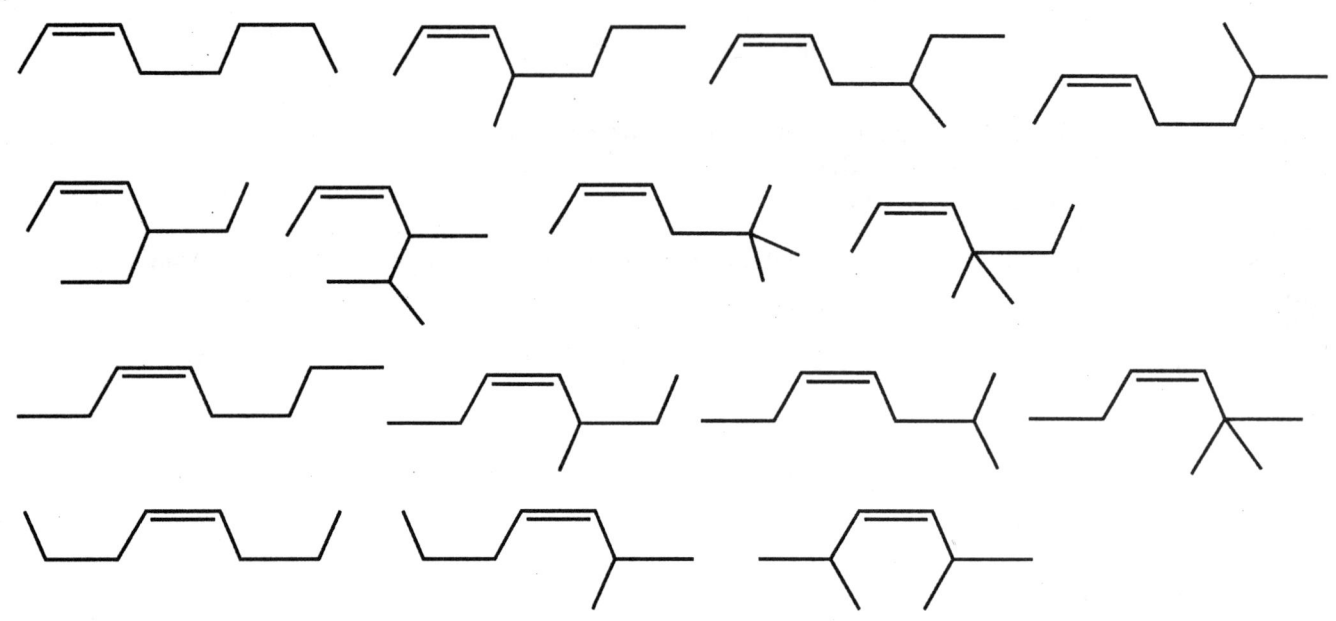

[Various alkene structures]

9.8

propene; 1-pentene
(Z)-2-butene; (E)-2-butene; 2-methylpropene; (Z)-cyclohexene
2-methyl-2-butene; (Z)-3-methyl-3-hexene
2,3-dimethyl-2-butene; (E)-3,4-dimethyl-3-hexene

9.9

(E)-2-pentene   (Z)-2-pentene   2-methyl-1-butene

9.10 Junctures of isoprene units are shown with dotted bonds.

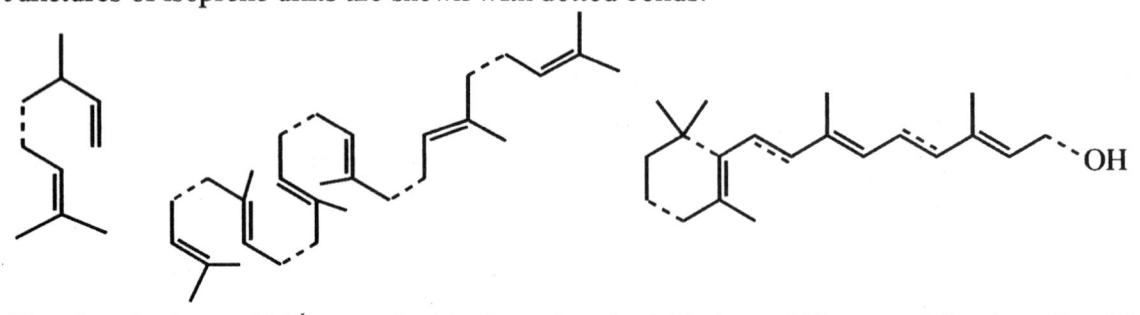

[Structures of myrcene, squalene-like, and vitamin A with OH]

9.11
The vitamin A would be treated with chromic anhydride in pyridine to oxidize it to the aldehyde without disturbing the olefinic linkage.

9.12
(a) propane; (b) 1-pentene; (c) 2-methyl-1-pentene

9.13

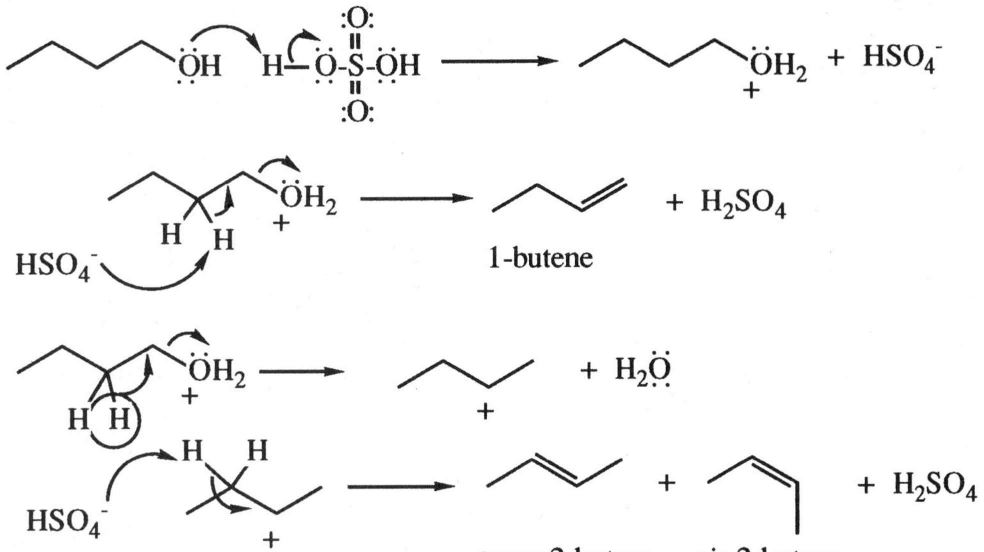

[mechanism showing protonation of 1-butanol by H₂SO₄ to give butyloxonium ion + HSO₄⁻; then E2-like elimination by HSO₄⁻ removing a β-H to give 1-butene + H₂SO₄; alternate pathway giving trans-2-butene + cis-2-butene + H₂SO₄]

1-butene

trans-2-butene   cis-2-butene

9.14 If a carbocation were involved, we would anticipate rapid rearrangement such that the positive charge would be located at an internal position leading to internal alkene products rather than only a terminal alkene.

9.15 1. Treatment with phosphorus tribromide to form 1-bromopentane; 2. Treatment with a strong base such as potassium *tert*-butoxide to cause the elimination to 1-pentene.

9.16 (CH₃)₃CBr

9.17. (a) (Z)-2,3-dimethyl-3-heptene
(b) (E)-2-pentene
(c) (2E,6Z)-2,3,6-trimethyl-2,6-octadiene
(d) 3-methylcyclopropene
(e) (R)-3-methyl-1-pentene
(f) (R)-(Z)-4,5-dimethyl-2-hexene
(g) (Z-3,5-dimethyl-4-propyl-4-octene

9.18 (a) (b) (c) (d) (e) (f) [structures]

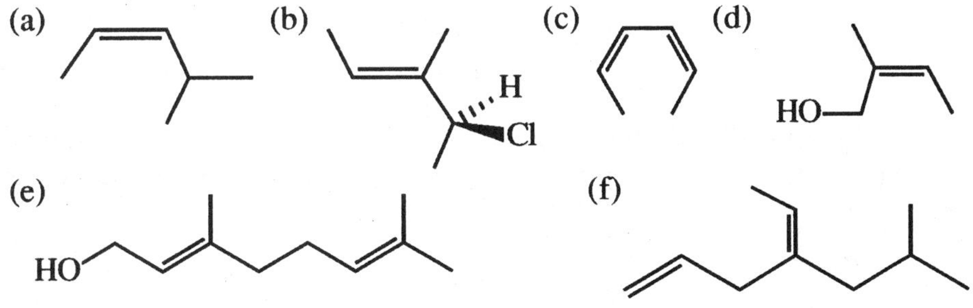

9.19 [structure with (E) labels marked on double bonds]   The remaining two carbon-carbon double bonds have two like groups attached to one of the carbon atoms and thus can not need be specified as to geometry.

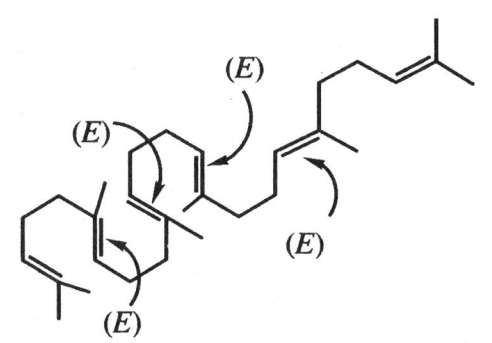

9.20 1-bromo-4-ethylhexane

9.21

(a) (Z)-2-hexene (E)-2-hexene (Z)-4-methyl-2-pentene (E)-4-methyl-2-pentene

(Z)-3-hexene (E)-3-hexene 2-methyl-1-pentene 2-ethyl-1-butene 2,3-dimethyl-1-butene

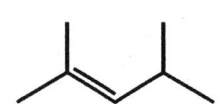

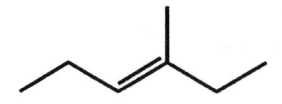

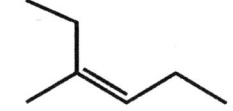

(b) 2-methyl-2-hexene 2,4-dimethyl-2-pentene (E)-3-methyl-3-hexene (Z)-3-methyl-3-hexene

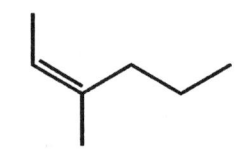

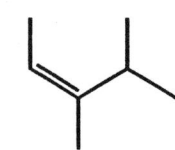

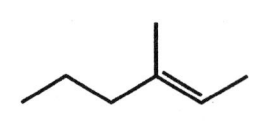

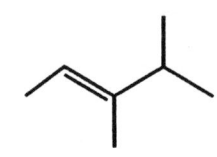

(Z)-3-methyl-2-hexene (Z)-3,4-dimethyl-2-pentene (E)-3-methyl-2-hexene (E)-3,4-dimethyl-2-pentene

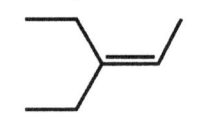

3-ethyl-2-pentene

9.22 (a) cis-2-butene   The effects of the two methyl groups add rather than oppose each other.
 (b) cis-1,2-dichloropropene   The methyl group effect adds to that of the chlorine attached to the same carbon atom.
 (c) (Z)-2,3-dichloro-2-butene   The effects of the two chlorine substituents are in the same direction.

9.23 The van der Waals effect is greater with cis-1-bromopropene, relatively increasing its boiling temperature.

9.24 The junctures between isoprene units are shown by dotted bonds.

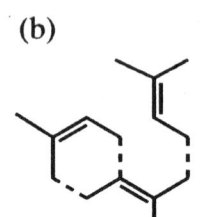

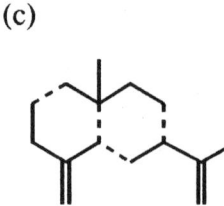

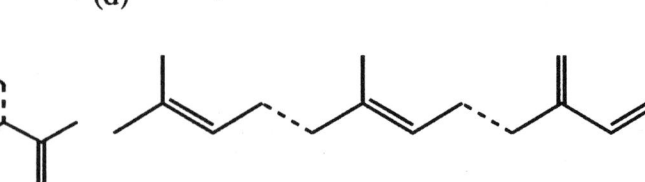

(a) (b) (c) (d)

9.25

9.26 (a) 2-ethyl-1-butene
 (b) (E)-1-bromopropene
 (c) 2-ethyl-3-buten-1-ol

9.27 (a) 2-methyl-3-buten-2-ol

(b) (3Z,5Z,7E)-3,5,7-dodecatrien-1-ol

9.28 (a)

[reaction scheme: HO:⁻ attacks H on carbon adjacent to C-Br, forming alkene + H₂O + :Br:⁻]

(b)

[reaction scheme: tert-butyl chloride ionizes to tert-butyl cation + :Cl:⁻]

[reaction scheme: tert-butyl cation + :OH⁻ → isobutylene + H₂O]

9.29 Dehydration of the alcohol using acid can proceed with rearrangement upon formation of a carbocation from the protonated alcohol, thus forming additional undesired products. A better approach is to use the two-step procedure: (1) treatment with phosphorus tribromide to form 1-bromohexane; (2) elimination using a strong base such as potassium *tert*-butoxide.

9.30 (a) 1-bromo-2,2-dimethylbutane
(b) 1-bromohexane; 1-bromo-4-methylpentane; 1-bromo-3-methylpentane; 1-bromo-3,3-dimethylbutane; 1-bromo-2-methylpentane; 1-bromo-2-ethylbutane; 2-bromo-3,3-dimethylbutane

9.31 The effect of the methyl groups which donate electron density toward an alkene linkage is to lower the ionization potential. Placement of a cyano group on an alkene linkage increases the ionization potential. Thus we infer that the cyano group withdraws electron density from the alkene linkage.

9.32 For an alkene reacting by using its π-HOMO with the LUMO from another molecule, the higher the energy of that HOMO the closer it will be to the interactring LUMO. The closer in energy the reacting HOMO and LUMO are, the more facile is the reaction. Thus, the greater the alkyl substitution is about the alkene linkage, the more facile is the alkene in this type of reaction.

9.33

[reaction scheme: carbocation intermediate with -OP₂O₅³⁻ group and base B: removes H to form alkene product + HB⁺]

Solution of Study Guide Practice Problems

9.1 (a) σ; (b) σ; (c) π; (d) σ

9.2 The weakest C-H bond in propene is that of the methyl group. The carbon uses an sp^3 hybrid orbital to form this bond; the sp^3 hybrid orbital has only 25% *s* character and thereby forms weaker bonds than the others, which involve carbon using sp^2 hybrid orbitals.

9.3 (a) 4-bromo-1-butene
(b) (Z)-3-penten-1-ol
(c) 3-bromo-2-isobutyl-4-methyl-1-pentene
(d) (Z)-4-ethyl-2,7-dimethyl-4-octene
(e) (2Z,4Z)-2,4-dibromo-3-methyl-2,4-hexadiene

9.4

(a) [structures shown]

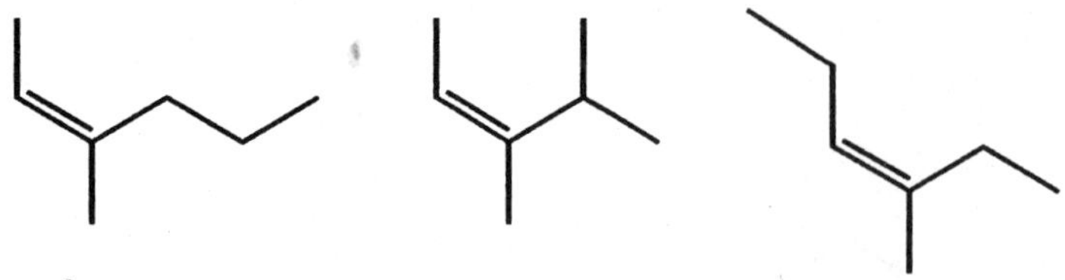

(b) [structure shown]

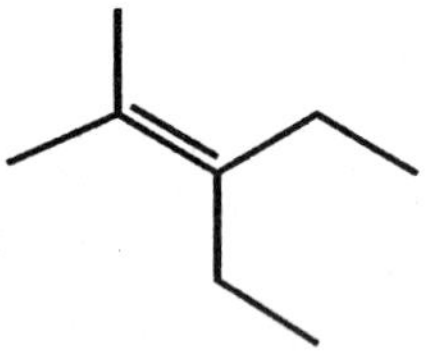

9.5 Structure B could not be caryophyllene as it is not an isoprenoid.

9.6 The mechanism of dehydration involves more than one step; the alcohol must first be protonated and then water is lost and the alkene generated. Dehydrohalogenation can occur in a single step.

9.7 2-Bromo-2-methylpropane will undergo dehydrohalogenation to yield only 2-methylpropene.
  1-Bromobutane will undergo dehydrohalogenation to yield only 1-butene.
  1-Bromo-2-methylpropane will undergo dehydrohalogenation to yield only 2-methylpropene.

9.8 Dehydration is not a good method as the reaction will proceed through a rearranged carbocation yielding a mixture of alkenes. A better approach involves the conversion of the 1-hexanol to 1-bromohexane using phosphorus tribromide followed by elimination using a strong base. No carbocation is involved in this approach and no rearrangement of the skeleton can occur.

CHAPTER 10
CARBON-CARBON DOUBLY BONDED SYSTEMS II. REACTIONS OF ALKENES

Key Points

This is a very important chapter as it introduces many reactions that you will encounter frequently in the remainder of the course.

• Know how to predict the products of reactions in which there is stereospecific *anti* or *syn* addition to a carbon-carbon double bond.

Practice Problem 10.1
A reaction is performed in which a reagent Y-Z (Y ≠ Z) adds to *cis*-2-butene. Assuming that each of Y and Z has a higher atomic number that does carbon, which product or products will form if the addition occurs in an *anti* manner?
    (a) a single *meso* product.
    (b) a racemic mixture of (2R,3S) and (2S,3R) $CH_3CH(Y)CH(Z)CH_3$.
    (c) a racemic mixture of (2R,3R) and (2S,3S) $CH_3CH(Y)CH(Z)CH_3$.
    (d) none of the above.

Practice Problem 10.2
Which of the possible answers in Practice Problem 10.1 would be correct if the addition occurred in a *syn* manner?

Practice Problem 10.3
Consider the alkenes and cycloalkenes indicated below. In which cases will the addition of a reagent Z-Z yield the same product regardless of whether the addition occurs in a *syn* or an *anti* manner?
    (a) 1-butene (b) *cis*-2-butene (c) *trans*-2-butene (d) cyclohexene (e) cyclopentylidene=CH_2

• Know how to correlate heats of hydrogenation with stability.

Practice Problem 10.4
The heat of hydrogenation of A is larger than that of B by ~10 kcal/mole.

    cyclopropyl—CH_3 cyclopropylidene=CH_2
        A   B

Which isomer, A or B, is the more strained and which is more stable? Suggest a reason for the relative stabilities of these two compounds (review Chapter 4).

• Learn the reagents used, the mechanisms, and the regiochemistry and stereochemistry of common addition reactions to alkenes. The following is a summary of these items.

Groups added to alkene	Reagents used	Stereochemistry	Regiochemistry
H, H	H$_2$, catalyst	usually *syn*	-
Br, Br	Br$_2$ in CCl$_4$	*anti*	-
H, Br	HBr, polar solvent	mixed	Markovnikov
H, Br	HBr, ether, peroxides	mixed	*anti*-Markovnikov
Br, OH	Br$_2$ in water	*anti*	Markovnikov
H, OH	aq. acid	mixed	Markovnikov
H, OH	1. H$_2$O, Hg(OAc)$_2$ 2. NaBH$_4$	mixed	Markovnikov
H, OH	1. BH$_3$, THF 2. H$_2$O$_2$, HO$^-$	*syn*	*anti*-Markovnikov
HO, OH	KMnO$_4$, KOH	*syn*	-
HO, OH	1. OsO$_4$ 2. H$_2$O, NaHSO$_3$	*syn*	-
HO, OH	H$_2$O$_2$, H$_2$O, HCO$_2$H	*anti*	-

Practice Problem 10.5
Which alkene (specify *E* or *Z* if necessary) and which reagents would you use to prepare the following as free as possible from regioisomers and diastereoisomers by way of addition reactions?
  (a) 2-bromobutane
  (b) *meso*-2,3-dibromobutane
  (c) *meso*-2,3-butanediol
  (d) 1-bromobutane
  (e) 1-butanol
  (f) *trans*-2-methylcyclopentanol
  (g) 1-iodo-1-methylcyclohexane

Practice Problem 10.6
Which of the following alcohols can be prepared cleanly (that is, free of regioisomers and diastereoisomers) by way of oxymercuration/demercuration of an alkene? Which alkene would you use for each preparation?
  (a) 1-hexanol
  (b) 3-hexanol
  (c) *cis*-2-methylcyclopentanol
  (d) 3-methyl-3-pentanol

Practice Problem 10.7
Which of the following alcohols can be prepared relatively cleanly (that is, free of regioisomers and diastereoisomers) by way of hydroboration/oxidation of an alkene? Which alkene would you use for each preparation?
  (a) 4-octanol
  (b) 1-methylcyclohexanol
  (c) *cis*-2-methylcyclohexanol
  (d) *trans*-2-methylcyclohexanol

Practice Problem 10.8
What are the chain propagating steps in the reaction of 1-butene with HBr in the presence of peroxides?

• Learn the uses of alkylboranes other than in simple hydroboration/oxidation.

Practice Problem 10.9
Explain how you could use boron chemistry to accomplish each of the following conversions.
  (a) *trans*-2-heptene to 1-heptanol
  (b) 1-pentene to 1-bromopentane
  (c) 2-ethyl-1-butene to (CH$_3$CH$_2$)$_2$CHCH$_2$D

- Learn how to predict the products of oxidative cleavage reactions of alkenes and be able to work out the structure of the starting alkene or polyene from a knowledge of its oxidative cleavage products.

Practice Problem 10.10
An optically active compound A, of formula C_8H_{16}, on ozonolysis followed by workup with hydrogen peroxide yields an optically active carboxylic acid B plus acetone. Suggest structures for A and B.

- Combine your knowledge of the new reactions learned in this chapter with earlier presented reactions to propose synthetic routes to designated products.

Practice Problem 10.11
Suggest preparative routes for the following conversions. (More than one step is generally required.)
    (a) 2-methylpropene to $(CH_3)_2CHCO_2H$
    (b) bromocyclopentane to *trans*-1,2-dibromocyclopentane (racemic)
    (c) 3-methyl-1-butene to 3-methyl-2-butanone
    (d) 1-butene to $CH_3CH_2CH_2CH=O$

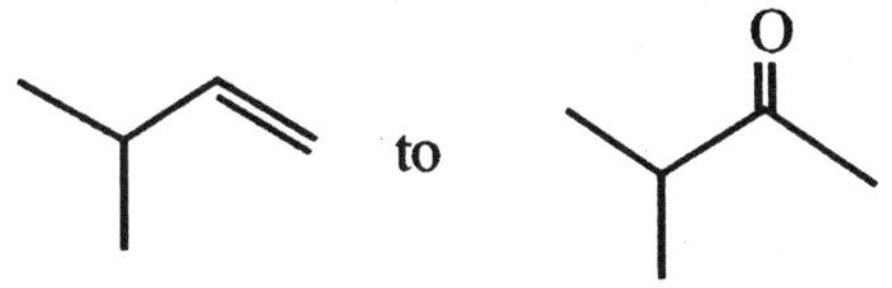

Solution of Text Problems

10.1 In order to distinguish *syn* from *anti* addition in a given reaction we need to use an alkene which would yield a different product from each possible route. With cyclohexene we would generate cyclohexane from either *syn* or *anti* addition. We need not only to have an alkene with a distinctive *E* or *Z* geometry, but one in which the groups present attached to the olefinic carbons differ from those being added. 1,2-Dimethylcyclohexene is such an alkene.

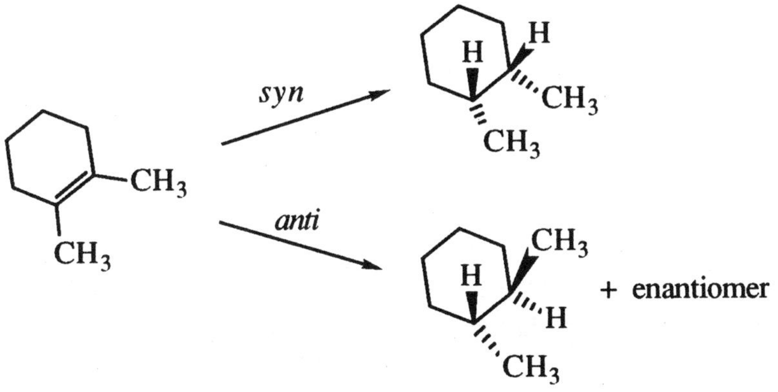

10.2 (a)

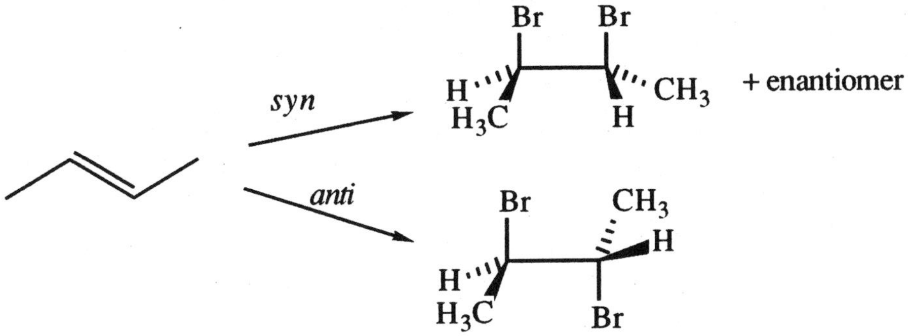

(b)

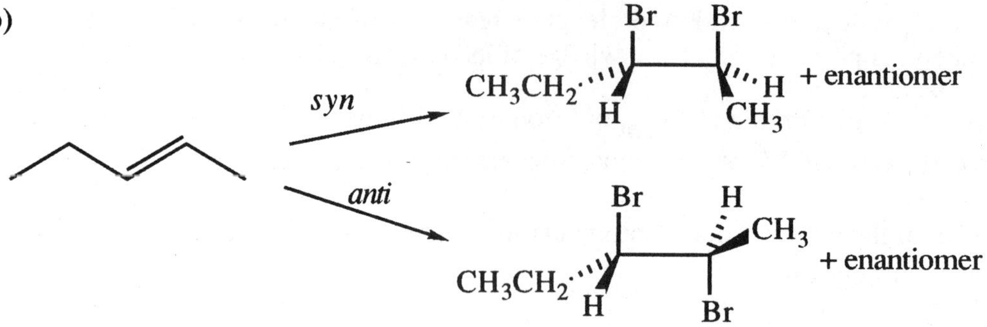

[alkene] —syn→ CH₃CH₂⋯C(Br)(H)–C(H)(CH₃)Br + enantiomer

—anti→ CH₃CH₂⋯C(Br)(H)–C(CH₃)(H)Br + enantiomer

(c)

[alkene] —syn→ CH₃CH₂⋯C(Br)(H)–C(H)(CH₃)Br + enantiomer

—anti→ CH₃CH₂⋯C(Br)(H)–C(CH₃)(H)Br + enantiomer

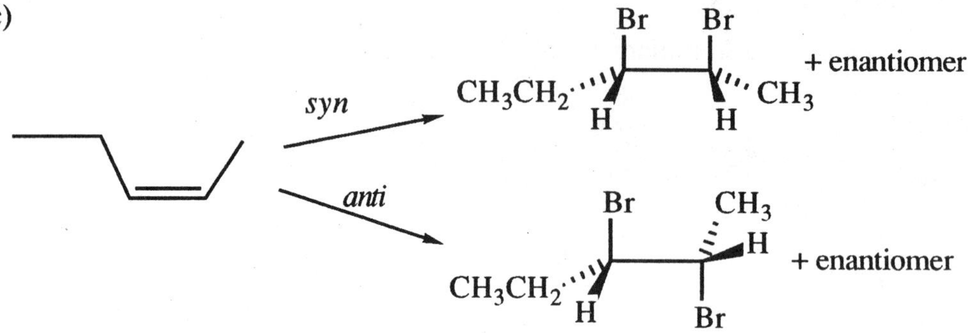

10.3

From *cis*-3,4-dimethyl-3-hexene:

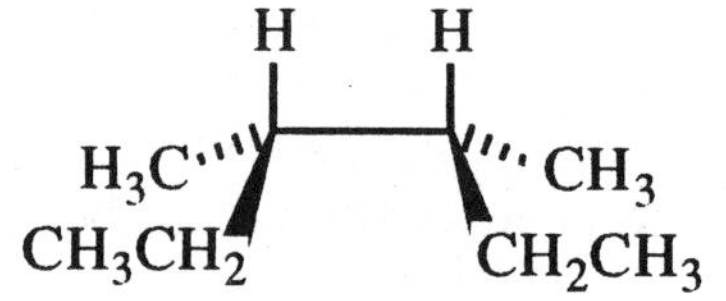

H₃C⋯C(H)(CH₃CH₂)–C(H)(CH₂CH₃)⋯CH₃
(3R,4S)-3,4-dimethylhexane

From *trans*-3,4-dimethyl-3-hexene:

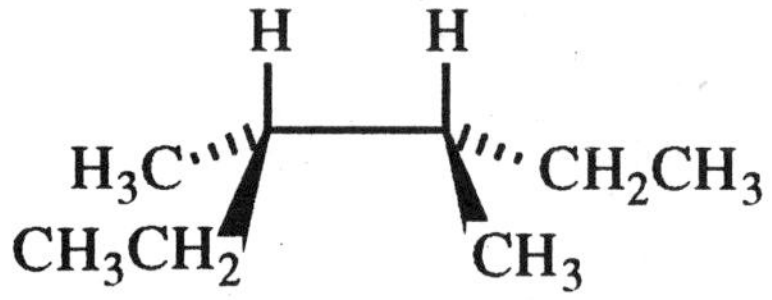

H₃C⋯C(H)(CH₃CH₂)–C(CH₃)(H)⋯CH₂CH₃   CH₃CH₂⋯C(H)(H₃C)–C(H)(CH₂CH₃)⋯CH₃
(3S,4S)-3,4-dimethylhexane   (3R,4R)-3,4-dimethylhexane

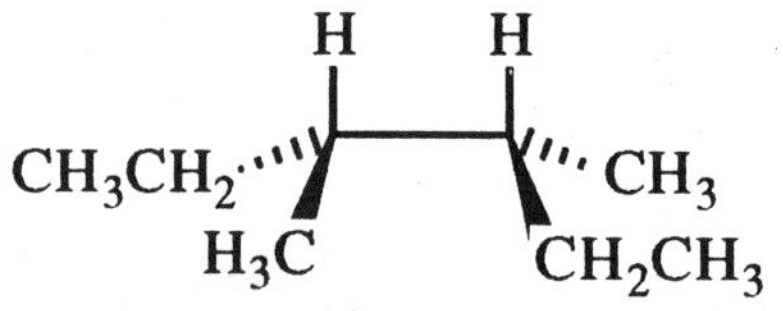

10.4 The role of a catalyst is to lower the activation energy between reactants and products, and thereby speed reaction. It does not change the energies of the reactants or products. With or without a catalyst, the energy difference between reactants and products is the same.

10.5 In general, disubstituted alkenes are more stable than monosubstituted alkenes.

10.6 This result is to be expected. For each set of reactions the products being compared are the same. The only difference for the reactions lies in the relative energies of the starting materials, which are the same for the two sets of reactions.

10.7 *trans*-Cyclooctene is the more strained, and thereby the higher energy alkene of the two. It thus liberates a greater amount of energy on hydrogenation than does its isomer, that is, it has a higher heat of hydrogenation.

10.8 (a) -110 kcal/mole; (b) -41 kcal/mole; (c) -28 kcal/mole; (d) -7 kcal/mole   Thus, we expect the relative reactivities to follow the order: $F_2 > Cl_2 > Br_2 > I_2$

10.9

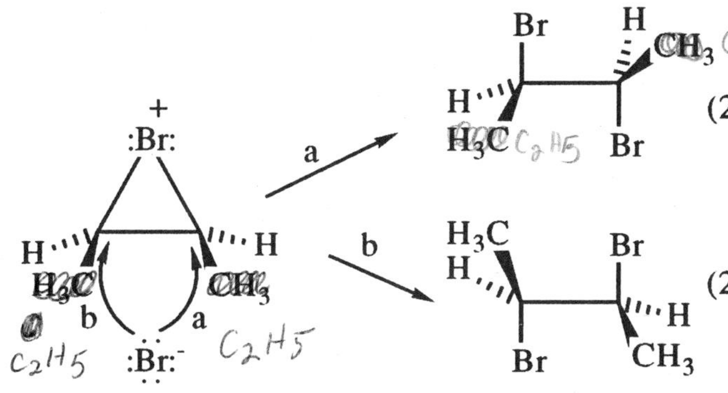

[Reaction scheme showing bromonium ion intermediate with H₃C, C₂H₅, H substituents and :Br:⁻ attacking at positions a and b]

- Path a → (2R,3R)-2,3-dibromobutane
- Path b → (2S,3S)-2,3-dibromobutane

10.10

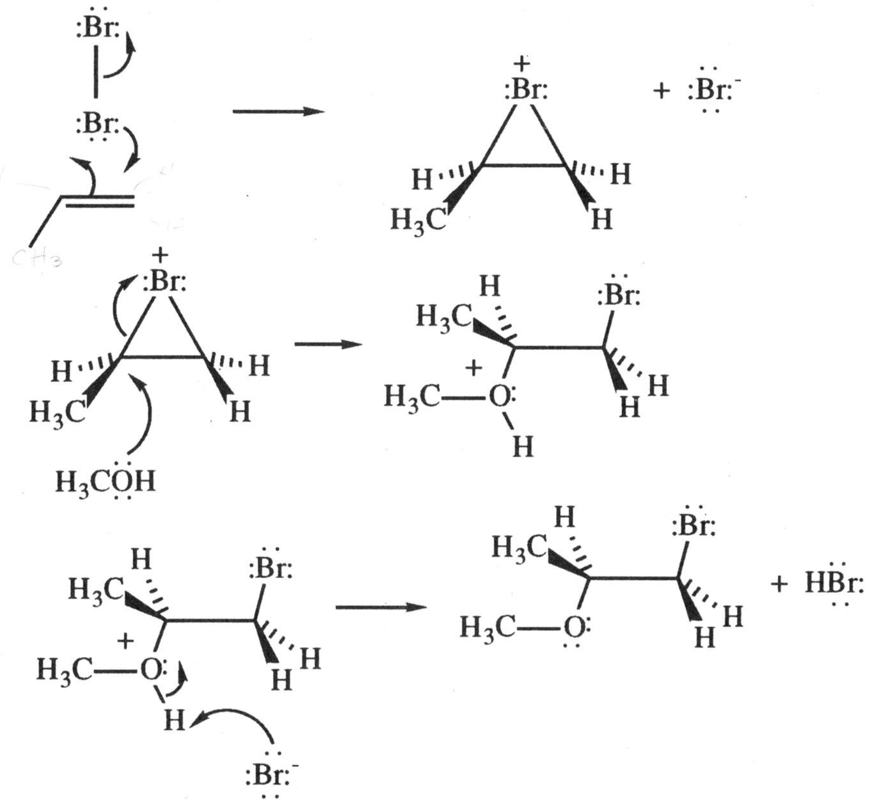

[Mechanism scheme: Br₂ adds to propene to form bromonium ion + :Br:⁻; methanol (H₃COH) attacks the bromonium ion to form protonated ether intermediate; deprotonation by :Br:⁻ yields the methoxy-bromo product + HBr]

10.11 We calculate k_1 to be 1.51 Å and k_2 to be 1.64 Å. Thereby $k_2/k_1 = 1.09$.

10.12 A $(CH_3)_2CHOSO_3H$ B $(CH_3)_2CHOH$

In the cold, sulfuric acid protonates propene at the terminal position to generate the 2-propyl cation. Bisulfate anion then combines with the carbocation to form the sulfate mono-ester, A. On addition of water the sulfate mono-ester is attacked *at the sulfur*, cleaving the sulfur-oxygen bond, and forming 2-propanol.

10.13 The bisulfate anion adds at the more highly substituted carbon atom of the alkene linkage; the protonation occurs on the least highly substituted carbon atom of the alkene. We would prepare *tert*-butyl alcohol rather than isobutyl alcohol by the reaction using methylpropene.

10.14 Treat the reaction mixture with cold, concentrated sulfuric acid, followed by the addition of water. The alkene impurity will react to form 2-hexanol, which will be soluble in the aqueous acid, but the hexane will remain insoluble.

10.15

(a) 2-methyl-1-butanol (b) 3-methyl-2-butanol (c) 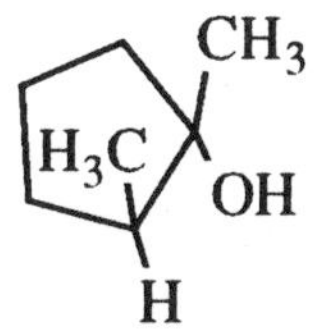 [cyclopentane with CH₃, H₃C, OH, H substituents] (d) 3-methyl-2-pentanol

84 Study Guide and Solutions Manual

10.16 (a) 1. BD$_3$; 2. CH$_3$CO$_2$H
(b) 1. BH$_3$; 2. CH$_3$CO$_2$D

10.17 (a) [cyclooctane-1,2-diol with H, H, HO, OH] (b) [meso 2,3-butanediol] (c) [racemic 2,3-butanediol]

10.18 [epoxide ring opening mechanism: protonated epoxide → water attack → loss of H$^+$ → trans-1,2-cyclohexanediol]

10.19 They are diastereoisomers. The product from permanganate hydroxylation is *meso*-1,2-cyclohexanediol while that from the peroxyformic acid reaction is racemic 1,2-cyclohexanediol.

10.20

$$2 \text{ RCH-CH}_2\text{Br} \longrightarrow \begin{array}{c}\text{RCH-CH}_2\text{Br} \\ | \\ \text{RCH-CH}_2\text{Br}\end{array}$$

$$\text{HO:} + \text{RCH-CH}_2\text{Br} \longrightarrow \begin{array}{c}\text{RCH-CH}_2\text{Br} \\ | \\ \text{:OH}\end{array}$$

$$\text{HO:} + \text{:Br:} \longrightarrow \text{HO-Br:}$$

10.21
(a) 1-bromo-1-methylcyclohexane
(b) 1-bromo-2-methylcyclohexane (*cis*- and *trans*)

10.22 (a) CH$_2$=CH- (b) methylenecyclobutane with ethyl (c) (CH$_3$)$_2$C=CHCH$_3$

10.23 (CH$_3$)$_2$C=O and HO$_2$CCH$_2$CH$_2$CO$_2$H

10.24 (a) (CH$_3$)$_2$C=O
(b) cyclohexanone and (CH$_3$)$_2$CHCH$_2$CH=O
(c) (CH$_3$)$_2$C=O, (CH$_3$)$_2$CHCH=O, and O=CHCH$_2$CH=O

10.25 (a) (b) (c) (d) and (e) i, ii

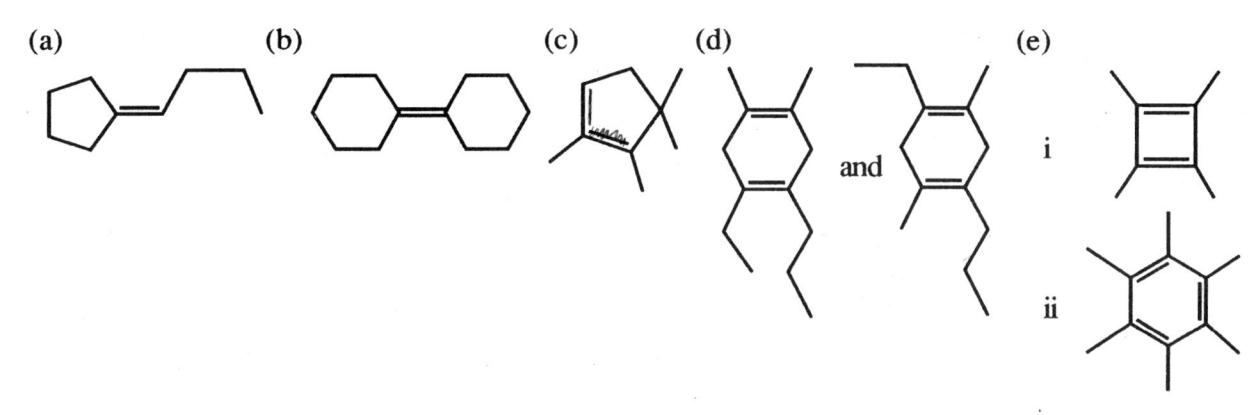

10.26 (a) two; (b) two; (c) three; (d) four; (e) seven; (f) five

10.27 There could be two double bonds or one triple bond. There is also a ring present in the compound.

10.28

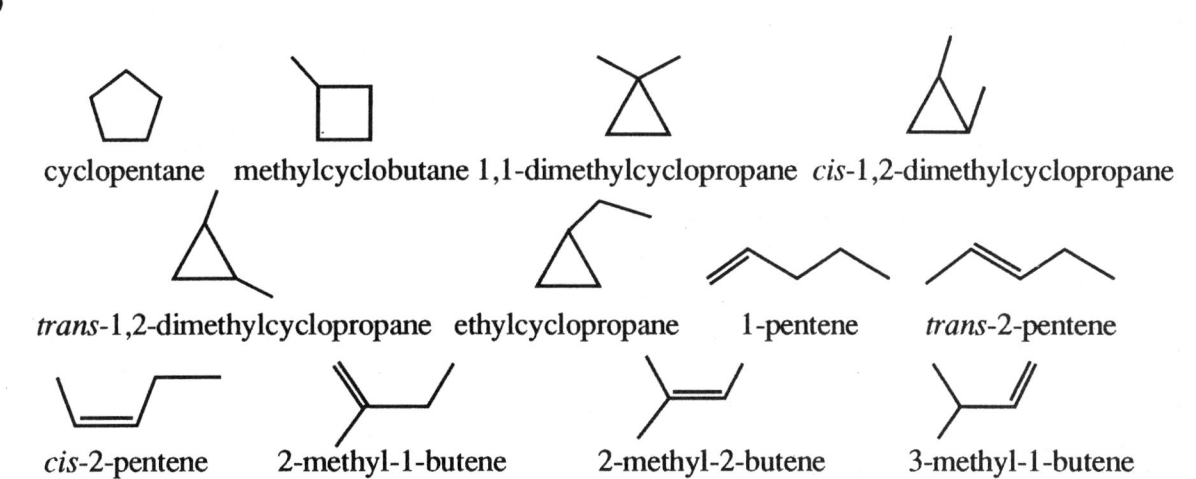

10.29

cyclopentane   methylcyclobutane 1,1-dimethylcyclopropane   cis-1,2-dimethylcyclopropane

trans-1,2-dimethylcyclopropane   ethylcyclopropane   1-pentene   trans-2-pentene

cis-2-pentene   2-methyl-1-butene   2-methyl-2-butene   3-methyl-1-butene

10.30

(a)–(n) [structures]

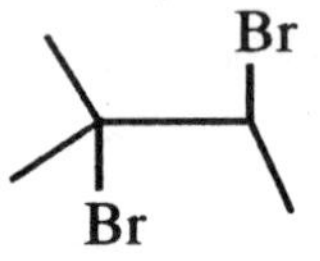

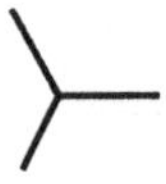

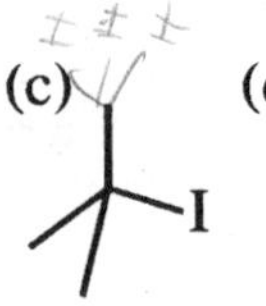

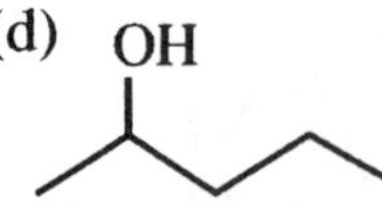

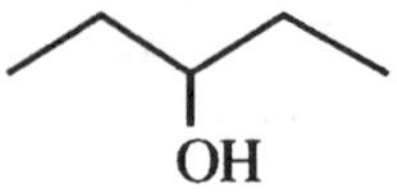

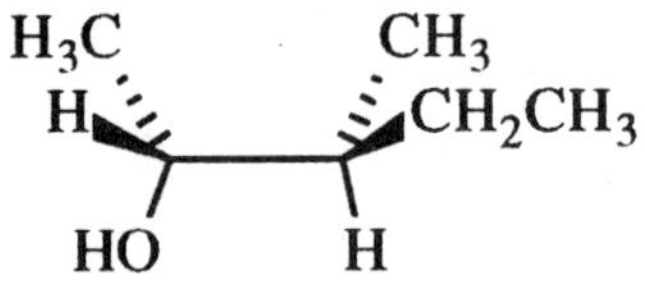

10.31
(a) formic acid, hydrogen peroxide, water
(b) formic acid, hydrogen peroxide, water
(c) osmium tetroxide
(d) potassium permanganate, potassium hydroxide, water, heat; aqueous acid workup
(e) ozone; zinc, acetic acid workup
(f) osmium tetroxide
(g) mercuric acetate, water; sodium borohydride
(h) borane; hydrogen peroxide, potassium hydroxide, water
(i) potassium *tert*-butoxide
(j) potassium permanganate, potassium hydroxide, water, heat

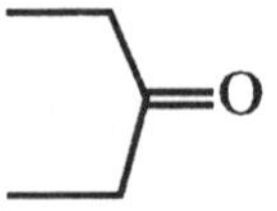

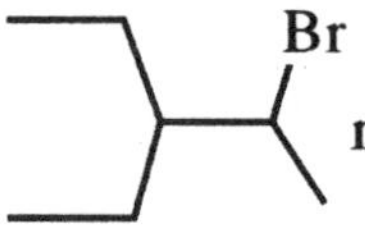

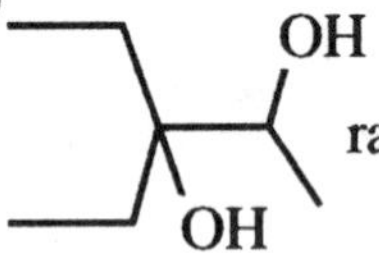

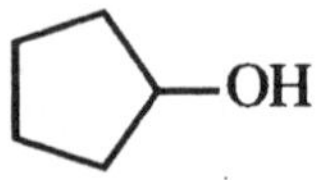

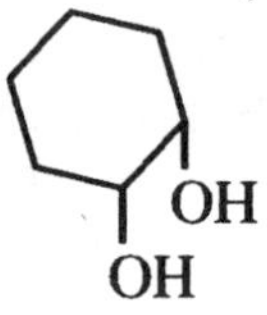

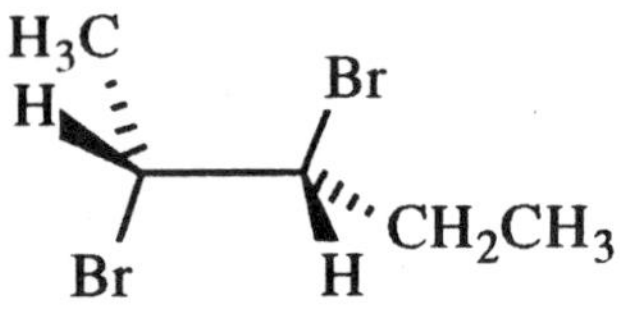

10.32 Since 2,3-dimethyl-2-butene leads to a more stable carbocation upon protonation, we would expect, on the basis of the Hammond postulate, that it would have the lower activation energy for protonation.

10.33 (Z)-2-pentene is of higher energy than is (E)-2-pentene.

10.34 (Z)-3-heptene

10.35 (R)-1-bromo-3-methylpentane

10.36 [Fischer projection structures]

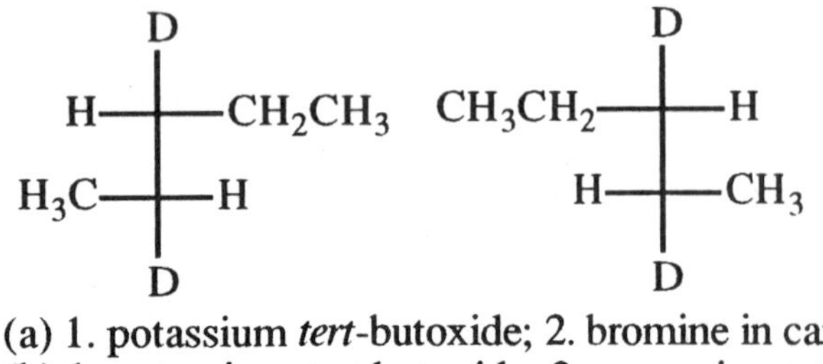

10.37
(a) 1. potassium *tert*-butoxide; 2. bromine in carbon tetrachloride
(b) 1. potassium *tert*-butoxide; 2. mercuric acetate, water; 3. sodium borohydride
(c) 1. potassium *tert*-butoxide; 2. mercuric acetate, water; 3. sodium borohydride
(d) 1. potassium *tert*-butoxide; 2. HBr
(e) 1. potassium *tert*-butoxide; 2. potassium permanganate, potassium hydroxide, water, heat; 3. aqueous acid
(f) 1. mercuric acetate, water; 2. sodium borohydride; 3. chromic anhydride, sulfuric acid
(g) 1. borane; 2. hydrogen peroxide, water; 3. chromic anhydride, sulfuric acid
(h) 1. potassium *tert*-butoxide; 2. hydrogen iodide
(i) 1. potassium *tert*-butoxide; 2. borane; 3. hydrogen peroxide, water
(j) 1. phosphoric acid, heat; 2. osmium tetroxide

(k) 1. borane; 2. hydrogen peroxide, water; 3. chromic anhydride
(l) 1. phosphoric acid, heat; 2. ozone; 3. hydrogen peroxide, water

10.38

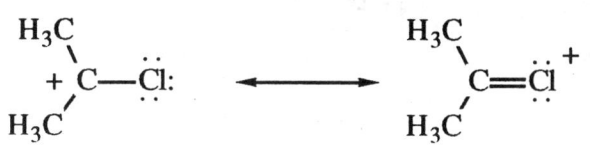

$$\overset{H_3C}{\underset{H_3C}{\diagdown}}\overset{+}{C}-\ddot{\underset{\cdot\cdot}{Cl}}: \longleftrightarrow \overset{H_3C}{\underset{H_3C}{\diagdown}}C=\overset{+}{\underset{\cdot\cdot}{Cl}}$$

Protonation occurs at the 1-position of 2-chloropropene preferentially to the 2-position since the former leads to a resonance stabilized carbocation.

10.39

A: (structure) or (structure)

10.40 1-bromo-4-ethylhexane

10.41 The bromine adds first. The second step of halogen addition occurs at the more highly substituted carbon site of the intermediate bridged ion, here the 2-position.

10.42 Based on bond energies, $\Delta H = 18$ kcal/mole. The product is $(CH_3)_2CBrCH_2I$.

10.43 (structure with Br)

10.44
(a) (structure) (b) (structure with Br) (c) (structure with O=O lactone)

10.45

B: (structure) and (structure)

C: (structure with Br) or (structure with Br)

D: (structure)

10.46

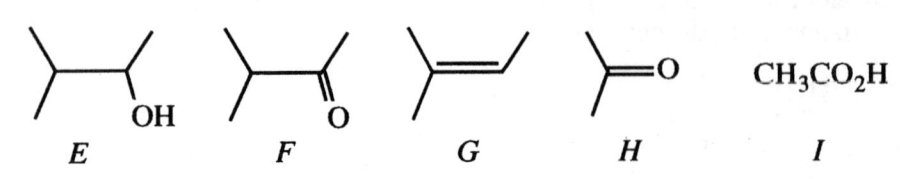

E (isopropyl carbinol with OH), F (isopropyl methyl ketone), G (2-methyl-2-butene type alkene), H (isobutyraldehyde, C=O), I: CH₃CO₂H

10.47

J

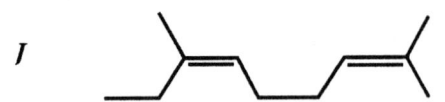

10.48 At equilibrium we expect the *trans*-2-butene to be the dominant material because it is the thermodynamically more stable material. The difference in energy between *cis*-2-butene and *trans*-2-butene is 1 kcal/mole. Thereby, ΔH for the conversion of *cis*-2-butene to *trans*-2-butene is -1 kcal/mole. The reaction progress diagram is as shown below.

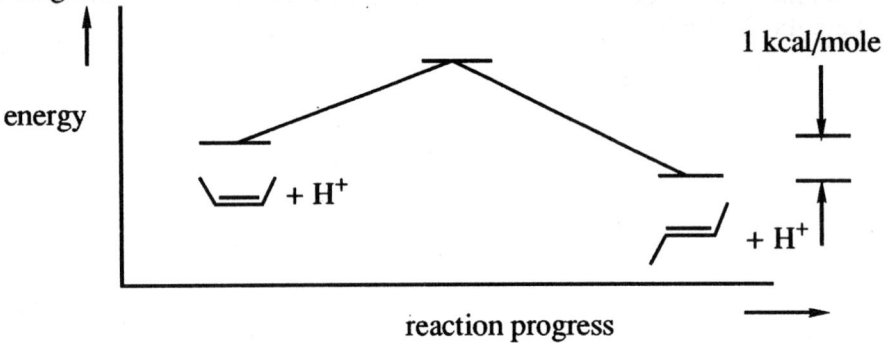

[Reaction progress diagram: energy vs reaction progress, showing cis-2-butene + H⁺ starting, going over a transition state, down to trans-2-butene + H⁺, with 1 kcal/mole difference]

10.49

[cyclohexane structure]

10.50 The maximum number of double bonds present is five. On the basis of 0.05 mole of the compound yielding 22.5 g of product in the bromination reaction, realizing that each double bond would add 8 g of bromine, we conclude that there are probably two double bonds in the molecule.

10.51

K

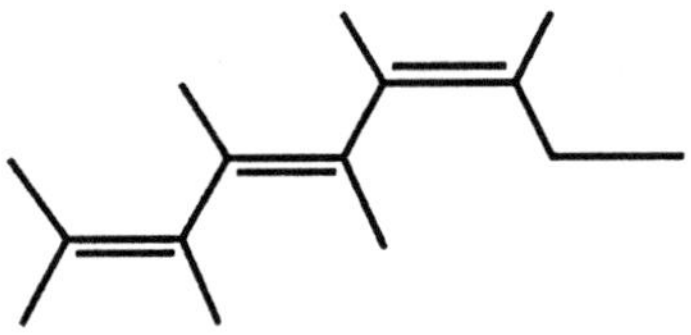

10.52 The tribromomethyl radical adds to the double bond to form the more stable of the possible organic radicals.

RO· + CBr₄ ⟶ R-Ö-Br + ·CBr₃

·CBr₃ + [alkene] ⟶ [alkyl radical with CBr₃]

Br₃C—Br: + [radical] ⟶ [product with CBr₃ and Br] + ·CBr₃

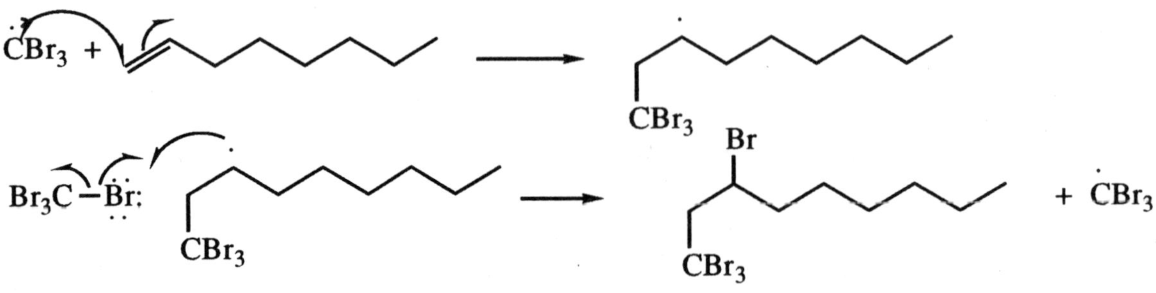

10.53

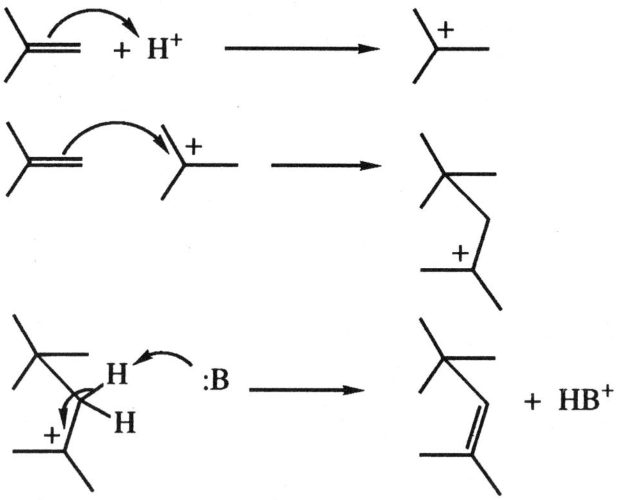

$$\text{(CH}_3\text{)}_2\text{C=CH}_2 + \text{H}^+ \longrightarrow (\text{CH}_3)_3\text{C}^+$$

$$(\text{CH}_3)_2\text{C=CH}_2 + (\text{CH}_3)_3\text{C}^+ \longrightarrow (\text{CH}_3)_3\text{C-CH}_2\text{-C}^+(\text{CH}_3)_2$$

$$(\text{CH}_3)_3\text{C-CH}_2\text{-C}^+(\text{CH}_3)_2 \;\; (\text{with H}) + :\text{B} \longrightarrow (\text{CH}_3)_3\text{C-CH=C(CH}_3)_2 + \text{HB}^+$$

10.54

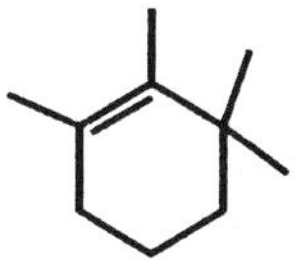

(1,2,3,3-tetramethylcyclohexene structure)

10.55

(a)

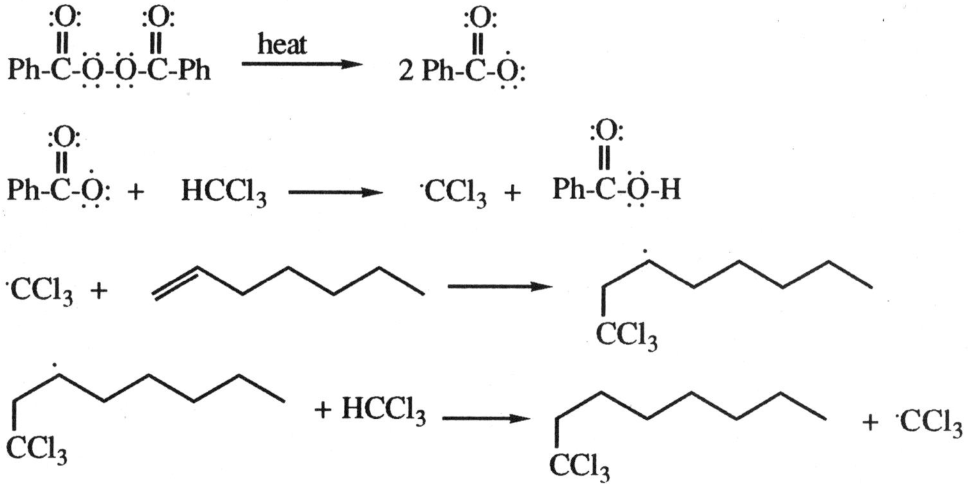

$$\text{Ph-C(=O)-O-O-C(=O)-Ph} \xrightarrow{\text{heat}} 2\;\text{Ph-C(=O)-O}\cdot$$

$$\text{Ph-C(=O)-O}\cdot + \text{HCCl}_3 \longrightarrow \cdot\text{CCl}_3 + \text{Ph-C(=O)-OH}$$

$$\cdot\text{CCl}_3 + \text{CH}_2\text{=CH-(CH}_2)_4\text{CH}_3 \longrightarrow \text{Cl}_3\text{C-CH}_2\text{-}\overset{\cdot}{\text{CH}}\text{-(CH}_2)_4\text{CH}_3$$

$$\text{Cl}_3\text{C-CH}_2\text{-}\overset{\cdot}{\text{CH}}\text{-(CH}_2)_4\text{CH}_3 + \text{HCCl}_3 \longrightarrow \text{Cl}_3\text{C-CH}_2\text{-CH}_2\text{-(CH}_2)_4\text{CH}_3 + \cdot\text{CCl}_3$$

(b)

$$\text{Ph-C(=O)-O-O-C(=O)-Ph} \xrightarrow{\text{heat}} 2\;\text{Ph-C(=O)-O}\cdot$$

$$\text{Ph-C(=O)-O}\cdot + \text{H}_3\text{C-C(=O)-H} \longrightarrow \text{H}_3\text{C-}\overset{\cdot}{\text{C}}\text{=O} + \text{Ph-C(=O)-OH}$$

$$\text{H}_3\text{C-C(=O)}\cdot + \text{CH}_2\text{=CH-(CH}_2)_4\text{CH}_3 \longrightarrow \text{H}_3\text{C-C(=O)-CH}_2\text{-}\overset{\cdot}{\text{CH}}\text{-(CH}_2)_4\text{CH}_3$$

$$\text{H}_3\text{C-C(=O)-CH}_2\text{-}\overset{\cdot}{\text{CH}}\text{-(CH}_2)_4\text{CH}_3 + \text{H}_3\text{C-C(=O)-H} \longrightarrow \text{H}_3\text{C-C(=O)-CH}_2\text{-CH}_2\text{-(CH}_2)_4\text{CH}_3 + \text{H}_3\text{C-C(=O)}\cdot$$

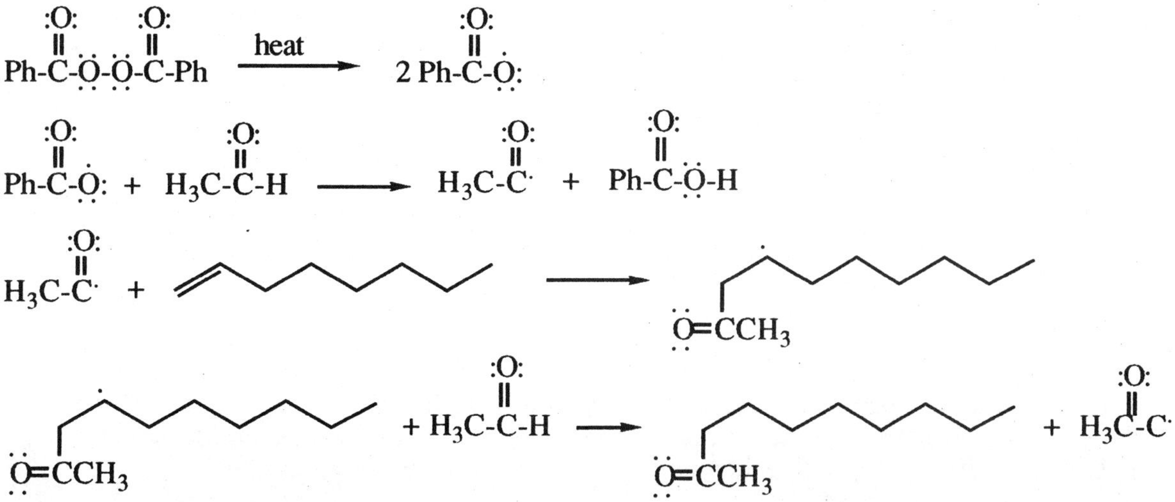

10.56 (a) While both lead to the same carbocation intermediate, the 1-butene initially is of higher energy than either of the 2-butenes. We thereby expect the 1-butene to have a lower activation energy for the protonation reaction.

(b) The chloride ion can attack the intermediate bromonium ion. However, in carbon tetrachloride solution there is no free chloride ion.
(c) correction sheet.

Solution of Study Guide Practice Problems

10.1 (c)

10.2 (b)

10.3 (a), (e)

10.4 Isomer A is more strained and isomer B is more stable. There are two sp^2 hybridized carbon atoms in the ring of A, but only one in the ring of B. Each sp^2 hybridized carbon atom in the ring causes strain because the ring bond angles of the cyclopropane ring are appreciably smaller than 120°, an angle associated with sp^2 hybridized carbon atoms.

10.5 (a) 2-Butene (either isomer) with HBr (presence or absence of peroxides).
  (b) *trans*-2-Butene with bromine in carbon tetrachloride.
  (c) *cis*-2-Butene with potassium permanganate in aqueous potassium hydroxide.
  (d) 1-Butene with HBr in the presence of peroxides.
  (e) 1-Butene treated with: 1. Borane-THF; 2. aqueous basic hydrogen peroxide.
  (f) 1-Methylcyclopentene treated with: 1. Borane-THF; 2. aqueous basic hydrogen peroxide.
  (g) 1-Methylcyclohexene with hydrogen iodide.

10.6 (a) No
  (b) Yes - either isomer of 3-hexene
  (c) No
  (d) Yes - 3-methyl-2-pentene (*E* or *Z*)

10.7 (a) Yes - either isomer of 4-octene
  (b) No
  (c) No
  (d) Yes - 1-methylcyclohexene

10.8

:Br: + /=\ ⟶ /\/\ :Br:

/\/\ :Br: + H-Br: ⟶ /\/\ :Br: + :Br:

10.9 (a) Heat the *trans*-2-heptene with borane followed by treatment with aqueous basic hydrogen peroxide.
  (b) After treatment with borane, allow to react with bromine in methanol/sodium methoxide solution.
  (c) After treatment with borane, add CH_3CO_2D.

10.10

A: CH_3CH_2, CH_3, H—C, C=C, H, CH_3, CH_3
or enantiomer

B: CH_3CH_2, CH_3, H—C, CO_2H
or enantiomer

10.11 (a) 1. BH_3-THF; 2. H_2O_2, KOH, water; 3. KOH, $KMnO_4$; 4. aqueous acid
  (b) 1. potassium *tert*-butoxide; 2. bromine in carbon tetrachloride
  (c) 1. mercuric acetate, water; 2. sodium borohydride; 3. chromic anhydride, sulfuric acid
  (d) 1. BH_3-THF; 2. H_2O_2, KOH, water; 3. chromic anhydride, pyridine

PRACTICE EXAMINATION TWO

A time limit of 90 minutes should be set for completion of this entire practice examination. Answers should be written out completely as they would be when presented for independent grading. No text or supplemental materials should be consulted during the testing period, and you should not check your answers until you have worked out the complete examination and the time limit has been reached.

1. For each part, choose the structure(s) which match the description given. None, one, or both possibilities given may match the description (2 points for each part).
   (a) is a tertiary alcohol: A) 3-pentanol; B) 2-propanol
   (b) the cycloalkene that, on hydroboration/oxidation, yields *trans*-2-methylcyclobutanol free of structural isomers and diastereoisomers: A) 1-methylcyclobutene; B) 3-methylcyclobutene
   (c) the alcohol that on protonation followed by loss of water yields a resonance stabilized carbocation:

   A) [cyclopentenol structure] B) [cyclopentenol structure]

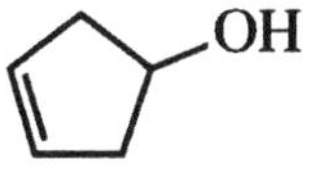

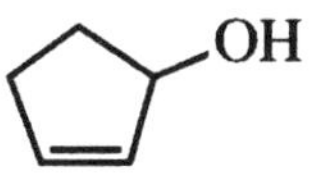

   (d) is a (Z) alkene:
   A) [alkene structure with Br]   B) [alkene structure]

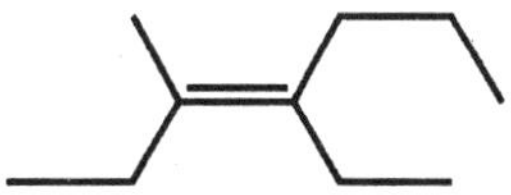

   (e) has the (R) configuration:
   A) [Newman projection with CH(CH₃)₂, H, H, Br, CH₂CH₃, CH₂CH₃]
   B) [Fischer-like projection with CH(CH₃)₂, Br, CH₂CH₃, CH₂CH₂CH₃]

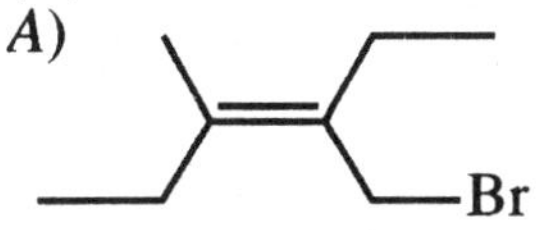

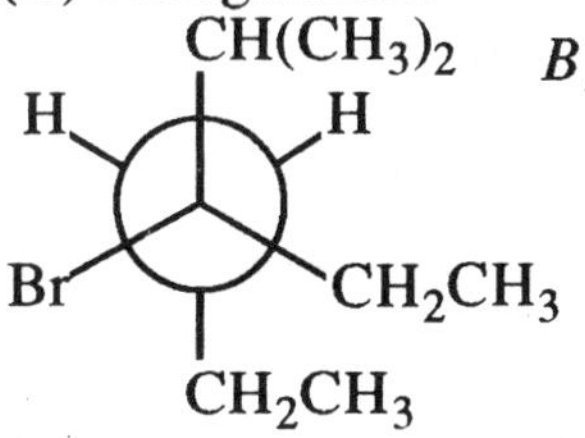

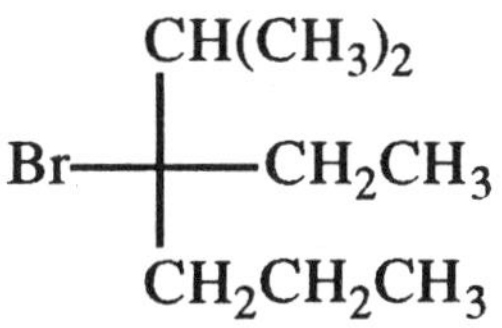

2. Give the required formulas or structures for each part (3 points for each part).
   (a) the reagent used in the demercuration step of the alkoxymercuration/demercuration procedure
   (b) the combination of reagents used to oxidize a primary alcohol to an aldehyde without continuing oxidation to the carboxylic acid stage
   (c) a reagent suitable for converting 2-methyl-2-propanol to 2-bromo-2-methylpropane
   (d) a bromoalkane of formula $C_5H_{11}Br$ that does not undergo the E2 reaction to form an alkene under any circumstances
   (e) the principal radical, $[C_3H_6Br]\cdot$, that is produced in the reaction of a bromine atom with propene

3. Give a suitable IUPAC name for each of the following structures (4 points for each name):

   (a) 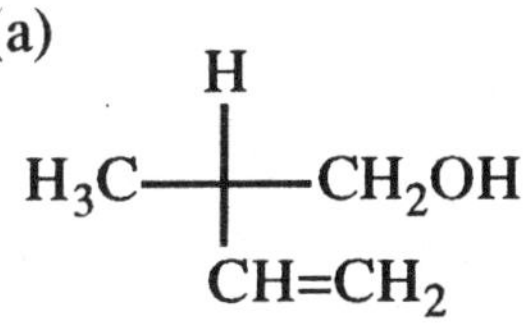 [structure: H₃C—C(H)(CH=CH₂)—CH₂OH]
   (b) [diene structure]
   (c) [Newman projection with Br, Br, H, H, CH₃, and C=C(CH₃)(CH₂CH₃)]
   (d) 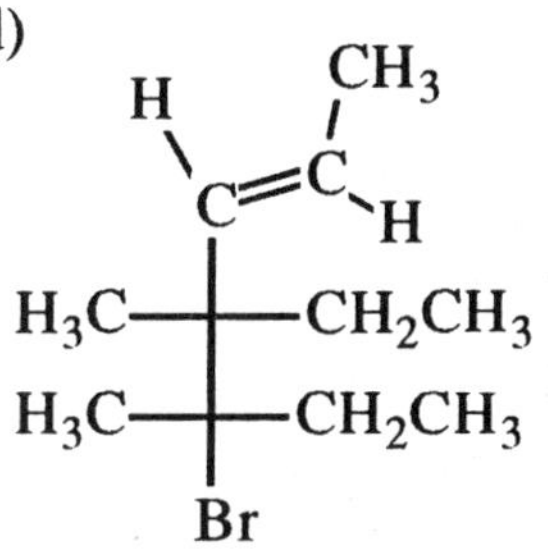 [structure with C=C(H)(CH₃), H₃C—C(CH₂CH₃)(CH₂CH₃)—Br]
   (e) 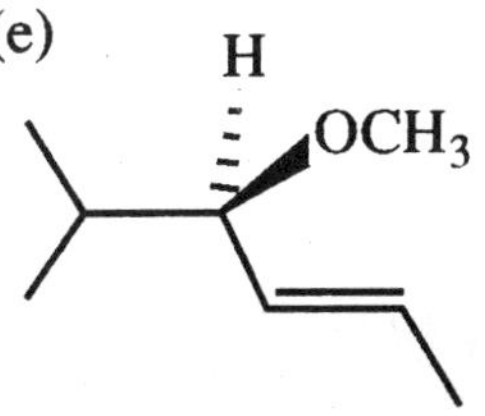 [structure with H, OCH₃, isopropyl and vinyl groups]

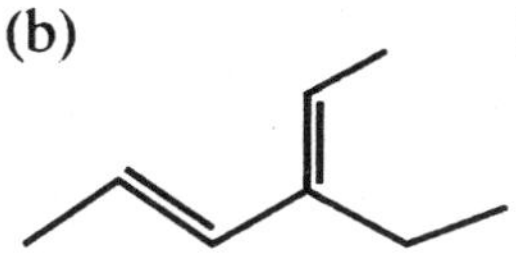

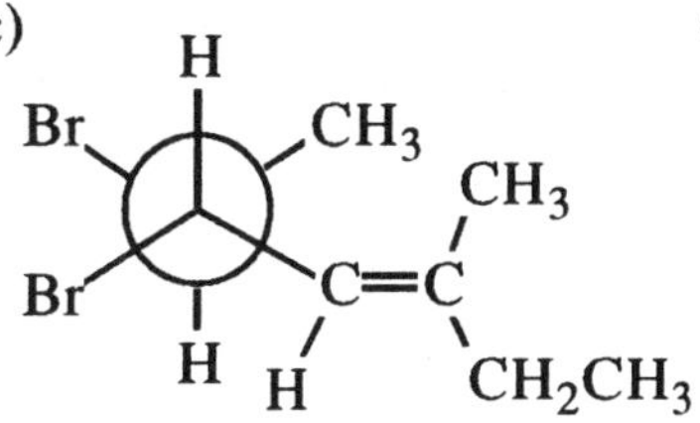

4. Provide IUPAC names or structures for compounds matching each of the following descriptions (4 points for each part):
   (a) formula $C_5H_{12}O$; reacts with sodium metal to form an alkoxide salt, but is inert to hot aqueous acidified potassium dichromate solution.
   (b) formula C_4H_8; yields the same product when treated with either cold aqueous basic potassium permanganate solution or with formic acid in aqueous hydrogen peroxide.
   (c) is formed by the application of the hydroboration/oxidation procedure to 1-methylcyclopentene.
   (d) formula C_6H_{12}; yields *meso*-3,4-dibromohexane on treatment with bromine in carbon tetrachloride solution.
   (e) formula $C_6H_{13}Br$; formed by reaction of 3,3-dimethyl-1-butene with HBr by a mechanism in which a carbocation rearrangement occurs.

5. Provide a reaction sequence to accomplish each of the indicated conversions (5 points for each part).
   (a) methylenecyclohexane → cyclohexanecarboxylic acid
   (b) 1-bromo-2-methylpropane → 2-methyl-2-propanol (Br-CH₂-CH(CH₃)-CH₃ → HO-C(CH₃)₃ with ethyl shown)
   (c) cyclopentene → H-C(=O)-(CH₂)₃-C(=O)-H
   (d) 1-methylcyclopentene → 2-hydroxy-2-methylcyclopentanone

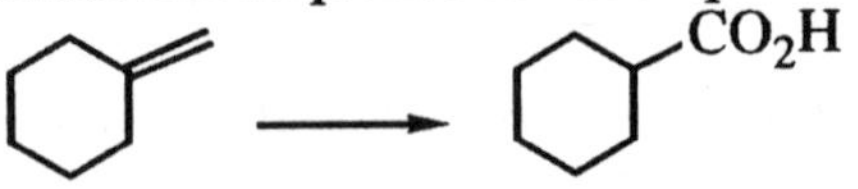

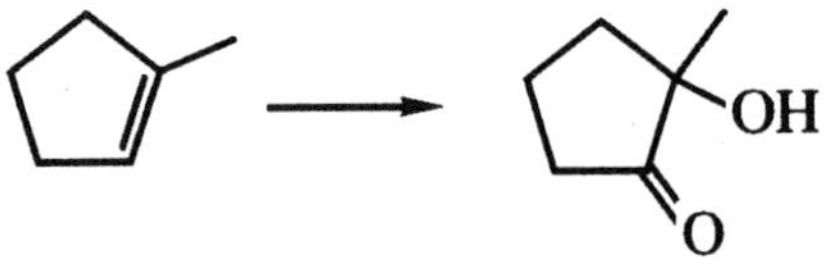

6. Give curved-arrow mechanisms for each of the transformations shown below. Each transformation may involve more than one step (5 points for each part).
   (a) (acyclic terpene alcohol) $\xrightarrow{H^+}$ (cyclized terpene)
   (b) 1-methylcyclohexanol $\xrightarrow{HBr}$ 1-bromo-1-methylcyclohexane
   (c) $(CH_3)_3C\text{-}Br + \overset{+}{K}\ {}^-OC(CH_3)_3 \longrightarrow (CH_3)_2C=CH_2$

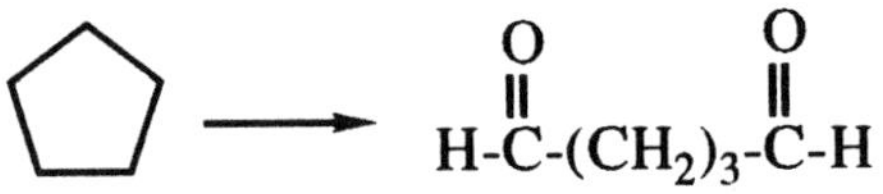

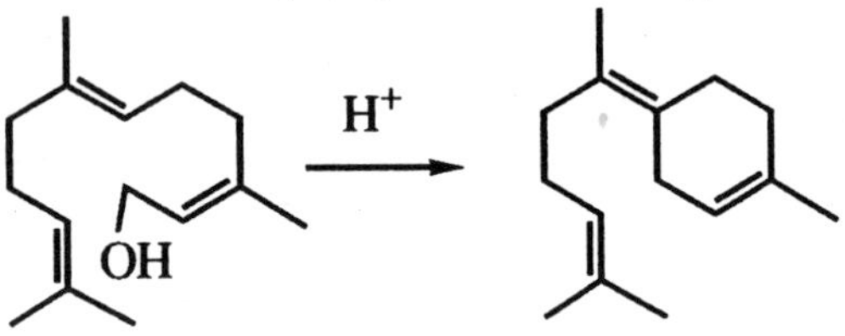

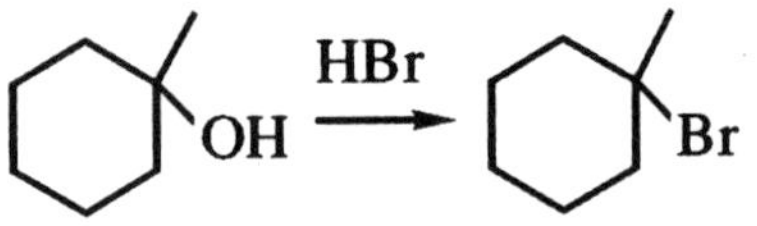

CHAPTER 11
CARBON-HALOGEN AND CARBON-METAL BONDS: TWO EXTREMES OF POLARITY

Key Points

• Know the reagents used for syntheses of haloalkanes and understand the mechanisms of those syntheses and any associated stereochemical aspects.

Practice Problem 11.1
Give the missing reagents or products, as appropriate, for each of the following synthetic conversions:

(a) 1-pentene $\xrightarrow{?}$ racemic 2-bromopentane

(b) 1-pentene $\xrightarrow{?}$ 1-bromopentane

(c) (R)-2-pentanol $\xrightarrow{?}$ (R)-2-chloropentane

(d) (R)-2-pentanol $\xrightarrow{?}$ (S)-2-chloropentane

(e) (R)-2-pentanol $\xrightarrow{Ph_3P, CCl_4}$?

• Know how to name halogen containing organic compounds.

Practice Problem 11.2
Give complete IUPAC names for each of the following compounds:

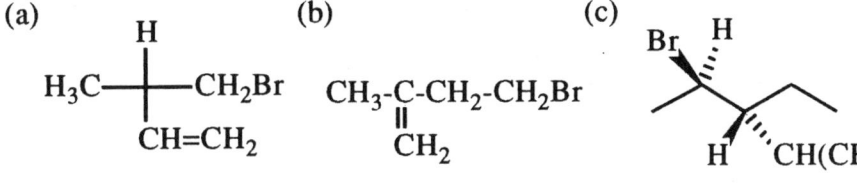

(a) H₃C-C(H)(CH=CH₂)-CH₂Br

(b) CH₃-C(=CH₂)-CH₂-CH₂Br

(c) (Br, H on one carbon)-C-C(H)(CH(CH₃)₂) with methyl and ethyl

Practice Problem 11.3
Name the bromoalkane which has all of the following characteristics.
(a) formula of $C_7H_{13}Br$; (b) is 3°; (c) is chiral; (d) can be prepared in good yield (in racemic form) by adding HBr to a terminal alkene; (e) on dehydrohalogenation with a strong base yields a mixture of disubstituted and trisubstituted alkenes.

• Know how to prepare Grignard reagents and organolithium reagents.

Practice Problem 11.4
Propose a series of reactions for the accomplishment of each of the following preparations:
(a) ethylmagnesium bromide from ethanol
(b) *tert*-butyllithium from 2-methylpropene

• Learn that, in their reactions, organometallic compounds behave as a source of R⁻ - a carbanion. This gives organometallic reagents the ability to react as strong bases or as potent nucleophiles. They behave as bases toward substances that can donate a proton, such as substances containing an O-H or an N-H bond.

Practice Problem 11.5
Write equations for the reactions of 2-propyllithium with each of the following:
(a) water
(b) methanol
Show a mechanism using the curved arrow formalism for the reaction indicated in part (a).

94 Study Guide and Solutions Manual

Practice Problem 11.6
The pure (+) enantiomer of a primary chiral chloroalkane of formula C_5H_9Cl is allowed to react with lithium in ether. Water is then added to the reaction mixture. Write structures for the organic reactant and product. Do you expect the product to be optically active? Explain your decision.

• Be able to predict the products obtained when organometallic compounds, behaving as nucleophiles, react with aldehydes and ketones. Be able to choose the correct combination of organometallic compound and carbonyl compound to prepare a desired alcohol product.

Practice Problem 11.7
Provide the missing reagents or starting materials.

(a) ? → 1. CH_3MgBr / 2. aq. acid → cyclobutane-OH

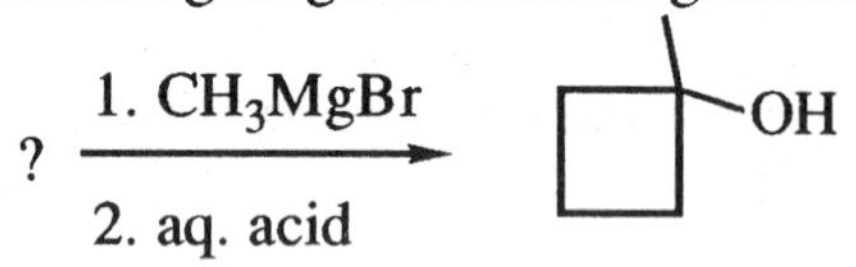

(b) cyclohexyl-MgBr → 1. ? / 2. aq. acid → cyclohexyl-CH_2OH

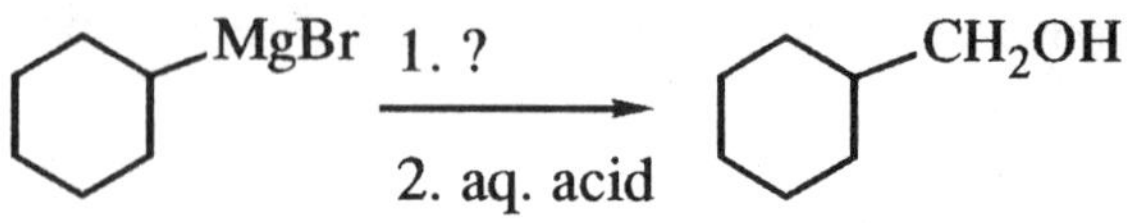

• Learn the use of lithium dialkylcuprate reagents (R_2CuLi) for the formation of carbon-carbon bonds.

Practice Problem 11.8
Which compound would you prepare if you allowed *sec*-butyl bromide to react with lithium in ether, treated it with CuI, and finally added bromoethane to the reaction mixture?

Practice Problem 11.9
Which haloalkane would you use to prepare a lithium dialkylcuprate reagent that would react with 1-bromobutane to generate (after suitable workup) each of the following:
(a) 2-methylheptane
(b) 3,4-dimethyloctane

Solution of Text Problems

11.1 (a) treatment with triphenylphosphine and carbon tetrachloride
(b) treatment with thionyl chloride using dioxane as the solvent

11.2 (a) treatment with HBr in the presence of benzoyl peroxide
(b) treatment with HCl

11.3
(a), (b), (c), (d), (e) — structures shown

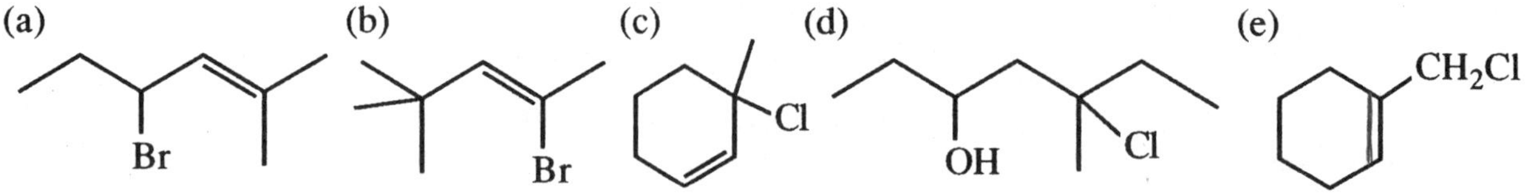

11.4 (a) 4-chloro-4-ethyl-3-methylheptane
(b) (Z)-1-bromo-2-methyl-2-heptene
(c) 3-bromo-2-ethyl-1-pentene
(d) 4-ethyl-3-iodo-4-methyl-1-hexene

11.5 (a) tertiary; (b) primary; (c) secondary; (d) secondary

11.6 Ammonia is more acidic than is an alkane. Thus, the Grignard reagent would abstract a hydrogen ion from ammonia to form the alkane and amide ion.

11.7 This would be a futile attempt to prepare propylsodium since ethanol is considerably more acidic than is propane. The equilibrium lies virtually completely to the side of propane and sodium ethoxide.

11.8 (a) 1-bromo-3-methylbutane
(b) bromocyclohexane
(c) 2-bromohexane

11.9 (a) 1. HBr; 2. Mg, ether; 3. D_2O
(b) 1. HBr, benzoyl peroxide; 2. Mg, ether; 3. D_2O

(c) 1. potassium *tert*-butoxide; 2. borane, heat; 3. aqueous H$_2$O$_2$; 4. thionyl chloride; 5. Mg, ether; 6. D$_2$O
(d) 1. thionyl chloride; 2. Mg, ether; 3. D$_2$O

11.10

(a) 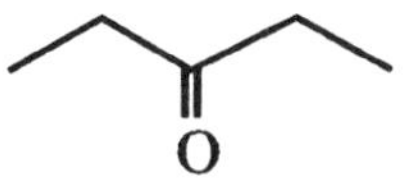 pentan-3-one

(b) H$_2$C=O

(c) 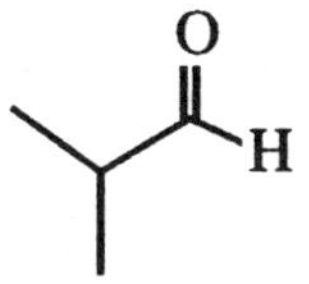 (CH$_3$)$_2$CHCHO (isobutyraldehyde)

(d) 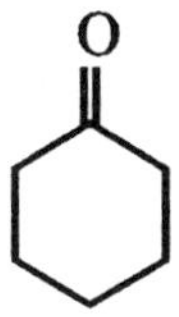 cyclohexanone

11.11

(a) Three routes to the same tertiary alcohol:
- nonan-3-one + butylMgBr
- nonan-5-one + CH$_3$CH$_2$MgBr
- heptan-3-one + pentylMgBr
→ 3-ethylnonan-3-ol (OH on tertiary carbon)

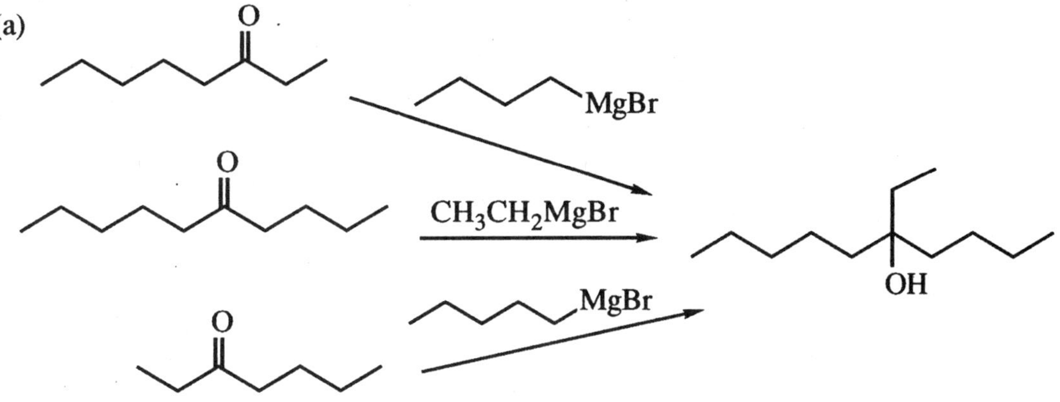

(b) sec-butylMgBr + CH$_3$CH=O → 3-methylpentan-2-ol
2-methylbutanal + CH$_3$MgI → 3-methylpentan-2-ol

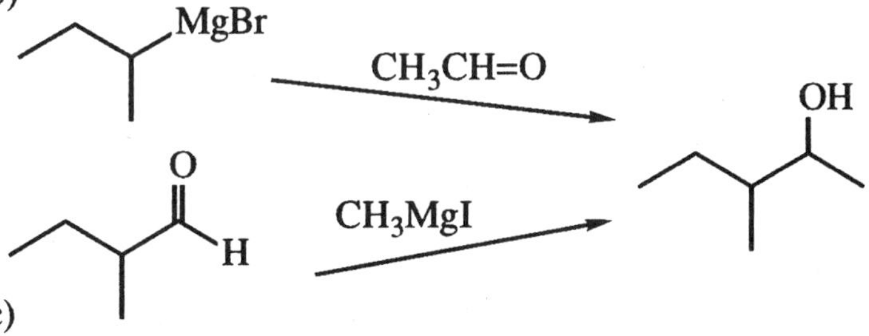

(c) (CH$_3$)$_2$CHCH(MgBr)CH$_3$ + H$_2$C=O → 2,3-dimethylbutan-1-ol (shown with CH$_2$OH)

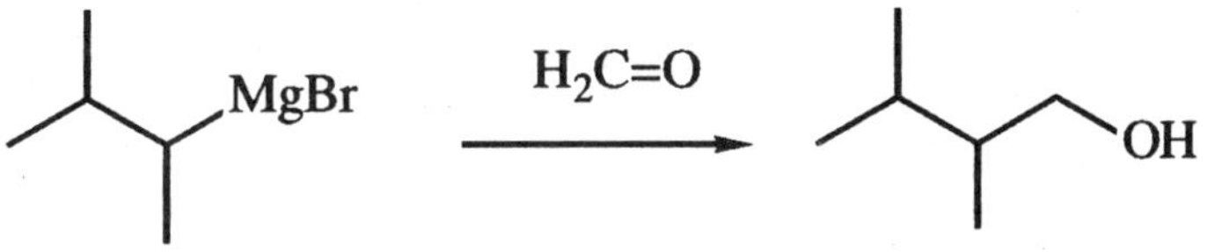

(d) butan-2-one + CH$_3$CH$_2$MgBr → 3-methylpentan-3-ol

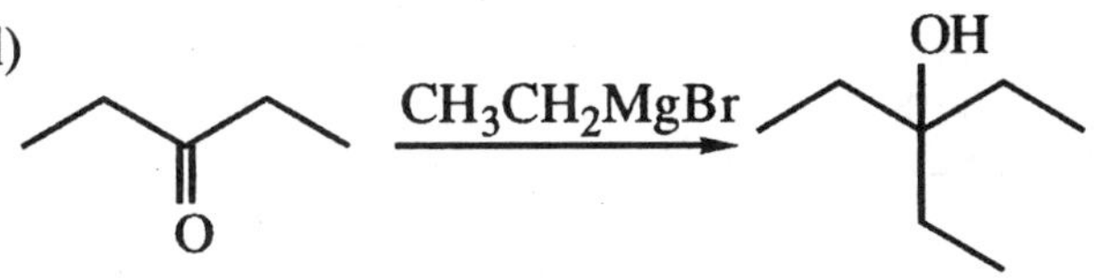

11.12

(a)

CH$_3$CH$_2$OH + PBr$_3$ → CH$_3$CH$_2$Br —Mg, ether→ CH$_3$CH$_2$MgBr

CH$_3$CH$_2$CH$_2$OH —CrO$_3$, pyridine→ CH$_3$CH$_2$CH=O —CH$_3$CH$_2$MgBr, aq. acid workup→ CH$_3$CH$_2$CH(OH)CH$_2$CH$_3$

(b)
$(CH_3)_2CHOH + PBr_3 \longrightarrow (CH_3)_2CHBr \xrightarrow[ether]{Mg} (CH_3)_2CHMgBr$

$(CH_3)_2CHOH \longrightarrow (CH_3)_2C=O \xrightarrow[\text{aq. acid workup}]{(CH_3)_2CHMgBr} (CH_3)_2CHC(CH_3)_2OH$

(c)
$CH_3CH_2CH=O + (CH_3)_2CHMgBr \xrightarrow{\text{aq. acid workup}} CH_3CH_2CH(OH)CH(CH_3)_2$
[prepared as in parts (a) and (b) above]

11.13

(a) (sec-butyl)Li + CH$_3$CO$_2$H   or   CH$_3$Li + (2-methylpropanoic acid with CO$_2$H)

(b) (cyclohexylmethyl)Li + CH$_3$CO$_2$H   or   CH$_3$Li + cyclohexyl-CH$_2$CO$_2$H

(c) $(CH_3)_2CHCO_2H + (CH_3)_2CHLi$

11.14 (a) 1-chloro-3-methylbutane
(b) 1-chloro-3,5-dimethylhexane
(c) chlorocyclohexane

11.15

(a) 3-chloro-2-methyl-1-butene structure
(b) 2-ethyl-1-bromo-alkene structure
(c) HO, H$_3$C, CH$_2$Cl, H chiral center
(d) H$_3$C, CH$_3$CH$_2$, Cl, H chiral center (+ enantiomer)

(e) CH$_3$CH$_2$CH$_2$, H$_3$C, CH$_2$CH$_3$, Cl chiral center (+ enantiomer)   and 4 stereoisomers of C-C-C-C-C-C with Cl

(f) HOCH$_2$, CH$_3$CH$_2$, Cl, CH$_3$ chiral center (+ enantiomer)

11.16
(a) 3-methyl-3-heptanol; (b) 2-methyl-2-hexanol; (c) 1-pentanol;
(d) lithium ethoxide and butane; (e) 1-cyclopentyl-2-ethyl-1-butanol;
(f) 2-deuteriobutane; (g) CH$_3$CO$_2$H

(h) $\underset{\text{CH-C-CH}_2\text{CH}_3}{\overset{O}{\|}}$

(i) hexane; (j) 3-methylpentane; (k) 1-deuterio-1-methylcyclopentane

(l) bicyclobutane   (m) cyclohexane with Cl and CH$_3$ (n) cyclohexane with H$_3$C and Cl

11.17 (a) 1. Mg, ether; 2. formaldehyde; 3. aq. acid workup
(b) 1. chromic anhydride, sulfuric acid; 2. two equivalent amounts of methyllithium
(c) 1. phosphorus tribromide; 2. Mg, ether; 3. formaldehyde; 4. aq. acid workup
(d) 1. HBr; 2. Mg, ether; 3. D$_2$O
(e) 1. aq. basic potassium permanganate, heat; 2. aq. acid workup; 3. four equivalent amounts of

ethyllithium
    (f) 1. Mg, ether; 2. formaldehyde; 3. aq. acid workup; 4. H$_2$, PtO$_2$
    (g) 1. Mg, ether, 2. 1/2 equivalent amount CdCl$_2$

11.18

$$CH_3CH_2CH_2CH_2-MgBr + \overset{\ddot{O}:}{\underset{:\ddot{O}:}{\overset{\|}{C}}} \longrightarrow CH_3CH_2CH_2CH_2\overset{:\ddot{O}:^-}{\underset{:\ddot{O}:}{\overset{\|}{C}}} \xrightarrow{H^+} CH_3CH_2CH_2CH_2CO_2H$$

11.19

[structures: 2-bromo-3,3-dimethylbutane type and 1-bromo-2,3-dimethylbutane type]

11.20

[structure: neopentyl bromide type, 1-bromo-2,2-dimethylpropane]

11.21

A: isopropyl bromide
B: 2,4-dimethyl-3-pentanone (diisopropyl ketone)
C: 2,4-dimethyl-3-isopropyl-3-pentanol
D: 2,4-dimethyl-3-isopropyl-2-pentene

11.22 The allyl Grignard reagent, H$_2$C=CH-CH$_2$MgBr, undergoes a free radical coupling reaction with another of its own kind. The product is H$_2$C=CH-CH$_2$CH$_2$-CH=CH$_2$.

11.23

E: 2-methylcyclohexanol (cis or trans)
F: 2-methylcyclohexanone
G: 1,2-dimethylcyclohexanol
H: 1,2-dimethylcyclohexene

11.24

(a) cyclohexene $\xrightarrow[\text{2. aq. acid workup}]{\text{1. KMnO}_4\text{, heat}}$ adipic acid (CO$_2$H, CO$_2$H) $\xrightarrow{\text{4 CH}_3\text{Li}}$ diketone $\xrightarrow{\text{H}_2, \text{Pd/C}}$ 2,7-dimethyl-2,7-heptanediol (OH, OH)

(b) propanol $\xrightarrow{\text{PBr}_3}$ propyl bromide $\xrightarrow{\text{Mg, ether}}$ propyl MgBr

isopropanol $\xrightarrow[\text{H}_2\text{SO}_4]{\text{CrO}_3}$ acetone $\xrightarrow{\text{aq. acid workup}}$ 2-methyl-2-pentanol

11.25 (a) 1. 1-butanol treated with PBr$_3$ followed by Mg in ether to yield the 1-butyl Grignard reagent; 2. ethanol oxidized with CrO$_3$ in pyridine to yield the aldehyde CH$_3$CH=O; 3. addition of materials from parts (1)

and (2) with aq. acid workup.

(b) 1. cyclopentanol treated with phosphorus tribromide followed by reaction with Mg in ether to yield the cyclopentyl Grignard reagent; 2. oxidation of methanol with CrO_3 in pyridine to yield formaldehyde ($H_2C=O$); 3. addition of materials from parts (1) and (2) with aq. acid workup.

(c) 1. 1-propanol treated with CrO_3 in pyridine to yield the aldehyde $CH_3CH_2CH=O$; 2. isobutyl alcohol treated with phosphorus tribromide followed by reaction with Mg in ether to yield the isobutyl Grignard reagent; 3. addition of the materials from parts (1) and (2) with aq. acid workup.

(d) 1. 3-methyl-1-butanol oxidized with CrO_3 in pyridine to yield the aldehyde $(CH_3)_2CHCH_2CH=O$; 2. addition of the materials from part (1) with isobutyl Grignard reagent [from part (2) of problem section (c)] and aq. acid workup.

(e) 1. addition of isobutyl Grignard reagent [from problem section (c)] with $CH_3CH_2CH=O$ [from problem section (c)] followed by oxidation with CrO_3 in sulfuric acid; 2. addition of cyclopentyl Grignard reagent [from problem section (b)] to the carbonyl compound formed in part (1) with aq. acid workup.

(f) 1. 2-methyl-1-butanol treated with phosphorus tribromide followed by reaction with Mg in ether to yield the Grignard reagent; 2. reaction of the Grignard reagent from part (1) with formaldehyde [from problem part (b)] and aq. acid workup to yield 3-methyl-1-pentanol; 3. treatment of 3-methyl-1-pentanol with phosphorus tribromide followed by reaction with Mg in ether to generate the Grignard reagent to which is added D_2O.

(g) 1. 2-butanol treated with phosphorus tribromide followed by reaction with Mg in ether to generate the 2-butyl Grignard reagent; 2. reaction of the 2-butyl Grignard reagent with the aldehyde $CH_3CH=O$ [from problem section (a)] with aq. acid workup to give 3-methyl-2-pentanol; 3. treatment of 3-methyl-2-pentanol with phosphorus tribromide followed by reaction with Mg in ether and addition of D_2O.

(h) 1. treatment of 3-pentanol with phosphorus tribromide followed by reaction with Mg in ether to yield the 3-pentyl Grignard reagent; 2. reaction of the 3-pentyl Grignard reagent with formaldehyde [from problem part (b)] and aq. acid workup to yield 2-ethyl-1-butanol; 3. treatment of 2-ethyl-1-butanol with phosphorus tribromide followed by reaction with Mg in ether and addition of D_2O.

(i) 1. oxidation of 2-butanol with CrO_3 in sulfuric acid to generate the ketone; 2.. formation of the cyclopentyl Grignard reagent [from problem part (b)]; 3. addition of the ketone from part (1) with the cyclopentyl Grignard reagent with aq, acid workup to generate the tertiary alcohol, 2-cyclopentyl-2-butanol; 4. dehydration of the 2-cyclopentyl-2-butanol with sulfuric acid followed by reduction with H_2 and PtO_2.

(j) 1. treatment of 1-butanol with phosphorus tribromide followed by reaction with lithium and the addition of CuBr to generate the lithium di(1-butyl)cuprate reagent; 2. reaction of the lithium di(1-butyl)cuprate reagent with 1-bromobutane [formed as in part (1)].

(k) 1. treatment of 1-bromo-2-methylpropane [from problem part (c)] with lithium followed by the addition of CuBr to generate the lithium diisobutylcuprate reagent; 2. treatment of ethanol with phosphorus tribromide to yield bromoethane; 3. reaction of the lithium diisobutylcuprate with the bromoethane.

(l) 1. treatment of 3-methyl-1-butanol with phosphorus tribromide followed by lithium to generate the 3-methyl-1-butyllithium reagent; 2. oxidation of 1-propanol with CrO_3 in sulfuric acid to yield the carboxylic acid $CH_3CH_2CO_2H$; 3. addition of two equivalent amounts of the 3-methyl-1-butyllithium reagent to the carboxylic acid $CH_3CH_2CO_2H$.

(m) 1. oxidation of isobutyl alcohol with CrO_3 in sulfuric acid to generate the carboxylic acid $(CH_3)_2CHCO_2H$; 2. reaction of the carboxylic acid with two equivalent amounts of the 2-methyl-1-propyllithium reagent [from problem part (k)].

Solution of Study Guide Practice Problems

11.1   (a) HBr, polar solvent
       (b) HBr, peroxides
       (c) thionyl chloride, dioxane
       (d) phosphorus trichloride or thionyl chloride, pyridine
       (e) (*S*)-2-chloropentane

11.2   (a) (*R*)-4-bromo-3-methyl-1-butene
       (b) 4-bromo-2-methyl-1-butene
       (c) (2*R*,3*R*)-2-bromo-3-ethyl-4-methylpentane

11.3 3-bromo-3-methylhexane

11.4 (a) treatment of ethanol with phosphorus tribromide followed by reaction with Mg in ether
(b) treatment of 2-methylpropene with HBr followed by reaction with lithium

11.5

(a) $(CH_3)_2CHLi + H_2O \longrightarrow CH_3CH_2CH_3 + Li^+ + HO^-$

(b) $(CH_3)_2CHLi + CH_3OH \longrightarrow CH_3CH_2CH_3 + Li^+ + H_3CO^-$

$$(CH_3)_2\overset{\overset{Li}{|}}{CH} \quad H\!-\!\ddot{O}H \longrightarrow (CH_3)_2\overset{\overset{Li^+}{|}}{\underset{H}{C}}H + {:}\ddot{O}H^-$$

11.6 The product of the overall reaction is 2-methylbutane which does not have a stereogenic center. Originally, the stereogenic center bore the four groups methyl, ethyl, chloromethyl, and hydrogen. In the reaction the chloromethyl group was converted to a methyl group leaving two identical substituents at the originally stereogenic carbon site.

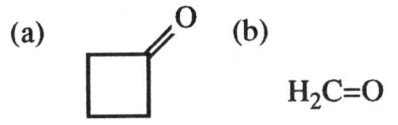

11.7 (a) cyclobutanone (b) $H_2C=O$

11.8 3-methylpentane

11.9 (a) 1-bromo-2-methylpropane
(b) 2-bromo-3-methylpentane

CHAPTER 12
SUBSTITUTION AND ELIMINATION REACTIONS OF HALOALKANES

Key Points

• Recognize that in a nucleophilic substitution reaction of a haloalkane, a nucleophile displaces a halogen atom as a halide ion from the haloalkane. The displaced group is said to be a leaving group. Groups other than halides can be displaced. In general, there is an inverse correlation between leaving group ability and basicity.

Practice Problem 12.1
Sodium iodide reacts faster with ethyl mesylate than with ethyl acetate to form ethyl iodide. Which acid do you thus infer to be stronger, acetic acid or methanesulfonic acid? Explain your decision. The pertinent structures are shown below.

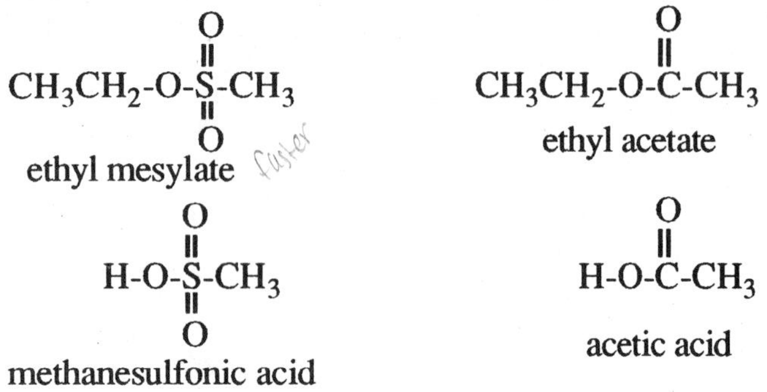

$$CH_3CH_2\text{-}O\text{-}\underset{\underset{O}{\|}}{\overset{\overset{O}{\|}}{S}}\text{-}CH_3 \qquad CH_3CH_2\text{-}O\text{-}\overset{\overset{O}{\|}}{C}\text{-}CH_3$$

ethyl mesylate (faster)   ethyl acetate

$$H\text{-}O\text{-}\underset{\underset{O}{\|}}{\overset{\overset{O}{\|}}{S}}\text{-}CH_3 \qquad H\text{-}O\text{-}\overset{\overset{O}{\|}}{C}\text{-}CH_3$$

methanesulfonic acid   acetic acid

• Understand how a rate law is obtained, and how it relates to the mechanism of a reaction.

Practice Problem 12.2
Consider a reaction:

$$A + B \rightarrow C$$

(a) In no intermediate forms in this reaction, that is, it is concerted, what is the form of the rate law?
(b) If the reaction does involve an intermediate X formed in a slow step by the transformation of A, and X then reacts in a rapid step with B to generate product C, what would be the form of the rate law?

Practice Problem 12.3
Suppose that a reaction of overall stoichiometry:

$$A + 2B \rightarrow C + D$$

has the rate law:

rate = k[A][B].

Suggest a possible sequence of steps for the mechanism that is consistent with the rate law.

Practice Problem 12.4
Which of the following remains constant as a reaction proceeds: (a) the rate; (b) the rate constant; (c) both the rate and the rate constant; (d) neither the rate nor the rate constant?

Practice Problem 12.5
Suppose a reaction between A and B has a rate law of the form:

rate = k[A]^2[B].

Predict the ratio of the rate of reaction when [A] = [B] = 1.0 M to the rate of the reaction when [A] = [B] = 3.0 M.

Practice Problem 12.6
Compare the rate constants for the reaction of cyanide ion with; (a) CH_3Cl; (b) CH_3I. Tell which of the following statements is true: (i) The rate constant is the same for the two reactions. (ii) The rate constant is larger for the CH_3Cl reaction. (iii) The rate constant is larger for the CH_3I reaction.

- Learn that nucleophilic substitution reactions of haloalkanes generally obey one of two common rate laws:
    rate = k[haloalkane]   (S_N1)
    rate = k[haloalkane][nucleophile]   (S_N2)

Practice Problem 12.7
Which reaction type, S_N1 or S_N2, can you infer is not concerted, based on the form of the rate law?

Practice Problem 12.8
Consider a nucleophilic substitution reaction:

$$Nuc:^- + R\text{-}X \rightarrow R\text{-}Nuc + X^-$$

Suppose that we double the concentrations of both the nucleophile and the haloalkane simultaneously and find that the reaction rate is doubled. What is the reaction type, S_N1 or S_N2?

- Recognize that the tendency for a given haloalkane to undergo S_N2 reaction with a given nucleophile decreases with increasing branching at the α-carbon atom, and also (but to a lesser extent) at the β-carbon atom.

Practice Problem 12.9
In each of the following cases choose the combination that will result in faster S_N2 reaction, assuming equal concentrations of substrate and nucleophile.
   (a) iodoethane or 1-iodopropane with NaCN in acetone solution
   (b) isobutyl bromide or *sec*-butyl bromide with NaCN in acetone solution

- Be able to recognize when to use S_N2 reactions as a part of a synthetic sequence.

Practice Problem 12.10
Suggest series of reactions that will accomplish each of the following synthetic conversions.
   (a) 1-butanol to 1-cyanobutane
   (b) 1-heptene to 1-aminoheptane

- Be aware that solvents can have a major influence on the rate of nucleophilic substitution reactions. In S_N2 reactions in which the nucleophile is an anion, the use of a polar aprotic solvent is desirable. These solvents leave the anion relatively unsolvated, and therefore more reactive.

Practice Problem 12.11
Which reaction would you expect to occur faster: KCN with 1-bromobutane in aqueous acetone, or in pure acetone?

Practice Problem 12.12
Name and show the structure of three common polar aprotic solvents.

- Learn that in S_N2 reactions there is an inversion of configuration at the site of nucleophilic attack, and that this suggests that S_N2 reaction can occur only if the nucleophile approaches from the backside of the carbon-halogen bond.

Practice Problem 12.13
Give full names for the products you would expect to form in each of the following S_N2 reactions.
   (a) (*R*)-1-deuterio-1-iodoethane with water in HMPT
   (b) (*R*)-2-bromo-1-methoxybutane with NaCN in acetone

- Learn the following aspects of S_N1 reactions:
    - The reaction rate is independent of the concentration of the nucleophile.
    - The reaction occurs in two steps. First, there is slow dissociation of the haloalkane to a carbocation and a halide ion. This is followed by the rapid reaction of the nucleophile with the carbocation.
    - In some cases the carbocation rearranges prior to reaction with the nucleophile.
    - S_N1 reactions of haloalkanes become more favorable (that is, more competitive with S_N2 reaction) as the stability of the carbocation produced in the first step increases. Thus, the general order of S_N1 reactivity is:

    3° RX > 2° RX > 1° RX > CH_3X

However, you should also be aware that haloalkanes that form resonance stabilized carbocations upon the loss of halide ion also tend to have high S_N1 reactivities (such as allylic and benzylic halides).

- Increasing polarity of the solvent speeds S_N1 reactions by helping to stabilize the ionic products (R^+ and X^-) of the slow step.
- If halide ion is displaced from a stereogenic carbon atom, the product will be largely racemic. This correlates with the loss of chirality in the rate-determining step - a carbocation is planar (flat) and achiral.

Practice Problem 12.14
In each of the following cases choose the compound you expect to have the greater S_N1 reactivity. Explain your choices.
- (a) 1-bromopropane or 2-bromopropane
- (b) 1-bromopropane or 3-bromo-1-propene

Practice Problem 12.15
When 4-iodocyclohexene is heated with acetic acid, a nucleophilic substitution reaction occurs that results in the formation of a mixture of 4-acetoxycyclohexene and an isomeric material that is formed as the result of the initially generated carbocation rearranging to a more stable carbocation. Suggest a mechanism for the formation of the product of rearranged carbon skeleton and rationalize its formation.

Practice Problem 12.16
Suppose that a given secondary haloalkane reacts with a particular nucleophile and that analysis of the product mixture indicates that 70% of the product is formed by way of an S_N1 mechanism and 30% is formed by way of an S_N2 mechanism. Would you expect the S_N1/S_N2 product ratio to increase, decrease, or to remain the same if the nucleophile were replaced by one that is more potent, assuming that all other conditions remain the same?

- Recognize that although any Lewis base has the potential to act as a nucleophile toward a haloalkane, it also has the potential to act as a Brønsted base, thereby bringing about an elimination reaction to generate a carbon-carbon double bond.

- In reactions with Lewis bases that are good Brønsted bases, tertiary haloalkanes give exclusively elimination products while secondary haloalkanes give a mixture of elimination and substitution products, with elimination being the dominant reaction route.

- A primary haloalkane in reaction with an unbranched Lewis base, such as an ethoxide ion, tends to undergo mainly nucleophilic substitution. Increasing amounts of elimination products are obtained if: (a) there is branching at the β-carbon atom of the haloalkane, (b) if the base is bulky, as with *tert*-butoxide ion, or (c) if the strength of the base is increased (such as changing from hydoxide ion to amide ion).

Practice Problem 12.17
For each of the following decide which you expect to proceed with the larger elimination/substitution ratio:
- (a) 2-bromo-2-methylbutane upon heating with water or with a 10% aqueous solution of NaOH
- (b) isobutyl bromide or *n*-butyl bromide heated with 10% sodium ethoxide in ethanol

- Be aware of the two common elimination mechanisms, E1 and E2. The E1 reaction occurs in two steps involving the formation of a carbocation intermediate. The E2 reaction is a concerted (one-step) reaction.

Practice Problem 12.18
When the salt $[PhCH_2CH_2N(CH_3)_3]^+ Cl^-$ is treated with sodium ethoxide in ethanol there is obtained a mixture of $PhCH=CH_2$ (styrene) and $(CH_3)_3N$ (triethylamine). The reaction rate is found to be proportional to the concentration of sodium ethoxide and to the concentration of the salt.
- (a) Explain why you think that a carbocation does or does not form in this reaction.
- (b) Use the curved-arrow formalism to depict the mechanism of the reaction.

- Realize that the E1 mechanism is relatively rare. Usually we observe it only in the reaction of a tertiary haloalkane with a weak base, such as in a solvolysis reaction where the solvent (generally water or an alcohol) plays the role of the weak base.

- In elimination reactions most haloalkanes (1°, 2°, or 3°) react with most bases by the E2 mechanism. This is unlike the situation for nucleophilic substitution where the mechanism (S_N1 or S_N2) depends heavily on the nature of the haloalkane.

- E1 and E2 reactions of haloalkanes are also known as dehydrohalogenations. Some haloalkanes can undergo dehydrohalogenation in different ways to yield a mixture of isomeric alkene products.

Practice Problem 12.19
Name all C_4H_9Br isomers that on undergoing a dehydrohalogenation reaction
 (a) yield a single alkene.
 (b) yield a mixture of two alkenes.
 (c) yield a mixture of three alkenes.

- If more than one alkene can form in an E2 reaction, the predominant product is the more stable (more highly substituted) alkene isomer, as long as an unhindered (unbranched) base is used. This is referred to as Zaitzev elimination. When a highly branched base, such as *tert*-butoxide ion, is used, a greater proportion of alkenes that are less highly substituted are obtained.

Practice Problem 12.20
Predict the major products formed when 1-bromo-1-methylcyclohexane reacts with each of the following:
 (a) potassium ethoxide in ethanol
 (b) potassium *tert*-butoxide in *tert*-butyl alcohol

- From experimental evidence we conclude that E2 reactions have a very specific stereochemical requirement. The bonds from carbon to the hydrogen and carbon to the halogen undergoing elimination must be periplanar. That is, they must lie in the same plane. This can happen only if the C-H and C-X bonds are either *anti* or *syn* relative to each other. Because *anti* conformations are generally more stable than *syn* conformations, most E2 reactions are stereospecific *anti* eliminations. You should be able to predict the products of such E2 reactions.

Practice Problem 12.21
Name the product (including the *E/Z* descriptor) that you would obtain from the elimination of HBr from (1*R*,2*R*)-1,2-dibromo-1,2-diphenylethane by an E2 mechanism.

- Chloroform, $CHCl_3$, and some related molecules undergo a reaction with strong bases known as α-elimination. In these reactions a hydrogen atom and a halogen atom are eliminated from the same carbon atom resulting in the generation of a highly reactive divalent carbon species. These species are too reactive to be isolated, but if generated in the presence of an alkene will react with it to form a cyclopropane derivative.

Practice Problem 12.22
Suggest a series of reactions that will accomplish the following conversion:
 1-bromobutane to 1,1-dichloro-2-ethylcyclopropane

Solution of Text Problems

12.1 (a) k = 2.4 x 10^{-5} $M^{-1}sec^{-1}$
 (b) rate = 1.35 x 10^{-5} $M\ sec^{-1}$
 (c) rate = 6 x 10^{-6} $M\ sec^{-1}$
 (d) k = 4.8 x 10^{-5} $M^{-1}\ sec^{-1}$

12.2 H-H (the hydrogen molecule)

12.3 (a) treatment of isobutyl alcohol with sodium hydride to generate the alkoxide salt, followed by reaction of the alkoxide salt with 1-bromopropane
 (b) treatment of 2,2-dimethyl-1-propanol with sodium hydride to generate the alkoxide salt, followed by reaction of the alkoxide salt with bromoethane
 (c) treatment of 2,2-dimethyl-1-propanol with sodium hydride to generate the alkoxide salt followed by reaction of the alkoxide salt with 2,3-dimethyl-1-bromobutane

104 Study Guide and Solutions Manual

12.4

$$H_3C-\overset{\overset{\displaystyle :\!O\!:}{\|}}{\underset{\cdot\cdot}{S}}-CH_3$$

12.5 The inversion of configuration occurred as normal for an S_N2 reaction. However, the product has the same *designation* as the starting material owing to the fact that the introduced group (CN) has a lower priority relative to the already attached $CH_2OCH_2CH_3$ than does the displaced Cl substituent.

12.6

$$H_2\ddot{O} < :NH_3 < CH_3\ddot{O}:^- < H\ddot{O}:^- < CH_3CH_2\ddot{O}:^- < H_2\ddot{N}:^- < :CH_3^- < CH_3\ddot{C}H_2^- < (CH_3)_2\ddot{C}H^-$$

12.7 (a) 180°; (b) 60°; (c) 60°
12.8 (a) 120°; (b) 0°; (c) 120°
12.9 (a) (*E*); (b) (*E*); (c) (*Z*)
12.10 (a) $(CH_3)_3C\text{-}O\text{-}CH_2CH_3$ - formed by the reaction of the intermediate *tert*-butyl cation with ethanol and the loss of a proton to the solution
    (b) $(CH_3)_3C\text{-}OH$ - formed by the reaction of the intermediate *tert*-butyl cation with water followed by the loss of a proton to the solution
    (c) $(CH_3)_2C=CH_2$ - formed by E1 reaction of the intermediate *tert*-butyl cation involving either water or ethanol
12.11 (a) iodomethane - Iodide ion is a better leaving group than is bromide ion.
    (b) iodomethane - There is less hindrance to S_N2 attack by hydroxide ion.
    (c) bromomethane - Bromide ion is a much better leaving group than is hydroxide ion.
    (d) 1-bromopropane - There is less hindrance to S_N2 attack by iodide ion.
    (e) 2-bromo-2-methylbutane - This substance forms a carbocation much more readily for E1/S_N1 reaction with ethanol.
    (f) aq. HBr - Displacement reaction on an alcohol requires initial protonation of the oxygen.
    (g) bromocyclohexane - Bromide ion is a better leaving group for reaction with water.
    (h) 2-chloro-3-methylbutane - This substance has the possibility of forming a more stable internal alkene in an E2 reaction.
    (i) 1:1 aq. ethanol - The higher polarity medium favors more rapid formation of the carbocation.
    (j) hydrogen sulfide - The sulfur is more nucleophilic than oxygen of water.
    (k) 1-chloropentane - This substance is less sterically hindered for S_N2 and is structurally capable of undergoing E2 reaction.
    (l) NaCN in DMSO - In DMSO the cyanide ion is relatively unsolvated and acts as a more potent nucleophile.
12.12 (a) $HO^- > HS^- > HSe^-$
    (b) $(CH_3)_3CO^- > CH_3CH_2O^- > HO^-$
    (c) $H_2N^- > HO^- > NH_3$
12.13 The substitution product is 2-methyl-1-aminopropane, and the elimination product is 2-methylpropene. The more basic is the reagent, the more prominent is the E2 route compared to the S_N2 route; amide ion has the greater basicity/nucleophilicity ratio.
12.14 With the stronger base, the significance of the E2 route increases. Thus, there will be a higher E2/E1 ratio with 0.1 *M* aq NaOH than with 0.1 *M* aq. ammonia.
12.15 (a) 2-methyl-2-butene - E1 and E2
    (b) $(CH_3)_3C\text{-}O\text{-}CH_3$ - S_N2
    (c) 2-methyl-2-butene - E1
    (d) *trans*-2-butene - E2
    (e) 3-chloro-1-cyanobutane - S_N2
    (f) (*S*)-2-iodobutane - S_N2
    (g) No reaction

(h) racemic 3-methyl-3-hexanol - S_N1
(i) 1,1-dichloro-2,2,3-trimethylcyclopropane - α-elimination
(j) 1-ethylcyclopentene - E2
(k) *cis*-1-iodo-3-methylcyclohexane - S_N2

12.16

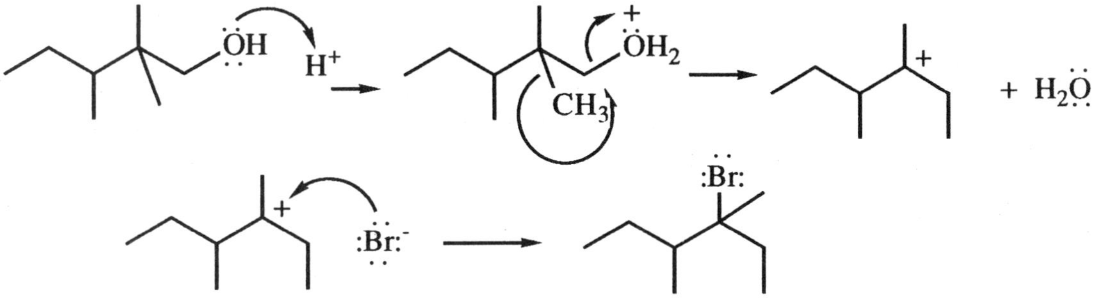

12.17

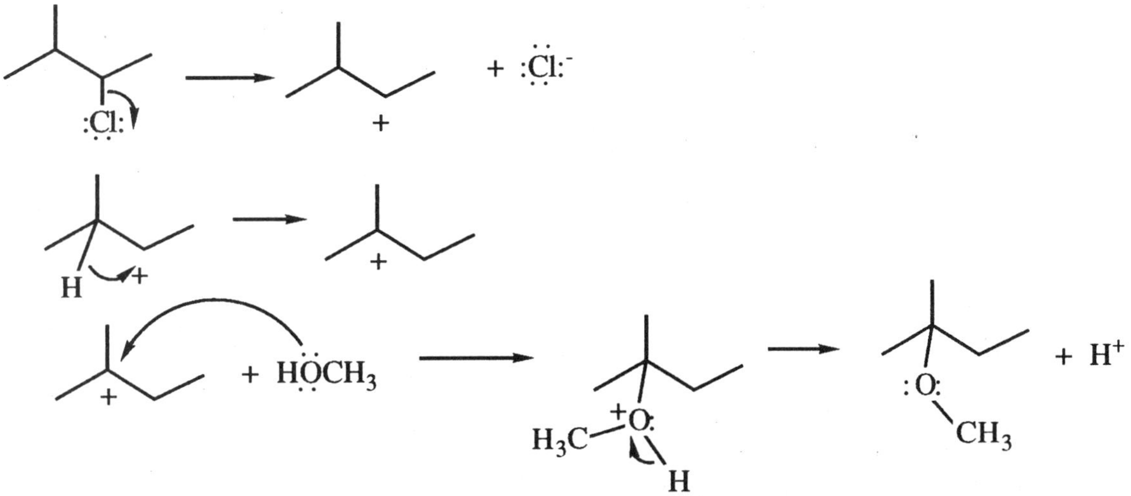

12.18

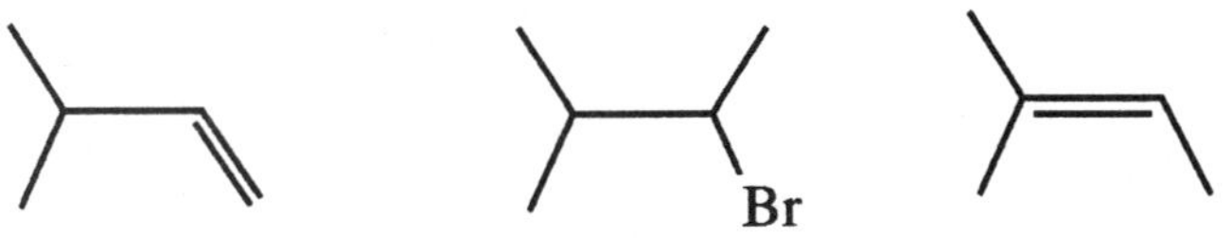

A, B, C

12.19

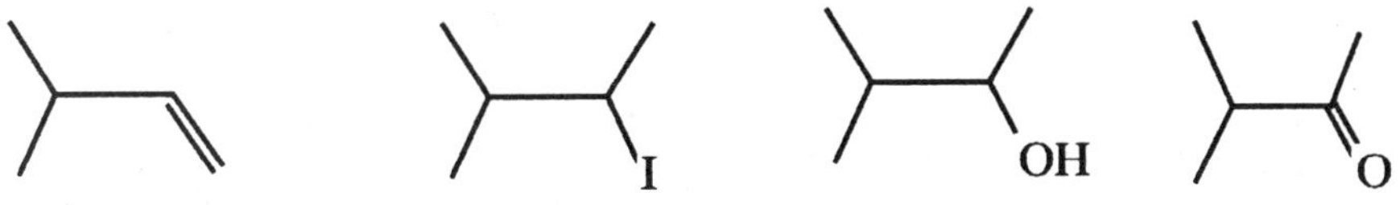

D, E, F, G

12.20

$$CH_3-\overset{..}{\underset{..}{O}}-\overset{+}{C}H_2 \longleftrightarrow CH_3-\overset{+}{\underset{..}{O}}=CH_2$$

The carbocation is resonance stabilized upon delocalization of the charge from carbon to oxygen. This stabilization lowers the actication energy for loss of chloride ion and generation of the carbocation.

12.21

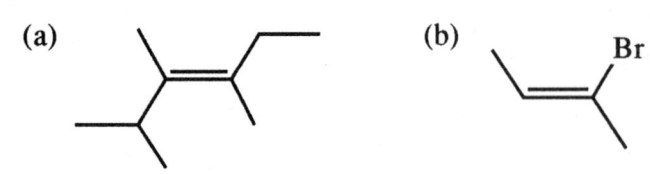

(a) [structure: 3,4-dimethyl-hex-3-ene with isopropyl substituent]
(b) [structure: 2-bromopropene with methyl]

12.22 (a) E1 rate = 1.2 x 10⁻⁶ $M\ sec^{-1}$; E2 rate = 8.0 x 10⁻⁷ $M\ sec^{-1}$
(b) E1 rate = 1.2 x 10⁻⁶ $M\ sec^{-1}$; E2 rate = 4.0 x 10⁻⁶ $M\ sec^{-1}$
E2 reaction is favored by high base concentrations.

12.23 (a) 2.9 x 10⁻⁵ $M\ sec^{-1}$; (b) 0.7 x 10⁻⁵ $M\ sec^{-1}$

12.24

$$[\ :\!\ddot{S}\!-\!C\!\equiv\!N\!: \longleftrightarrow \ \ddot{S}\!=\!C\!=\!\ddot{N}\ ]^-$$

$CH_3CH_2\text{-}S\text{-}C\!\equiv\!N$ $CH_3CH_2\text{-}N\!=\!C\!=\!S$

12.25

CH_3CH_2O—[cis-alkene structure]   [trans-alkene]—OCH_2CH_3 [branched alkene with OCH_2CH_3]

12.26 (a) 1-halo-4-ethylhexane
(b) 3-halo-3-~~ethyl~~ methyl pentane
(c) 2-chlorobutane

12.27

[structures labeled H (with Br, br), I, J, K]

12.28

[structure L: bromocyclopentane with methyls]   [structure M: methylcyclopentene]   [structure N: keto-acid with CO₂H and O]

L   M   N

12.29 (a) Cyclohexanol is treated with thionyl chloride to yield chlorocyclohexane, followed by reaction with sodium cyanide in DMSO solution.
(b) *trans*-2-Butene is trated with NBS under irradiation with light to give the 1-bromo-2-butene, followed by nucleophilic substitution using sodium cyanide in DMSO and reduction of the olefinic linkage with hydrogen and a Pd/C catalyst.
(c) (*S*)-2-Chlorobutane is allowed to react with sodium iodide in acetone solution to give (*R*)-2-iodobutane, followed by reaction with sodium cyanide in DMSO to give (*S*)-2-cyanobutane.
(d) Cyclohexanol is treated with sodium hydride to generate the alkoxide salt, followed by treatment with iodomethane to form cyclohexyl methyl ether.
(e) The 2-bromo-2,3-dimethylbutane is treated with potassium *tert*-butoxide to cause an elimination to form 2,3-dimethyl-2-butene, which is then treated with chloroform in the presence of potassium *tert*-butoxide to form the 1,1-dichloro-2,2,3,3-tetramethylcyclopropane.

12.30 Iodide ion attacks the alkyl iodide, performing a displacement reaction yielding the enantiomer of the original alkyl iodide. Thus racemization has occurred, generating a (±)-mixture from an originally optically active material.

12.31 Both substrates proceed through the same intermediate, the *tert*-butyl cation.

12.32 [structures] Protonation of the original alkene yields a tertiary carbocation which can lose a proton from either of two adjacent atoms to give the isomerized alkene products.

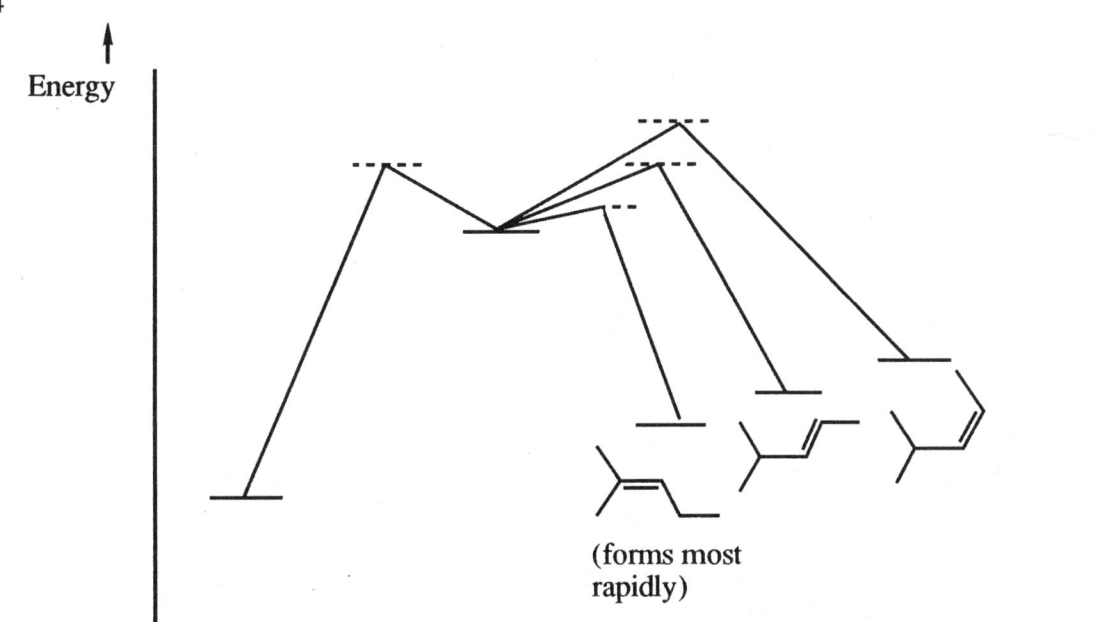

12.33 [mechanism shown: S²⁻ 2Na⁺ attacks carbon bearing Br, displacing Br⁻; intramolecular closure gives thietane (four-membered S ring) + NaBr]

12.34 [energy diagram with Reaction Progress on x-axis, Energy on y-axis; branching to three alkene products, with the middle one labeled "(forms most rapidly)"]

12.35 A formula of C₄H₈ for compound O indicates there is one ring or one site of unsaturation present. As O can add HCl and have it removed to regenerate O, there must be one site of unsaturation present. Thus, O is $(CH_3)_2C=CH_2$, P is $(CH_3)_3C\text{-}Cl$, and Q is $(CH_3)_3CH$.

12.36 2-bromo-2-methylpropane < 2-chloro-3-methylbutane < 2-bromobutane < 1-chlorobutane < 1-bromobutane

12.37 1-bromobutane < 1-chlorobutane < 2-bromobutane < 2-chloro-3-methylbutane < 2-bromo-2-methylpropane

12.38

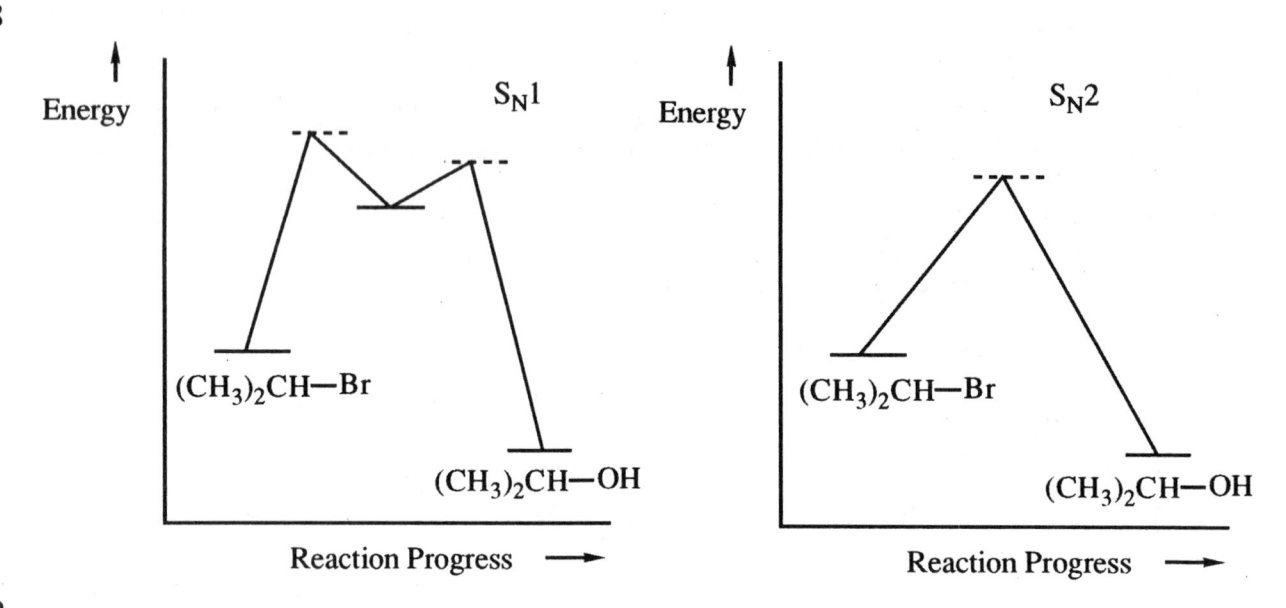

SN1 energy diagram: (CH₃)₂CH—Br → [transition state] → intermediate → [transition state] → (CH₃)₂CH—OH (two-hump profile with carbocation intermediate)

SN2 energy diagram: (CH₃)₂CH—Br → [single transition state] → (CH₃)₂CH—OH (one-hump profile)

12.39

From *meso*:

$$\underset{(CH_3)_3C}{\overset{H}{\diagdown}}C=C\underset{C(CH_3)_3}{\overset{Br}{\diagup}}$$

From racemic:

$$\underset{(CH_3)_3C}{\overset{H}{\diagdown}}C=C\underset{Br}{\overset{C(CH_3)_3}{\diagup}}$$

As the activated complex for reaction of the racemic material is less crowded than that for reaction of the *meso* compound, we expect the racemic compound to react faster.

12.40

(Cyclic piperidine: six-membered ring with NH)

The cyclic material is formed by a sequence of two nucleophilic substitution reactions, the first occurring by ammonia attack on the 1,5-dibromopentane, and the second an intramolecular attack by nitrogen of the intermediate 1-bromo-5-aminopentane to displace the second bromine.

12.41 The relatively small ring (six-membered) is not able to accommodate either a carbon-carbon triple bond or an allenic linkage as they would require an energetically unreasonable stretching and twisting of the remaining bonds of the ring.

Solution of Study Guide Practice Problems

12.1 From the data in the problem, mesylate is a better leaving group than is acetate. We infer that mesylate is a weaker base than is acetate, and that the conjugate acid of mesylate is a stronger acid than is the conjugate acid of acetate. Thus, methylsulfonic acid is a stronger acid than is acetic acid.

12.2 (a) rate = k[A][B]
(b) rate = k[A]

12.3 A + B → X   (slow; X is an intermediate)

X + B → C + D   (rapid)

Overall: A + 2 B → C + D

12.4 Only the rate constant remains constant as a reaction proceeds (at constant temperature).

12.5 The ratio of rates would be 1:27. Tripling the concentration of A would speed the reaction nine-fold since the reaction is second-order in A, and tripling the concentration of B would speed the reaction an additional three-fold, since the reaction is first-order in B.

12.6 Iodomethane will have the larger rate constant as iodide is a better leaving group than is chloride in an S_N2 reaction.

12.7 An S_N1 reaction cannot be concerted. The rate law for a concerted reaction contains concentration terms relating to each reactant.

12.8 The reaction is S_N1. If the reaction mechanism were S_N2, the rate would quadruple upon doubling the concentration of each of the nucleophile and haloalkane.

12.9 (a) Iodoethane
(b) Isobutyl bromide

12.10 (a) treatment of 1-butanol with phosphorus tribromide to generate 1-bromobutane, followed by nucleophilic substitution using sodium cyanide in DMSO to yield the 1-cyanobutane
(b) treatment of 1-heptene with HBr in the presence of peroxides to form 1-bromoheptane, followed by nucleophilic substitution reaction with ammonia.

12.11 In pure acetone, a completely aprotic solvent, the cyanide ion is more free to perform a nucleophilic substitution reaction.

12.12

propanone   dimethyl sulfoxide   diemthyl formamide

12.13 (a) (S)-1-deuterioethanol
(b) (R)-2-cyano-1-methoxybutane

12.14 (a) 2-Bromopropane will have the greater S_N1 reactivity as it leads to the more stable 2° carbocation.
(b) 3-Bromo-1-propene will have the greater S_N1 reactivity as it leads to a resonance stabilized carbocation.

12.15

Rearrangement leads to a resonance stabilized allylic carbocation.

12.16 The product ratio would decrease. If the potency of the nucleophile is increased, the S_N2 reaction route is favored (the nucleophile is involved in the rate determining step), but the S_N1 reaction rate is not affected (the nucleophile is not involved until after the rate determining step). Therefore, the S_N2 rate will increase relative to the S_N1 rate and the ratio will decrease.

12.17 (a) The 10% aq. NaOH is a stronger basic medium than is water, and reaction in it will produce more elimination.
(b) Since isobutyl bromide is the more highly branched haloalkane, it will undergo elimination more readily and substitution less readily than the less highly branched system.

12.18 (a) We exclude the possibility of involvement of a carbocation since the experimental rate law data suggests a second-order reaction is involved.
(b)

PhCH—CH$_2$ → PhCH=CH$_2$ + (CH$_3$)$_3$N: + CH$_3$CH$_2$ÖH

12.19 (a) 1-bromobutane; 2-bromo-2-methylpropane; 1-bromo-2-methylpropane
(b) No compound of formula C$_4$H$_9$Br will yield a mixture of two alkenes on dehydrohalogenation.
(c) 2-bromobutane

12.20

(a) 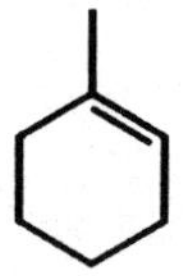 1-methylcyclohex-1-ene

(b) 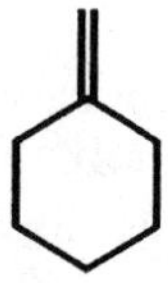 methylenecyclohexane

12.21 (Z)-1-bromo-1,2-diphenylethene

12.22 1. potassium *tert*-butoxide in *tert*-butyl alcohol; 2. chloroform with potassium *tert*-butoxide

CHAPTER 13
ALKANES AND CYCLOALKANES II. REACTION MECHANISMS AND CONFORMATIONAL ANALYSIS

Key Points

• Learn the nomenclature of bicyclic compounds
Practice Problem 13.1
Norbornane is the common name given to the structure shown below. Provide a systematic name for this compound.

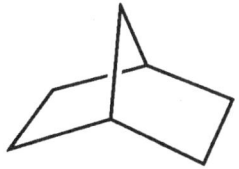

Practice Problem 13.2
Name the bicyclic starting material that on ozonolysis followed by oxidative workup would be expected to yield the molecule illustrated below.

• Learn the various methods (section 13.3 of text) for converting haloalkanes to alkanes, and be able to use these reactions in conjunction with other reactions you have met in earlier chapters to propose synthetic routes to specified target compounds.
Practice Problem 13.3
How would you convert 1-bromohexane to each of the following compounds. More than one step may be necessary for each synthesis.
    (a) hexane
    (b) 1-deuteriohexane
    (c) heptane
    (d) 2-ethylhexane

• Learn the mechanism of free radical halogenation reactions of alkanes.
Practice Problem 13.4
What are the two chain-propagating steps in the reaction:

$$C_2H_6 + Br_2 \xrightarrow{h\nu} C_2H_5Br$$

• Understand that mixtures of products often arise from free radical halogenation reactions of alkanes, and that the relative amounts the different substances formed depend on both statistical factors and the relative reactivities of different hydrogen atoms toward abstraction.
Practice Problem 13.5
In a free radical reaction of propane with a halogen molecule, X_2, a chemist finds that 20% of the

monohalogenated product is substituted at the 1-position and 80% at the 2-position. Calculate the relative reactivity of the 1° and 2° hydrogen atoms in propane toward X atoms.

• Review the use of divalent carbon for the synthesis of cyclopropane derivatives, and learn the Simmons-Smith reaction.

Practice Problem 13.6
What combination of reagents is used in the Simmons-Smith reaction to generate a carbene-equivalent? Name the product expected when cyclopentene is subjected to the Simmons-Smith reaction.

• Learn that the cyclopropane ring, unlike most other rings, undergoes addition reactions with hydrogen, with halogens, and with hydrogen halides, and that these reactions result in an opening of the ring. These special reactions correlate with the strain present in a three-membered ring. The strain is removed when the ring opens, and this ring-opening contributes to the driving force of the reaction.

Practice Problem 13.7
Which cyclopropane derivative, on catalytic hydrogenation, will yield 4,4-dimethylheptane?

Practice Problem 13.8
1-Ethylcyclopropane reacts with hydrogen bromide to form 3-bromopentane in 79% yield. Which bond of the cyclopropane ring is cleaved in this reaction? Rationalize the cleavage of this particular bond. Why does Br add to the same carbon atom that bears the ethyl group?

• Study the conformations of three, four, five, and six membered rings. Give particular attention to six membered rings. Make sure that you can draw chair structures neatly and accurately with all axial and equatorial bonds properly located. Understand the general preference for a large substituent to occupy an equatorial position, but be aware of exceptions.

Practice Problem 13.9
Which would have the least angle strain *if it were planar*: cyclopropane, cyclobutane, cyclopentane, or cyclohexane. However, this compound does not actually exist in a planar conformation. Explain why it does not exist in a planar conformation.

Practice Problem 13.10
Draw the most stable chair conformations for each of the following.
(a) [structure: cyclohexane with (CH₃)₃C and CH₃ and Cl substituents]

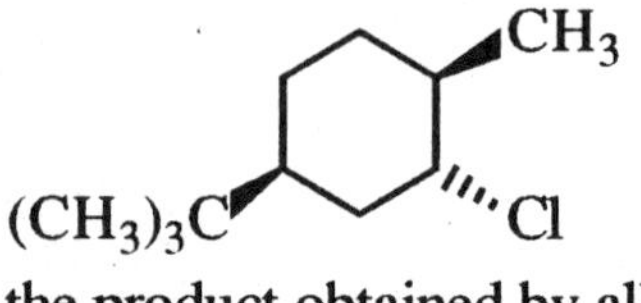

(b) the product obtained by allowing 1-methylcyclohexene to react with bromine in carbon tetrachloride
(c) the product obtained by allowing 1-methylcyclohexene to react with H₃B-THF followed by treatment with aqueous alkaline hydrogen peroxide.

Practice Problem 13.11
Compare the two chair conformations of the molecule shown below.
[structure: cyclohexane with H₃C, CH₃, CH₃ substituents]

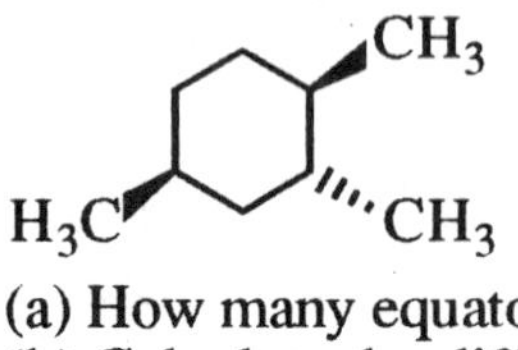

(a) How many equatorial methyl substituents are present in the more stable chair conformation?
(b) Calculate the difference in stabilities of the two chair conformations using 0.9 kcal/mole as the strain energy associated with a butane *gauche* interaction or with a CH₃-H 1,3-diaxial interaction.

Practice Problem 13.12
What is the dihedral angle between the C2-H and C3-H bonds in *cis*-2,3-dibromobicyclo[2.2.1]heptane? What is the magnitude of the same dihedral angle for the *trans* isomer?

• Be able to predict the products of reactions in which ring expansion or contraction results from a carbocation rearrangement.

Practice Problem 13.13
Each of the following reactions involves a ring expansion or ring contraction. Name the products in each reaction.

(a) [structure] →acid→ (C₉H₁₆) [expansion]

(b) [structure] →acid→ (C₈H₁₄) [contraction followed by hydride migration]

Solution of Text Problems

13.1 There are a maximum of two rings, and no sites of unsaturation.

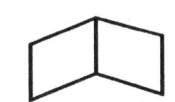

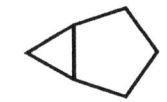

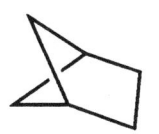

bicyclo[2.2.0]hexane   bicyclo[3.1.0]hexane   bicyclo[2.1.1]hexane

13.2 (a) 2-methylbicyclo[2.1.0]pentane
(b) 7,7-dichlorobicyclo[4.1.0]heptane
(c) bicyclo[3.2.2]nonane
(d) bicyclo[2.1.1]hexane

13.3 6,6-dichlorobicyclo[3.1.0]hexane

13.4 All syntheses begin with 1-pentanol.
(a) 1. NaI, sulfuric acid; 2. Zn, HCl, acetic acid
(b) 1. NaI, sulfuric acid; 2. Mg, ether; 3. D₂O
(c) 1. NaI, sulfuric acid; 2. potssium *tert*-butoxide, *tert*-butyl alcohol; 3. HCl; 4. Mg, ether; 5. D₂O
(d) 1. NaI, sulfuric acid; 2. Li; 3. CuBr; 4. 1-iodobutane (as formed in step 1)
(e) 1. NaI, sulfuric acid; 2. Li; 3. CuBr; 4. 2-chloropentane (as formed in part c above)

13.5 (a) We need to look at the chain propagating steps for each reaction. The net energy changes for the chain propagating steps are: F₂, -102 kcal/mole; Cl₂, -24.5 kcal/mole; Br₂, -7.5 kcal/mole.

(b) For the bromination of methane: step 2, +16.5 kcal/mole; step 3, -24 kcal/mole.

13.6 1°:2°:3° = 0.20:0.77:1.00

13.7 1°:2°:3° = 0.00067:0.053:1.00

13.8
(a) Cl₂CHCH₂CH₂CH₃; ClCH₂CH(Cl)CH₂CH₃; ClCH₂CH₂CH(Cl)CH₃; ClCH₂CH₂CH₂CH₂Cl; CH₃CCl₂CH₂CH₃; CH₃CH(Cl)CH(Cl)CH₃ [*meso* and racemic pair]
(b) Cl₂CHCH₂CH₂CH₂CH₃; ClCH₂CH(Cl)CH₂CH₂CH₃; ClCH₂CH₂CH(Cl)CH₂CH₃; ClCH₂CH₂CH₂CH(Cl)CH₃; ClCH₂CH₂CH₂CH₂CH₂Cl; CH₃CCl₂CH₂CH₂CH₃; CH₃CH(Cl)CH(Cl)CH₂CH₃ [two racemic pairs]; CH₃CH(Cl)CH₂CH(Cl)CH₃ [*meso* and racemic pair]; CH₃CH₂CCl₂CH₂CH₃
(c) Cl₂CHCH(CH₃)CH₂CH₃; (ClCH₂)₂CHCH₂CH₃; ClCH₂CCl(CH₃)CH₂CH₃; ClCH₂CH(CH₃)CH(Cl)CH₃; ClCH₂CH(CH₃)CH₂CH₂Cl; (CH₃)₂C(Cl)CH(Cl)CH₃; (CH₃)₂C(Cl)CH₂CH₂Cl; (CH₃)₂CHCCl₂CH₃; (CH₃)₂CHCH(Cl)CH₂Cl; (CH₃)₂CHCH₂CHCl₂
(d) Cl₂CHC(CH₃)₃; (ClCH₂)₂C(CH₃)₂
(e)

[structures: cyclopentane derivatives]

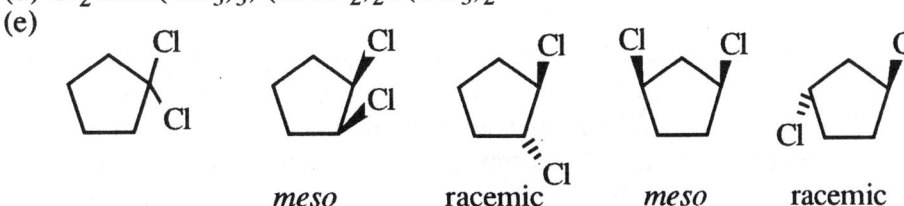

meso   racemic   meso   racemic

13.9 These are chain termination coupling products of alkyl radicals.

CH₃CH₂CH₂CH₂CH₂CH₃; (CH₃)₂CHCH₂CH₂CH₃; (CH₃)₂CHCH(CH₃)₂

13.10 For Cl₂ the product distribution is 1.2/1 for 1°/3°, whereas for sulfuryl chloride the product distribution is 1.7/1 for 1°/3°. As the sulfuryl chloride reaction result is closer to the statistical product distribution, it is less selective than the Cl₂ reaction.

13.11

$$RO\!-\!OR \xrightarrow{\text{heat}} 2\,RO\cdot$$

$$RO\cdot + :Cl\!-\!S(=O)_2\!-\!Cl: \longrightarrow RO\!-\!Cl: + \cdot SO_2Cl:$$

$$\cdot SO_2Cl: + H\!-\!CH_3 \longrightarrow HSO_2Cl: + \cdot CH_3$$

$$\cdot CH_3 + :Cl\!-\!S(=O)_2\!-\!Cl: \longrightarrow :ClCH_3 + \cdot SO_2Cl:$$

13.12

$$\overset{H}{\underset{H}{\diagdown}}C=\overset{+}{N}=\overset{..}{\underset{..}{N}}{:}^{-} \longleftrightarrow \overset{H}{\underset{H}{\diagdown}}\overset{..}{\underset{-}{C}}-\overset{+}{N}\equiv N:$$

13.13 (a) *cis*
(b) *cis*

13.14

The difference in dihedral angles for the two rotational forms is 120°, the amount of rotation possible about the intervening bond.

13.15

Viewing along pairs of bonds: a   b   c

13.16 With both the ethyl and isopropyl groups there is possible a rotational form in which there is a C-H bond pointing across the ring to interact with cross-ring hydrogens rather than having a methyl group pointing across the ring. However, with a *tert*-butyl group there is no hydrogen substituent which can point across the ring, only methyl groups, and the more crowded interaction of a methyl group with cross-ring hydrogens cannot be avoided.

13.17 (a) axial
(b) equatorial

13.18 The two methyl groups attached to the ring participate in *gauche* butane type interactions with each other, rather than with bonds of the ring.

Chapter 13 115

13.19 Since both materials yield the same combustion products, the difference in energy for each process is the difference in energy between the starting materials. Thus, the *cis* compound liberates 1.8 kcal/mole more energy.

13.20 The *p* orbitals necessary for the π bond in such compounds cannot be parallel. Thereby, the π bond is greatly weakened.

13.21
(a)

(b)

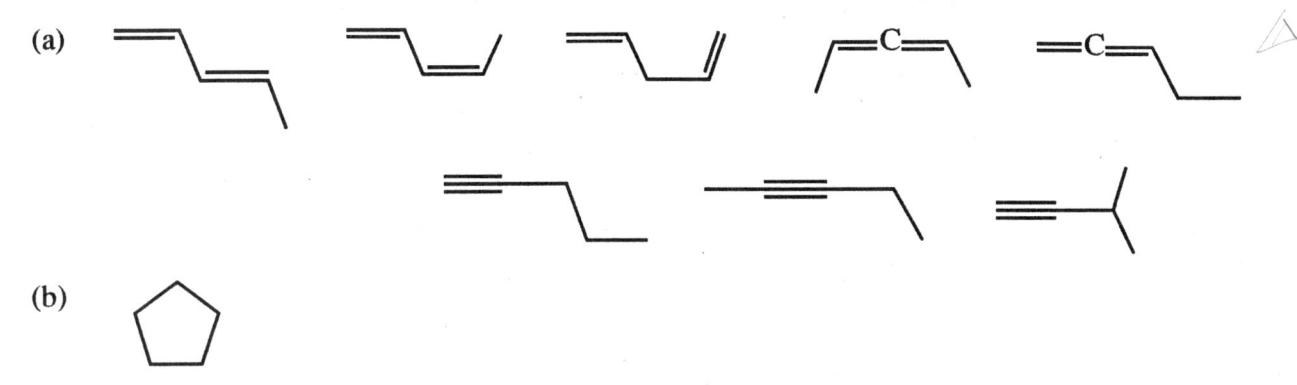

13.22 (a) bicyclo[3.2.2]nonane
(b) *trans*-bicyclo[4.4.0]decane
(c) 8-(1-methylpropyl)bicyclo[4.2.0]oct-2-ene
(d) *trans*-2,3-difluorobicyclo[2.2.2]octane

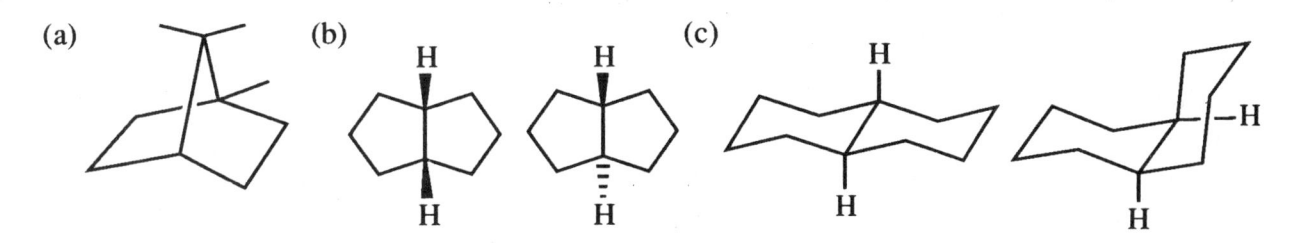

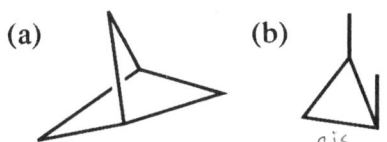

13.23
(a)   (b)   (c)

13.24
(a) bicyclo[1.1.1]pentane   (b) *cis*-1,2-dimethylcyclopropane

13.25 CH₃CH₂CH₂CH₂CH₃ and CH₃CH₂CH(CH₃)CH₂CH₃
        heptane

13.26 The observed dipole is consistent with the puckered form. A planar form would have zero dipole moment.

13.27 The dominant product is CH₃CH₂CH(Br)CH₃ and the minor product is BrCH₂CH(CH₃)₂. The reaction proceeds to place the halide ion at a site which is better stabilized as a carbocation.

13.28 (a)–(h) [structural drawings]

(f) ClCH₂CH₂CH₂CH₂Cl
   CH₃CHCl CHClCH₃

(h) major product (Br, OCH₃ on cyclopentane)

13.29 2°:1° = 47.5:0.83

13.30 In addition to the 1.0 g of 1-chlorohexane, each of 2-chlorohexane and 3-chlorohexane are formed to the extent of 1.5 g.

13.31 (a) 2,2-dichlorobutane and (2R,3R)-2,3-dichlorobutane
       (b) *meso*-2,4-dichloro-3-methylpentane

13.32 There is no change in the stereogenic center or in the relative designation of the groups attached to it. Thus, the stereogenic center remains *R*.

13.33 (a) 1. Mg, ether; 2. D₂O
       (b) osmium tetroxide, or potassium permanganate in aqueous potassium hydroxide solution
       (c) bromine, irradiation with light
       (d) potassium permanganate, aqueous potassium hydroxide solution, heat

13.34 (a) 1. sulfuric acid; 2. D₃B-THF; 3. acetic acid
       (b) 1. bromine in carbon tetrachloride; potassium *tert*-butoxide
       (c) 1. potassium *tert*-butoxide; 2. CHCl₃; potassium *tert*-butoxide
       (d) 1. sulfuric acid; 2. CH₂I₂, Zn-Cu alloy
       (e) 1. bromine, light; potassium *tert*-butoxide
       (f) 1. bromine, light; 2. Li; 3. CuI; 4. CH₃I
       (g) 1. Mg/ether; 2. H₂C=O; 3. acid w.u.; 4. PBr₃
       (h) 1. bromine, light; 2. potassium *tert*-butoxide; 3. ozone; 4. hydrogen sulfide

13.35 (a) 2-iodo-2,3-dimethylbutane
       (b) 2-iodobutane

13.36 (a) 2; (b) 1; (c) 2

13.37 (a) *cis*-1,2-Dimethylcyclohexane and *trans*-1,2-dimethylcyclohexane each have two stereogenic carbon atoms. While the *trans* isomer is in principle resolvable, the *cis* isomer is a *meso* compound. However, neither *cis*- nor *trans*-1,4-dimethylcyclohexane molecules have any stereogenic carbon atoms.
       (b) The two chair forms of *cis*-1,2-dimethylcyclohexane are not superimposable. They are conformational enantiomers, interconvertible by rotation about single bonds. With the facile interconversion, we do not expect these conformational enantiomers to be resolvable.
       (c) These are true enantiomers which should be resolvable.
       (d) The situation is the same as with the *cis*- and *trans*-1,2-dimethylcyclohexanes.

13.38 The *cis*-1,2-dimethylcyclopropane is less stable than the *trans* isomer owing to repulsive interactions of the eclipsing methyl groups.

13.39 We can estimate the difference in the heat of combustion for the two compounds by looking at the different types of methylene groups for each, and using the values of heat of combustion *per* methylene group of various sized rings from Table 4.1 of the text. Using this approach, we expect the bicyclo[2.2.0]hexane to liberate 13.3 kcal/mole more heat on combustion than the bicyclo[3.1.0]hexane.

13.40

(a) DCH₂CH₂CH₂D

(b) [chair cyclohexane with H, Cl axial-equatorial at one carbon and Br, H at adjacent carbon] (plus enantiomer)

(c) [chair cyclohexane with D, CH₃ at one carbon and OD, H at adjacent carbon] (plus enantiomer)

(d) [cyclopentene]

13.41

[Mechanism: epoxide protonation by H⁺, ring opening by Br⁻, giving trans-2-bromocyclopentanol shown in two equivalent depictions]

13.42 Using Newman projections of the two chair conformations, we find that there are a total of four more *gauche* butane type interactions in the diaxial structure as compared to the diequatorial structure. Thereby, the diequatorial structure is more stable by ~3.6 kcal/mole.

13.43

[Chair conformations comparing 1,2-dimethylcyclohexane and 1,4-dimethylcyclohexane diequatorial vs diaxial forms]

The 1,4-dimethylcyclohexane system has the greater difference between the two chair forms. With the 1,2-dimethylcyclohexane system the diequatorial conformation has a *gauche* butane interaction between the methyl groups which is absent in the diequatorial conformation of the 1,4-dimethylcyclohexane system.

13.44 The energy difference between the two chair conformations of *cis*-1,3-dimethylcyclohexane calculated using this approach is 5.5 kcal/mole.

13.45 These two chair conformations are of equal energy.

13.46

[Four chair conformation structures with C(CH₃)₃ and Cl substituents]

13.47 The *cis*-4-bromo-*tert*-butylcyclohexane will yield the *trans*-1-*tert*-butyl-4-iodocyclohexane, and the *trans*-4-bromo-1-*tert*-butylcyclohexane will yield the *cis*-1-*tert*-butyl-4-iodocyclohexane. The *cis* starting material will react faster owing to a less hindered approach of the incoming nucleophile.

13.48 For 3-methylcyclohexene to be the *only* product, there must be only one hydrogen in the 2-chloro-1-methylcyclohexane which can exist with a dihedral angle of 180° relative to the chlorine, that being on the 3-position. This requires the methyl and the chlorine to be in a 1,2-relationship and both axial, meaning that the compound has a *trans* configuration. Moreover, as hydroboration/oxidation is a *syn* process, the intermediate alcohol must also have been *trans*, requiring the replacement of the hydroxyl group by chlorine to occur with retention of configuration at carbon. The thionyl chloride reaction thereby must have *not* been performed in HMPT

solution.
13.49

cis and *trans*

13.50 Statement (b) describes the actual situation.

13.51

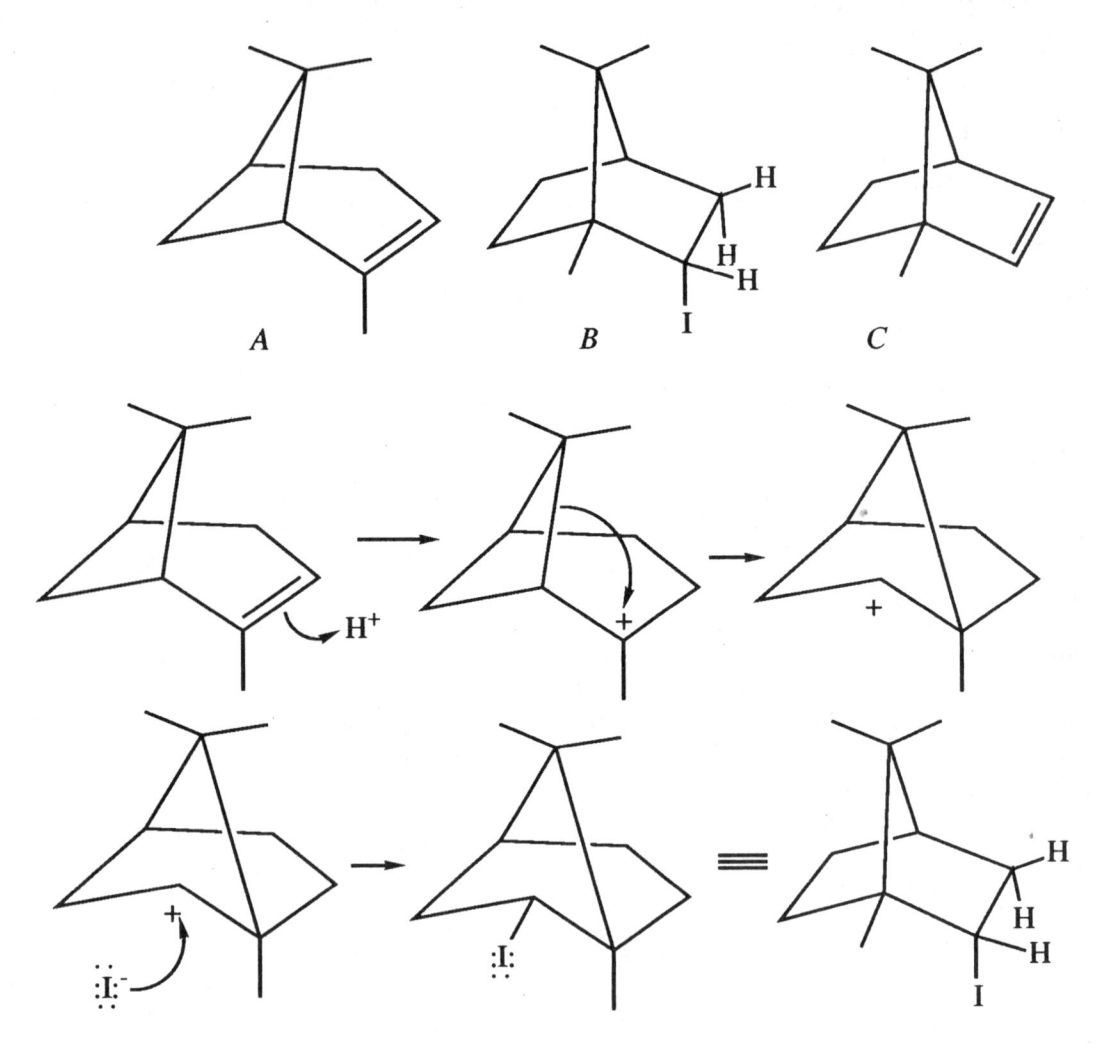

A   B   C

Solution of Study Guide Practice Problems
13.1   bicyclo[2.2.1]heptane
13.2   1,6-dimethylbicyclo[4.2.0]oct-7-ene
13.3   (a) 1. Mg, ether; 2. water
       (b) 1. Mg, ether; 2. D$_2$O
       (c) 1. Li; 2. CuI; 3. iodomethane
       (d) 1. potassium *tert*-butoxide; 2. HBr; 3. Li; 4. CuI; 5. iodoethane

13.4

H_3C-CH_3 + :$\ddot{B}r\cdot$ ⟶ $H_3C-\dot{C}H_2$ + $H\ddot{B}r$:

$H_3C-\dot{C}H_2$ + :$\ddot{B}r-\ddot{B}r$: ⟶ $H_3C-CH_2\ddot{B}r$: + :$\ddot{B}r\cdot$

13.5 reactivity ratio 2°:1° = 33.3:1

13.6 methylene iodide (CH_2I_2) and zinc-copper alloy; bicyclo[3.1.0]hexane

13.7 1,1-di(1-propyl)cyclopropane

13.8 The bond between C1 (carbon bearing the ethyl group) and C2 of the cyclopropane ring is cleaved. The reaction proceeds by initial protonation of the cyclopropane ring, and bromide ion adds to the site which best stabilizes a carbocation, the most highly substituted position.

13.9 Cyclopentane would have the least angle starin if it were planar. It is not planar, however, owing to the fact that hydrogens on adjacent carbon atoms would be completely eclipsed in the planar structure, adding an element of instability to such a structure which can be eliminated by slight twisting from planarity.

13.10
(a) (b) (c)

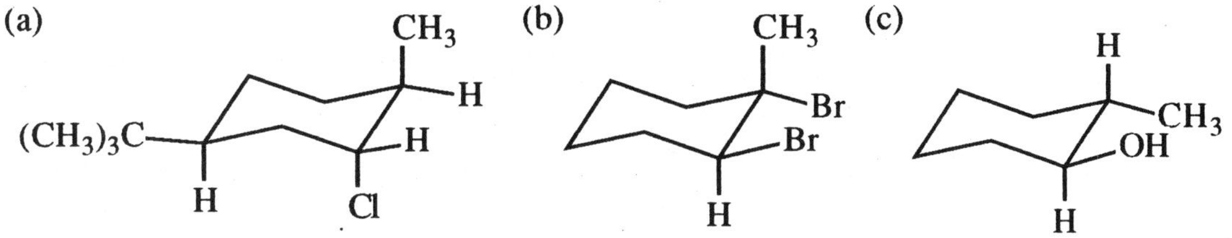

13.11 (a) 2 methyl groups
(b) 0.9 kcal/mole

13.12 For the *cis* compound, the dihedral angle is 0°, while for the *trans* compound the dihedral angle is 120°.

13.13 (a) 3,3-diethylcyclopentene
(b) 1-isopropylcyclopentene

CHAPTER 14
ETHERS AND EPOXIDES

Key Points

• Know how to name open-chain and cyclic ethers, and be able to draw the structure of an ether if you are given its name.

Practice Problem 14.1
Name each of the following structures:

(a) C₆H₅—O—CH₂—CH=CH₂

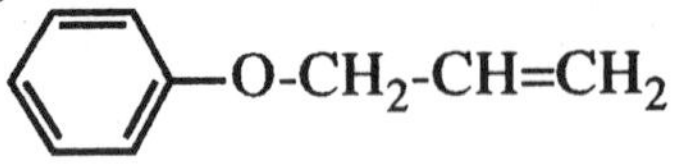

(b) *trans*-cyclohexane with CH₃ and OCH(CH₃)₂ substituents

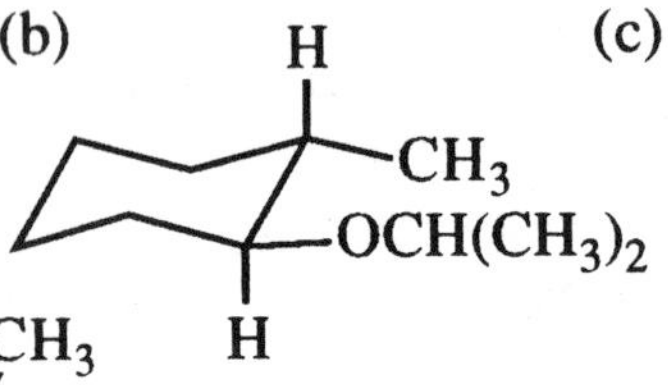

(c) (CH₃)₂C=CHCH₂CH₂OCH₂CH(CH₃)₂

(d) tetrahydropyran

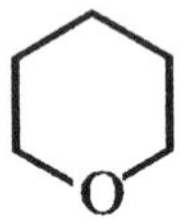

(e) 1,3-dioxane with two CH₃ groups

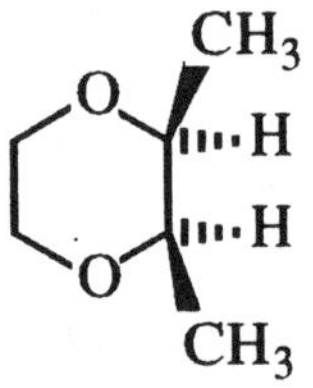

Practice Problem 14.2
Draw skeletal structures for the compounds described by each of the following names:
  (a) *sec*-butyl vinyl ether
  (b) *trans*-2-butene oxide
  (c) 2-isopropoxypentane
  (d) *n*-propyl 2-propenyl ether
  (e) 2-isobutoxy-2-methylpropane

• Learn that although simple ethers are relatively inert, they can be cleaved by concentrated HI. (See Equation 14.1 in the text, and study the mechanism for this reaction.)

Practice Problem 14.3
In the reaction of Equation 14.1 in the text, rationalize why the products are iodoethane and 2-propanol rather than 2-iodopropane and ethanol.

Practice Problem 14.4
When *tert*-butyl methyl ether is treated with aqueous HI, cleavage occurs to give mainly methanol and *tert*-butyl iodide. Why does this imply that the mechanism of this reaction is somewhat different from that of the reaction shown in Equation 14.1 in the text? Suggest an alternative mechanism for this reaction that is in accord with the observed products.

• Know the main methods for preparing ethers, the Williamson synthesis and alkoxymercuration/demercuration. Be aware that the haloalkane used in the Williamson synthesis should be methyl or primary for best results. (Tertiary alkyl halides undergo elimination when treated with alkoxides.) Also, be aware that the oxygen atom of the alcohol used in alkoxymercuration adds to the more highly substituted carbon atom of the olefinic linkage.

Practice Problem 14.5
Outline an efficient synthesis of each of the ethers listed below. You may use the Williamson synthesis only if it will provide the target ether in good yield.
  (a) cyclohexyl methyl ether
  (b) *sec*-butyl 1-propyl ether

Practice Problem 14.6
What starting alkene and alcohol should you use to prepare *tert*-pentyl isopentyl ether by the alkoxymercuration/demercuration procedure?

Chapter 14 121

• Be aware that intermolecular dehydrative coupling of primary and secondary alcohols can be used for the synthesis of symmetric ethers. Unsymmetrical ethers can also be prepared by intermolecular dehydration if one of the alcohols is tertiary and the other is secondary or primary.
Practice Problem 14.7
Write a mechanism for the reaction shown in Equation 14.8 of the text.

• Recognize that vinyl ethers undergo cleavage reactions under much milder conditions (for example, dilute aqueous sulfuric acid) than do most other ethers (which require concentrated HI solution).
Practice Problem 14.8
Explain why vinyl ethers undergo cleavage so readily in dilute aqueous acid.
Practice Problem 14.9
Suggest a structure for a product of formula C_3H_6O formed when 1-propenyl 1-propyl ether is cleaved using dilute aqueous sulfuric acid. What is the other organic product of this cleavage reaction?

• Understand the use of particular types of ether linkages as protecting groups for alcohols.
Practice Problem 14.10
Tell what reagents you would use to convert a primary hydroxyl group (-OH) to each of the following:
(a) $-O-C(CH_3)_3$
(b) $-O-Si(CH_3)_2C(CH_3)_3$
Practice Problem 14.11
How would you accomplish the synthetic conversion shown below, recognizing that protection of the hydroxyl group is required?

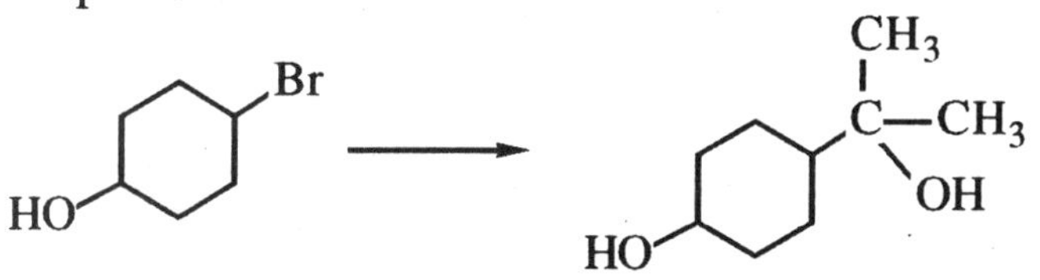

• Know the main methods for the preparation of epoxides - from halohydrins by way of intramolecular nucleophilic displacement of the halogen, and from alkenes by way of reaction with a peroxycarboxylic acid.
Practice Problem 14.12
Identify the compounds indicated A to C.

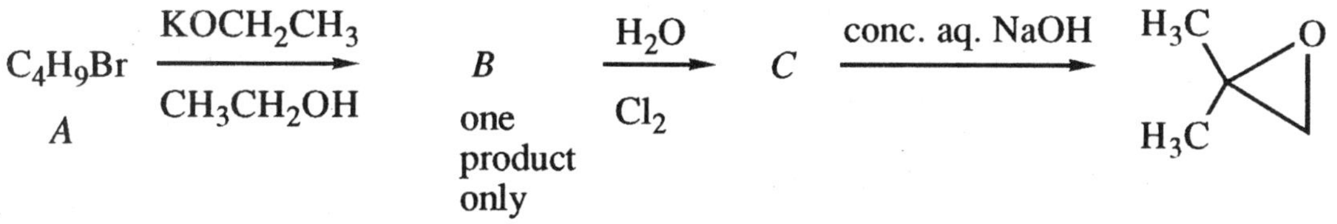

C_4H_9Br (A) $\xrightarrow{KOCH_2CH_3, CH_3CH_2OH}$ B $\xrightarrow[Cl_2]{H_2O}$ C (one product only) $\xrightarrow{conc.\ aq.\ NaOH}$ [epoxide with H_3C, H_3C groups]

Practice Problem 14.13
In the reaction of 1,2-dimethyl-1,4-cyclohexadiene with one equivalent amount of *m*-chloroperoxybenzoic acid, one of the two double bonds is epoxidized preferentially to the other. Suggest a structure for the major product in this epoxidation and account for the selectivity in the reaction.

• Know the stereochemistry of the intramolecular Williamson procedure for preparing cyclic ethers. The alkoxy anion site attacks the backside of the carbon-halogen bond resulting in inversion of configuration at the reaction center.
Practice Problem 14.14
Suppose that (2R,5R)-5-bromo-2-hexanol is treated with sodium hydroxide solution so that an intramolecular reaction occurs leading to a cyclic ether. Fully name the product cyclic ether. Do you expect this product to be optically active? Explain your conclusion.

• Recognize that epoxides readily undergo ring-opening reactions. Under neutral or basic conditions, the epoxide ring is attacked preferentially at the less substituted carbon atom of the oxirane ring. The epoxide is protonated

122 Study Guide and Solutions Manual

under acidic conditions and there is a tendency for nucleophilic attack to occur at the more highly substituted carbon atom of the oxirane ring.

Practice Problem 14.15
Give the major organic products in each of the following reactions.

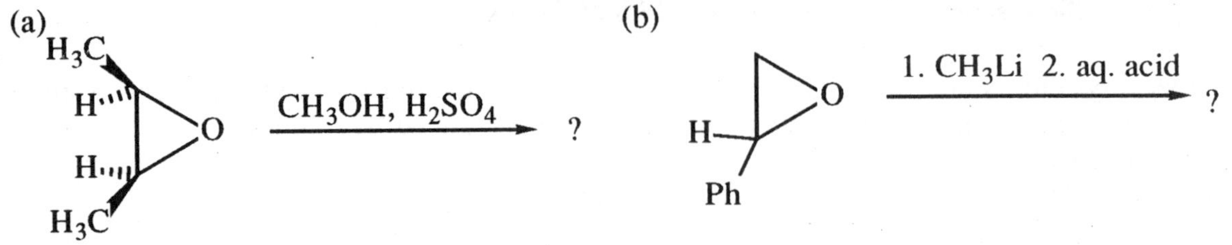

(a) 2,2,3,3-tetramethyloxirane (H₃C, H, H, H₃C substituents with stereochemistry shown) + CH₃OH, H₂SO₄ → ?

(b) 2-phenyl-2-methyloxirane (with H and Ph) 1. CH₃Li 2. aq. acid → ?

• Be able to use the reactions introduced in this chapter along with reactions from previous chapters to devise syntheses of target compounds.

Practice Problem 14.16
Propose efficient synthetic routes for each of the following transformations:

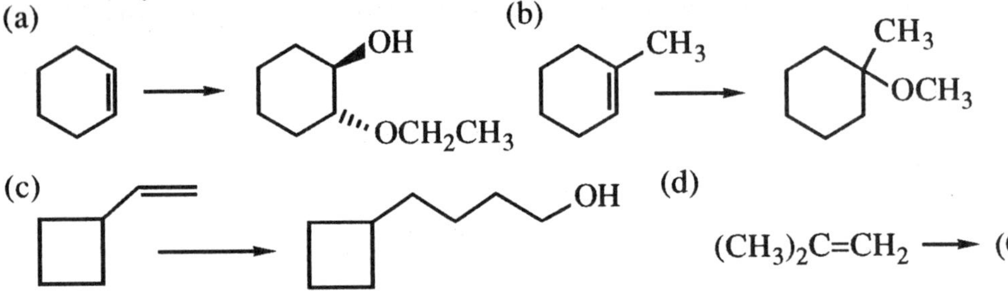

(a) cyclohexene → trans-2-ethoxycyclohexanol (OH and OCH₂CH₃)

(b) 1-methylcyclohexene → 1-methyl-1-methoxycyclohexane (CH₃ and OCH₃)

(c) vinylcyclobutane → 4-cyclobutylbutan-1-ol (cyclobutane-CH₂CH₂CH₂CH₂OH)

(d) (CH₃)₂C=CH₂ → (CH₃)₃CCH₂CHO

Solution of Text Problems

14.1 ethoxyethane; 2-methyl-1-propoxypropane; 1-butoxy-4-methyl-3-pentene

14.2

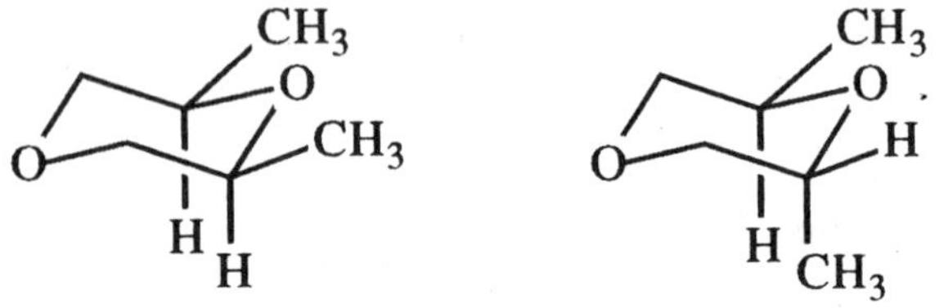

cis   trans

14.3 (a) 2-ethyl-2-methyloxirane
(b) *cis*-2-chloro-3-ethyloxirane

14.4

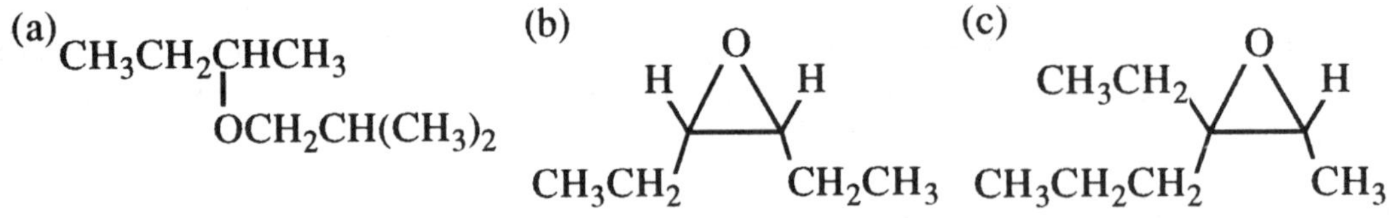

(a) CH₃CH₂CHCH₃
        |
        OCH₂CH(CH₃)₂

(b) oxirane with H, CH₃CH₂ and H, CH₂CH₃ substituents

(c) oxirane with CH₃CH₂, H and CH₃CH₂CH₂, CH₃ substituents

14.5 The bicyclic amine is more nucleophilic than triethylamine because the alkyl groups attached to nitrogen in the bicyclic amine cannot interfere with its approach toward an electrophilic site.

14.6

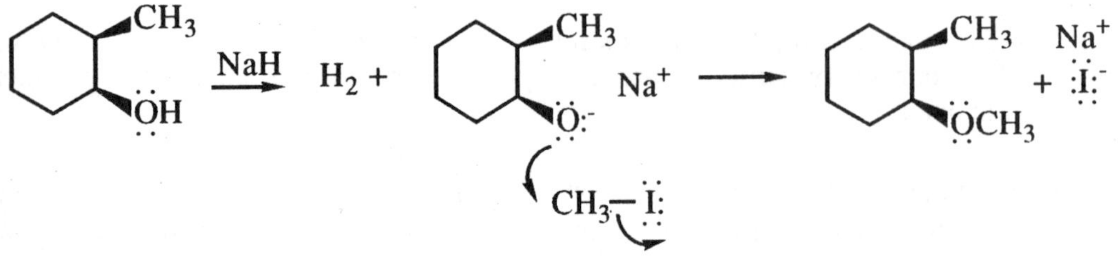

cyclohexane-CH₃, OH + NaH → H₂ + cyclohexane-CH₃, O⁻ Na⁺ → cyclohexane-CH₃, OCH₃ + Na⁺ + :I:⁻

            CH₃—I:

14.7
(a)

CH₃OCH₂CH₂CH₃ →1 CH₃I + NaOCH₂CH₂CH₃
CH₃OCH₂CH₂CH₃ →2 NaOCH₃ + ICH₂CH₂CH₃

1 is favored as no elimination is possible

(b)

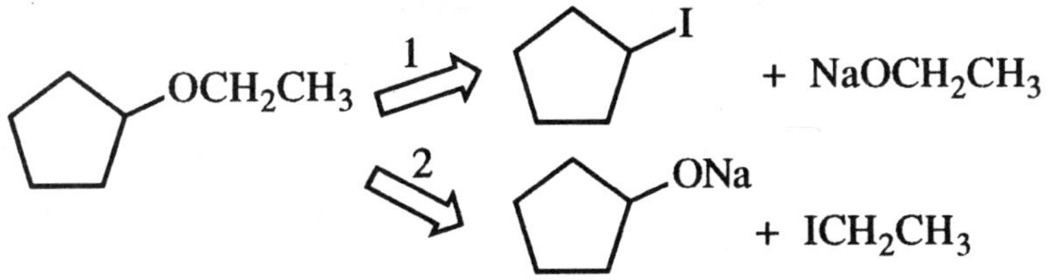

cyclopentyl-OCH₂CH₃ →1 cyclopentyl-I + NaOCH₂CH₃
cyclopentyl-OCH₂CH₃ →2 cyclopentyl-ONa + ICH₂CH₃

2 is favored as elimination would be minimized

14.8 (a) treatment of iodoethane with sodium isopropoxide
(b) treatment of a mixture of propene and ethanol first with mercuric trifluoroacetate and then with sodium borohydride under basic conditions

14.9 Elimination reaction would occur to the exclusion of substitution.

14.10

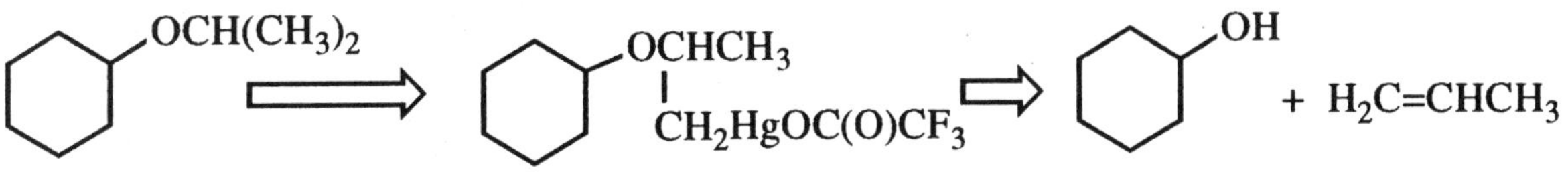

cyclohexyl-OCH(CH₃)₂ ⇒ cyclohexyl-OCHCH₃(CH₂HgOC(O)CF₃) ⇒ cyclohexyl-OH + H₂C=CHCH₃

14.11 (a) Williamson
(b) Both are satisfactory.
(c) alkoxymercuration/demercuration

14.12

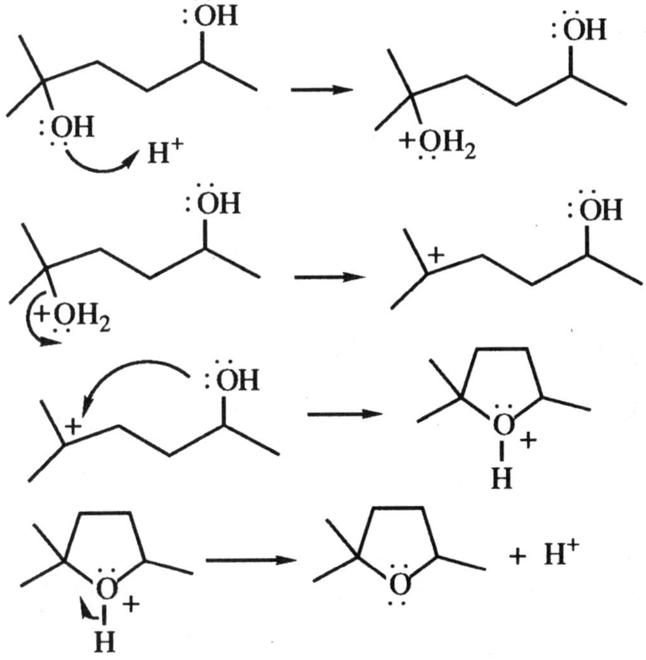

[mechanism showing protonation of one OH to +OH₂, loss of water to form carbocation, intramolecular attack by remaining OH to form cyclic oxonium ion, and deprotonation to give cyclic ether]

14.13

[tetrahydrofuran with OH substituent]

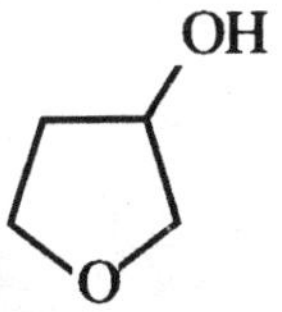

14.14 Rearrangement occurs readily, along with significant amounts of elimination.

124 Study Guide and Solutions Manual

14.15

(a) (CH$_3$)$_2$CHOCH$_2$CH$_2$OCH$_3$ (b) CH$_3$OCH$_2$CH$_2$OCH$_3$ (c) [tetrahydrofuran structure] (d) [1,4-dioxane structure]

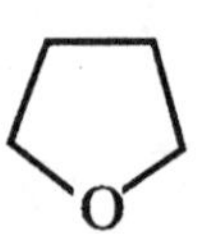

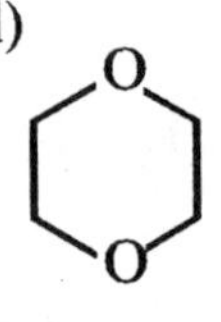

14.16 Protonation at the β-carbon atom yields a cation which is resonance stabilized. Protonation at either of the other two possible sites yields only charge-localized cations.

14.17

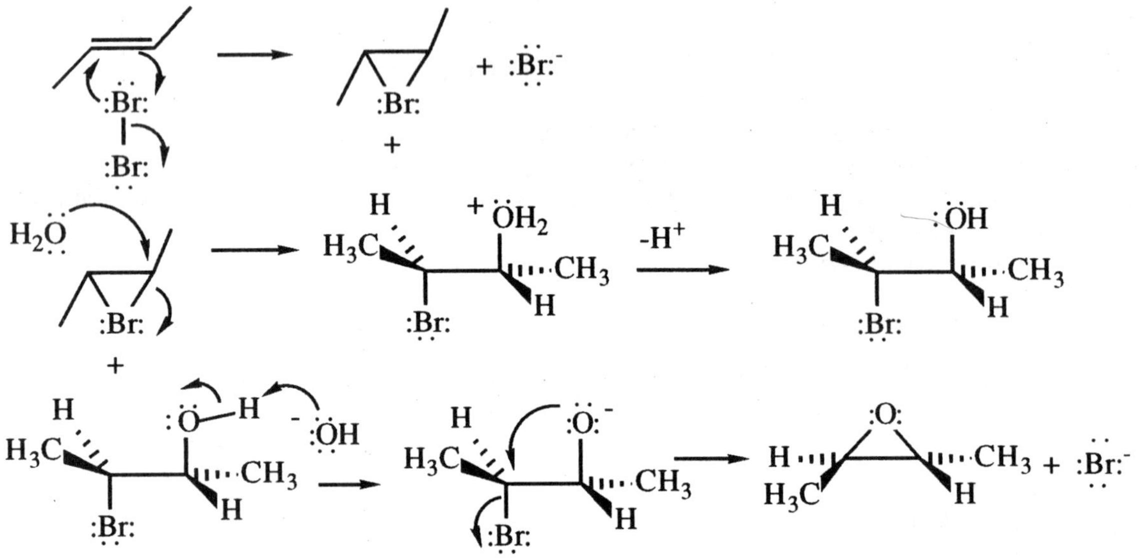

[reaction mechanism scheme showing bromonium ion formation, water attack, and subsequent epoxide formation]

14.18

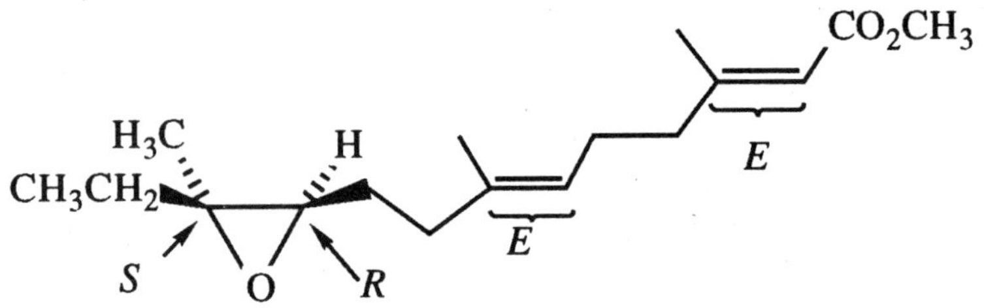

[structure showing epoxide with S and R configurations, ethyl and methyl substituents, with E-configured alkenes and CO$_2$CH$_3$ group]

14.19

[structure similar to 14.18 but without stereochemistry indicated, with CO$_2$CH$_3$ group]

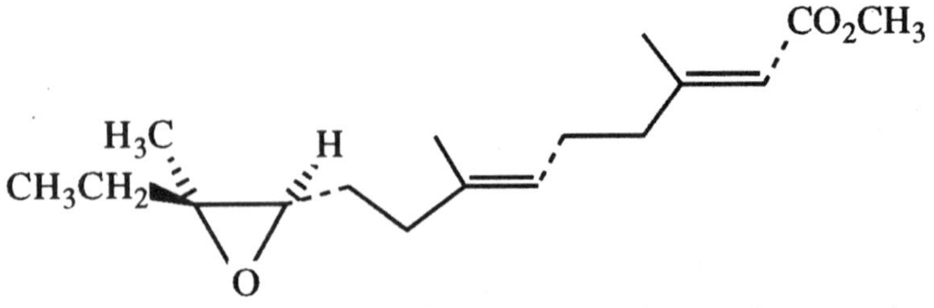

The bonds joining individual isoprene units are shown as broken lines.

14.20 (a) treatment of 2-methyl-1-butene with *m*-chloroperoxybenzoic acid in methylene chloride solution
(b) treatment of 1-bromo-2-methyl-2-butanol with sodium carbonate

14.21 2-Propanol is formed.

14.22 The two products formed initially (the 1,2-propanediols) would be enantiomeric. However, the material formed in the acid catalyzed reaction would slowly undergo racemization.

14.23 *trans*-2-cyanocyclohexanol (racemic mixture)

14.24
(a) (CH₃)₂CHCH₂CH₂OCH₃ (b) HOCH₂CH₂CH₂OCH(CH₃)₂ (c) H₂C=CHOCH₂CH=CH₂

(c) [cis-CH=CH-CH₂-OCH₃] (d) CH₃OCHCH₃
                                      |
                                     CH₂CH₃

14.25 (a) *trans*-2,3-dimethyloxirane
       (b) isopropyl propyl ether
       (c) cyclopropyl ethyl ether
       (d) isopropyl vinyl ether
       (e) *trans*-2,3-dimethyl-1,4-dioxane
       (f) dicyclobutyl ether
       (g) allyl isopentyl ether

14.26 (a) 1. *tert*-butyl alcohol treated with potassium to generate the alkoxide salt; 2. treatment of the alkoxide salt with iodomethane
       (b) 1. 2-propanol treated with mercuric trifluoroacetate in the presence of propene; 2. reduction with sodium borohydride in aqueous sodium hydroxide solution
       (c) 1. 2-butanol treated with sodium metal to generate the alkoxide salt; 2. treatment of the alkoxide salt with 2-methyl-1-iodopropane
       (d) 1. treatment of 2-propanol with sodium metal to generate the alkoxide salt; treatment of the alkoxide salt with 1-iodohexane: and, 1. treatment of 1-hexanol with mercuric trifluoroacetate in the presence of propene; 2. reduction with sodium borohydride in aqueous sodium hydroxide solution
       (e) treatment of *tert*-butyl alcohol with mercuric trifluoroacetate in the presence of cyclohexene; 2. reduction with sodium borohydride in aqueous sodium hydroxide solution

14.27

[Mechanism scheme showing protonation of ether by H⁺ followed by nucleophilic attack by Br⁻ on the methyl group, yielding the alcohol and CH₃Br]

14.28 (a) 2-iodopropane
       (b) no reaction
       (c) no reaction
       (d) 1-methyl-*trans*-1,2-cyclohexanediol
       (e) racemic 3,4-hexanediol
       (f) racemic *trans*-2-chlorocyclopentanol

14.29 *cis*-2-butene

14.30 (a) 1-propanol
       (b) 2-methyl-1,2-propanediol
       (c) 1,3-diiodopropane
       (d) methylpropene
       (e) pyran

14.31

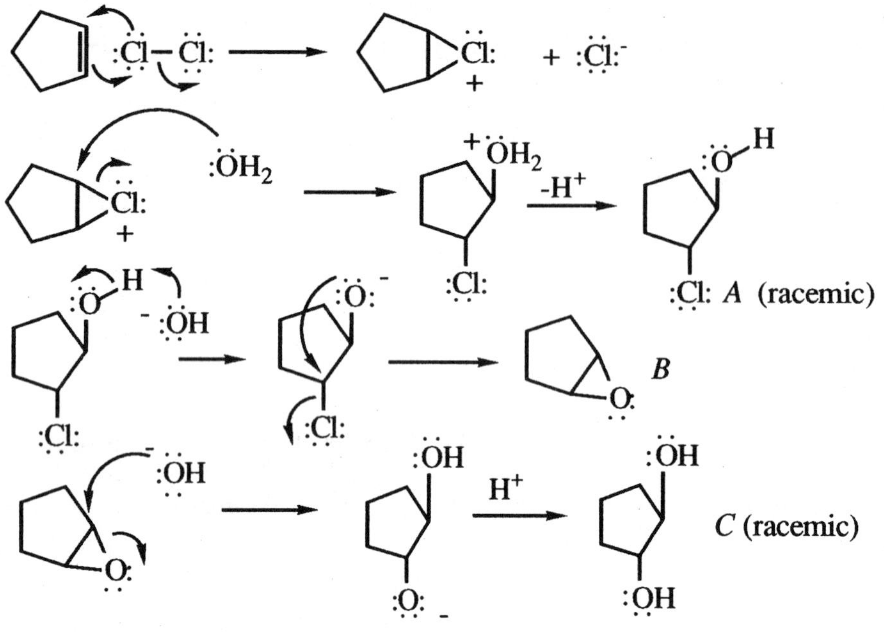

[Mechanism scheme showing chlorination of cyclopentene leading to chloronium ion opening by water to give trans-2-chlorocyclopentanol A (racemic); then base-induced cyclization via alkoxide displacing chloride to form epoxide B; followed by hydroxide opening of epoxide to give trans-1,2-cyclopentanediol C (racemic).]

14.32 (CH₃)₂CHOCH(CH₃)₂

14.33 The compound is 2,4-dimethylfuran. This material forms by protonation of the alkene at the terminal position followed by intramolecular attack of the alcohol oxygen on the carbocation site.

14.34 (a) ethylene glycol (1,2-ethanediol)
(b) ethyl 2-hydroxyethyl ether
(c) 2-aminoethanol
(d) diethanolamine [HN(CH₂CH₂OH)₂]
(e) acrylonitrile (H₂C=CHCN)

14.35

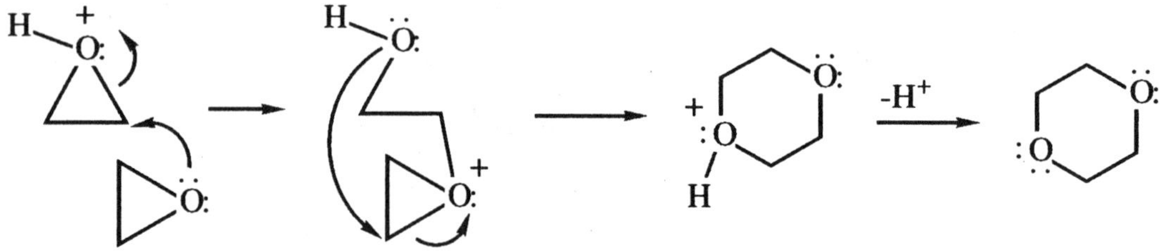

[Mechanism showing acid-catalyzed dimerization of ethylene oxide: protonated epoxide attacked by second epoxide oxygen, forming protonated dioxane, then loss of H⁺ to give 1,4-dioxane.]

14.36

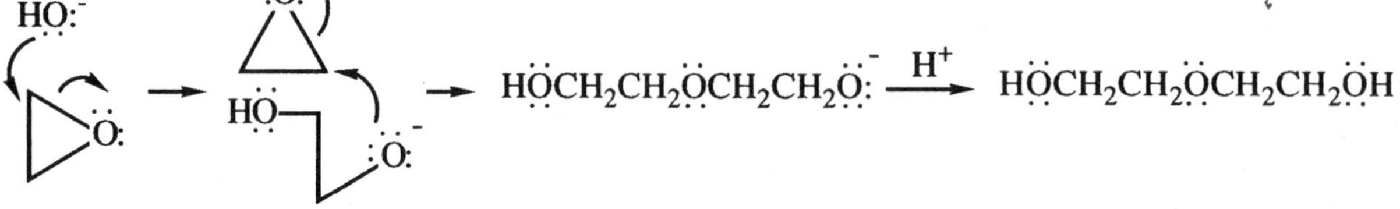

[Mechanism showing hydroxide opening of ethylene oxide to give HOCH₂CH₂O⁻, which attacks a second epoxide to give HOCH₂CH₂OCH₂CH₂O⁻, then H⁺ workup to HOCH₂CH₂OCH₂CH₂OH.]

14.37

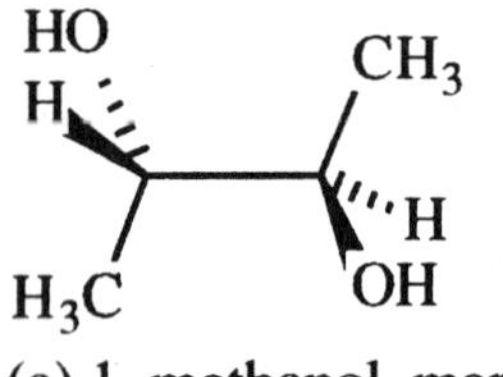

[Structure of (2R,3R)- or meso-2,3-butanediol drawn with wedge/dash: HO and H on left carbon (with H₃C), CH₃ and H on right carbon (with OH).]

14.38 (a) 1. methanol, mercuric trifluoroacetate; 2. sodium borohydride, aqueous sodium hydroxide
(b) 1. *m*-chloroperoxybenzoic acid, methylene chloride, heat; 2. sodium methoxide
(c) 1. 1-butanol treated with thionyl chloride to form 1-chlorobutane; 2. isobutyl alcohol treated with sodium metal to generate the alkoxide salt; 3. reaction of the alkoxide salt with the 1-chlorobutane

Chapter 14 127

(d) 1. trimethylchlorosilane, pyridine; 2. magnesium, ether; 3. D$_2$O; 4. tetrabutylammonium fluoride, water

(e) 1. potassium *tert*-butoxide; 2. potassium permanganate, potassium hydroxide, water

(f) 1. potassium *tert*-butoxide; 2. formic acid, hydrogen peroxide, water

14.39 CH$_3$CH$_2$OCH$_2$CH$_2$OCH$_2$CH$_3$

14.40

14.41 (a) 1. 2-propanol, mercuric trifluoroacetate; 2. sodium borohydride, aqueous sodium hydroxide
(b) 1. phosphoric acid; 2. acetic acid, hydrogen peroxide; 3. methylmagnesium iodide, acid workup
(c) 1. bromine, light; 2. potassium *tert*-butoxide; 3. formic acid, hydrogen peroxide, water
(d) 1. sodium metal; 2. propylene oxide

14.42

14.43 R'$_3$Si-F

14.44

Solution of Study Guide Practice Problems

14.1 (a) allyl phenyl ether
(b) *trans*-1-isopropoxy-2-methylcyclohexane
(c) isobutyl 4-methyl-3-pentenyl ether
(d) tetrahydrofuran
(e) *cis*-2,3-dimethyl-1,4-dioxane

14.2

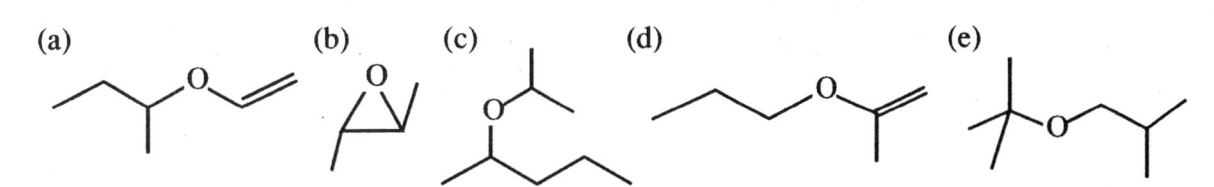

14.3 After the ether is protonated, iodide ion attacks the alkyl group that is least hindered toward backside attack. That is, it attacks the ethyl rather than the isopropyl group. The mechanism of the reaction is S$_N$2.

14.4 The mechanism must be different from that of the reaction in the previous problem since iodine is bonded to the *more* bulky group in the products. A mechanism consistent with these findings involves the following steps:

1. protonation of the ether at the oxygen; 2. loss of methanol leaving behind a *tert*-butyl cation; 3. combination of the iodide ion with the *tert*-butyl cation. The mechanism is first order.

14.5 (a) 1. treatment of cyclohexanol with sodium hydride to generate the alkoxide salt; 2. reaction of the alkoxide salt with iodomethane

(b) 1. treatment of 1-propanol with 1-butene in the presence of mercuric trifluoroacetate; 2. reduction with sodium borohydride in aqueous sodium hydroxide solution

14.6 We need to use 2-methyl-2-butene and 3-methyl-1-butanol.

14.7

14.8 Vinyl ethers undergo acid catalyzed cleavage quite readily because protonation (on *carbon*) leads to a resonance stabilized cation.

14.9 The products are CH_3CH_2CHO and $CH_3CH_2CH_2OH$.

14.10 (a) *tert*-butyl alcohol and acid, or methylpropene with acid
(b) $(CH_3)_2Si(Cl)C(CH_3)_3$

14.11 1. $(CH_3)_2Si(Cl)C(CH_3)_3$; 2. Mg, ether; 3. $(CH_3)_2C=O$; 4. aqueous acid workup

14.12 A 2-bromo-2-methylpropane; B 2-methylpropene; C 1-chloro-2-methyl-2-propanol

14.13 The more electron-rich (tetra-substituted) double bond is the one which is more reactive toward the electrophilic peroxyacid reagent.

14.14 *cis*-2,5-dimethyltetrahydrofuran This is a *meso* compound, and thus is optically inactive.

14.15 (a) 3-methoxy-2-butanol
(b) 1-phenyl-1-propanol

14.16 (a) 1. *m*-chloroperoxybenzoic acid, methylene chloride, heat; 2. sodium ethoxide

(b) 1. mercuric acetate, water; 2. sodium borohydride, aqueous sodium hydroxide; 3. sodium hydride; 4. iodomethane

(c) 1. H_3B-THF; 2. hydrogen peroxide, water, sodium hydroxide; 3. phosphorus tribromide; 4. Mg, ether; 5. aqueous acid

(d) 1. HBr; 2. Mg, ether; 3. ethylene oxide, acid workup; 4. chromic anhydride, pyridine

CHAPTER 15
ALKADIENES AND ALKYNES

Key Points

• Learn how to name dienes and alkynes. The systematic naming of these compounds should be easy to master as the approach used is very much like the approach used to name alkenes. Be careful, however, to specify E or Z, when appropriate, for *each* double bond of a diene.

Practice Problem 15.1
Name each of the following compounds:
   (a)   H$_3$C≡C-CH$_2$-CH$_2$Br

   (b)   [skeletal structure]

   (c)   [skeletal structure]

   (d) the lowest molecular weight conjugated diene for which it is necessary to specify E or Z designators for each double bond

Practice Problem 15.2
Draw skeletal structures for each of the following compounds:
   (a) 4-bromo-2-pentyne
   (b) (2E,4Z)-5-methyl-2,4-heptadiene

• Understand the application of bonding models to alkynes, and be familiar with how they are used to help us to rationalize many aspects of the chemistry of alkynes. Be aware of the enhanced acidity of terminal alkynes.

Practice Problem 15.3
In each case, choose the substance with the indicated property, and give an explanation for your choice.
   (a) more strained: *cis*-cyclooctene or cyclooctyne
   (b) reacts with CH$_3$MgBr to produce methane: dimethylacetylene or ethylacetylene
   (c) gives a red precipitate when treated with cuprous chloride in aqueous ammonia: 1-pentyne or 2-pentyne
   (d) contains the shorter C-H bond: 1-nonyne or cyclononyne
   (e) is a stronger acid than ammonia: 1-pentyne or 2-pentyne

• Learn the synthetic methods that are useful for the preparation of alkynes.

Practice Problem 15.4
Show a three-step synthesis of acetylene from ethanol.

Practice Problem 15.5
Suggest a synthesis of diethylacetylene from ethylacetylene.

Practice Problem 15.6
Suggest structures for compounds *A* to *C* as indicated below.

$$C_5H_{12}O \xrightarrow[H_2SO_4, \text{ heat}]{K_2Cr_2O_7} C_5H_{10}O \xrightarrow{PCl_5} C_5H_{10}Cl_2 \xrightarrow[\text{liquid NH}_3]{\text{excess NaNH}_2} \text{2-pentyne (only alkyne product)}$$

A, B, C

• Learn the synthetic methods that are used for the synthesis of conjugated dienes.

Practice Problem 15.7
Suggest a series of reactions that will convert 2-methyl-3-phenyl-2-butene to 2-methyl-3-phenyl-1,3-butadiene.
Practice Problem 15.8
How could you convert cycloheptene to 1,3-cycloheptadiene?

- Learn the important addition reactions of alkynes and of conjugated dienes. Particular points to study are:
    - We use poisoned catalysts for selective hydrogenation of alkynes to alkenes.
    - A different stereochemistry of reduction results depending on whether catalytic hydrogenation or dissolving metal reduction were used.
    - The hydration of alkynes produces enols that tautomerize to carbonyl compounds.
    - The hydration reaction can be made to proceed with either of two orientations depending on the reaction conditions used.
    - 1,2- and 1,4-addition can occur when conjugated dienes undergo addition reactions.

Practice Problem 15.9
Name the alkyne corresponding to each of the following descriptions:
    (a) the alkyne of formula C_4H_6 that will give the *same* product (of formula C_4H_8) by either catalytic hydrogenation or dissolving metal reduction (Li/liquid ammonia).
    (b) the alkyne of formula C_4H_6 that will give the *same* product (of formula C_4H_8O) following *either* treatment with disiamylborane followed by hydrogen peroxide/aqueous sodium hydroxide workup *or* by treatment with mercuric sulfate with aqueous sulfuric acid.

Practice Problem 15.10
Name the products resulting from 1,2- and 1,4-addition of HBr to (2Z,4E)-3,4-dimethyl-2,4-hexadiene. Which experimental conditions will favor the formation of the 1,4-addition product?

Practice Problem 15.11
Consider the addition of HBr to (2E,4E,6E)-2,4,6-octatriene. In principle, 1,2-, 1,4-, or 1,6-addition could occur. What mode of addition does *not* result in the formation of a conjugated diene?

- Learn the synthetic utility and stereochemical aspects of the Diels-Alder reaction.

Practice Problem 15.12
What are the products of each of the following Diels-Alder reactions?

(a)

(b)

Practice Problem 15.13
Which diene and which dienophile would you use for the synthesis of each of the following compounds by way of a Diels-Alder reaction?
    (a)   (b)

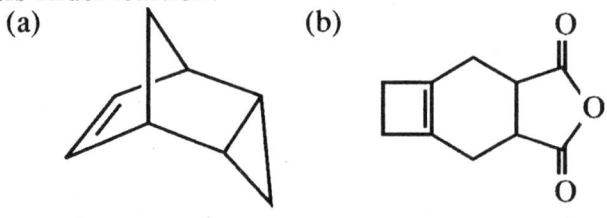

Solution of Text Problems

15.1 (a), 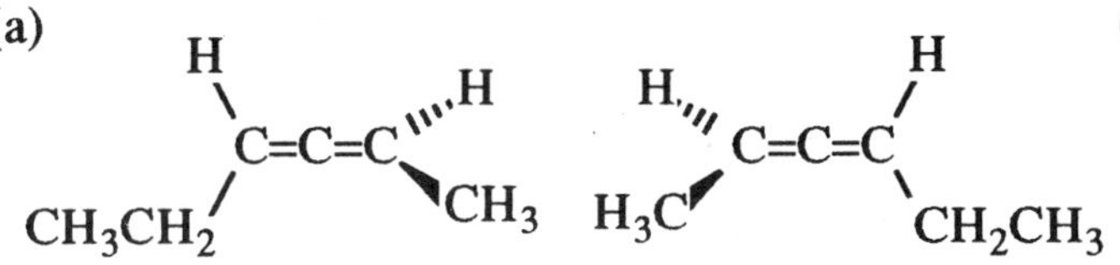 (b), 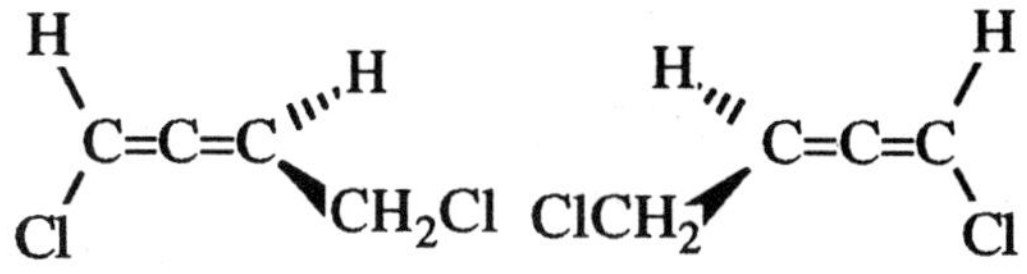 (c) 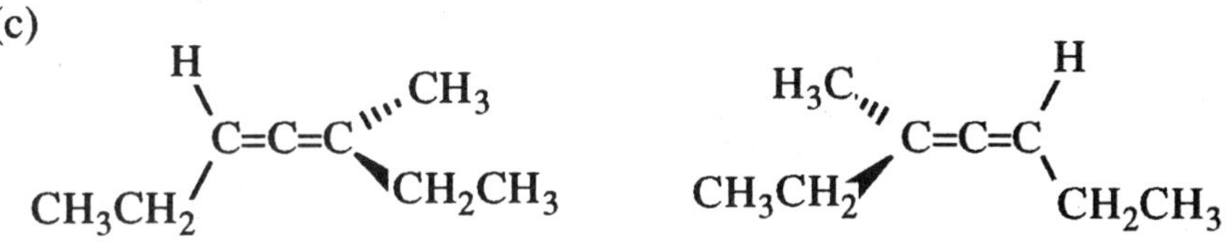 [structures shown]

15.2 (a), 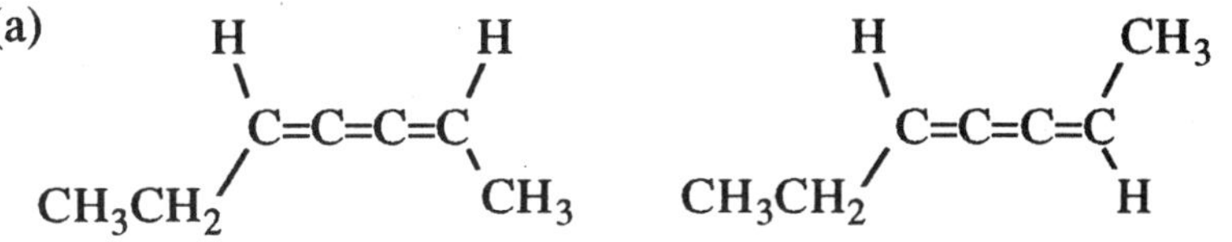 (b) 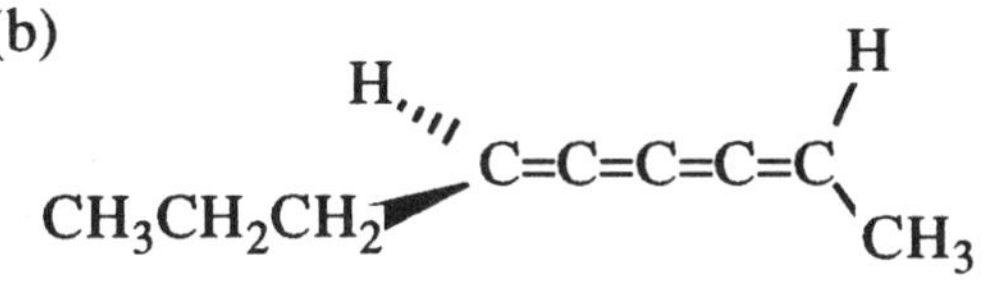 + enantiomer, (c) 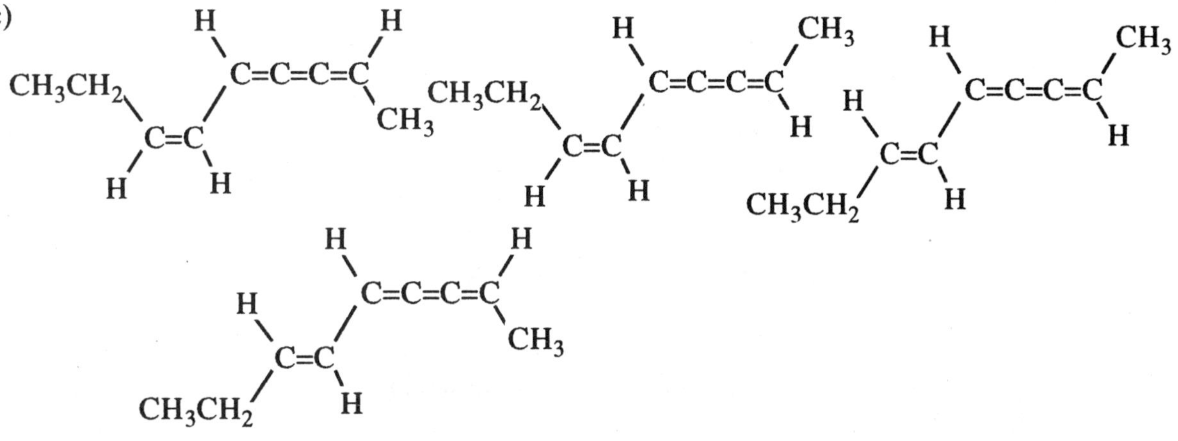 [structures shown]

15.3 $H_2O > NH_3 > CH_3C\equiv CH > CH_3C\equiv CCH_3$

15.4 The sodium amide would immediately react with the water or alcohol to generate hydroxide or alkoxide ion and ammonia.

15.5 1. propene is treated with bromine in carbon tetrachloride; 2. treatment with sodium amide

15.6 2,2-dimethyl-3-hexyne and 5,5-dimethyl-2-hexyne

15.7 1. 3-pentanone treated with phosphorus pentachloride to generate 3,3-dichloropentane; 2. treatment with sodium amide

15.8 The major products are 2-methylpropene and acetylene (ethyne).

15.9 (a) 1. treatment of propyne with sodium amide; 2. addition of 1-bromopropane

(b) 1. treatment of acetylene with sodium amide; 2. addition of iodomethane; 3. treatment with sodium amide; 4. addition of 1-bromopropane

(c) 1. treatment of acetone with phosphorus pentachloride; 2. treatment with sodium amide to generate propyne; 3. treatment with sodium amide to generate the methylacetylide salt; 4. addition of 1-bromopropane

15.10

CH₃CH₂C≡C—H + ⁻:NH₂ → CH₃CH₂C≡C:⁻ + :NH₃

CH₃CH₂C≡C:⁻ + [epoxide :O:] → CH₃CH₂C≡CCH₂CH₂Ö:⁻ —H⁺→ CH₃CH₂C≡CCH₂CH₂ÖH

15.11 Only the conjugated diene (1,3-cyclohexadiene) is obtained. Neither the alkyne nor the allene can form because of ring strain which would be present in the six-membered ring.

15.12 (a) 1,3-butadiene, 1,2-butadiene, and 2-butyne
(b) 1,2-butadiene, 2-butyne, and 1-butyne

15.13 3,4-dimethyl-2,4-hexadiene (mixture of three geometric isomers)

15.14 (E)-2-methyl-1,3-pentadiene, (Z)-2-methyl-1,3-pentadiene, and 4-methyl-1,3-pentadiene

15.15 acetone

15.16

Ph	Ph	Ph
C₃H₇——OH	C₃H₇——OH	HO——C₃H₇
C₃H₇——OH	HO——C₃H₇	C₃H₇——OH
Ph	Ph	Ph
meso	racemate

We would not expect these products to be all formed in equal amounts. As the energies of the conformations leading to each of the *meso* and racemic product are different, there will be different activation energies for the formation of each and the product distribution will not be equal. Three dienes would be formed in the double dehydration of the mixture of diols, as shown below.

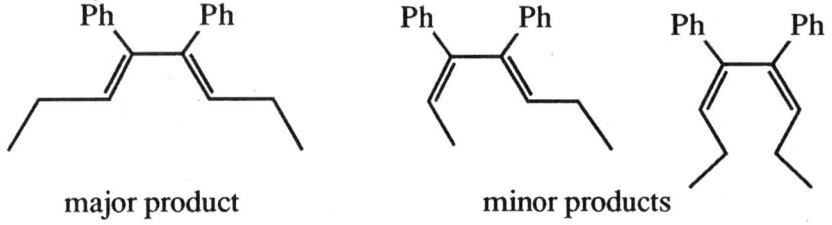

major product   minor products

15.17 There would be a high concentration of bromine present which would undergo ordinary addition to the carbon-carbon double bond.

15.18 The three products we isolate are: 3-bromo-1-hexene, (E)-1-bromo-2-hexene, and (Z)-1-bromo-2-hexene. The 3-bromo-1-hexene will be formed as the major product. The three products result as a consequence of the resonance stabilized intermediate allylic radical which has radical character at both the 1- and 3-positions. With cyclohexene the alkene is symmetrical and only 3-bromocyclohexene is formed from the reaction at either allylic site.

15.19 1. propene treated with bromine in carbon tetrachloride; 2. treatment with sodium amide to generate propyne; 3. treatment with sodium amide to generate the methylacetylide anion; 4. addition of 3-bromopropene (The 3-bromopropene is generated by the treatment of propene with NBS under irradiation.)

15.20 One product is generated: (3R,4S)-3,4-dibromo-3,4-dichlorohexane.

15.21

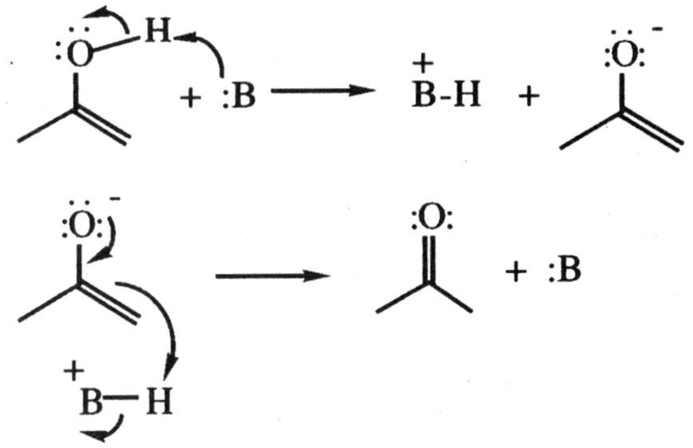

15.22 A mixture of 2-methyl-4-heptanone and 6-methyl-3-heptanone results in this reaction.

15.23 (Z)-2-chloro-2-butene

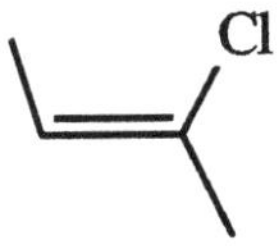

15.24 [cyclohexenyl-CN structure]

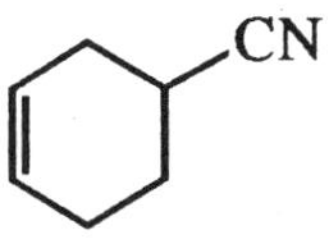

15.25 $H_2C=CH-C\equiv N: \longleftrightarrow H_2\overset{+}{C}-CH=C=\ddot{N}:^-$

15.26 [Diels-Alder reaction of hexachlorocyclopentadiene with norbornadiene → isodrin]

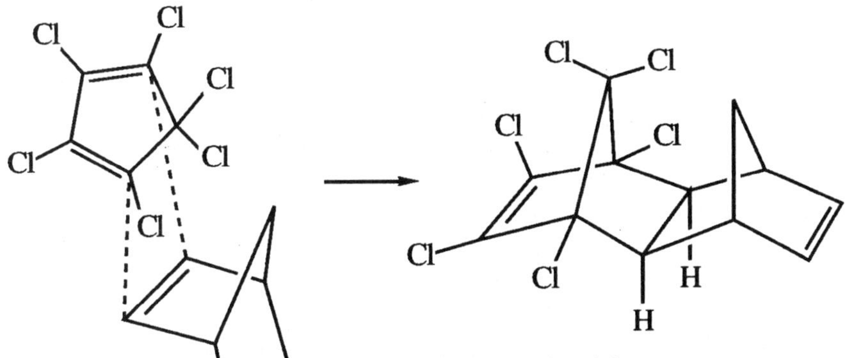

15.27

(a) [1-ethoxy-1,3-butadiene] + [methyl acrylate CO₂CH₃]

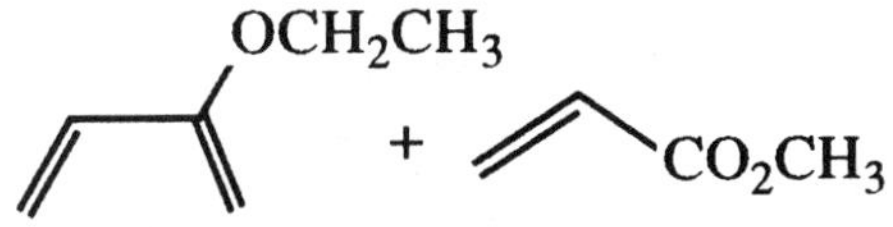

(b) [cyclopentadiene with CH₂OCH₃] + $H_2C=CHNO_2$

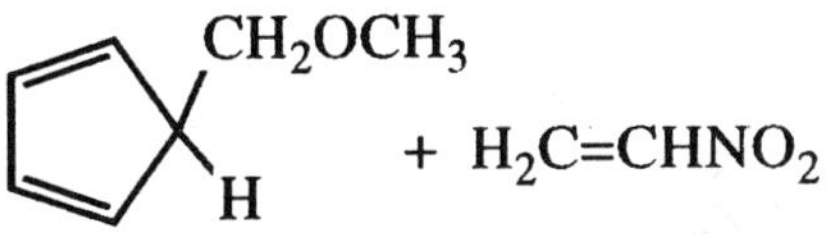

15.28 $CH_3O_2C-C\equiv C-CO_2CH_3$

15.29 [1,3-butadiene + 1,3-butadiene → 4-vinylcyclohexene] $\xrightarrow[\text{2. }H_2O_2]{\text{1. }O_3}$ $HO_2C-CH(CO_2H)-CH_2-CO_2H$ (HO₂C, HO₂C) + CO₂

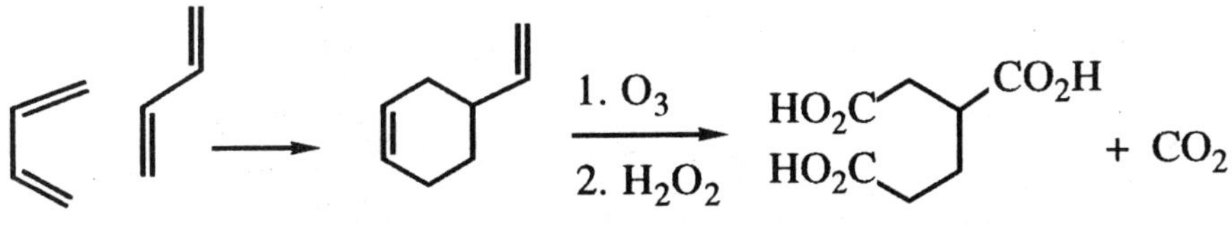

15.30

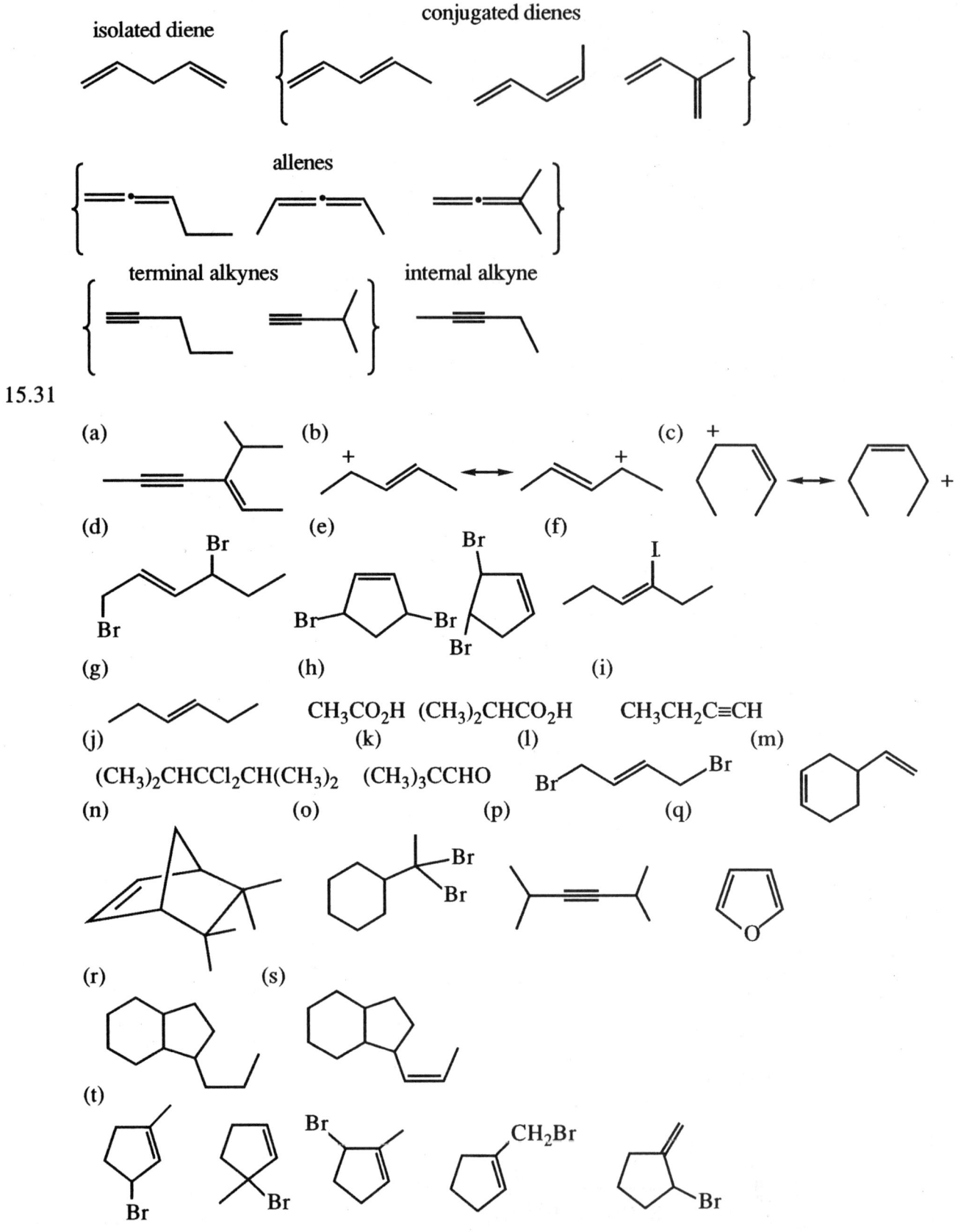

isolated diene / conjugated dienes / allenes / terminal alkynes / internal alkyne

15.31

(a)–(t) [structures]

15.32 Product *B*, thermodynamically more stable than product *A*, is favored at high temperature, but not at low temperature owing to the higher activation energy for its formation from *C*.

15.33 The three isomers are: (*E,E*)-2,4-hexadiene, (*Z,Z*)-2,4-hexadiene, and (*E,Z*)-2,4-hexadiene. The (*E,E*)-2,4-hexadiene would be most reactive in a Diels-Alder process because it can most easily adopt the required *cisoid* conformation.

15.34 (a) [structure: H₂C=C(H)(Cl) ↔ H₂C-C(H)(Cl)⁺ resonance] (b) [structure: oxocarbenium ion resonance in tetrahydropyranyl cation]

15.35 (a) (*E*)-4-methyl-1,3-hexadiene treated with HBr at low temperature
(b) 1,3-butadiene heated with H₂C=CH-CO₂CH₂CH₃

15.36 (a) treatment of 1-pentyne with mercuric sulfate in aqueous sulfuric acid
(b) 1. treatment of 1-pentyne with (Sia)₂BH; 2. NaOH, water, hydrogen peroxide
(c) treatment of 1-butyne with methylmagnesium bromide

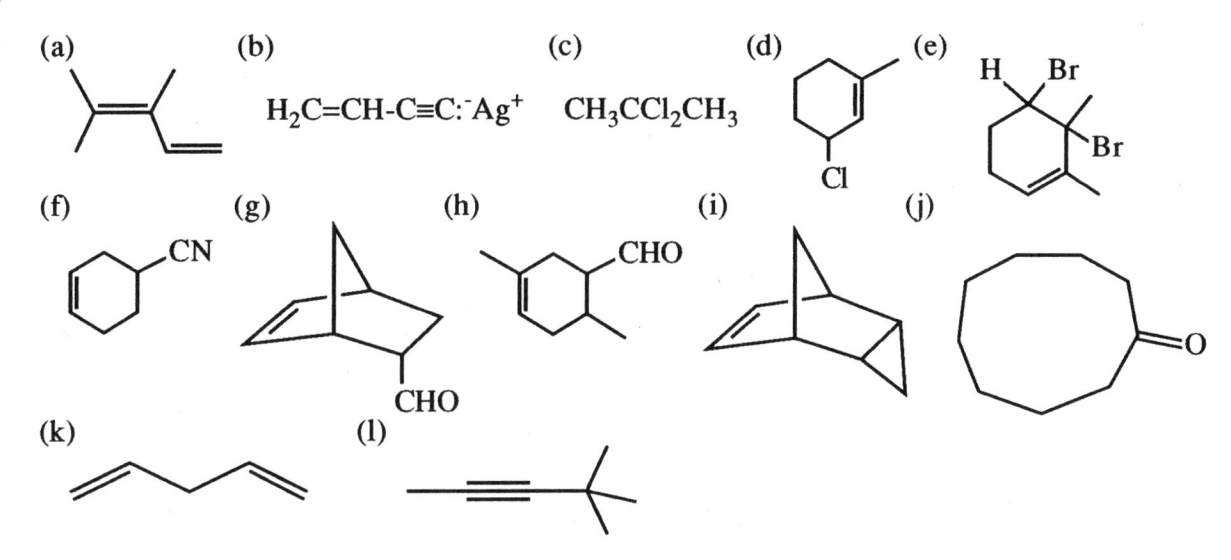

15.37 (a) [structure] (b) H₂C=CH-C≡C:⁻Ag⁺ (c) CH₃CCl₂CH₃ (d) [3-chloro-1-methylcyclohexene] (e) [1,2-dibromo-cyclohexene structure]
(f) [cyclohexenyl-CN] (g) [norbornene with CHO] (h) [methylcyclohexene with CHO] (i) [norbornene derivative] (j) [cyclononanone]
(k) [1,4-pentadiene] (l) [4,4-dimethyl-2-pentyne]

15.38 (a) HBr
(b) 1. sodium amide; 2. iodomethane; 3. sodium amide; 4. iodomethane; 5. sodium in liquid ammonia
(c) 1. sodium amide; 2. iodoethane; 3. sodium amide; 4. iodoethane
(d) 1. sodium amide; 2. iodoethane; 3. mercuric sulfate, sulfuric acid, water
(e) *N*-bromosuccinimide; 2. sodium amide
(f) 1. heat wiyh sodium amide; 2. sodium amide; 3. 1-iodopropane
(g) 1. sodium amide; 2. iodoethane; 3. hydrogen, platinum oxide
(h) 1. chromic anhydride, pyridine; 2. phosphorus pentachloride; 3. sodium amide
(i) 1. hydrogen, Lindlar's catalyst; 2. potassium permanganate, potassium hydroxide, water
(j) 1. phosphorus pentachloride; 2. sodium amide; 3. borane; 4. sodium hydroxide, hydrogen peroxide, water; 5. phosphorus tribromide
(k) 1. thionyl chloride; 2. sodium amide; 3. bromine; 4. heat with sodium amide; 5. sodium amide; 6. chloroethane (from treatment of ethanol with thionyl chloride); 7. borane; 8. sodium hydroxide, hydrogen peroxide, water; 9. phosphorus tribromide
(l) 1. bromine; 2. heat with sodium amide; 3. sodium in liquid ammonia
(m) for (Z): deuterium, Lindlar's catalyst; for (E): sodium in liquid ammonia
(n) 1. sodium amide; 2. iodomethane; 3. sodium in liquid ammonia
(o) 1. iodomethane; 2. water
(p) 1. disiamylborane; 2. hydrogen peroxide, potassium hydroxide, water

15.39 1. treatment of acetylene with sodium amide; 2. addition of 1-bromooctane (from the treatment of 1-octanol with phosphorus tribromide); 3. sodium amide; 4. addition of 1-bromotridecane (from the reaction of 1-tridecanol with phosphorus tribromide; sodium in liquid ammonia

15.40 Treatment of 1,4-pentadien-3-ol with acid generates a resonance stabilized carbocation, delocalizing the positive charge over positions 1, 3, and 5 of the pentadienyl chain. Bromide ion adds to the terminal position of this carbocation generating the more stable of the possible adducts.

136 Study Guide and Solutions Manual

15.41

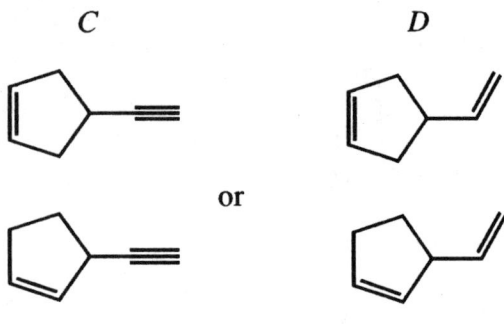

C or D

15.42

15.43 E, F, G

15.44

(a) H, I, J, K

(b) (CH₃)₂CHCHO   HCOCH₂CHO   (CH₃)₂C=O
(c) cyclobutene
(d) 2-methylpropene   The remaining carbon is lost as carbon dioxide.

15.45

Q →(base) [cyclopentadienone] →(Diels-Alder dimerization) R

15.46

(a)   (b)

15.47

(a) S, U

(b) T, V

Solution of Study Guide Practice Problems

15.1
   (a) 3-bromo-1-propyne
   (b) (E)-2,7-dimethyl-2,5-nonadiene
   (c) (3Z,6E)-3-methyl-3,6-nonadiene
   (d) 2,4-hexadiene

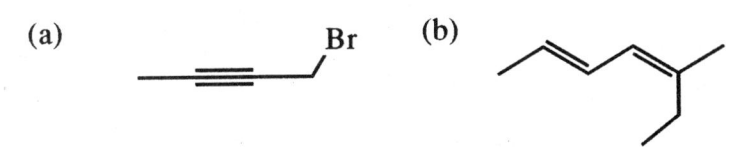

15.2 (a) [structure: HC≡C–CH₂–Br] (b) [structure: CH₃–CH=CH–C(CH₃)=CH–CH₃ type diene]

15.3 (a) cyclooctyne A triple bond introduces more strain into a small ring than does a double bond because with an alkyne linkage four atoms are constrained to be along a straight line.
   (b) ethylacetylene Terminal alkyne linkages react with strong bases, such as Grignard reagents, but internal alkyne linkages do not so react.
   (c) 1-pentyne Terminal alkynes react with Ag(I) and Cu(I) species to form salts, but internal alkynes do not so react.
   (d) 1-nonyne The C-H bond can be considered to arise from the overlap of a hydrogen 1s orbital with a carbon sp hybridized orbital. The large fraction of s character in the hybrid orbital correlates with a stronger bond.

15.4 1. Ethanol is dehydrated to ethene by treatment with sulfuric acid; 2. Ethene is treated with bromine in carbon tetrachloride to generate 1,2-dibromoethane; 3. The 1,2-dibromoethane is subjected to treatment with an excess of sodium amide to eliminate two equivalent amounts of hydrogen bromide and form acetylene.

15.5 1. treatment of ethylacetylene with sodium amide; 2. addition of iodoethane

15.6 A $CH_3CH_2CH(OH)CH_2CH_3$ B $CH_3CH_2C(O)CH_2CH_3$ C $CH_3CH_2CCl_2CH_2CH_3$

15.7 1. treatment with bromine in carbon tetrachloride; 2. treatment with two equivalent amounts of a base such as sodium carbonate

15.8 1. treatment with N-bromosuccinimide under light irradiation; 2. treatment with sodium *tert*-butoxide

15.9
   (a) 1-butyne
   (b) 2-butyne

15.10 1,2-Addition will lead to an E/Z mixture of 4-bromo-3,4-dimethyl-2-hexene. 1,4-Addition will lead to an E/Z mixture of 2-bromo-3,4-dimethyl-3-hexene

15.11 1,4-Addition will not lead to a conjugated diene product.

15.12 (a) [cyclohexene with CO₂H substituent] (b) [decalin-type structure with NO₂ substituent]

15.13
(a) [cyclopentadiene] + [cyclopropene (triangle with double bond)]

(b) [1,2-dimethylenecyclobutane] + [maleic anhydride]

PRACTICE EXAMINATION THREE

A time limit of 90 minutes should be set for completion of this entire practice examination. Answers should be written out completely as they would be when presented for independent grading. No text or supplemental materials should be consulted during the testing period, and you should not check your answers until you have worked out the complete examination and the time limit has been reached.

1. For each part provide structures *and* systematic IUPAC names (3 points for each part):
   (a) the product obtained by treating cyclohexanone with ethylmagnesium bromide, followed by workup with aqueous acid.
   (b) a compound of formula $C_6H_{13}Cl$ that on E2 dehydrohalogenation yields the thermodynamically most stable C_6H_{12} isomer.
   (c) the product obtained by treating *trans*-1,2-dimethylcyclopropane with bromine in carbon tetrachloride solution.
   (d) a compound of formula $C_7H_{14}O$ that on treatment with an excess of concentrated HI yields 1,5-diiodo-3,3-dimethylpentane.
   (e) the open-chain C_6H_{10} isomer that reacts *most readily* with methyl acrylate, $H_2C=CH-CO_2CH_3$, to produce a cyclohexene derivative.

2. For each part, give the *numerical* value requested (2 points for each part):
   (a) the C-C-C bond angle in cyclohexane.
   (b) the dihedral angle between the adjacent C-Cl bonds in *trans*-2,3-dichlorobicyclo[2.2.1]heptane
   (c) the number of equatorial methyl substituents in the more stable chair conformation of the structure shown below.

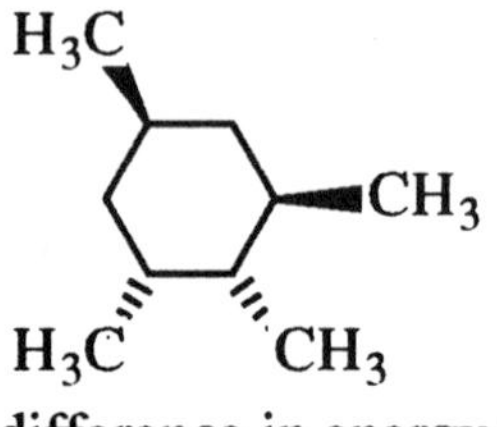

   (d) the difference in energy between the two chair conformations of *trans*-1,2-dimethylcyclohexane.
   (e) the ratio of the rate of reaction A to reaction B (shown below).
   reaction A: a reaction solution in DMSO which is 0.30 M in ICH_3 and 0.080 M in NaCN
   reaction B: a reaction solution in DMSO which is 0.15 M in ICH_3 and 0.020 M in NaCN

3. In each case, choose which reaction (if any) is the faster, and give a rationalization of your choice (4 points for each part):
   (a) 2,2-dimethyl-1-iodobutane *or* 1-iodobutane reacting with KCN in acetone.
   (b) 2-bromo-2-phenylpropane reacting with 0.010 M KI in ethanol *or* with 0.020 M KI in ethanol.
   (c) 1-propanol reacting with KI in water *or* with HI in water.
   (d) 1-chlorobutane reacting with KCN in aqueous acetone *or* with KCN in DMSO.
   (e) *cis*- or *trans*-1-chloro-4-*tert*-butylcyclohexane reacting with potassium ethoxide in ethanol to give 4-*tert*-butylcyclohexene.

4. Compound A, of formula C_5H_8, is converted to compound B, of formula C_5H_{10}, when hydrogenated in the presence of Lindlar's catalyst. Upon treatment with cold aqueous basic potassium permanganate solution compound B yields compound C, of formula $C_5H_{12}O_2$, but compound B yields compound D, an isomer of C, when treated with hydrogen peroxide and formic acid in aqueous solution. Compound A also yields compound E, of formula $C_5H_{10}O$ upon treatment with mercuric sulfate in aqueous sulfuric acid. Give structures for each of the compounds A-E and tell which can be resolved into optically active forms (4 points for each structure).

5. The reactions of a substance F, of formula $C_5H_{11}Br$ but unknown structure, are investigated. It is found that under no circumstances will it undergo E2 elimination when treated with sodium ethoxide. Rather, it is slowly

converted into compound G, of formula $C_7H_{16}O$. When compound F is treated with ethyl alcohol it is slowly converted into compound H (a structural isomer of compound G) along with 2-methyl-2-butene. In light of these data, answer the following questions (5 points for each part).
    (a) What mechanistic type is the conversion of compound F to compound G?
    (b) What are the structures and names of compounds F-H?
    (c) What are the mechanistic steps in the conversion of compound F to the mixture of compound H and 2-methyl-2-butene?

6. Propose reaction routes for each of the conversions shown below (4 points for each part).
    (a) cyclohexene to *trans*-1-ethoxy-2-methoxycyclohexane
    (b) any alcohol and any haloalkane to *tert*-butyl ethyl ether
    (c) $PhCH=CH_2$ to $Ph-C\equiv C-CH_2CH_3$
    (d) ethene (only organic reagent permitted) to $CH_3CH_2CH_2CO_2H$
    (e) 2-butyne to *meso*-2,3-dibromobutane (free of other isomers, including stereoisomers)

CHAPTER 16
MOLECULAR ORBITAL CONCEPTS IN ORGANIC CHEMISTRY: CONJUGATED π SYSTEMS AND AROMATICITY

Key Points

• Review the application of the molecular orbital model to π bonding in simple alkenes.

Practice Problem 16.1
Offer an explanation, based on the molecular orbital model, for the observation that a *trans* alkene is converted into a mixture of *cis* and *trans* isomers on irradiation with light of the appropriate wavelength.

Practice Problem 16.2
Would you expect an electron-donating substituent bonded to a carbon atom of an alkene double bond to raise or lower the HOMO energy relative to an unsubstituted alkene? Explain your choice.

• Know how to combine *p* orbitals to construct π molecular orbitals of conjugated dienes, trienes, and so forth.

Practice Problem 16.3
In each of the following choose the item or items with the indicated property.
   (a) has the higher energy HOMO: 1,3-butadiene or 1,3,5-hexatriene
   (b) has a LUMO containing three vertical nodal planes: 1,3-butadiene or 1,3,5-hexatriene
   (c) has a node at its central carbon atom: the HOMO of the allyl cation or the LUMO of the allyl cation
   (d) contains the most stable occupied π molecular orbital: ethylene, 1,3-butadiene, or 1,3,5-hexatriene
   (e) contains the least stable unoccupied π molecular orbital: ethylene, 1,3-butadiene, or 1,3,5-hexatriene

• Be able to compare the energies of the π molecular orbitals of cyclic polyenes with those of the corresponding open-chain polyenes containing the same number of double bonds by consideration of the relative numbers of bonding and antibonding interactions. Appreciate that this leads to a rationalization of the Hückel ($4n + 2$) rule.

Practice Problem 16.4
Two π electrons occupy a molecular orbital of ground state 1,3-cyclobutadiene that is more stable than any π molecular orbital of 1,3-butadiene. How many vertical nodes are in this orbital of 1,3-cyclobutadiene? Compute the number of net π bonding interactions in this orbital and compare this number with the number of net π bonding interactions in the lowest energy molecular orbital of 1,3-butadiene. What is the percentage increase in the net number of π bonding interactions between these orbitals that occurs on orbital transformation from 1,3-butadiene to 1,3-cyclobutadiene?

Practice Problem 16.5
In the ground state of 1,3-cyclobutadiene there are two degenerate SOMO that are less stable than the HOMO of ground state 1,3-butadiene. How many vertical nodes are there in each of these SOMO? Compute the number of net π bonding interactions in each SOMO and compare this number with the number of net π bonding interactions in the HOMO of ground state 1,3-butadiene. What is the percentage decrease in the net π bonding interaction that occurs on transformation of the HOMO of the ground state 1,3-butadiene to the SOMO of ground state 1,3-cyclobutadiene?

Practice Problem 16.6
Assuming that the increase or decrease in stability of a π molecular orbital on orbital transformation from an open-chain to a cyclic form correlates with the percentage increase or decrease in the number of net bonding interactions, would you predict a net π stabilization or destabilization on orbital transformation from 1,3-butadiene

to 1,3-cyclobutadiene?

- Understand how to count π electrons to be able to predict aromatic, antiaromatic, or non-aromatic character of ions and molecules, and appreciate the consequences of aromatic and antiaromatic character.

Practice Problem 16.7
Consider the molecule shown below. Would you expect this molecule to be planar? Would you expect this molecule to be aromatic? Explain your choices.

Practice Problem 16.8
For each of the following pairs, predict which would be the stronger base. Explain your choices.

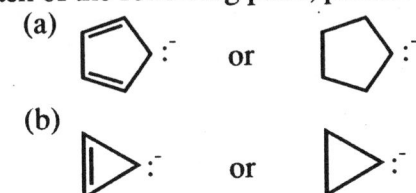

(a)

(b)

Practice Problem 16.9
Explain why pyrimidine and pyridine have comparable basicities, while imidazole is a much stronger base than is pyrrole.

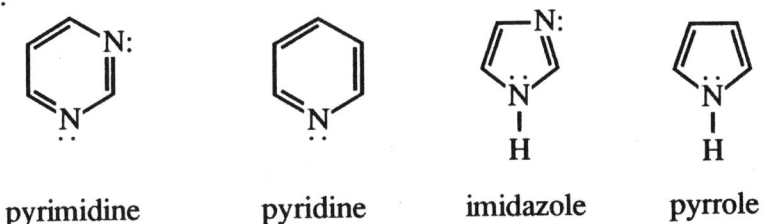

pyrimidine   pyridine   imidazole   pyrrole

Solution of Text Problems

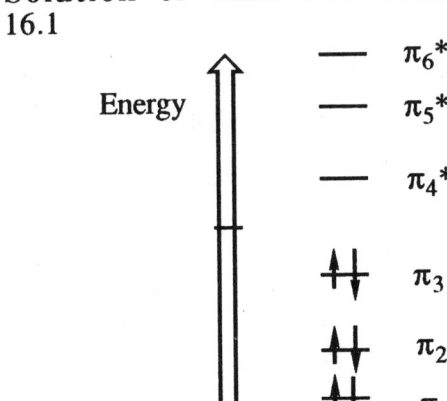

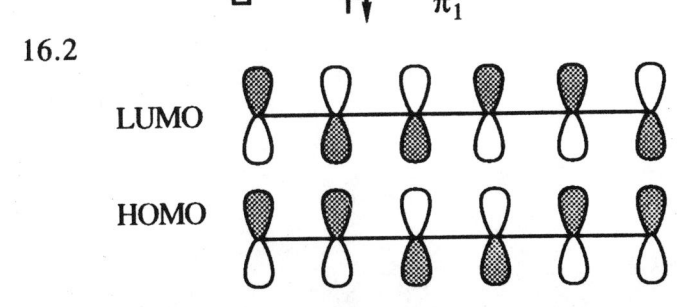

16.1

Energy ↑

— π_6^*
— π_5^*
— π_4^*
↑↓ π_3
↑↓ π_2
↑↓ π_1

16.2

LUMO

HOMO

142 Study Guide and Solutions Manual

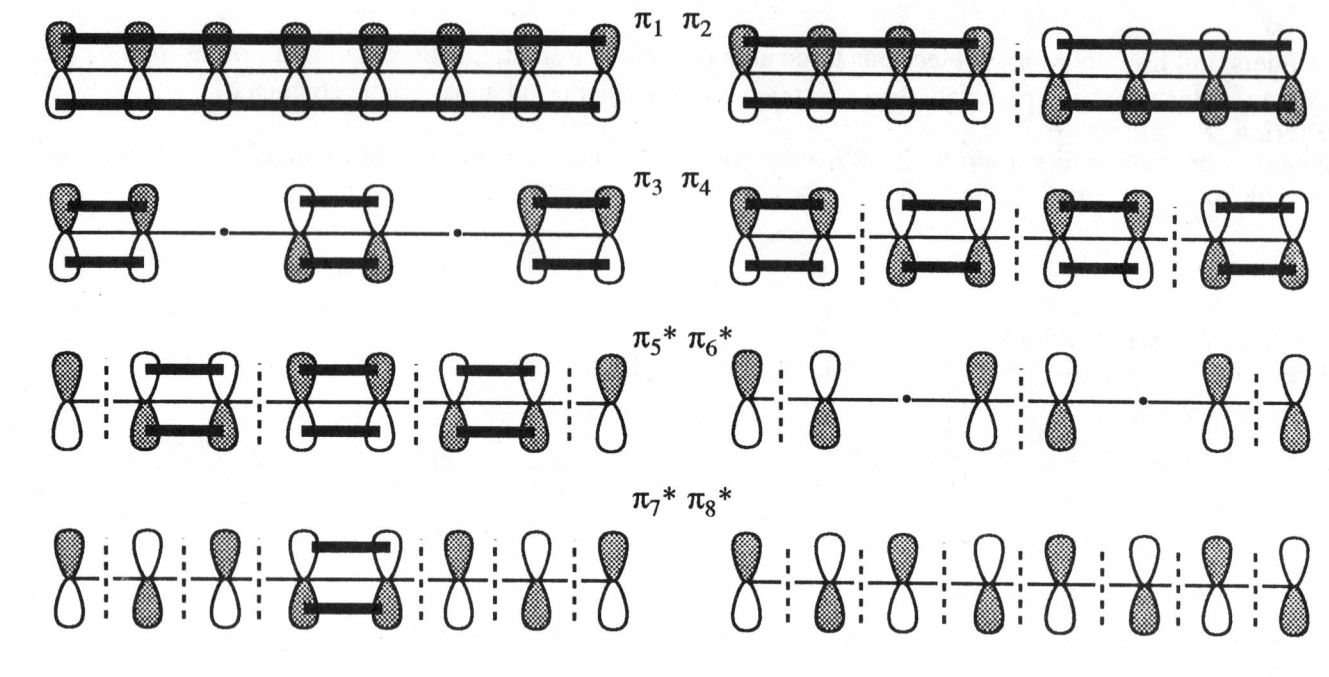

16.3

[Molecular orbital diagrams showing π₁, π₂, π₃, π₄, π₅*, π₆*, π₇*, π₈*]

16.4

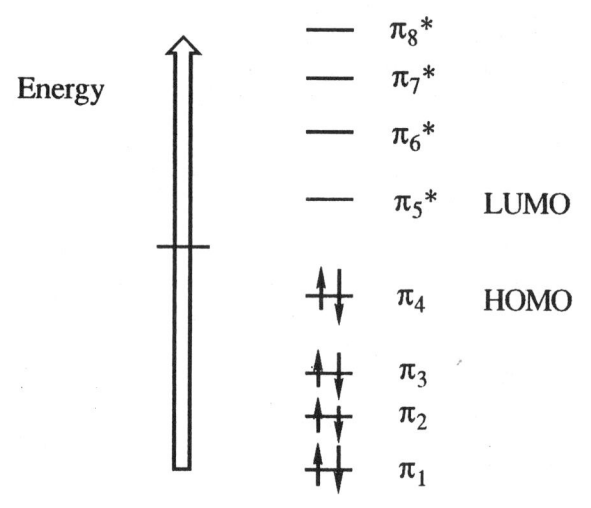

Energy ↑

— π₈*
— π₇*
— π₆*
— π₅*   LUMO

⇅ π₄   HOMO

⇅ π₃
⇅ π₂
⇅ π₁

16.5

(a)

$$H_2C=C(Br)-\overset{+}{C}(H)-CH_3 \longleftrightarrow H_2\overset{+}{C}-C(Br)=C(H)-CH_3$$

(b)

$$CH_3-\overset{\cdot}{C}(H)-CH=CH_2 \longleftrightarrow CH_3-C(H)=CH-\overset{\cdot}{C}H_2$$

16.6 $H_2{}^{14}C=CHCH_2Br$ and $H_2{}^{14}C(Br)-CH=CH_2$

16.7

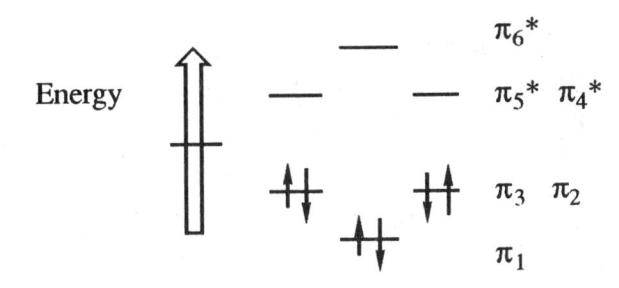

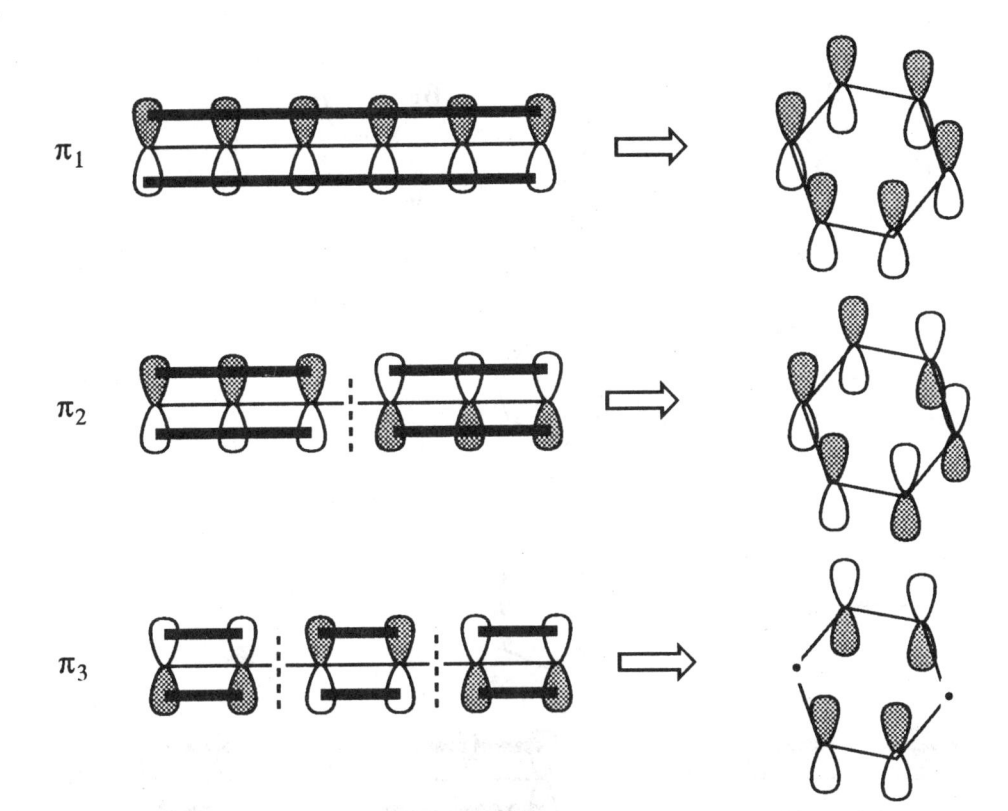

[Energy diagram: π₁ (↑↓) lowest; π₂, π₃ (both ↑↓) degenerate above; π₄*, π₅* degenerate higher; π₆* highest]

16.8

π₁ ⇒ [all p-orbitals in phase around hexagon]

π₂ ⇒ [one nodal plane]

π₃ ⇒ [two nodal planes, nodes through atoms]

16.9

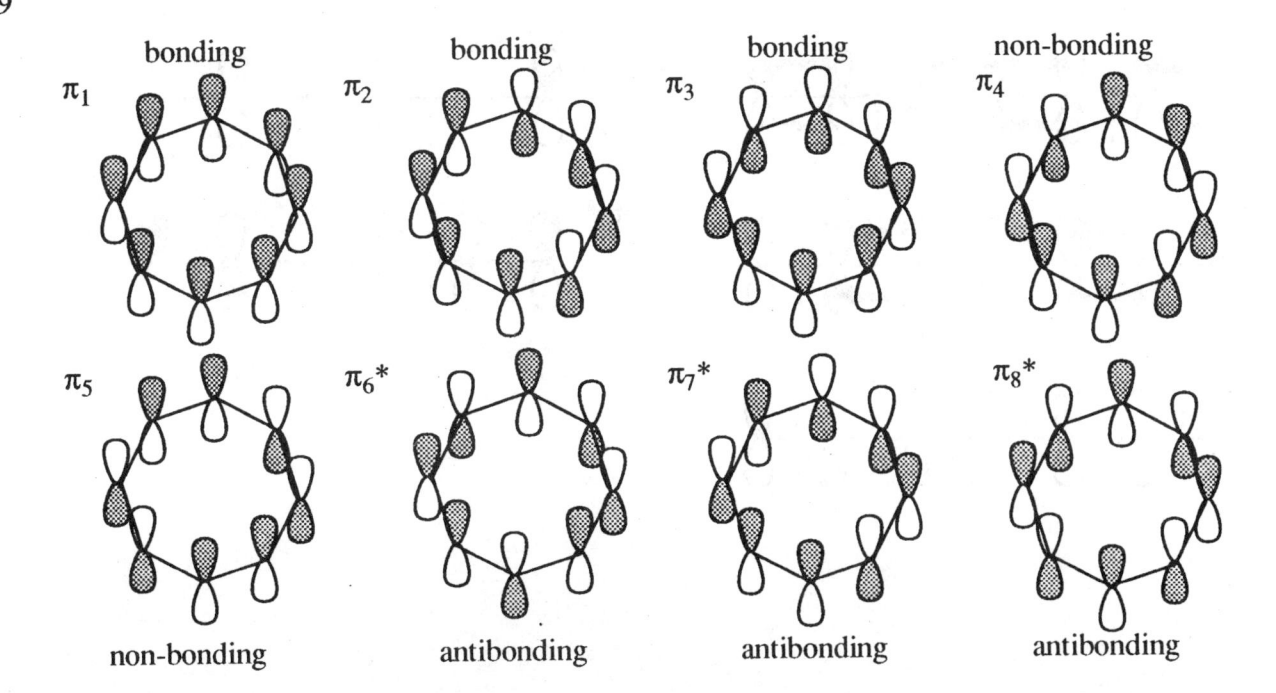

π₁ bonding   π₂ bonding   π₃ bonding   π₄ non-bonding

π₅ non-bonding   π₆* antibonding   π₇* antibonding   π₈* antibonding

144 Study Guide and Solutions Manual

16.10

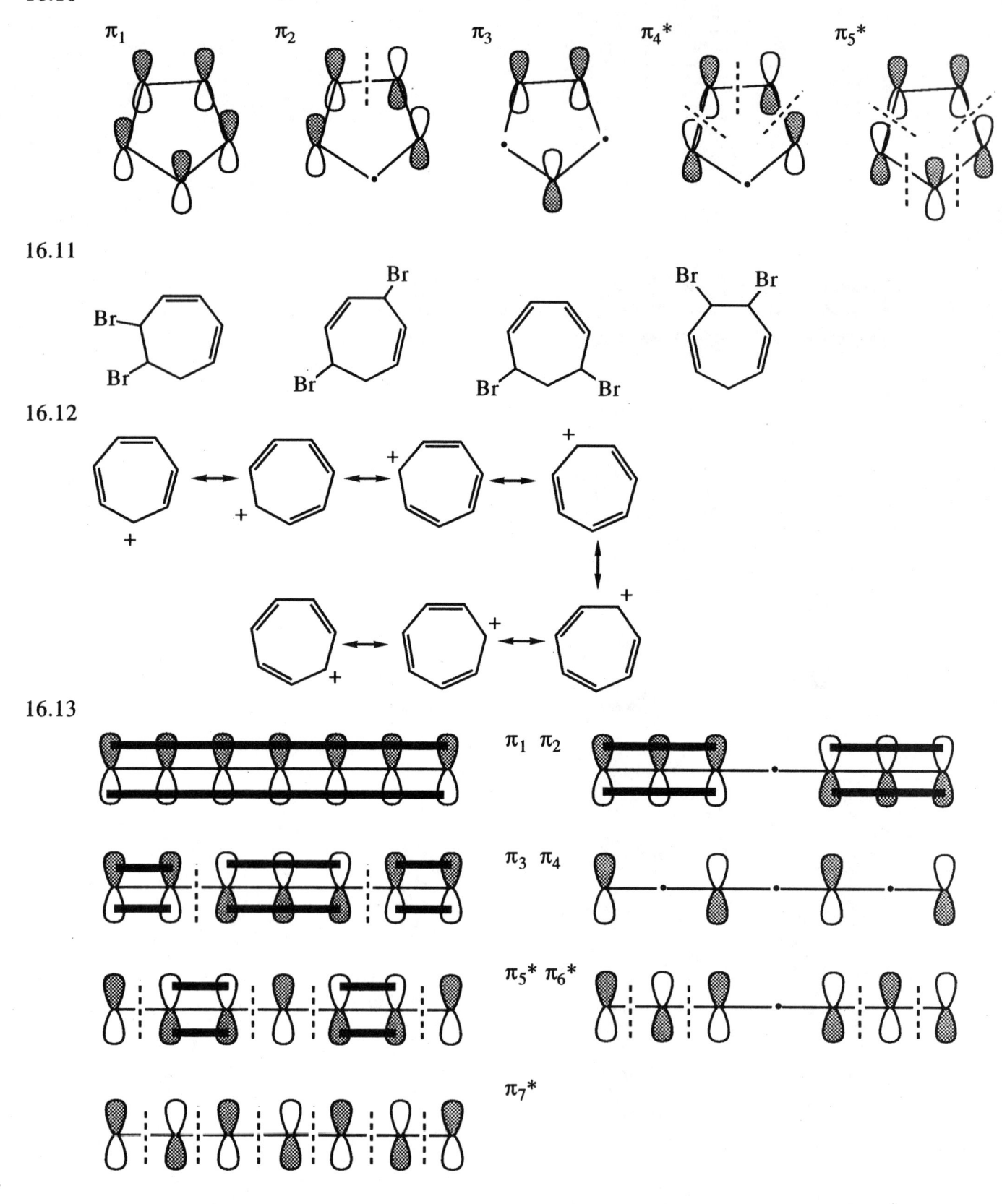

π_1 π_2 π_3 π_4^* π_5^*

16.11

16.12

16.13

π_1 π_2

π_3 π_4

π_5^* π_6^*

π_7^*

16.14

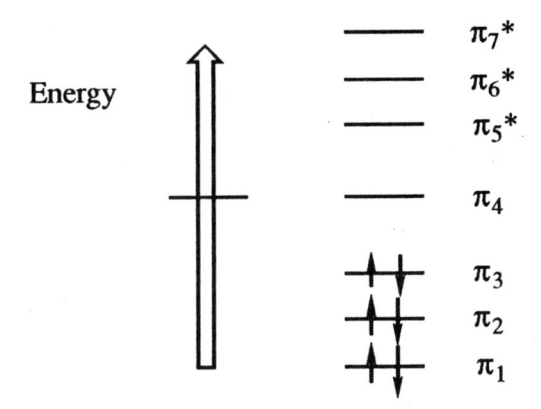

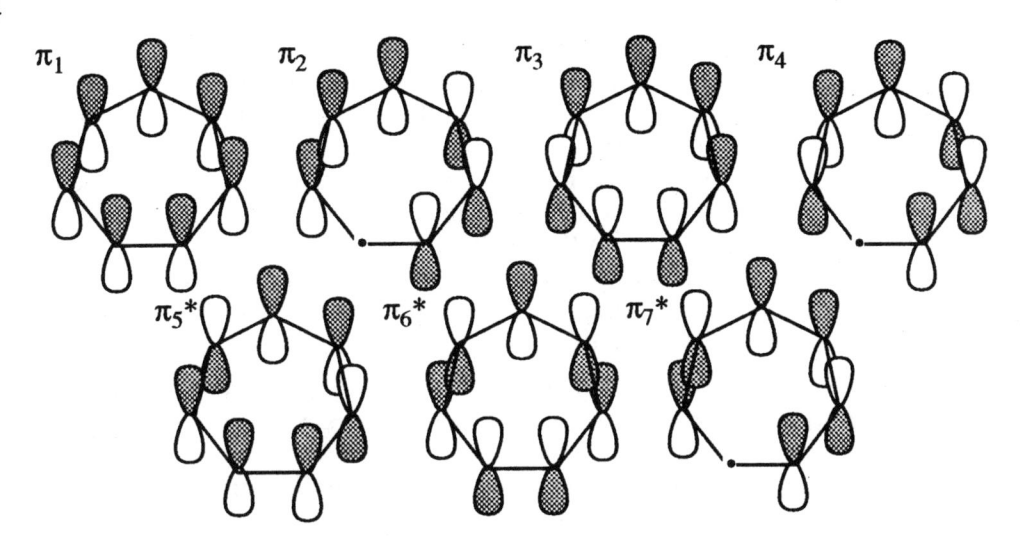

There is extra stabilization (an extra bonding interaction) in π_1 and π_3, although there is one greater antibonding interaction in π_2.

16.15

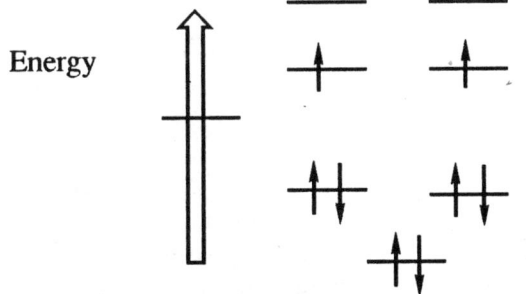

The cycloheptatrienyl anion is destabilized relative to the cycloheptatrienyl cation. The two highest energy electrons in the cycloheptatrienyl anion, those in SOMO, are in antibonding orbitals, whereas no electrons are in antibonding levels with the cycloheptatrienyl cation. Similarly, the cycloheptatrienyl anion is higher in energy than the 1,3,5-heptatrienyl anion, for which the two highest energy electrons are in a non-bonding orbital.

16.16 We would not expect the removal of a proton from cycloheptatriene to be a facile reaction as it would generate a cycloheptatrienyl anion (see the previous problem).

16.17

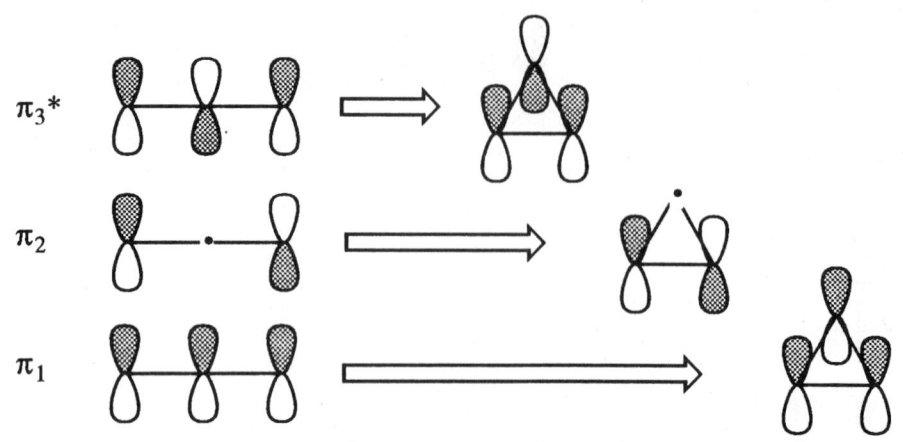

In the cyclopropenyl cation, only the lowest energy molecular orbital is occupied, and both of the antibonding molecular orbitals are empty. That is, all bonding levels are filled and all antibonding levels are empty. The maximum stabilization for the cyclic system is obtained.

16.18 The planar cyclobutadiene dianion has all of its bonding molecular orbitals filled, and no electrons are in any antibonding orbitals. In the 1,3-butadiene anion, a pair of electrons are in an antibonding orbital, as well as having the bonding molecular orbitals completely filled. For the cyclobutadiene dianion there is an extra bonding interaction compared to the 1,3-butadiene dianion, thus we conclude that the cyclobutadeine dianion is aromatic.

16.19 The hydrogens pointing toward the center of the ring system bump into each other and force the structure away from the planarity required for the system to be aromatic.

16.20 In pyridine the unshared electron pair on nitrogen is in an sp^2 orbital, orthogonal to the aromatic π system. In pyrrole, the unshared electron pair on nitrogen is in a p orbital which is parallel with the other p orbitals comprising the aromatic system.

16.21 A greater number of nodes indicates a greater energy for an orbital. A node represents a break between possible bonding interactions. When a node is present between adjacent atoms it actually results in a repulsive (antibonding) interaction.

16.22 (a) 3; (b) 1; (c) 2; (d) 1; (e) 3; (f) 4

16.23

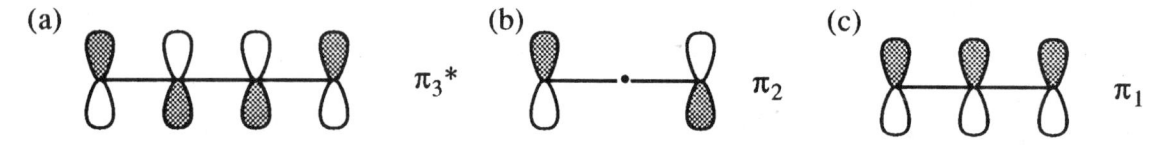

16.24 (a) 2 (1 vertical); (b) 1; (c) 1; (d) 0

16.25 (a) 8; (b) 5; (c) 2; (d) 2; (e) 4; (f) 6

16.26 The HOMO of the 1,3,5-heptatrienyl cation has 4 bonding and 2 antibonding interactions, providing a net 2 bonding interactions for the orbital. With the cycloheptatrienyl cation the pair of degenerate HOMO have a net of 3 bonding interactions each, indicating a greater stability than with the open-chain counterpart. For the 1,3,5-heptatrienyl anion the HOMO is a net non-bonding orbital whereas with the cycloheptatrienyl anion each of the pair of degenerate HOMO is net antibonding. Thereby, for the anion, the open-chain system is the more stable.

16.27 (a) Each of π_1 and π_3 contain two electrons and are stabilized, whereas π_2 contains two electrons and is destabilized for the transformation of the 1,3,5-heptatrienyl cation to the tropylium ion.

(b) The π_1 contains two electrons and is slightly stabilized, but π_2 also contains two electrons and is destabilized. Overall, there is a net destabilization in the transformation of the allyl anion to the cyclopropenyl anion.

(c) With the allyl cation being transformed to the cyclopropenyl cation electrons are present only in π_1 which is slightly stabilized. Thus there is net stabilization upon transformation.

16.28 C > B > A

16.29 (a) no; (b) no; (c) no; (d) no; (e) no; (f) yes; (g) no; (h) no; (i) yes; (j) yes; (k) no; (l) yes; (m) yes

16.30 The second reaction requires a lower pH (stronger acid). The first reaction is favored by having the product ion being stabilized, and occurs readily with a less acidic medium.

16.31 The charge separated structure here involves an aromatic ring system, whereas that without charge separation does not involve aromatic stabilization.

16.32
(a) [cyclopropenyl cation–O⁻ structure] (b) [cyclopropenyl cation–cyclopentadienyl anion structure] (c) [azulene-like cation structure]

16.33

[pyridine resonance structures showing three forms, with energy arrow]

[pyrrole resonance structures showing three forms, with energy arrow]

16.34
(a) [cyclopropenyl-H, Cl structure]   (b) [cycloheptatrienyl-H, Cl structure]

16.35 The reaction is endothermic. There is greater π stabilization for the cyclopropenyl cation than there is for the allyl cation.

16.36 Cyclopentadiene; the cyclopentadienyl anion is a resonance stabilized aromatic system whereas the cyclopropenyl anion is antiaromatic.

16.37 Pyridine; protonation does not disturb the aromatic electronic system with pyridine as it does with pyrrole.

16.38 For A the monocation is aromatic whereas for B the dication is aromatic.

16.39

[mechanism showing protonation of tertiary alcohol on cycloheptatriene, loss of water to form tropylium cation, and loss of (CH₃)₂C=CH₂]

when H–O–H is lost, a ⊕ is formed, loses a proton to form ↘

16.40 The reaction producing *p*-dihydroxybenzene occurs readily. The larger ring system involves cross-ring hydrogen-hydrogen steric interaction which keeps the system from becoming planar and aromatic.

16.41 Linear conjugated systems exhibit greater stabilization than do cross-conjugated systems.

Solution of Study Guide Practice Problems

16.1 Upon irradiation with light one electron from the (bonding) π_1 molecular orbital of *trans*-2-butene is excited to the (antibonding) π_2 molecular orbital. There is zero π bonding in this excited species and rotation about the carbon-carbon linkage is possible, occurring to generate a mixture of the *cis* and *trans* isomers upon loss of energy and reformation of the π bond.

16.2 An electron-donating substituent will raise the energy of the alkene π HOMO. One useful way to think about this is to realize that an electron-donating substituent will stabilize the cation remaining after an electron has been removed from the double bond. Thus, the ionization energy is *decreased*, and the the HOMO energy must have been increased.

16.3 (a) 1,3,5-hexatriene
(b) 1,3,5-hexatriene
(c) LUMO
(d) 1,3,5-hexatriene
(e) 1,3,5-hexatriene

16.4 There are zero vertical nodes in the π_1 of 1,3-cyclobutadiene. There is a 33 1/3% increase in the net number of π bonding interactions for these orbitals realized upon transformation.

16.5 There are two vertical nodes in each SOMO. There is a 100% decrease in the net π bonding interactions for these orbitals realized upon transformation.

16.6 There will be a net π stabilization.

16.7 This molecule would not be expected to be aromatic since it has 8 π electrons about the ring and would distort from planarity.

16.8 (a) The cyclopentyl anion would be the stronger base since it does not have aromatic stabilization.
(b) The cyclopropenyl anion is antiaromatic. Thus, we expect it to be a stronger base than the cyclopropyl anion.

16.9 Pyrimidine and pyridine are both readily protonated without disturbing the aromatic system, thus providing them with similar basic characteristics. Imidazole may be protonated at the nitrogen with the unshared electron pair orthogonal to the aromatic ring, but pyrrole has no such nitrogen and can not be protonated without disturbing the aromatic ring system.

CHAPTER 17
PHYSICAL METHODS OF STRUCTURAL ELUCIDATION: INFRARED AND NUCLEAR MAGNETIC RESONANCE SPECTROMETRY

Key Points

• Understand how an IR spectrum is measured, and also make sure you have a basic understanding of the fundamental theory.

Practice Problem 17.1
Use Hooke's law to predict which vibration in each group has the higher frequency.
  (a) N-H or N-D
  (b) C-N or C=N
  (c) C≡C or C≡N
  (d) C-O or C-S

Practice Problem 17.2
Write equations expressing:
  (a) the velocity of light (c) in terms of the wavelength (λ) and the frequency (v) of the light.
  (b) the energy of a photon in terms of Planck's constant, the velocity of light (c), and the wavelength (λ) of the light.

• Learn the characteristic IR absorptions associated with common functional groups (-OH, -NH, >C=O, -CO$_2$H, -C≡C-, -C≡N, >C=C<, *etc.*). Know the significance of *differences* in various frequencies, (*e.g.* C-H stretch above or below 3000 cm^{-1}), and know what factors contribute to an absorption band being strong or weak.

Practice Problem 17.3
From *memory* give the (approximate) IR absorbances characteristic of each of the following groups.
  (a) -O-H
  (b) >C=O
  (c) -CO$_2$H
  (d) -C≡C-
  (e) >N-H
  (f) C-O
  (g) H-C(sp^2)
  (h) H-C(sp^3)

Practice Problem 17.4
Which compound from each pair do you expect to give the larger peak for the indicated vibration?
  (a) -C≡C- stretch: 1-pentyne or 3-pentyne
  (b) C=C stretch: *cis*-1,2-dibromoethene or *trans*-1,2-dibromoethene

Practice Problem 17.5
In each case below an "unknown" compound is one of the two listed compounds. Specify a feature in the IR spectrum that would allow you to make a distinction between the possibilities.
  (a) cyclohexanone and 2-hexanone
  (b) methyl benzoate and propiophenone
  (c) ethylamine and trimethylamine
  (d) cyclohexene and 1-hexene
  (e) benzoic acid and benzaldehyde

(f) *cis*-3-hexene and *trans*-3-hexene
(g) cyclohexane and cyclohexene

Practice Problem 17.6
A chemist is performing a reaction in the laboratory. The reaction flask is filled with an orange solution that slowly turns green in color. Lying on the desktop is an IR spectrum labeled "reactant". This IR spectrum shows a strong band at ~3400 cm^{-1}. The chemist has also extracted a small amount of material from the progressing reaction and the IR spectrum of this material, also lying on the desktop, shows an absorption at ~3400 cm^{-1}, but which is decreased in intensity from that in the spectrum labeled "reactant", as well as a new band at ~1715 cm^{-1}. What type of reaction is the chemist performing?

Practice Problem 17.7
We are given an unknown substance of formula C_3H_6O for which the structure is one of those shown below. How could we use IR spectroscopy to distinguish among the possibilities?

$$CH_3CH_2\underset{\underset{O}{\|}}{C}H \quad CH_3\underset{\underset{O}{\|}}{C}CH_3 \quad H_2C=CHCH_2OH \quad H_2C=CH-OCH_3$$

• Understand the basic theory of NMR spectroscopy, and learn its terminology. Also, learn the fundamental experimental aspects of the technique.

Practice Problem 17.8
Suppose we obtained an NMR spectrum by keeping the magnetic field *constant*, while sweeping through a range of *radiofrequencies*. Would the frequency needed to bring a "downfield" proton into resonance be larger or smaller than that needed to bring an "upfield" proton into resonance?

Practice Problem 17.9
The proton NMR spectrum of $Br_2CHC\equiv C-CH_2Br$ consists of two peaks whose areas are in the ratio 1:2. Does the peak of larger area appear upfield or downfield relative to the peak of smaller area?

• Learn the approximate δ values (chemical shifts) associated with the different types of environments of H atoms in molecules.

Practice Problem 17.10
From *memory*, give the (approximate) NMR chemical shifts for the indicated protons (circled).

(a) \[vinyl H on C=C\]

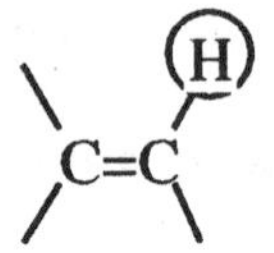

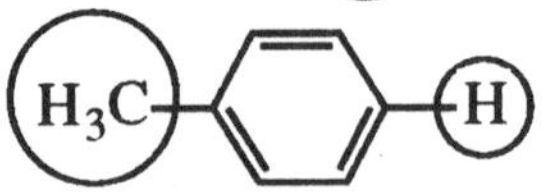

(b) -CH$_2$-CH$_2$-(CH$_3$)

(c) (H$_3$C)-C$_6$H$_4$-(H) (both)

(d) (H$_3$C)-C(=O)-O-(CH$_3$) (both)

(e) CH$_3$CH$_2$-C≡C-(H)

(f) C$_6$H$_5$-C(=O)-O-(H)

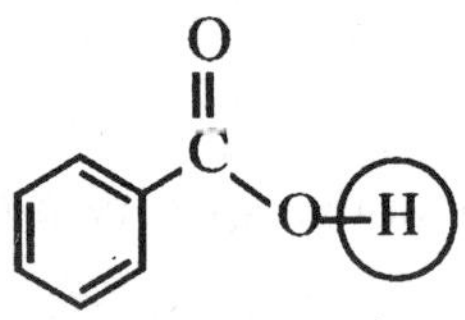

(g) CH$_3$CH$_2$-C(=O)(H)

- Understand how spin-spin coupling leads to the splitting of signals in NMR spectra.

Practice Problem 17.11
Predict the multiplicities of the signals for the indicated (circled) protons.
(a) Cl$_2$C-CHBr$_2$
    (H is circled)
(b) CH$_3$(CH$_2$)C(=O)-CH$_3$
    (CH$_2$ is circled)
(c) cyclopentyl-CH$_3$ (CH$_3$ is circled)
(d) dioxolane with two H's on adjacent carbons (both H's circled)
(e) BrCH$_2$(CH$_2$)Br (middle CH$_2$ circled)

Practice Problem 17.12
Predict the general appearance of the proton spectrum for each of the following compounds.
(a) CH$_3$CH(Cl)CO$_2$H
(b) CH$_3$CH$_2$CH$_2$NO$_2$
(c) PhC(CH$_3$)$_3$
(d) PhCH$_2$CH$_3$
(e) (CH$_3$)$_2$CHCO$_2$H

- Know how to use ^{13}C NMR spectroscopy; remember that the number of ^{13}C signals *generally* indicates the number of different environments for carbon atoms in the compound.

Practice Problem 17.13
Predict the number of ^{13}C signals you would expect to see in the ^{13}C spectrum of each of the following compounds.
(a) benzene
(b) toluene
(c) *p*-xylene
(d) di-*tert*-butyl ether
(e) benzophenone
(f) ethyl acetate

- The most important skill you need to develop in connection with NMR spectroscopy is the assignment of structure from spectral data. Try the (relatively simple) examples here, and then look at the end-of-chapter problems in the text.

Practice Problem 17.14
Suggest structures for each of the following indicated compounds.

(a) C$_7$H$_{14}$O The IR spectrum shows a strong band at 1715 cm^{-1}. The ^{1}H NMR shows singlets at δ = 2.3, 2.1, and 1.0 with peak areas in the ratio 2:3:9.

(b) C$_{10}$H$_{12}$O$_2$ The ^{1}H NMR spectrum shows signals at δ = 7.3 (singlet), 3.2 (quartet), 1.9 (singlet), and 0.9 (triplet).

(c) C$_9$H$_9$BrO The ^{1}H NMR spectrum shows bands at δ = 7.5 (singlet, 5 H), 5.3 (quartet, 1 H), and 1.9 (doublet, 3 H). The IR spectrum shows an intense band at ~1700 cm^{-1}.

(d) A compound isomeric with that in part (a) above exhibits peaks at 15, 19, 45, and 211 ppm downfield from TMS in the ^{13}C NMR spectrum.

(e) A compound isomeric with those in parts (a) and (d) above exhibits just *three* peaks in its

Solution of Text Problems

17.1 The sp^2 carbon-hydrogen bond is the stronger. The greater s character of the carbon contribution holds the electrons closer to the nucleus and thereby strengthens the bond.

17.2 (a) sp^2C-H 3010 cm^{-1}; sp^3C-H 2950 cm^{-1}; C=C 1650 cm^{-1}; C-H(bend) 1400cm^{-1}

(b) C=O 1710 cm^{-1}; sp^3C-H 2950 cm^{-1}; C-H(bend) 1400 cm^{-1}

(c) sp^2C-H 3010 cm^{-1}; sp^3C-H 2950 cm^{-1}; C≡N 2250 cm^{-1}; aromatic ring 700, 800 cm^{-1}

(d) sp^3C-H 2950 cm^{-1}; C-H(bend) 1400^{-1}

17.3 Both dissolve significant amounts of NaCl or KBr and would destroy the sample cells.

17.4 There is intramolecular hydrogen bonding between hydroxyl groups which is present even when individual molecules are far apart. For this type of hydrogen bonding all that is required is that the bonding groups lie in proper positions within a particular molecule, as they do with *trans*-1,2-dihydroxycyclohexane.

17.5 phenylacetaldehyde [PhCH$_2$CHO]

17.6 cyclopentanol

17.7 trichloroacetonitrile [Cl$_3$C-C≡N]

17.8 1-heptene [H$_2$C=CHCH$_2$CH$_2$CH$_2$CH$_2$CH$_3$]

17.9 propanoic acid [CH$_3$CH$_2$CO$_2$H]

17.10 1,5-cyclooctadiene

17.11 ethyl acetate [CH$_3$CO$_2$CH$_2$CH$_3$]

17.12 2-methyl-1,3-butadiene [H$_2$C=C(CH$_3$)CH=CH$_2$]

17.13 ethyl benzoate [PhCO$_2$CH$_2$CH$_3$]

17.14 anisole [PhOCH$_3$]

17.15 (a) 2 in the ^{1}H and 3 in the ^{13}C

(b) 2 in the ^{1}H and 5 in the ^{13}C

(c) 2 in the ^{1}H and 2 in the ^{13}C

(d) 2 in the ^{1}H and 3 in the ^{13}C

(e) 3 in the ^{1}H and 5 in the ^{13}C

17.16 (a) 2 H δ ~4.3 and 3 H δ ~3.3

(b) 2 H δ ~7.0 and 12 H δ ~2.2

(c) 4 H δ ~7.0 and 2 H δ ~3.0

(d) 4 H δ ~7.0 and 6 H δ ~4.0

(e) 4 H δ ~3.7 and 6 H δ ~3.3

17.17

For the central CH₂   For the terminal CH₂'s

1   4   6   4   1   1   2   1

↑↑↑↑ ↑↑↑↓ ↑↑↓↓ ↑↓↓↓ ↓↓↓↓   ↑↑ ↑↓ ↓↓
       ↑↑↓↑ ↑↓↑↓ ↓↑↓↓   ↓↑
       ↑↓↑↑ ↓↑↑↓ ↓↓↑↓
       ↓↑↑↑ ↑↓↓↑ ↓↓↓↑
                ↓↑↓↑
                ↓↑↑↓

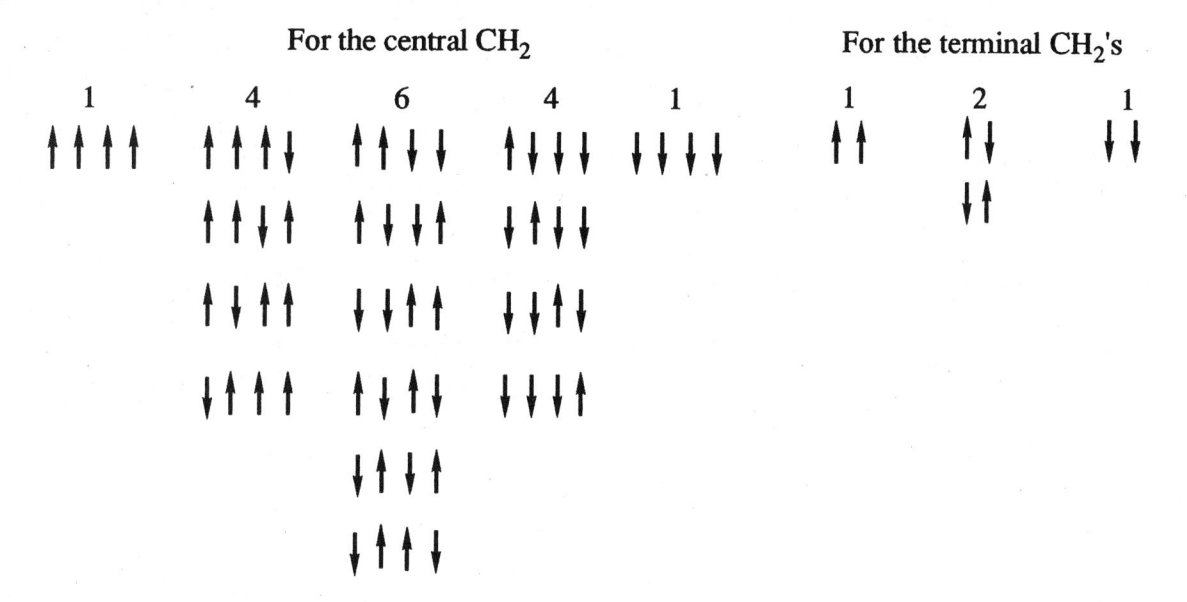

17.18 (a) OCH₃, 3 H, δ 3.3, singlet; OCH₂, 2 H, δ 3.6, quartet; C-CH₃, 3 H, δ 1.0, triplet.

(b) CH₂Cl, 2 H, δ 3.6, doublet; Br₂CH, 1 H, δ 4.2, triplet.

(c) CH₃, 6 H, δ 1.0, triplet; CH₂, 4 H, δ 2.3, quartet; ArH, 4 H, δ 7.0, singlet.

(d) CH₃, 6 H, δ 1.0, doublet; CH, 1 H, δ 3.9, septet; OH, 1 H, δ 2.0, singlet

(e) CH₃, 6 H, δ 1.2, doublet; CH₂O, 2 H, δ 3.4, triplet; CH₂Cl, 2 H, δ 3.8, triplet; CH, 1 H, δ 4.0, septet.

17.19 (a) CH₃, 6 H, δ 0.9, doublet; CH, 1 H, δ 2.3, septet; CH₂, 2 H, δ 2.1, quartet; CH₃, 3 H, δ 0.9 triplet.

(b) CH₃, 3 H, δ 1.0, doublet; CHBr, 1 H, δ 3.4, triplet of doublets; CH₂, 2 H, δ 2.4, doublet; ArH, 5 H, δ 7.0, singlet.

(c) C*H*₃CH, 3 H, δ 1.0, doublet; CH, 1 H, δ 3.5, triplet of quartets; C*H*₂CH, 2 H, δ 1.5, doublet of triplets; C*H*₂CH₃, 2 H, δ 1.2, triplet of quartets; CH₃, 3 H, δ 0.9, triplet; OH, 1 H, δ 4.0, singlet.

(d) CH₃, 6 H, δ 1.0, triplet; CH₂, 4 H, δ 1.4, doublet of quartets; CH, 1 H, δ 4.0, singlet; OH, 1 H, δ 3.5, singlet.

(e) CH₃, 3 H, δ 1.0, triplet; CH₂, 2 H, δ 1.7, doublet of quartets; C*H*CH₂, 1 H, δ 4.6, triplet of doublets; CHCl, 1 H, δ 5.6, doublet.

17.20 CH₃CH₂C(O)CH₂CH₃

17.21 *p*-bromobenzyl bromide

17.22 *trans*-1-bromopropene

17.23 H₂C=C(CH₃)CO₂CH₃

17.24 benzyl alcohol

17.25 (a) signals at 15, 40, and 60 ppm downfield from TMS
(b) signals at 15, 110, 120, 125, and 160 ppm downfield from TMS
(c) signals at 15, 45, and 200 ppm downfield from TMS
(d) signals at 15, 45, 110, and 125 ppm downfield from TMS

17.26 (a) signals at 15 (quartet), 40 (doublet), and 60 (triplet) ppm downfield from TMS
(b) signals at 15 (quartet), 110 (singlet), 120 (doublet), 125 (doublet), and 160 (singlet) ppm downfield from TMS
(c) signals at 15 (quartet), 45 (triplet), and 200 (singlet) ppm downfield from TMS

(d) signals at 15 (quartet), 45 (triplet), 110 (doublet), and 125 (doublet of doublets) ppm downfield from TMS

17.27 1,3,5-tribromobenzene
17.28 3-methyl-2-pentanol
17.29 cyclohexene
17.30 methyl isopropyl ketone
17.31 (a) the C=C and the vinylic C-H stretching absorption bands in cyclohexene
  (b) the -OH absorption band in cyclohexanol
  (c) the aldehydic C-H stretching absorption band in butanal
  (d) the carbon-carbon triple bond and the acetylenic C-H bond absorption bands in 1-butyne
  (e) the C=O absorption band in butanoic acid
  (f) the carbon-nitrogen triple bond stretching absorption band in benzonitrile
17.32 (a) $CH_3CH_2C(O)CH_2CH_3$
  (b) methyl isopropyl ketone
  (c) oxetane
  (d) 3-hexyne
  (e) di-*tert*-butyl ether
  (f) $CH_3CO_2CH_3$
  (g) 1,1,2-trichloroethane
  (h) $CH_3CH(Cl)CO_2H$
  (i) $(Ph)_2CHCl$
17.33 (a) 3600 (s), 2950 (s), and 1450 (m) cm^{-1}
  (b) 3010 (w), 2950 (s), 1650 (w), and 1450 (m) cm^{-1}
  (c) 2950 (s), 1450 (m), and 550 (s) cm^{-1}
  (d) 2950 (s), 1715 (s), and 1450 (m) cm^{-1}
  (e) 3010 (m), 2950 (m), 1450 (m), 1550 (s), and 750 (s) cm^{-1}
  (f) 3300 (m), 2950 (s), 2200 (m), and 1450 (m) cm^{-1}
  (g) 2950 (s), 1450 (m), 1100 (s) cm^{-1}
17.34 (a) CH_3, 3 H, δ 4.0, singlet; ArH, 5 H, δ 7.0, singlet
  (b) CH_3, 6 H, δ 1.0, triplet; CH_2 4 H, δ 1.8, doublet of quartets; CH, 2 H, δ 4.8, triplet
  (c) CH_3, 6 H, δ 1.0, doublet; CH, 2 H, δ 3.4, triplet of quartets; CH_2, 2 H, δ 1.3, triplet
  (d) CH_3, 6 H, δ 1.0, doublet of doublets; CH, 1 H, δ 4.2, doublet of septets
  (e) CH_3, 3 H, δ 2.2, singlet; ArH, 4 H, δ 7.0, AA'BB'
  (f) CH_2O, 4 H, δ 3.3, triplet; OH, 2 H, δ 4.0, singlet; C-CH_2-C, 4 H, δ 1.3, triplet
  (g) $CH_3C(O)$, 3 H, δ 2.1, singlet; $CH_2C(O)$, 2 H, δ 2.3, triplet; C-CH_2-C, 2 H, δ 1.3, triplet of quartets; CH_3-C, 3 H, δ 1.0, triplet
17.35 (a) 3 peaks: 20, 30, and 75 ppm downfield from TMS
  (b) 6 peaks: 10, 15, 20, 25 65, and 70 ppm downfield from TMS
  (c) 5 peaks: 30, 110, 120, 130, and 160 ppm downfield from TMS
  (d) 7 peaks: 30, 110, 120, 125, 130, and 135 ppm downfield from TMS
  (e) 4 peaks: 20, 25, 30, and 110 ppm downfield from TMS
  (f) 3 peaks: 15, 30, and 65 ppm downfield from TMS
  (g) 4 peaks: 15, 20, 65, and 80 ppm downfield from TMS
17.36 $PhCH_2CH_2OH$
17.37 $CH_3C(O)CH_2C(O)OCH_2CH_3$
17.38 cyclopentene
17.39 $PhN(CH_2CH_3)H$

17.40 $H_2C=CHCH_2OC(O)CH_3$

17.41 4-nitroanisole

17.42 A is α-methylstyrene and B is *trans*-β-methylstyrene

17.43 1-iodopropane

17.44 $CH_3CH_2CH(Br)CO_2H$

17.45 $ClC(O)CH_2CH_2CH_2CH_2C(O)Cl$

17.46 *C* is 2-methyl-1-propanol; *D* is 2-butanol; *E* is 2-methyl-2-propanol; *F* is 2-bromobutane

17.47 *G* is tetrahydrofuran

17.48 (a) 1-Hexyne would show acetylenic C-H and carbon-carbon triple bond stretching absorptions.
(b) Cyclohexanone would show a C=O absorption band whereas 1-hexen-3-ol would show the OH absorption band.
(c) *trans*-3-Hexene would show vinylic C-H stretching absorption bands.

17.49 (a) Ethyl isopropyl ether cleanly shows a 1 H septet and a 6 H doublet along with a 2 H quartet and a 3 H triplet. 1-Pentanol does not exhibit such clean splittings.
(b) 3-Pentanone shows only two signals, a triplet and a quartet.
(c) 4-Chlorotoluene shows a clean AA'BB' quartet in the aromatic region.

17.50 (a) 1-Bromobutane shows 4 signals while 1-bromo-2-methylpropane shows only 3 signals.
(b) Pentanal shows a signal far downfield.
(c) 1,1-Dibromocyclohexane shows 4 signals whereas 1,2-dibromocyclohexane shows only 3 signals.

17.51 (a) The methyl singlet in the proton NMR would be farther downfield with $CH_3CH_2CO_2CH_3$ than it would with $CH_3CO_2CH_2CH_3$.
(b) The 4-bromoanisole would show a clean AA'BB' quartet in the proton NMR spectrum.
(c) 2-Ethyl-1-butene would show a clean triplet and quartet pattern in the proton NMR.

17.52 (a) 1.363
(b) 1.003
(c) 0.9356

17.53 *H* is 1-chloro-3-methyl-2-butene; *I* is 3-methyl-2-buten-1-ol; *J* is 2-methyl-3-buten-2-ol

17.54 The intermediate ion is the same bridged bromonium ion as would be generated by bromine addition to cyclopentene. From the dibromide, this bridged bromonium ion is formed by the SbF_5-assisted loss of bromide ion. From the 4-bromocyclopentene, the bridged bromonium ion is formed *via* initial protonation of the alkene linkage, a hydride shift rearrangement to place the positive charge at a carbon adjacent to that bearing the bromine atom, and closure of the ring of the bridged bromonium ion. These are illustrated below.

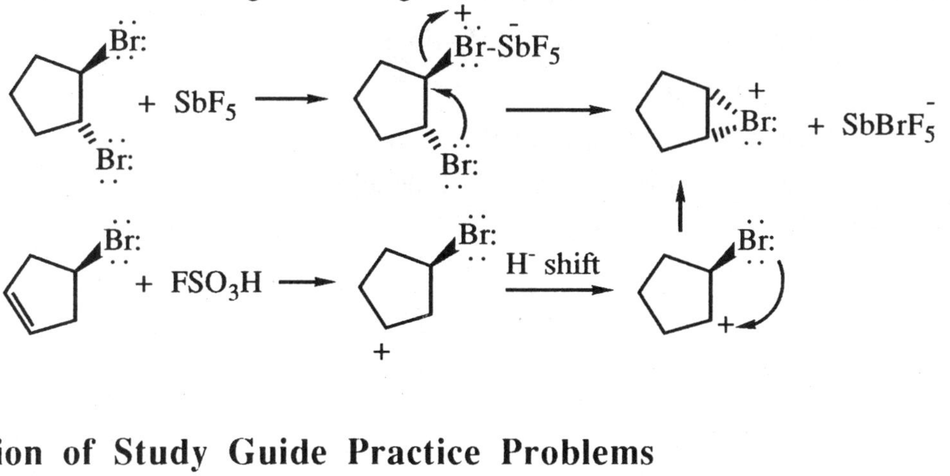

Solution of Study Guide Practice Problems

17.1 (a) N-H; (b) C=N; (c) C≡C; (d) C-O

17.2 (a) $c = \nu\lambda$; (b) $E = hc/\lambda$

17.3 (a) 3300-3500 cm^{-1}
(b) 1650-1800 cm^{-1}
(c) 2500-3400 cm^{-1}
(d) 2200 cm^{-1}

(e) 3300-3500 cm^{-1}

(f) 1050-1175 cm^{-1}

(g) 3000-3100 cm^{-1}

(h) 2850-2950 cm^{-1}

17.4 (a) 1-pentyne; (b) *cis*-1,2-dibromoethene

17.5 (a) 2-Hexanone has a methyl group, so we should observe a signal at ~1370 cm^{-1}.

(b) Methyl benzoate produces a peak at ~1200 cm^{-1} (C-O) whereas propiophenone does not since it has no C-O single bond.

(c) Ethylamine will exhibit N-H stretching absorption bands at 3300-3500 cm^{-1} whereas triethylamine will not since it has no N-H bonds.

(d) The C=C absorption band will be appreciably stronger in the spectrum of 1-hexene as compared to that of cyclohexene.

(e) The IR spectrum of benzoic acid will show absorption in the region 2500-3400 cm^{-1}.

(f) The C=C absorption band will be stronger in the spectrum of *cis*-3-hexene.

(g) Cyclohexene will show absorption at frequencies > 3000 cm^{-1}.

17.6 The reaction would seem to be the oxidation of a secondary alcohol to a ketone using a Cr(VI) reagent.

17.7 The CH_3CH_2CHO and $CH_3C(O)CH_3$ would both show a band at ~1715 cm^{-1} in the IR spectrum; the other two compounds would not. To distinguish CH_3CH_2CHO and $CH_3C(O)CH_3$ we would look for a band at 2750 cm^{-1} characteristic of aldehydes. To distinguish $H_2C=CHOCH_3$ from $H_2C=CHCH_2OH$ we should look for a band at 3400 cm^{-1} characteristic of alcohols.

17.8 The frequency would need to be larger. A "downfield" proton is less shielded so the energy gap between the two spin states is larger.

17.9 The peak of larger area appears upfield from the peak of smaller area. The Br_2CH-group proton will be farther downfield than those of the CH_2Br-group because of the presence of *two* electronegative substituents associated with the latter.

17.10 (a) δ 5-6.5; (b) δ 1.0; (c) δ 2.3; (d) δ 2.0 and 3.6; (e) δ 2.5; (f) δ 11-12; (g) δ 9-10

17.11 (a) doublet; (b) quartet; (c) doublet; (d) quintet; (e) singlet

17.12 (a) δ 1.8 (3 H, doublet), 4.5 (1 H, quartet), 12.0 (1 H, singlet)

(b) δ 1.0 (3 H, triplet), 2.0 (2 H, triplet of quartets), 4.5 (2 H, triplet)

(c) δ 1.3 (9 H, singlet), 7.2 (5 H, broad singlet)

(d) δ 1.2 (3 H, triplet), 2.5 (2 H, quartet), 7.2 (5 H, singlet)

(e) δ 1.0 (6 H, doublet), 2.7 (1 H, doublet of septets), 9.5 (1 H, doublet)

17.13 (a) 1; (b) 5; (c) 3; (d) 2; (e) 5; (f) 4

17.14 (a) methyl neopentyl ketone

(b) *p*-ethoxyanisole

(c) PhC(O)CH(Br)CH$_3$

(d) di-*n*-propyl ketone

(e) di-isopropyl ketone

CHAPTER 18
REACTIONS OF BENZENE AND ITS DERIVATIVES

Key Points

• Learn the basis of nomenclature for benzene derivatives.
Practice Problem 18.1
From *memory*, draw the structures of each of the following.
    (a) *p*-aminophenol
    (b) benzophenone
    (c) *m*-xylene
    (d) *o*-nitrobenzoic acid
    (e) 4-bromobenzonitrile
    (f) 2,4-dinitroanisole
    (g) *p*-nitrobenzyl bromide
    (h) (Z)-1-phenylbutene

• Commit to memory the basic mechanism for aromatic electrophilic aromatic substitution, and learn the reaction conditions (*i.e.* the reagents needed) to effect, bromination, nitration, alkylation, acylation, and sulfonation.
Practice Problem 18.2
From *memory*, give the reagents needed for the several electrophilic aromatic substitution reactions noted above.
Practice Problem 18.3
It is possible to perform an electrophilic aromatic substitution of hydroxyl for hydrogen on certain benzene derivatives (generating phenols) using H_2O_2 and fluorosulfonic acid, FSO_3H (a very strong acid). Propose a mechanism for the conversion of benzene to phenol using this reagent system.
Practice Problem 18.4
Benzene reacts with 2,2,5,5-tetramethyltetrahydrofuran in the presence of sulfuric acid to yield a compound of formula $C_{14}H_{20}$. Suggest a structure for the product and propose a mechanism for its formation.

• Learn the activating/deactivating and *ortho,para*-directing/*meta*-directing influence of substituents on electrophilic aromatic substitution reactions.
Practice Problem 18.5
Choose the *ortho,para*-directing groups from the following list:
    $-NO_2$, $-NH_2$, $-C\equiv N$, $-OCH_3$, $-SO_3H$, $-CHO$, $-CO_2H$, $-CH_2CH_3$, $-Cl$
Practice Problem 18.6
Which groups from the list given in the preceding problem are activating groups toward electrophilic aromatic substitution reactions?
Practice Problem 18.7
Predict the major products expected on mononitration of each of the following compounds.
    (a)   (b)   (c)

Practice Problem 18.8
Draw a Lewis structure for the intermediate arenonium ion $C_6H_5BrCl^+$ formed in the *p*-bromination of chlorobenzene in which all atoms have a noble gas electronic configuration.

• Learn how to convert alkyl groups on benzene rings to carboxyl groups, and how to convert nitro groups to amino groups.

Practice Problem 18.9
Propose series of reactions to accomplish each of the following transformations.
 (a) benzene to *m*-bromoaniline
 (b) benzene to *p*-nitrobenzoic acid

• Be aware of the limitations in synthetic methodology involving benzene derivatives that may flaw a synthetic scheme.

Practice Problem 18.10
Which reaction or reactions of each pair will *not* be synthetically viable?
 (a) *tert*-butylbenzene or toluene being converted to benzoic acid by oxidation with aqueous basic potassium permanganate
 (b) aniline or nitrobenzene being subjected to Friedel-Crafts alkylation using chloromethane with aluminum chloride
 (c) chlorobenzene reacting with potassium cyanide in acetone to generate benzonitrile, or chlorobenzene reacting with nitric acid in the presence of sulfuric acid to generate a mixture of *ortho-* and *para-*chloronitrobenzenes
 (d) 4-methoxyphenylmagnesium bromide reacting with formaldehyde to yield 4-methoxybenzyl alcohol after aqueous workup, or 4-nitrophenylmagnesium bromide reacting with formaldehyde to yield 4-nitrobenzyl alcohol after aqueous workup

• Appreciate the special stability of benzylic radicals and cations, and know the special chemistry associated with the benzylic position.

Practice Problem 18.11
What are the two chain propagating steps in the reaction of toluene with bromine under light irradiation?

Practice Problem 18.12
Propose a synthetic route for the preparation of PhCH$_2$-C≡C-H starting with toluene and acetylene.

Practice Problem 18.13
What major product would you expect from the reaction of PhCH=CHCH$_3$ with HBr if:

 (a) peroxides are present?
 (b) peroxides are excluded?

• Learn the special chemistry associated with the generation and reaction of benzyne species.

Practice Problem 18.14
From memory, show the mechanism for the production of benzyne from chlorobenzene and amide ion.

Practice Problem 18.15
One type of experiment that supports the notion that benzyne is an actual reaction intermediate in the reaction of chlorobenzene with amide ion involves "trapping" the benzyne. In a trapping experiment a substance is added to the reaction mixture that can readily react with the postulated intermediate but not with either the reactant or products. Because benzyne has a very reactive triple bond, it should behave as an excellent dienophile. What product would you expect to be formed if furan were added as the trapping agent?

Practice Problem 18.16
What do you predict to be the major product from the reaction of *o*-bromoanisole with potassium amide in liquid ammonia?

• Learn the mechanism and uses of reactions in which a halogen is displaced from an electron-poor benzene ring *via* an addition-elimination reaction.

Practice Problem 18.17
Which would you predict to be more readily hydrolyzed by aqueous sodium hydroxide solution, 3-chloronitrobenzene or 4-chloronitrobenzene? Explain your choice.

Solution of Text Problems

18.1

(a) 1,2,3-trinitrobenzene; 1,2,4-trinitrobenzene; 1,3,5-trinitrobenzene

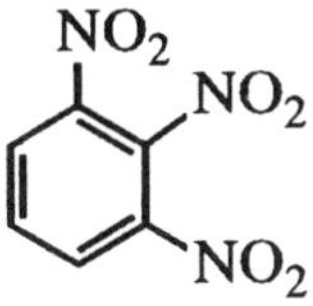

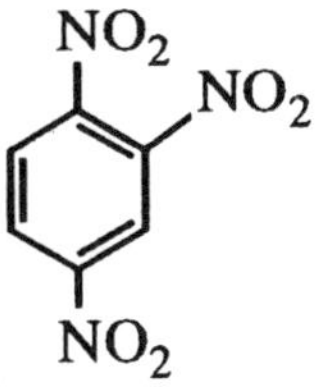

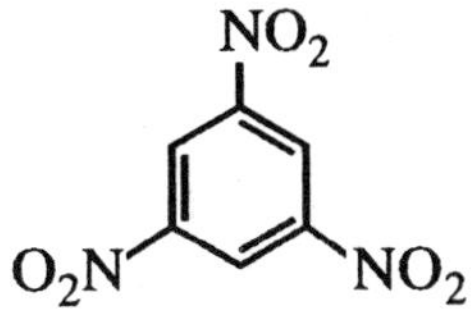

(b) 2,3-dichloronitrobenzene; 3,4-dichloronitrobenzene; 2,4-dichloronitrobenzene; 3,5-dichloronitrobenzene; 2,5-dichloronitrobenzene; 2,6-dichloronitrobenzene

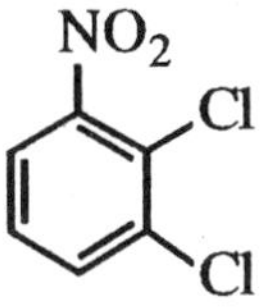

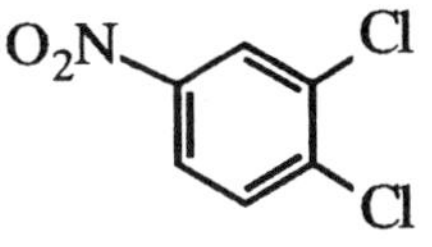

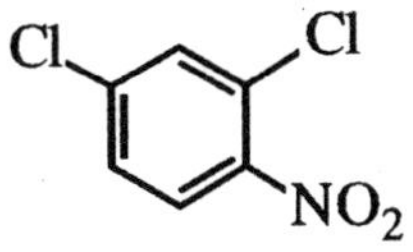

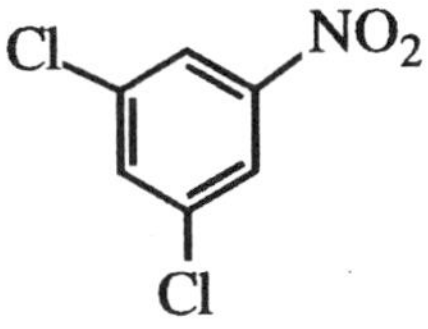

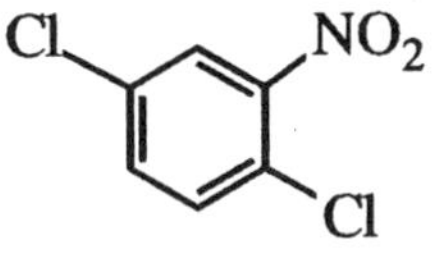

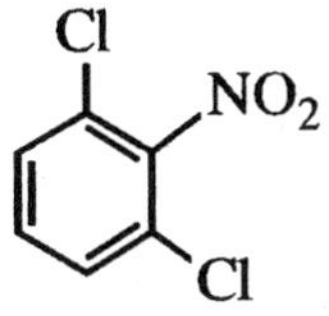

(c) 3-bromo-2-chloroethylbenzene; 2-bromo-3-chloroethylbenzene; 3-bromo-4-chloroethylbenzene; 4-bromo-3-chloroethylbenzene; 2-bromo-4-chloroethylbenzene; 4-bromo-2-chloroethylbenzene; 3-bromo-5-chloroethylbenzene; 5-bromo-2-chloroethylbenzene; 2-bromo-5-chloroethylbenzene; 2-bromo-6-chloroethylbenzene

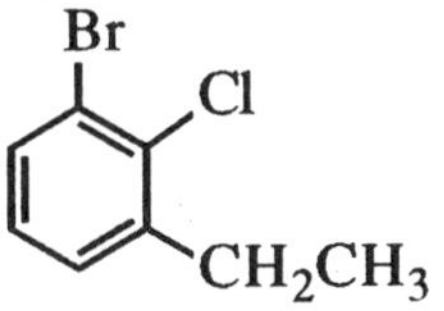

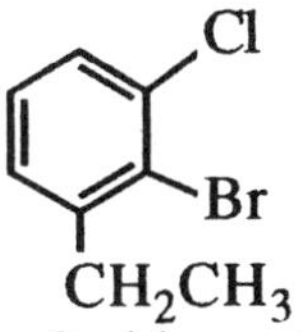

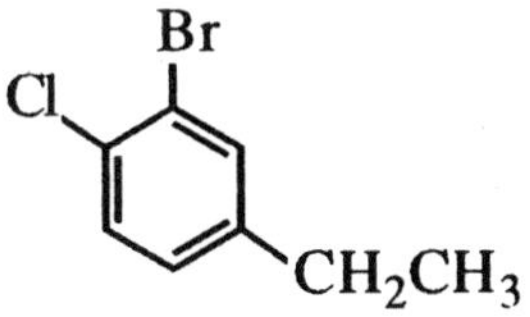

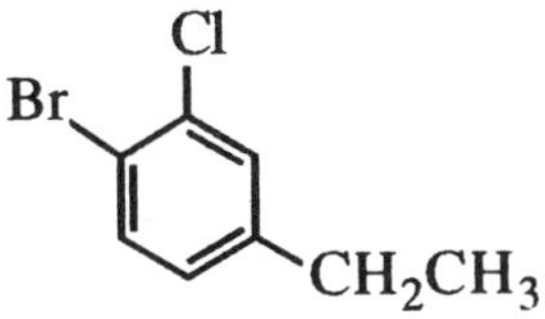

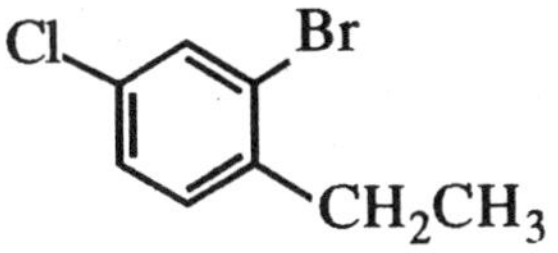

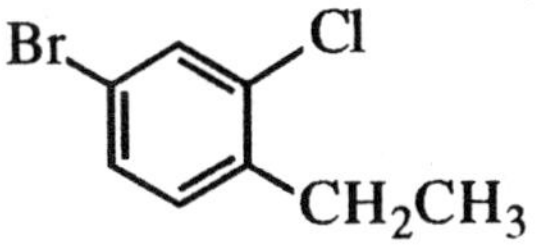

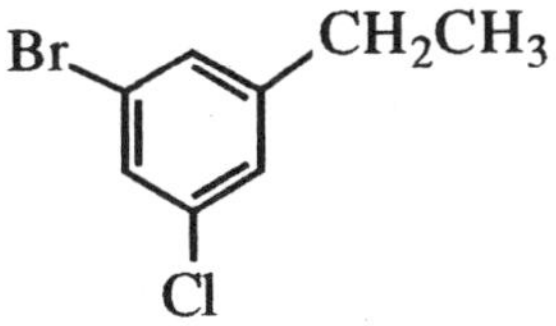

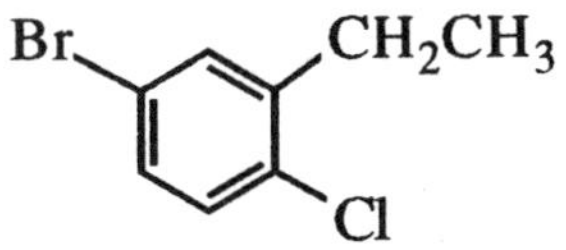

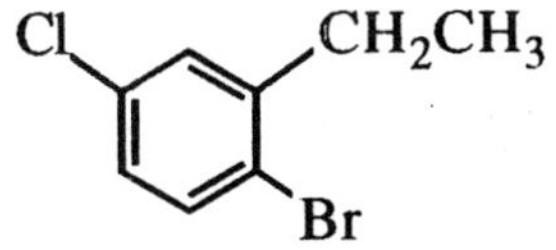

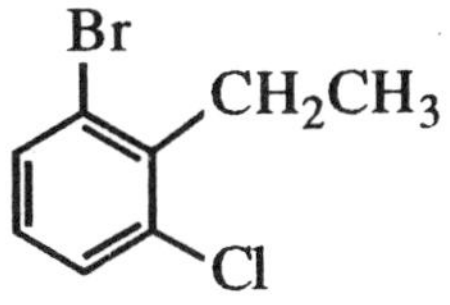

18.2 (a) methoxybenzene; (b) isopropylbenzene; (c) methylbenzene

18.3

(a) 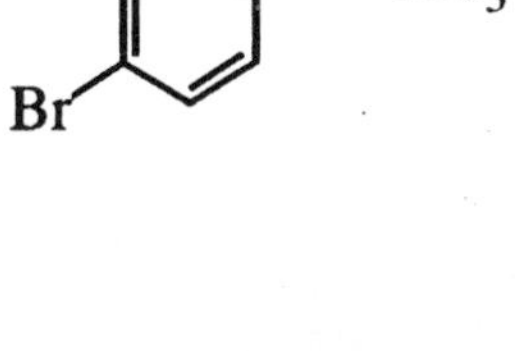 4-bromoacetophenone; (b) 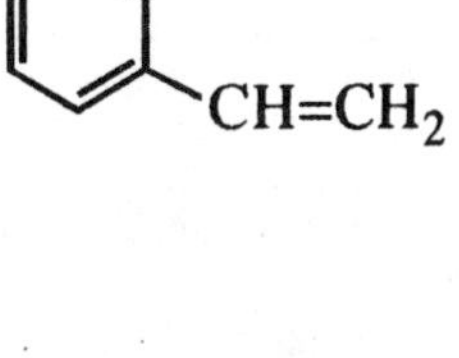 1,2-divinyl-3-methylbenzene (structure shown); (c) 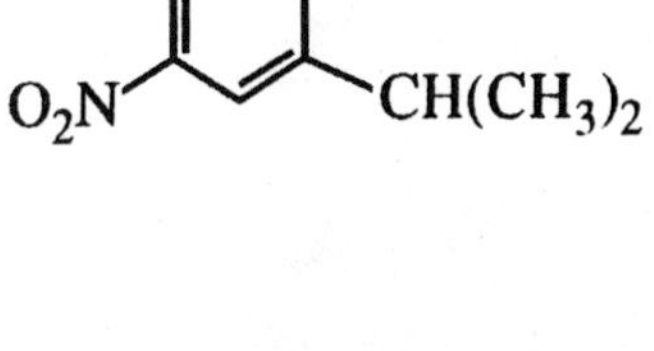 3,5-dinitroisopropylbenzene

160 Study Guide and Solutions Manual

18.4 In each case, numbering should be done to provide the smaller of the possible numbers.
(a) 2-bromotoluene; (b) 3,4-dichloroanisole

18.5
(a) 1,3-dichlorobenzene
(b) 2-chlorotoluene (Cl and CH₃ ortho)
(c) 4-nitroaniline (O₂N and NH₂ para)
(d) 1-ethyl-3-isopropylbenzene
(e) 1,3-divinylbenzene

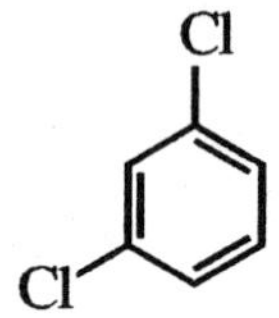

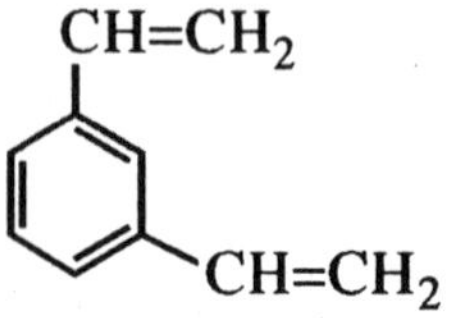

18.6
(a) benzyl alcohol (C₆H₅CH₂OH)
(b) 1,3-diphenylpropa-1,2-diene (stilbene-like structure with CH=C=CH-phenyl)
(c) (E)-propenylbenzene (C₆H₅-CH=CH-CH₃)
(d) 4-bromo-(bromomethyl)benzene (Br–C₆H₄–CH₂Br)
(e) phenyllithium (C₆H₅Li)

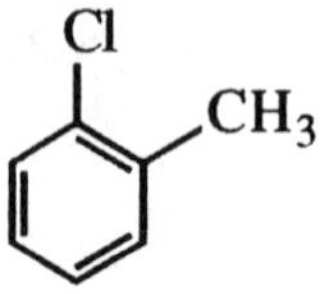

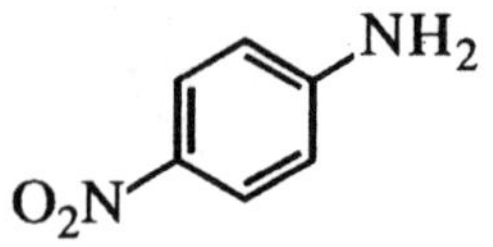

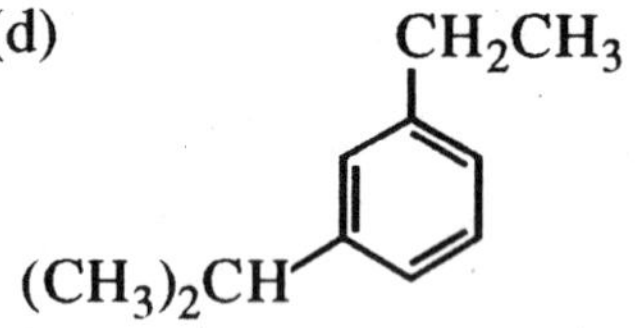

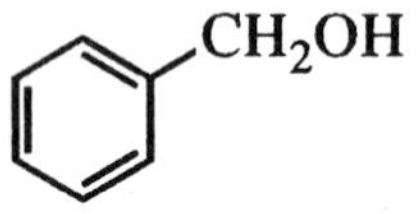

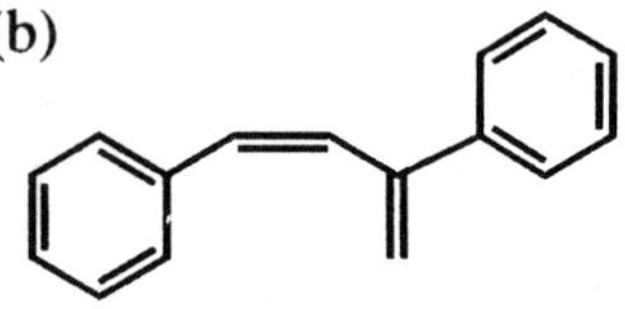

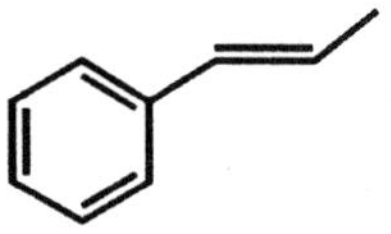

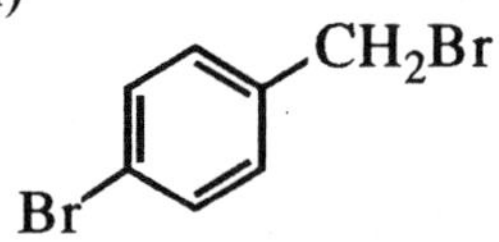

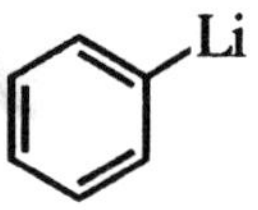

18.7

(mechanism showing protonation of HNO₃ by a second HNO₃, giving H₂O–NO₂⁺ and NO₃⁻, then loss of H₂O to give NO₂⁺ (O=N⁺=O) + H₂O)

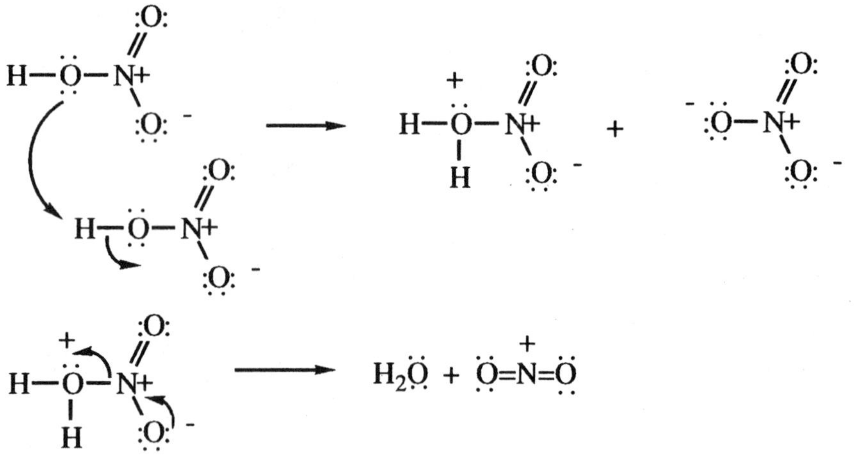

18.8 The nitrogen is considered to be *sp* hybridized and the ion is linear.

18.9

For Eqn. 18.3:

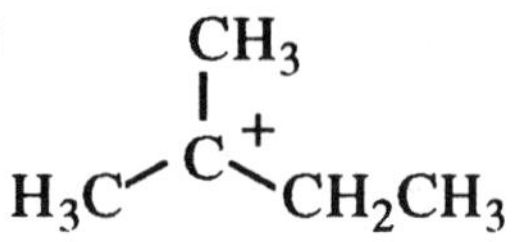

$$H_3C-\overset{CH_3}{\underset{CH_2CH_3}{\overset{+}{C}}}$$

For Eqn. 18.4: CH₃CH₂CH₂⁺   and   (CH₃)₂CH⁺
              (minor)   (major)

Chapter 18   161

18.10 The formation of the electrophile is shown here for each case. Once formed, the reaction of the electrophile (E$^+$) continues as is illustrated in section 18.4 of the text. *After n-butyl cation forms, it rearranges to sec-butyl cation via a hydride shift b-4 rxn with the benzene ring.*

Eqn. 18.5

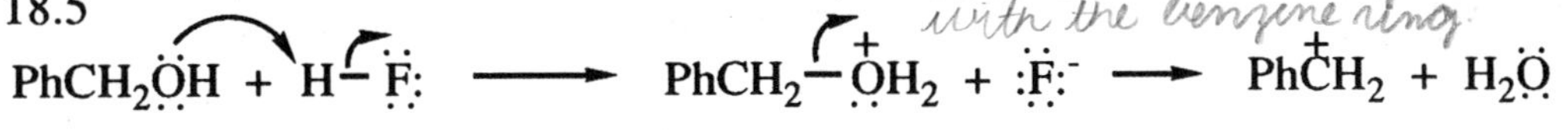

$$PhCH_2\ddot{O}H + H-\ddot{F}: \longrightarrow PhCH_2-\overset{+}{\ddot{O}}H_2 + :\ddot{F}:^- \longrightarrow Ph\overset{+}{C}H_2 + H_2\ddot{O}$$

Eqn. 18.6

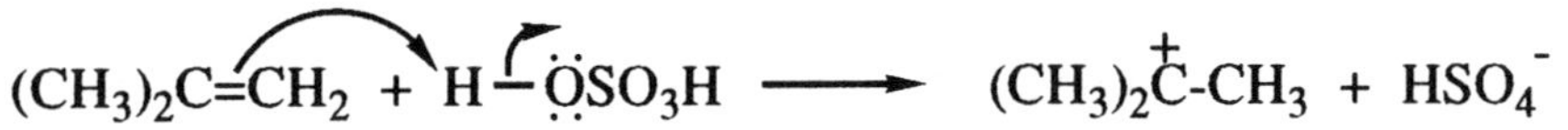

$$(CH_3)_2C=CH_2 + H-\ddot{O}SO_3H \longrightarrow (CH_3)_2\overset{+}{C}-CH_3 + HSO_4^-$$

Eqn. 18.7

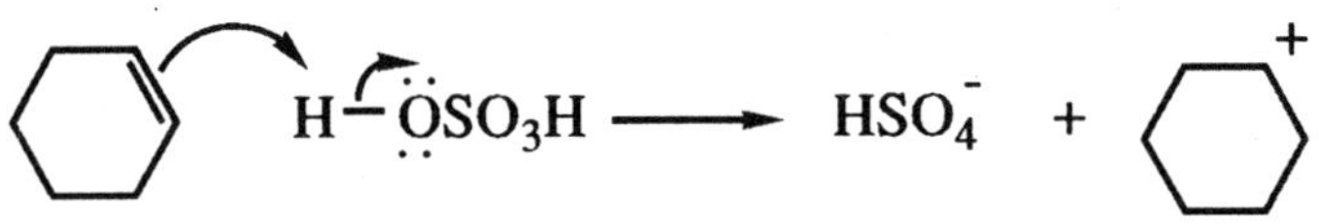

(cyclohexene) + H−ÖSO$_3$H ⟶ HSO$_4^-$ + (cyclohexyl cation)

Eqn. 18.8

$$CH_3CH_2CH_2CH_2-\ddot{O}H + BF_3 \longrightarrow CH_3CH_2CH_2CH_2-\overset{+}{\underset{-BF_3}{\ddot{O}H}} \longrightarrow CH_3CH_2CH_2\overset{+}{C}H_2 + H\ddot{O}BF_3^-$$

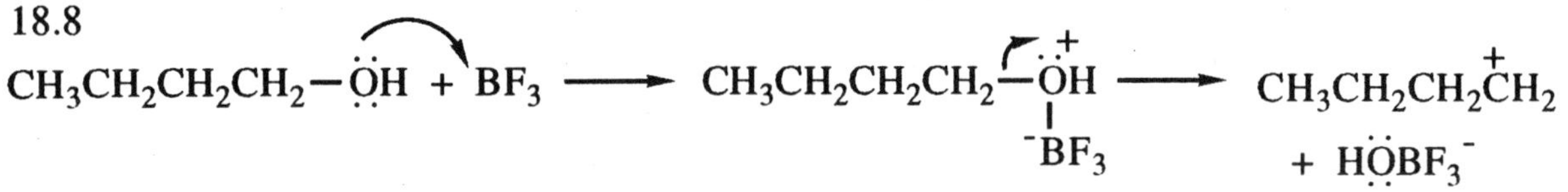

18.11 An equilibrium mixture of two secondary carbocations is generated from each of the starting alcohols. The carbocations are the "E$^+$" of the electrophilic aromatic substitution mechanism illustrated in section 18.4.

$$CH_3CH_2CH_2\underset{OH}{C}HCH_3 \xrightarrow{BF_3} CH_3CH_2CH_2\overset{+}{C}HCH_3$$

$$\updownarrow \sim H^-$$

$$CH_3CH_2\underset{OH}{C}HCH_2CH_3 \xrightarrow{BF_3} CH_3CH_2\overset{+}{C}HCH_2CH_3$$

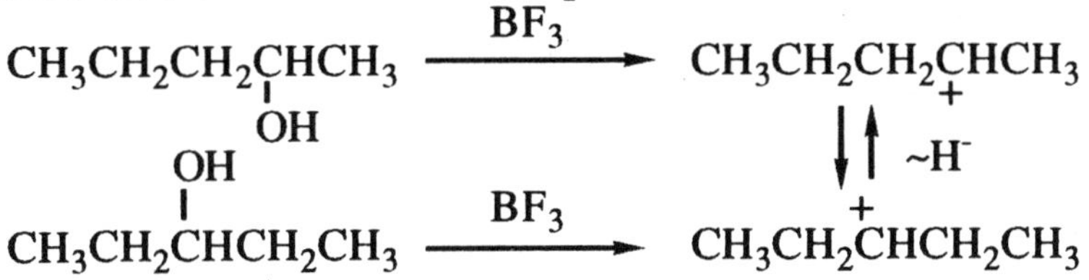

18.12

$$(CH_3)_2CHCH_2OH + BF_3 \longrightarrow (CH_3)_2CH\overset{+}{C}H_2 \xrightarrow{\sim H^-} (CH_3)_3\overset{+}{C} \longrightarrow \longrightarrow PhC(CH_3)_3$$

"E$^+$" for continuing electrophilic aromatic substitution reaction

18.13 The product is Ph$_3$CCl. It is formed by three successive electrophilic aromatic substitution reactions proceeding through PhCCl$_3$ and Ph$_2$CCl$_2$ as intermediates. The carbocations serving as "E$^+$" in the standard electrophilic aromatic substitution mechanism are $^+$CCl$_3$, $^+$CPhCl$_2$, and $^+$CPh$_2$Cl.

18.14

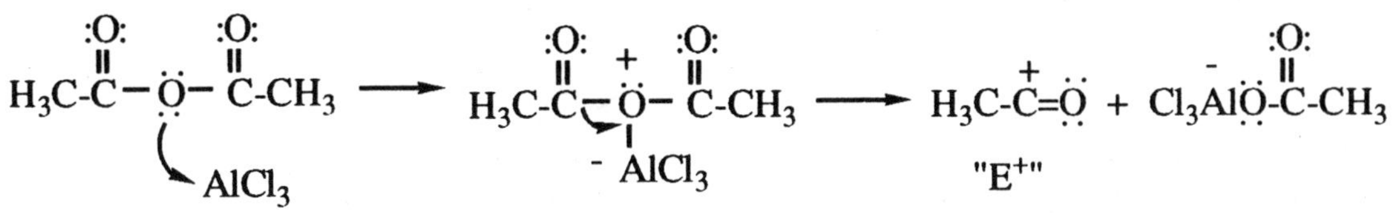

$$H_3C-\overset{\overset{\ddot{O}:}{\|}}{C}-\ddot{O}-\overset{\overset{\ddot{O}:}{\|}}{C}-CH_3 \longrightarrow H_3C-\overset{\overset{\ddot{O}:}{\|}}{\underset{-AlCl_3}{C}}-\overset{+}{\ddot{O}}-\overset{\overset{\ddot{O}:}{\|}}{C}-CH_3 \longrightarrow H_3C-\overset{+}{C}=\ddot{O} + Cl_3Al\overset{-}{\ddot{O}}-\overset{\overset{\ddot{O}:}{\|}}{C}-CH_3$$

$$\quad \quad \searrow AlCl_3 \quad \quad \quad \quad \quad \quad \quad \quad \quad \quad \quad \quad \quad \quad \quad \text{"E}^+\text{"}$$

The species "E$^+$" serves as the electrophile in the continuing electrophilic aromatic substitution reaction as illustrated in section 18.4 in the text.

18.15

[Structure: PhCH₂CH₂C(=O)Cl with AlCl₃ attacking Cl → PhCH₂CH₂-C⁺=O:]

"E⁺" for continuing intramolecular electrophilic aromatic substitution as illustrated in section 18.4 of the text

18.16 Benzene is treated with PhC(O)Cl (benzoyl chloride) in the presence of aluminum chloride.

18.17 In each case the formation of the electrophile "E⁺" is shown which continues *via* the standard mechanism for electrophilic aromatic substitution (section 18.4 of the text) to the product shown.

(a) [succinic anhydride + AlCl₃ → ring-opened acylium → after EAS with benzene → PhC(=O)CH₂CH₂COOH]

(b) [phthalic anhydride + AlCl₃ → acylium → EAS with benzene → o-benzoylbenzoic acid (PhC(=O)-C₆H₄-COOH)]

18.18

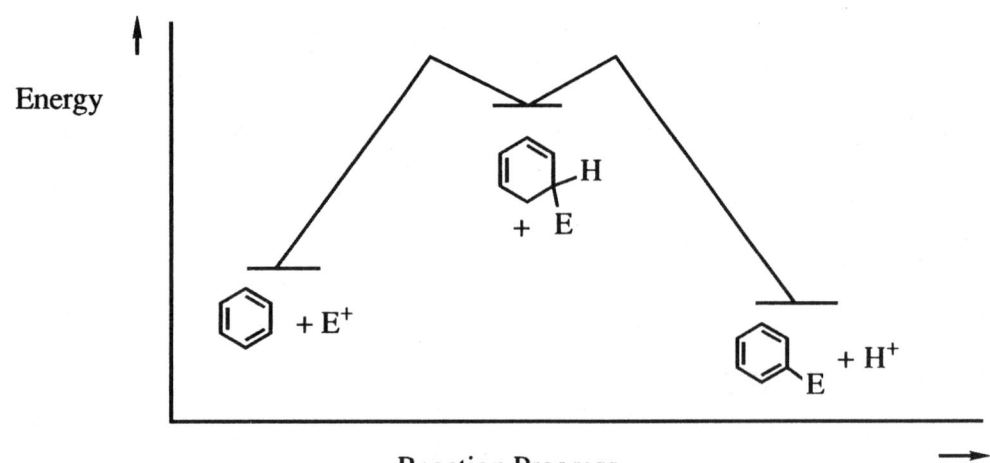

[Energy diagram showing benzene + E⁺ → arenium ion intermediate (cyclohexadienyl cation with H and E) → Ph-E + H⁺, with two transition states]

18.19 (a) *ortho,para* - Electron donation from the nitrogen to the π-system activates preferentially the *ortho* and *para* positions.

(b) *meta* - Electron withdrawl from the ring by the electron deficient boron specifically retards reaction at the *ortho* and *para* positions.

18.20

[Structure: cyclohexadienyl cation with Br and H on sp3 carbon, and Cl+ leaving]

18.21

[Resonance structures of aniline reacting with E+ showing the arenium ion intermediates with positive charge delocalized onto the nitrogen]

18.22 (a) 2-bromo-4-nitrophenol;   (b) 4-methyl-2-nitrophenol

18.23 2-ethyltoluene

18.24

[Structure of indane]

18.25 (a) 1. toluene treated with iodomethane in the presence of aluminum chloride; 2. heating of the resultant *p*-xylene with potassium permanganate in aqueous base, followed by aqueous acid workup

(b) 1. heating benzene with nitric acid in sulfuric acid to generate *m*-dinitrobenzene; 2. treatment with tin and hydrochloric acid to reduce both nitro groups to amino groups

(c) 1. treatment of toluene with nitric acid in sulfuric acid to generate *p*-nitrotoluene; 2. treatment with bromine in the presence of iron; 3. reduction of the nitro group to an amino group using tin and hydrochloric acid

18.26 (a) 1. heating toluene with aqueous basic potassium permanganate, followed by workup with aqueous acid; 2. treatment with bromine in the presence of iron

(b) 1. treatment of toluene with bromine in the presence of iron; 2. oxidation by heating with aqueous basic potassium permanganate, followed by workup with aqueous acid

(c) treatment of *p*-bromobenzoic acid [as prepared in part (b) above] with nitric acid in sulfuric acid

(d) 1. treatment of toluene with nitric acid in sulfuric acid to give *p*-nitrotoluene; 2. treatment with bromine in the presence of iron; 3. reduction of the nitro group to an amino group using tin and hydrochloric acid

18.27 (a) 1. treatment of toluene with bromine in the presence of iron to form *p*-bromotoluene; 2. reaction with Mg metal in ether to generate the Grignard reagent; 3. addition of 2-propanone (acetone), with aqueous workup; 4. heating with aqueous basic potassium permanganate, followed by aqueous acid workup (Only the methyl group attached to the aromatic ring is oxidized in this last step; the hydroxyl group is not oxidized as it is tertiary, nor is the remaining alkyl group attached to the aromatic ring oxidized as it lacks any benzylic hydrogens.)

(b) 1. treatment of the Grignard reagent from *p*-bromotoluene [prepared as in part (a)] with D$_2$O; 2. heating with aqueous basic potassium permanganate, followed by aqueous acid workup.

18.28 The intermediate radical leading to the major product is benzylic, that is it is stabilized by the delocalization of the unpaired electron about the aromatic ring. The intermediate radical leading to the minor product is isolated from the aromatic ring by an sp^3-hybridized carbon atom and thereby is not resonance stabilized. The Hammond postulate teaches us that for competing reactions, that reaction involving the more stable intermediate will have the lower activation energy and thus will proceed most rapidly.

18.29 We expect a ratio > 30:1. The bromination reaction exhibits greater selectivity owing to the fact that it involves less exothermic individual reaction steps.

18.30 (a) 1. treatment of ethylbenzene with bromine under irradiation with light; 2. treatment with potassium *tert*-butoxide; 3. reaction with HBr in the presence of peroxide free radical initiators; 4. formation of the Grignard reagent by reaction with Mg metal in ether; 5. addition of formaldehyde, with aqueous acid workup

(b) 1. treatment of toluene with bromine under irradiation with light; 2. formation of the Grignard reagent by treatment with Mg metal in ether; 3. addition of formaldehyde, with aqueous acid workup; 4. heating with sulfuric acid to perform dehydration of the alcohol

18.31

[Structure: biphenyl–CH(Br)CH₂–phenyl]

18.32 There is greater stabilization for the carbocation which is in a benzylic position relative to the *p*-methoxy group than for the carbocation which is in a benzylic position relative to the *p*-nitro group.

18.33 The first step of the reaction presumably leading to benzyne occurs, *i.e.* removal of the hydrogen (deuterium) adjacent to fluorine, but fluoride ion loss, that step required for the formation of benzyne, does not occur. Rather, the anion picks up a hydrogen ion from the solvent. Fluoride ion is sufficiently poor as a leaving group that it does not come off readily under these conditions.

18.34

[Mechanism: o-chloroanisole + PhLi → benzyne with OCH₃ → aryl anion with Ph substituent (Li⁺) → m-methoxybiphenyl via H₂O]

18.35

[Mechanism: 3-chlorophenethylamine + PhLi → benzyne intermediate → cyclization to protonated tetrahydroquinoline → –H⁺ → 1,2,3,4-tetrahydroquinoline]

18.36

[Structure: cyclohexadienide Meisenheimer-like intermediate with :Br, HO, and =N⁺(O:⁻)(O:⁻) (nitro) substituents]

18.37 A nitro group at the *ortho* or *para* position would speed the reaction by stabilizing the intermediate carbanion.

18.38 The 4-chloropyridine is favored for reaction over the 3-chloropyridine. The attack by a nucleophile at the 4-position of 4-chloropyridine leads to an intermediate in which the negative charge is delocalized to nitrogen, whereas the attack of a nucleophile at the 3-position of 3-chloropyridine leads to an intermediate in which the negative charge can be delocalized only to carbon. As nitrogen is more electronegative than is carbon and can better accommodate a negative charge, reaction of the 4-chloropyridine is favored.

18.39

[Three alkene structures: (Ph)(Ph)C=C(CH₃)(H); (H₃C)(Ph)C=C(Ph)(H); (Ph)(H₃C)C=C(Ph)(H)]

18.40 There is no activated H adjacent to the iodide position.

18.41

(a)
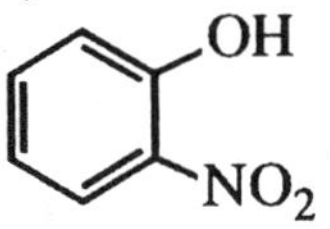

(b)
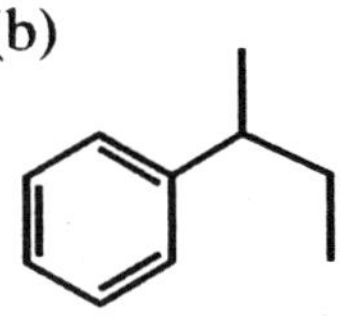

(c)
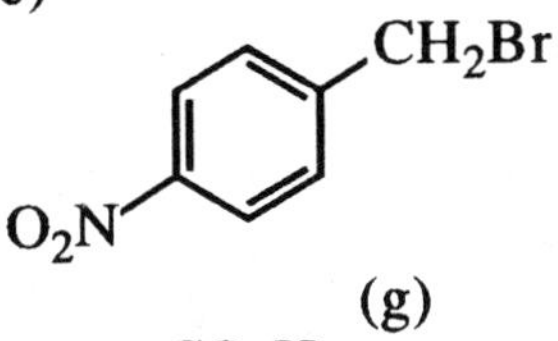

(d)
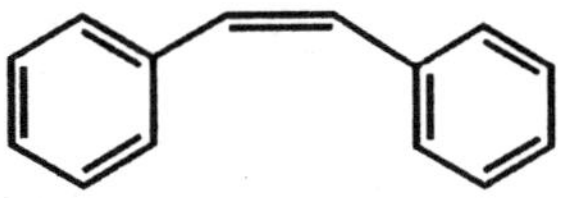

(e)
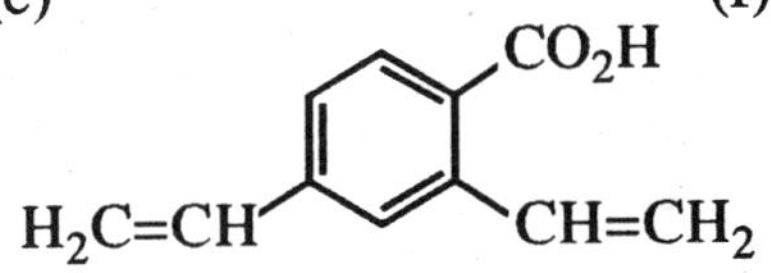

(f)
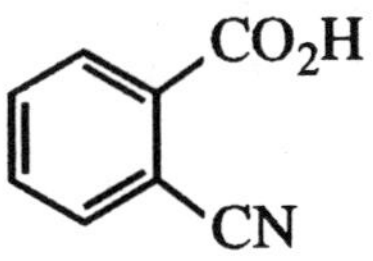

(g)
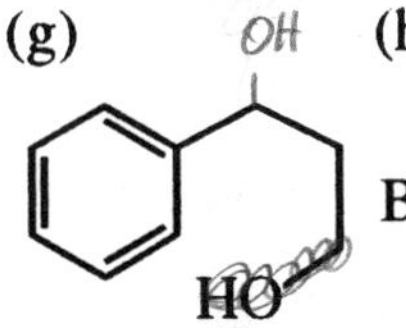

(h)
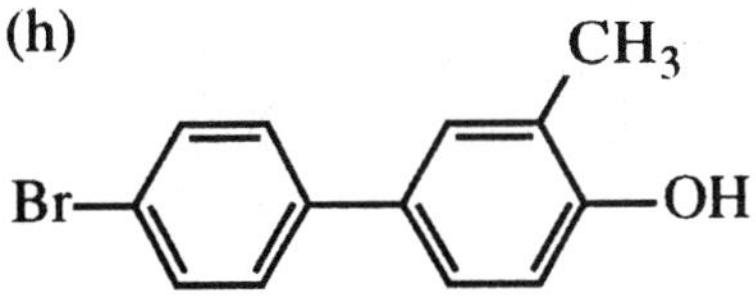

(i)
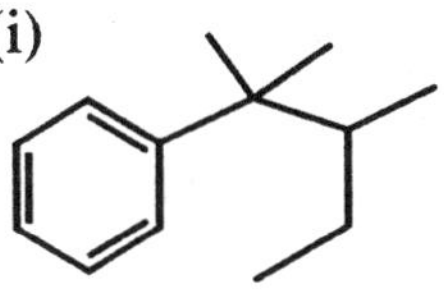

(j)
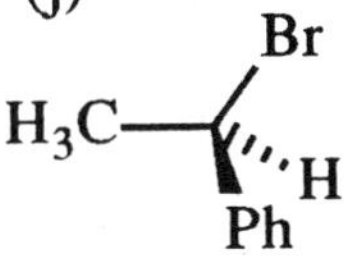

(k)
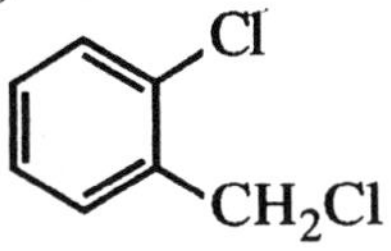

(l)
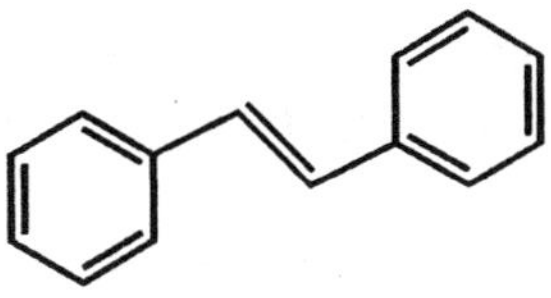

(m)
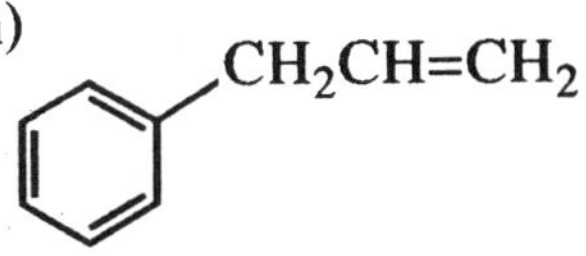

(n)
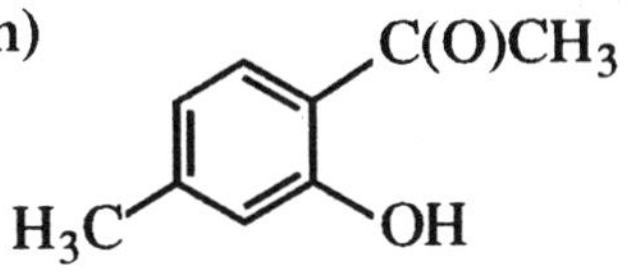

(o)
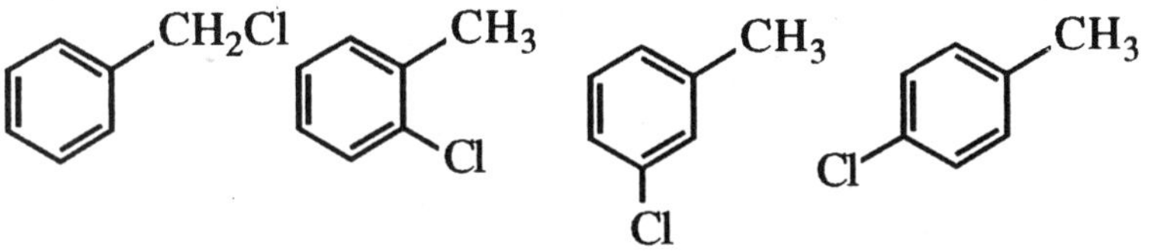

(p)
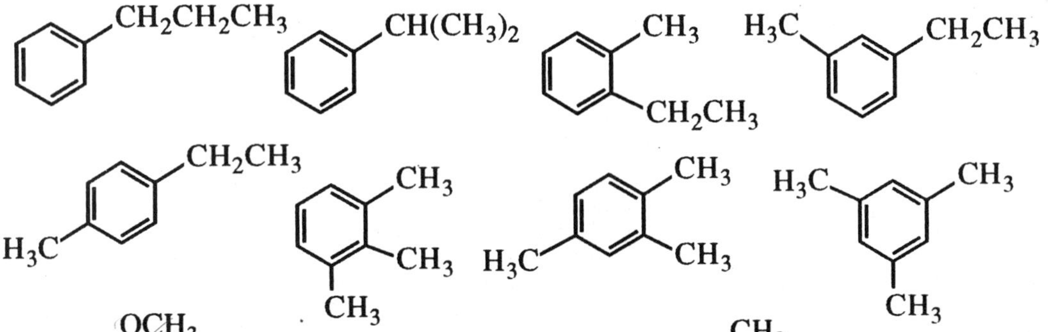

(q)
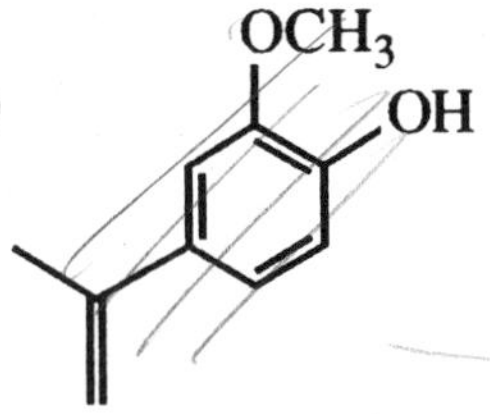

(r)
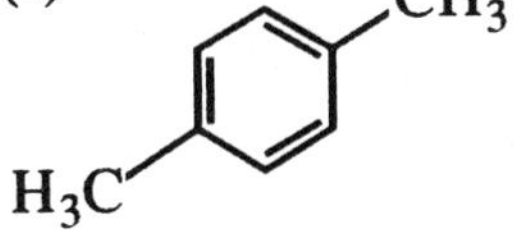

(s)
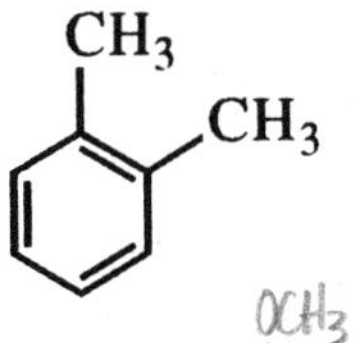

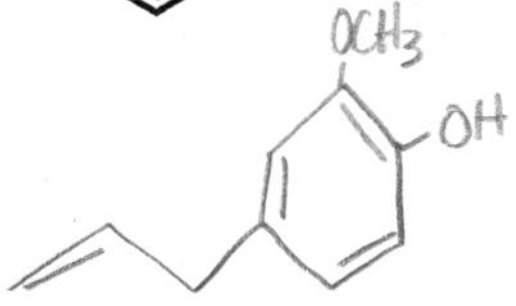

18.42 (a) 2-nitroethylbenzene
(b) 3-bromobenzenesulfonic acid
(c) 4-(bromomethyl)acetophenone anisole
(d) 2,4-dichlorobenzoic acid
(e) 2,6-dimethylethylbenzene
(f) 4-bromobenzoic acid
(g) 2,4-dinitroanisole Styrene
(h) *p*-divinylbenzene
(i) phenylcyclopentane

166 Study Guide and Solutions Manual

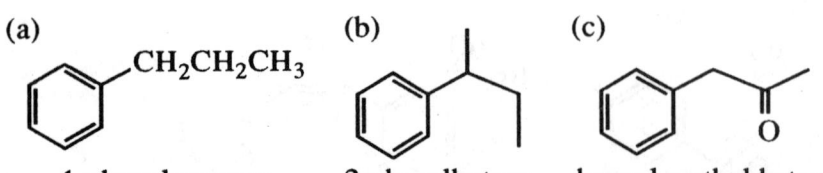

18.43
(a) PhCH₂CH₂CH₃ — 1-phenylpropane
(b) Ph-CH(CH₃)CH₂CH₃ — 2-phenylbutane
(c) PhCH₂C(O)CH₃ — benzyl methyl ketone

18.44 (a) With the hydroxyl group the activating effect through the π system is more important (electron donation) than the inductive electron withdrawl. Although halogens have an electron donating effect through the π system, it is less important than their inductive (electronegativity) effect withdrawing electron density from the ring.

(b) Once a nitro group has been introduced to one ring, that ring is particularly deactivated toward further nitration.

(c) The nitroso group has an unshared valence level electron pair on nitrogen which can interact with the ring π system to donate electron density to the ring; the nitro group does not have such an available electron pair.

18.45

[Resonance structures of chlorobenzene showing :Cl: donating into ring, and CH₂=CHCl showing :Cl: donation]

Chloroethane has the higher dipole moment.

18.46
(a) 1-Br, 2-NO₂, 4-OCH₃ benzene (ortho — another possible)
(b) 6-nitrochroman (O₂N on chroman)
(c) 3,5-dinitroacetophenone (C(O)CH₃ with two NO₂ meta)
(d) 2-methoxy-4-hydroxy-... (OH, OCH₃, NO₂)
(e) 2-hydroxy-3,5-dinitro... (OH with two O₂N)
(f) 1-Cl, 2-CH₃, 4-NO₂, 5-NO₂ (Cl, CH₃, O₂N, NO₂)

18.47
(a) 2-chloro-4-nitrotoluene (CH₃, Cl, NO₂)
(b) 2-chloro-4-nitroaniline (O₂N, Cl, NH₂)
(c) 2,4-dibromoacetophenone (C(O)CH₃, Br, Br)
(d) 2-methyl-3,4-dinitro-...-sulfonic acid (CH₃, HO₃S, NO₂, NO₂)
(e) phenol (OH)
(f) 2-phenyl-2-propanol (C(CH₃)₂OH)
(g) terephthalic acid (HO₂C, CO₂H)

18.48

(a) 3-bromo-4'-nitro-4-hydroxybiphenyl (HO—[ring]—[ring]—NO₂ with Br on the HO ring)

(b) PhCH₂O—C₆H₄—Br (para)

(c) Ph—CO₂—C₆H₄—Br (para)

(d) 2-hydroxy-5-bromobenzaldehyde (phenol with OH, CHO ortho, Br para to OH)

18.49 (a) 1. treatment of benzene with iodomethane and aluminum trichloride; 2. sulfuric acid and nitric acid

(b) 1. treatment of benzene with acetic anhydride and aluminum trichloride; 2. iodomethane and aluminum trichloride

(c) 1. treatment of benzene with iodomethane and aluminum trichloride; 2. bromine with light irradiation

(d) 1. treatment of benzene with bromine in the presence of iron; 2. magnesium metal in ether; 3. ethylene oxide; 4. dehydration with sulfuric acid

(e) 1. treatment of benzene with iodomethane and aluminum trichloride; 2. aqueous basic potassium permanganate, followed by acidic workup; 3. bromine in the presence of iron

(f) 1. treatment of benzene with iodomethane and aluminum trichloride; bromine with light irradiation [Fe]; 3. aqueous basic potassium permanganate, followed by acidic workup

(g) 1. treatment of benzene with nitric acid and sulfuric acid; 2. bromine in the presence of iron; 3. tin and hydrochloric acid

(h) 1. heating of benzene with nitric acid and sulfuric acid to generate *m*-dinitrobenzene; 2. chlorine in the presence of iron

(i) 1. treatment of benzene with bromine in the presence of iron; 2. magnesium metal in ether; 3. ethylene oxide

(j) 1. treatment of styrene [from the end of part (d)] with bromine in carbon tetrachloride; 2. heating with sodium amide

(k) 1. treatment of benzene with bromine in the presence of iron; 2. magnesium metal in ether; 3. D₂O

18.50 Of the two halogens, I and Cl, the Cl is more electronegative and ICl reacts to produce "I⁺" as the electrophile rather than "Cl⁺".

18.51

HÖ—Br: + H—ÖSO₃H → H₂Ö⁺—Br: + HSO₄⁻

H₂Ö⁺—Br: → H₂Ö + :Br:⁺

The protonated HOBr serves to donate an electron deficient bromine species as the electrophile.

18.52

(a) PhCH₂OH; (b) PhCH₂CH(OH)Ph; (c) Ph₂CH₂; (d) PhCH=CPh₂; Ph₂C(OH)CH₂Ph

(e) 2-mercapto-1,3-dinitrobenzene (SH with O₂N and NO₂ groups) (f) no reaction; (g) no reaction; (h) Ph₂C=O;

2,6-dimethylaniline

(i) 1-methyltetralin   (j) *p-tert*-butyltoluene (H₃C—C₆H₄—C(CH₃)₃)   (k) phenylcyclopentane

18.53 1. Treatment of toluene with bromine under light irradiation; 2. magnesium metal in ether; 3. D₂O; 4. sulfuric acid, nitric acid

18.54 (a) S$_N$1

(b) Rate = k_1[*p*-methoxybenzyl bromide]

(c) Doubling the concentration of the *p*-methoxybenzyl bromide doubles the rate. However, doubling the concentration of the hydroxide ion has no effect on the rate of the reaction.

(d) [Structure: CH₃-O⁺=C₆H₄=CH₂ quinoid resonance form]

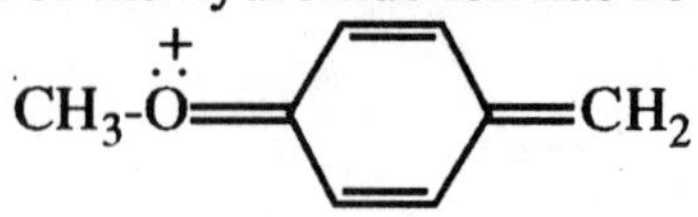

(e) The rate would decrease.

18.55

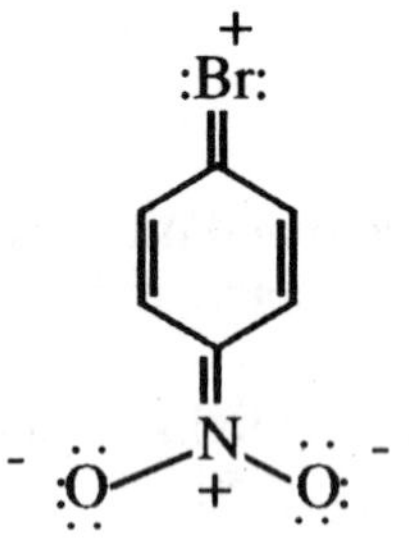

[Structure 1: quinoid form with +Br: at top and NO₂ at bottom]
less important (smaller dipole than nitrobenzene)

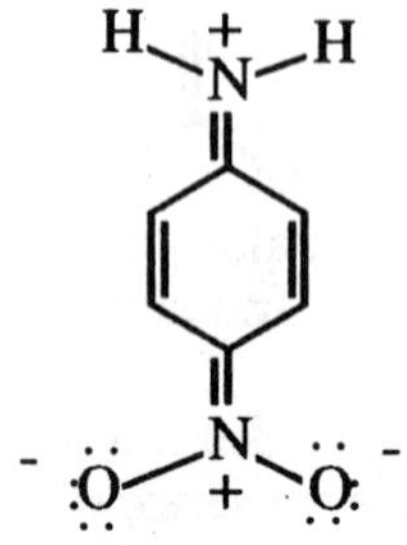

[Structure 2: quinoid form with +NH₂ at top and NO₂ at bottom]
more important (larger dipole than nitrobenzene)

18.56 The *p*-nitrophenol has a smaller dipole moment than does *m*-nitrophenol, and *p*-methylphenol has a larger dipole moment than does *m*-methylphenol.

18.57 Phenol is more acidic than is cyclohexanol. We judge this on the basis of resonance delocalization of the negative charge of the conjugate base of phenol (phenoxide ion), in which the negative charge is spread over the ring, whereas the charge is localized on oxygen in the conjugate base of cyclohexanol.

18.58

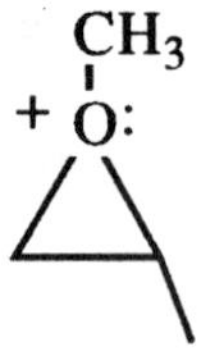

[Structure: protonated methyloxirane (oxonium ion) with CH₃ on oxygen]

Attack by the aromatic ring occurs at the less hindered site of the oxonium ion.

18.59

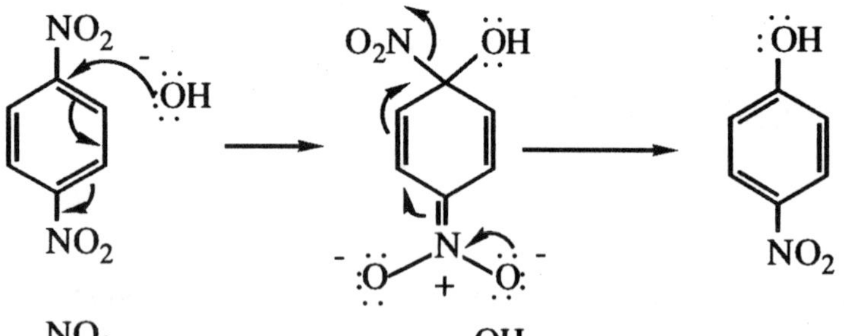

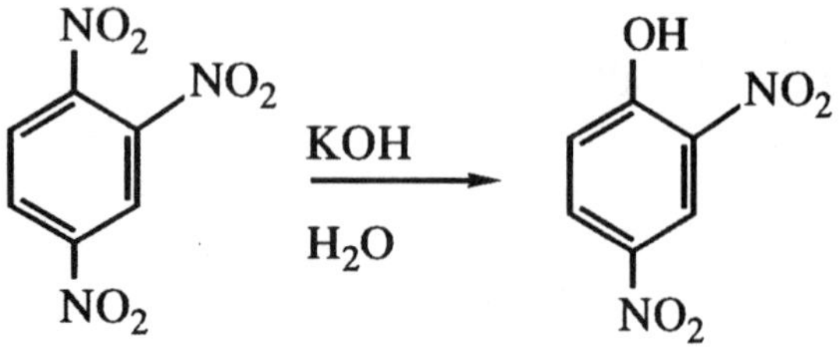

[Mechanism: hydroxide attacks 1,4-dinitrobenzene at carbon bearing NO₂, forming Meisenheimer intermediate, then loss of nitrite gives 4-nitrophenol]

[Reaction: 1,2,4-trinitrobenzene + KOH/H₂O → 2,4-dinitrophenol (with NO₂ at 2 and 4 positions)]

18.60

[Structure: 1,2,3,4-tetrahydronaphthalene]

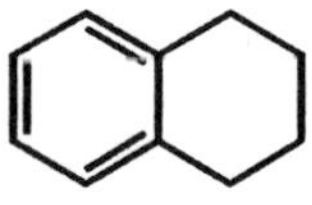

Two separate stepwise Friedel-Crafts alkylations occur, the second one intramolecularly.

18.61

18.62 A: indene; B: indan-1,2-diol; C: indane

18.63
(a) PhCH₂CH₂CH₂Ph
(b) PhCHBr-CH₂-C₆H₄-NO₂
(c) 2,4,6-trinitrophenol with Cl (picryl-type), structure as drawn
(d) α-tetralone
(e) 3-ethylbenzonitrile
(f) PhC(O)CH₂-C₆H₄-SO₃H

18.64 (a) 1. treatment of toluene with bromine under light irradiation; 2. magnesium metal in ether; 3. ethanal (CH₃CHO); 4. workup with water

(b) 1. treatment of toluene with sulfuric acid and nitric acid; 2. bromine under light irradiation; 3. sodium ethoxide in a Williamson ether synthesis; 4. reduction with tin and hydrochloric acid

(c) 1. treatment of ethylbenzene with bromine under light irradiation; 2. elimination using potassium *tert*-butoxide; 3. reaction with bromine in carbon tetrachloride solution 4. double elimination upon heating with sodium amide

(d) 1. treatment of toluene with sulfur trioxide and sulfuric acid; 2. heating with potassium hydroxide; 3. NaOH; 4. bromoethane; 5. heating with potassium permanganate in aqueous base, followed by acidic workup

(e) 1. bromobenzene heated with sulfuric acid and nitric acid to give dinitration; 2. treatment with aqueous potassium hydroxide; 3. reduction with tin and hydrochloric acid

(f) 1. treatment of benzene with bromine in the presence of iron; 2. sulfur trioxide in sulfuric acid; 3. heating with potassium hydroxide, followed by workup with aqueous acid

(g) treatment of phenylacetylene with mercuric sulfate in aqueous sulfuric acid

(h) 1. treatment of benzyl alcohol with phosphorus tribromide; 2. magnesium metal in ether; 3. ethylene oxide, followed by aqueous workup

(i) 1. treatment of 1-phenylpropane with bromine under light irradiation; 2. elimination using potassium *tert*-butoxide; 3. *N*-bromosuccinimide; 4. bromine in carbon tetrachloride

(j) 1. treatment of benzenesulfonic acid with bromine in the presence of iron; 2. heating with potassium hydroxide, followed by acidic workup

170 Study Guide and Solutions Manual

18.65

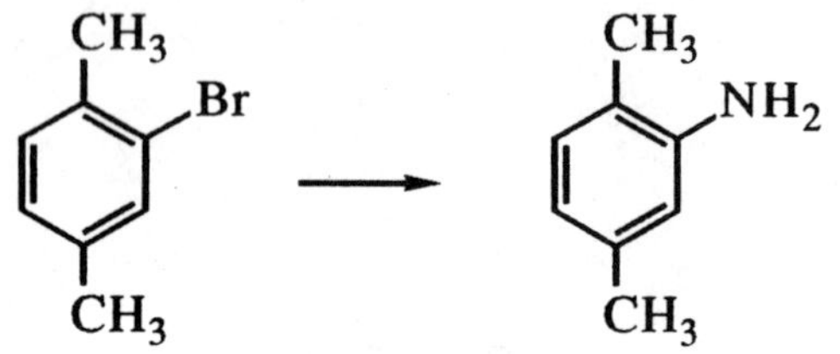

2-bromo-1,4-dimethylbenzene → 2-amino-1,4-dimethylbenzene (CH₃ ortho, NH₂, CH₃ para)

18.66

CH₂CH₃	CH₂CH₃	CH(Br)CH₃	CH(Br)CH₂Br	C≡CH	C(O)CH₃
(no NO₂)	p-NO₂	p-NO₂	p-NO₂	p-NO₂	p-NO₂
D	E	F	G	H	I

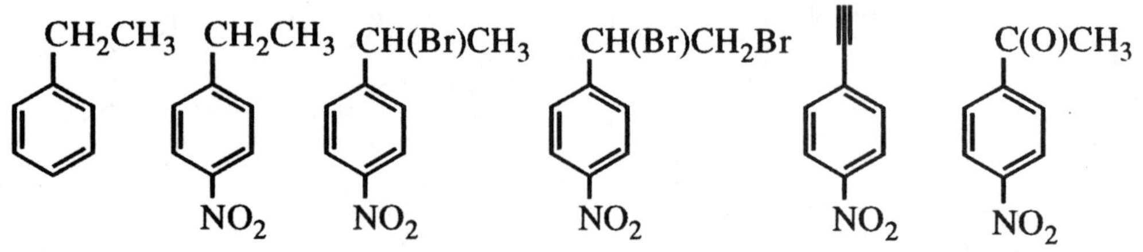

Solution of Study Guide Practice Problems *Correct!*

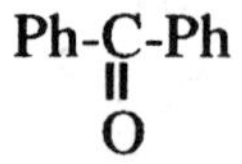

18.1

(a) 4-hydroxyaniline (HO–C₆H₄–NH₂)

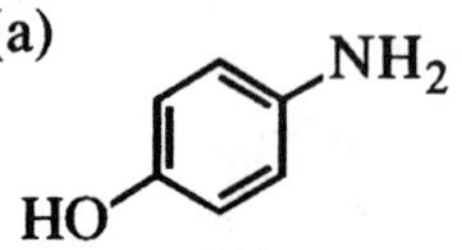

(b) Ph–C(=O)–Ph

(c) 1,3-dimethylbenzene (H₃C–C₆H₄–CH₃, meta)

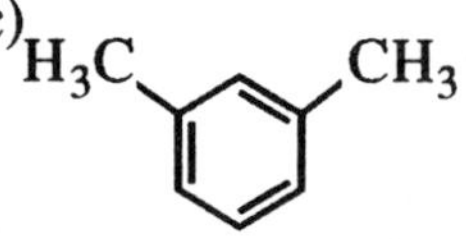

(d) 2-nitrobenzoic acid (o-NO₂, CO₂H)

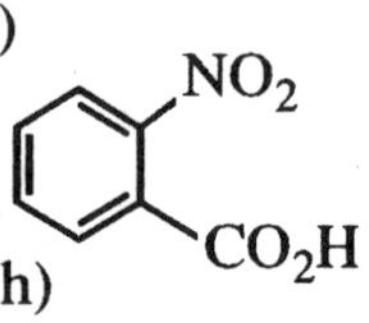

(e) 4-bromobenzonitrile (Br–C₆H₄–CN)

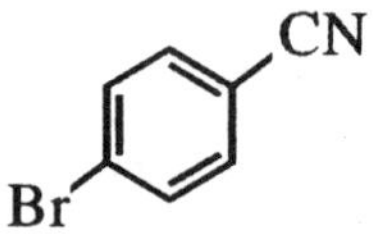

(f) 1-methoxy-2,4-dinitrobenzene (OCH₃ with O₂N at 4 and NO₂ at 2)

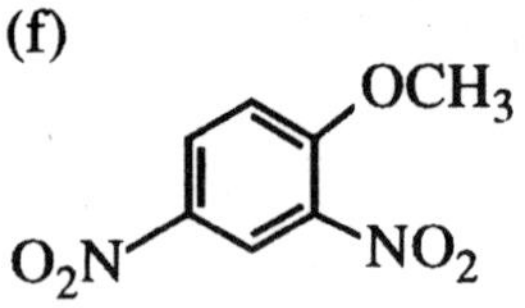

(g) 4-nitrobenzyl bromide (O₂N–C₆H₄–CH₂Br)

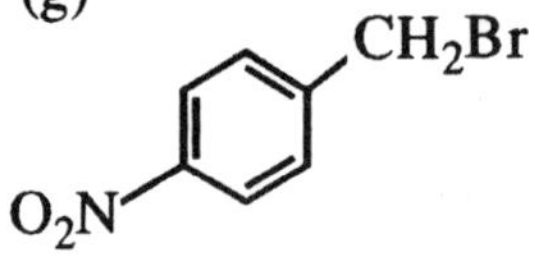

(h) Ph–CH=CH–CH₂CH₃ (cis)

18.2 See Table 18.1 of the text.

18.3

$$H\!-\!\ddot{O}\!-\!\ddot{O}\!-\!H + HSO_3F \longrightarrow H\!-\!\overset{+}{\underset{H}{O}}\!-\!\ddot{O}\!-\!H + FSO_3^-$$

[Benzene attacks H–O⁺(H)–O–H → cyclohexadienyl cation with H and :OH substituents + H₂Ö]

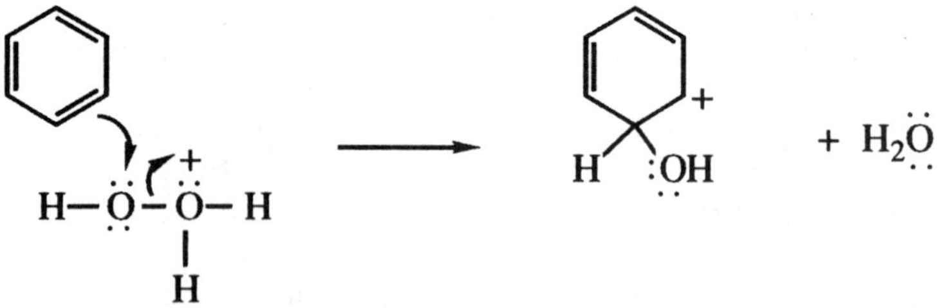

The reaction continues *via* the usual steps of electrophilic aromatic substitution reactions as shown in section 18.4 of the text.

18.4

[Reaction mechanism scheme showing protonation of cyclic ether, ring opening to carbocation with OH group, electrophilic aromatic substitution on benzene, then acid-catalyzed loss of water to form a second carbocation, which undergoes intramolecular electrophilic aromatic substitution to generate the tetrahydronaphthalene product.]

This carbocation undergoes electrophilic aromatic substitution on benzene.

This carbocation undergoes electrophilic aromatic substitution intramolecularly to generate the product.

18.5   -NH$_2$; -OCH$_3$; -CH$_2$CH$_3$; -Cl

18.6   -NH$_2$; -OCH$_3$; -CH$_2$CH$_3$

18.7

(a) [3-nitro-4-methoxybiphenyl structure]

(b) [N-(4-nitrophenyl)benzamide structure]
(plus *ortho*)

(c) [bis(4-nitrophenyl) ether structure]
(plus *ortho*)

18.8

[Sigma complex intermediate with Br and H on sp3 carbon, and =Cl$^+$ on the ring]

18.9 (a) 1. treatment of benzene with nitric acid and sulfuric acid; 2. reaction with bromine in the presence of iron; 3. reduction with tin and hydrochloric acid
     (b) treatment of benzene with iodomethane in the presence of aluminum chloride; 2. reaction with nitric acid and sulfuric acid; 3. heating with potassium permanganate in aqueous base, followed by acidic workup

18.10 (a) *tert*-Butylbenzene will not undergo side-chain oxidation since there is no benzylic hydrogen present.
     (b) Neither reaction will work. Nitrobenzene is too deactivated to undergo the reaction, and aniline forms an acid-base complex with the aluminum chloride which renders the ring highly deactivated.
     (c) Chlorobenzene will not undergo the reaction with potassium cyanide. Without particularly electron-withdrawing groups present, aryl halides are generally inert to nucleophilic substitution reactions.
     (d) The reaction with *p*-nitrophenylmagnesium bromide will not work. The nitro group will react to destroy the Grignard reagent.

18.11

Ph—CH₃ + :Br· ⟶ Ph—CH₂· + H–Br:

Ph—CH₂· + Br₂ ⟶ Ph—CH₂Br + :Br·

18.12 1. treatment of toluene with bromine under light irradiation; 2. addition of sodium acetylide (formed by the reaction of acetylene with sodium amide in liquid ammonia)

18.13 (a) PhCH(Br)CH₃ (b) PhCH₂CH₂Br

18.14 See section 18.16 of text.

18.15

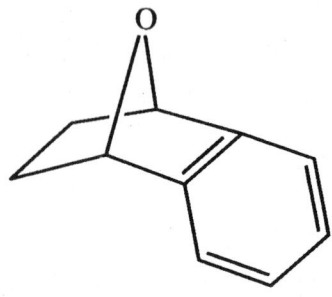

[Structure: bicyclic oxabridge fused to benzene ring, with O bridging two carbons]

18.16 *m*-bromoaniline

18.17 4-nitrochlorobenzene - Attack by hydroxide ion at the chlorine-bearing carbon site leads to a resonance stabilized carbanion.

CHAPTER 19
ALDEHYDES AND KETONES: PREPARATION, PROPERTIES AND NUCLEOPHILIC ADDITION REACTIONS

Key Points

• Learn the classification and nomenclature of aldehydes and ketones. Pay particular attention to systematic nomenclature, and learn the uses of the terms -carbaldehyde, -phenone, formyl and oxo.

Practice Problem 19.1
Draw structures for each of the following compounds:
- (a) 4-formylbenzoic acid
- (b) 5-oxohexanenitrile
- (c) propiophenone
- (d) β-bromobutyraldehyde
- (e) 3,3-diethylcyclohexanecarbaldehyde

• Be familiar with the preparation of aldehydes and ketones by a variety of methods, including:
  - oxidation of alcohols
  - hydration of alkynes
  - Friedel-Crafts acylation of aromatics
  - Reimer-Tiemann reaction
  - pinacol-pinacolone rearrangement
  - reaction of acid halides with organometallics (for ketones) or with reducing agents (for aldehydes)

Practice Problem 19.2
Give the reagents or combination of reagents that will effect the following synthetic conversions:

(a) 3,4-dimethoxybenzoyl chloride → 3,4-dimethoxybenzaldehyde

(b) 1-methylcyclopentanecarbonyl chloride → 1-acetyl-1-methylcyclopentane

(c) 1-ethynylcyclohexanol → 1-acetylcyclohexanol

(d) p-cresol → 2-hydroxy-5-methylbenzaldehyde

(e) n-C$_5$H$_{11}$-C≡C-H → n-C$_5$H$_{11}$CH$_2$CHO

(f) 5-isopropyl-3-methylcyclohexanol → 5-isopropyl-3-methylcyclohexanone

(g) 2-(hydroxymethyl)-2-norbornanol → 2-norbornanone

• Know the general mechanism for nucleophilic addition to the carbonyl group under neutral or acidic conditions.

174 Study Guide and Solutions Manual

Practice Problem 19.3
Write a complete mechanism for the reaction of cyclohexanone with KCN and aqueous sulfuric acid.

• Learn the main features relating to the addition of *oxygen* nucleophiles to aldehydes and ketones (*e.g.* water and alcohols). Learn the use of acetals and ketals as protecting groups.

Practice Problem 19.4
Write a mechanism for the overall conversion shown below.

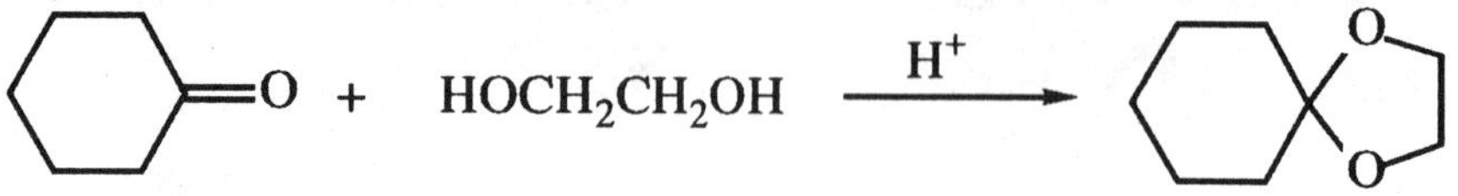

cyclohexanone + HOCH$_2$CH$_2$OH $\xrightarrow{H^+}$ cyclohexanone ethylene ketal

Practice Problem 19.5
Propose synthetic routes for the following transformations.
(a) 3-bromocyclopentanecarbaldehyde → 3-(2-hydroxypropan-2-yl)cyclopentanecarbaldehyde
(b) 3-ethynylcyclopentanone → 3-(prop-1-yn-1-yl)cyclopentanone

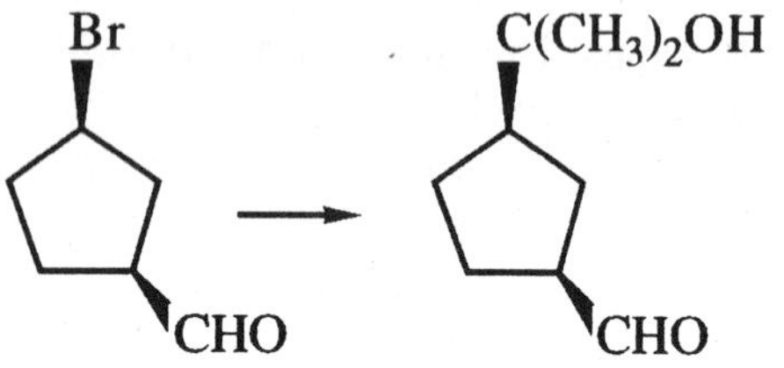

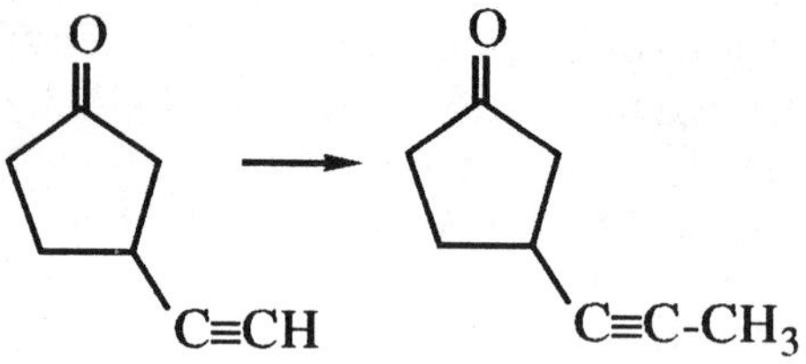

• Learn the reactions that occur between amines (and other nitrogen nucleophiles) and aldehydes and ketones.

Practice Problem 19.6
Using the mechanistic principles you have learned, propose a structure for the product C$_{13}$H$_9$N$_3$O$_5$ that you would expect to form initially in the reaction:

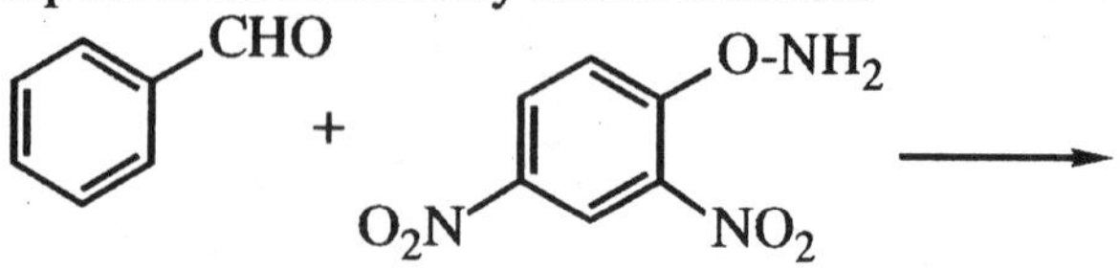

benzaldehyde (PhCHO) + 2,4-dinitrophenyl-O-NH$_2$ →

Actually, the observed products are benzonitrile and the 2,4-dinitrophenoxide ion. Propose a mechanism for the formation of these products, proceeding from the structure you proposed. The reaction does *not* proceed to benzonitrile and phenoxide ion if the nitro groups are absent from the reactant. Offer an explanation for this result (review Chapter 18).

Practice Problem 19.7
Give the products of each of the following reactions, each performed in the presence of a catalytic amount of acid.
(a) PhNH$_2$ + 2-pyridinecarbaldehyde →
(b) PhNH$_2$ + (CH$_3$)$_2$CHCH$_2$CHO →
(c) PhNHCH$_3$ + (CH$_3$)$_2$CHCH$_2$CHO →

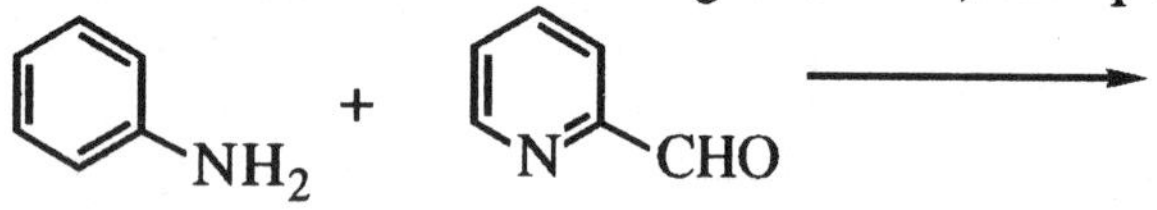

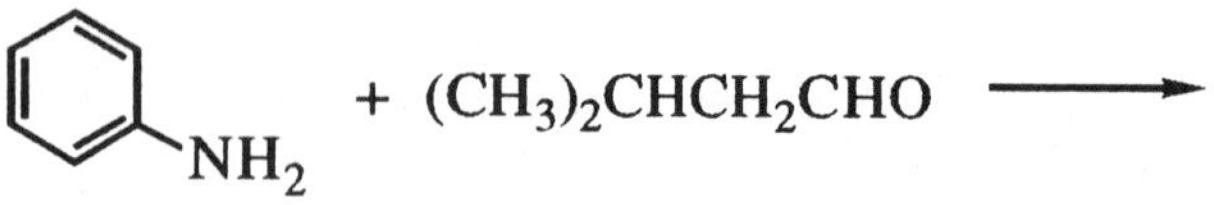

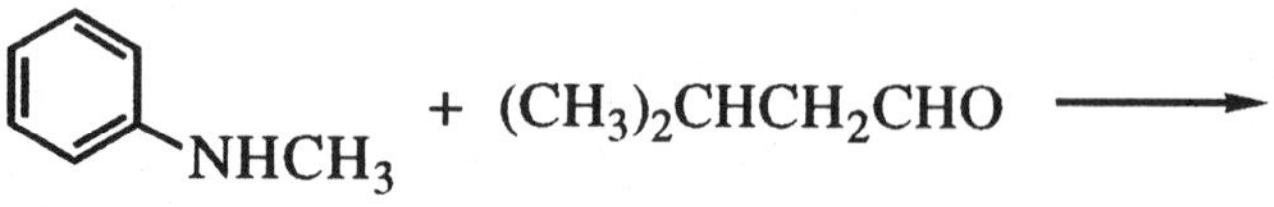

• Learn the different types of reactions between aldehydes (or ketones) and substances containing nucleophilic carbon. In particular, review the reactions of organometallics (Grignard reagents, alkyllithiums) with carbonyl compounds and learn the Wittig reaction.

Practice Problem 19.8
Propose a synthesis of 2-phenyl-2-propanol using benzene as your aromatic starting material.

Practice Problem 19.9
Which aldehyde and which haloalkane would you begin with in a Wittig synthesis of the compound shown below? Show all steps in the synthesis.

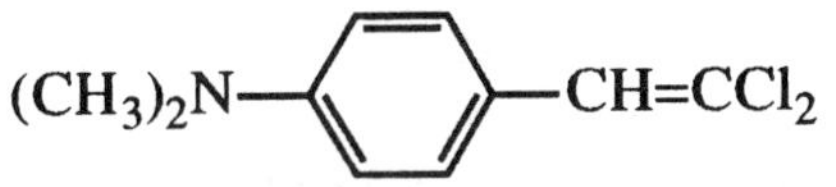

(CH₃)₂N—C₆H₄—CH=CCl₂

Practice Problem 19.10
Propose structures for the compounds indicated A-E.

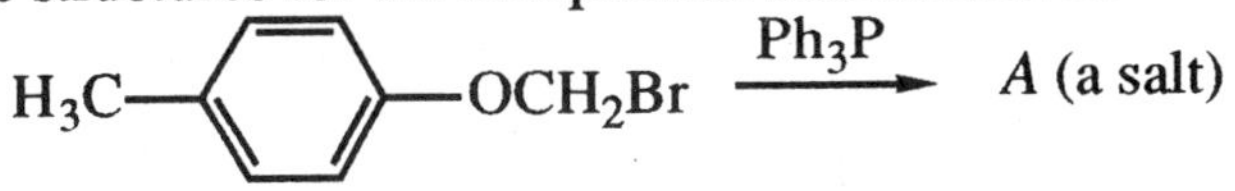

H₃C—C₆H₄—OCH₂Br $\xrightarrow{Ph_3P}$ A (a salt)

A $\xrightarrow{CH_3CH_2CH_2CH_2Li}$ B (an ylid)

B + PhCHO ⟶ C (C₁₅H₁₄O)

C $\xrightarrow{\text{acid hydrolysis}}$ D (C₈H₈O) + E

Give a mechanism for the conversion of C to D and E (see Chapter 14).

• Learn the chemistry of thioacetals and thioketals. Pay particular attention to their mode of preparation, their reduction with Raney nickel, the enhanced acidity of certain members of their class, and their hydrolysis.

Practice Problem 19.11 *Many corrections*
Provide structures for each of the compounds indicated A-D.

H₂C=O + 1,3-propanedithiol $\xrightarrow[\text{ethanol}]{BF_3 \quad KOH}$ A (C₄H₈S₂)

A $\xrightarrow[THF]{CH_3CH_2CH_2CH_2Li}$ B (C₄H₇S₂Li)

B + (propylene oxide, CH₃-substituted epoxide) ⟶ C (C₆H₁₃S₂OLi)

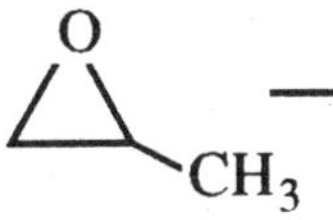

C $\xrightarrow{\text{aqueous acid}}$ D (C₆H₁₄S₂O)

Practice Problem 19.12
How would you effect the following synthetic transformations?

(a) cyclohexanone ⟶ cyclohexanol with D substituents at 2,2,6,6 positions and OH

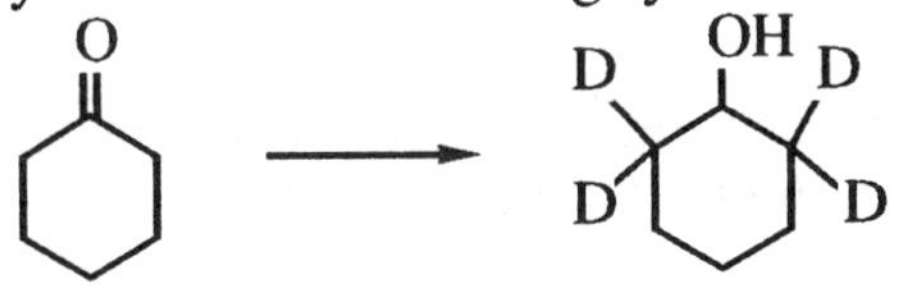

(b) cyclopentyl-CHO ⟶ 1-bromo-1-formylcyclopentane (CHO and Br on same carbon)

(c) (CH₃)₃CC(O)CH₃ ⟶ (CH₃)₃CCO₂H

Practice Problem 19.13
Which of the following compounds would you expect to racemize in aqueous basic solution?
  (a) (R)-2-ethyl-2-methylcyclopentanone
  (b) (R)-2-ethylcyclopentanone
  (c) (R)-3-ethylcyclopentanone
  (d) (R)-3-ethyl-3-methylcyclopentanone

Practice Problem 19.14
Which of the following will give a positive iodoform test?
  (a) cyclopentanol; (b) 2-butanol; (c) 2-butanone; (d) 3-hexanone

- Know the reagents used to oxidize and to reduce aldehydes and ketones.

Practice Problem 19.15

What reagents or combination of reagents would you use for each of the following conversions?

(a) Ph-C(=O)-Ph ⟶ Ph-CH(OH)-Ph

(b) HC(=O)–C₆H₄–CH₂OH ⟶ HO₂C–C₆H₄–CH₂OH

(c) HC(=O)–C₆H₄–OH ⟶ H₃C–C₆H₄–OH

(d) HO–C₆H₄–C(O)CH₃ ⟶ HO–C₆H₄–CO₂H

Solution of Text Problems

19.1 Although the position number for the carbonyl group is often used, it is not necessary as there is only one position at which the carbonyl group could be located in each size ketone.

19.2 methyl phenyl ketone *for* acetophenone
p-bromophenyl ethyl ketone *for* *p*-bromopropiophenone
2-naphthyl methyl ketone *for* acetonaphthone
diphenyl ketone *for* benzophenone

19.3 (a) methanal; (b) propanal; (c) methylpropanal; (d) 2,4-dimethyl-3-pentanone; (e) 3,5-dimethyl-4-heptanone

19.4 The methyl groups of acetone are electron donating toward the carbonyl carbon, whereas the hydrogens of formaldehyde are not electron donating. The methyl groups of acetone supplement the dipole of the C=O linkage.

19.5 There is a greater dipole-dipole interaction in acetone than there is in propanal (propionaldehyde).

19.6 (a) *A* is 3-pentanone
(b) *B* is methylpropanal (isobutyraldehyde)
(c) *C* is butanone (methyl ethyl ketone)
(d) *D* is 1-methoxy-2-propanone [methoxyacetone or CH₃OCH₂C(O)CH₃]
(e) *E* is 4,4-dimethyl-2-pentanone
(f) *F* is isopropyl methyl ketone

19.7 *A* is *p*-nitroacetophenone and *B* is *p*-methoxyacetophenone. Electron donation from the methoxy group in *B* through the π system gives the carbon-oxygen linkage greater single-bond character and less double-bond character.

19.8 Borane itself would result in further reduction of the alkyne.

19.9

[Mechanism: HC(=O)F + BF₃ ⟶ HC⁺=O + BF₄⁻; then addition of HC⁺=O to mesitylene gives arenium ion intermediate, then –H⁺ gives the aryl aldehyde product]

19.10 (a) PhCH₂CHO
(b) 3-phenyl-2-butanone [CH₃C(O)CHPhCH₃]
(c) butanal
(d) 2,2-diphenylcyclohexanone

(e) [spiro ketone structure]

19.11 Cyclohexanone is the limiting reagent (0.20 mole). Thereby, 0.20 mole of water, or 3.6 mL, would be generated in this reaction.

19.12 The change in entropy is negative. Two particles are being formed starting with three. [margin note: enthalpy Δ must be – if K>1 because ΔG<0.]

19.13 (a) 1. treatment of 3-bromopropanal with ethylene glycol (ethane-1,2-diol) in the presence of acid to form the ethylene acetal; 2. addition of the sodium salt of 1-butyne (formed from 1-butyne and sodium amide); 3. aqueous acid deprotection

(b) treatment of 4-bromobenzaldehyde with ethylene glycol in the presence of acid to form the ethylene acetal; 2. Mg in ether; 3. D$_2$O; 4. aqueous acid

19.14 [mechanism showing hydrolysis of imine PhCH=NH via protonation, water addition, and loss of NH$_3$ to give benzaldehyde]

19.15 [mechanism showing reaction of benzaldehyde with phenylhydrazine PhNH-NH$_2$ via protonation, addition, proton transfers, and loss of water to give phenylhydrazone]

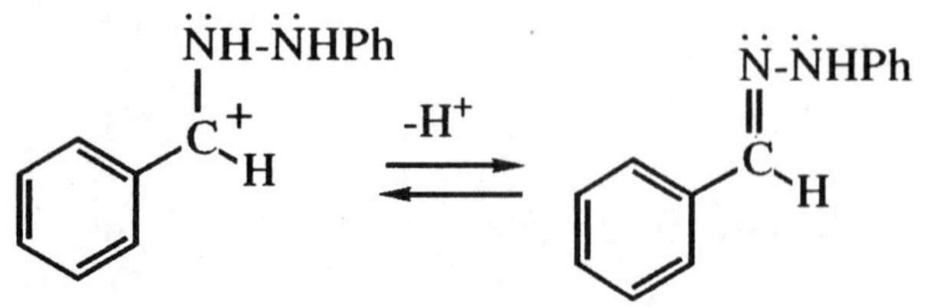

178 Study Guide and Solutions Manual

[structure: PhCH(NH-NHPh)+ ⇌ (-H+) PhCH=N-NHPh]

19.16 This would require the (positively charged) nitrogen atom to form five bonds. A nitrogen atom does not have the orbitals available to do this.

19.17

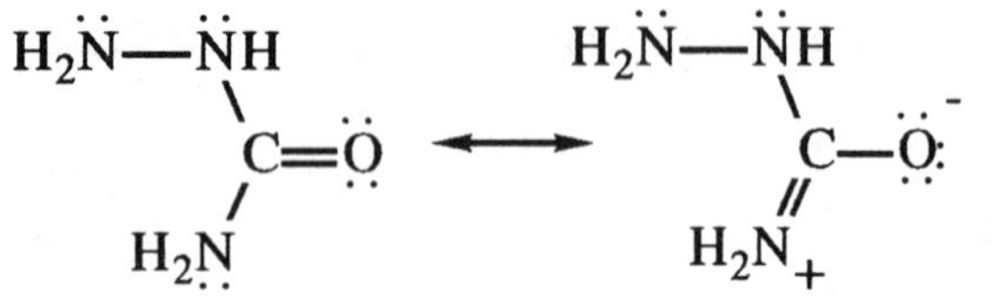

[structure: (H2N)(H2N-NH)C=O ↔ (H2N-NH)(H2N+)C-O:–]

The unshared electron pair on the one nitrogen is delocalized making that nitrogen diminished in nucleophilicity.

19.18 $(CH_3)_2C=N-OH$

19.19 The reaction stops at the intermediate iminium ion since no α-hydrogen is available to be lost.

19.20

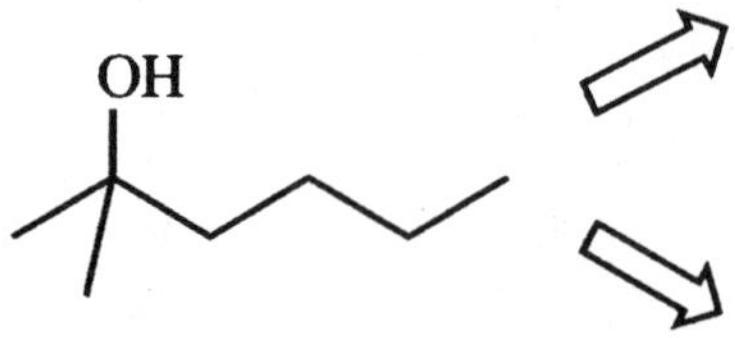

[retrosynthesis of tertiary alcohol (CH3)2C(OH)CH2CH2CH2CH3 →
 $(CH_3)_2C=O + CH_3CH_2CH_2CH_2MgX$
 or $CH_3MgX + CH_3CH_2CH_2CH_2C(O)CH_3$]

19.21 1. treatment of toluene with bromine under light irradiation; 2. Mg in ether to generate the benzyl Grignard reagent; 3. addition of isobutyraldehyde (formed from the oxidation of isobutyl alcohol with chromic anhydride in pyridine); 4. aqueous acid workup

19.22 (a) pentanal taken in reaction with $Ph_3P=CH_2$
(b) benzaldehyde taken in reaction with $Ph_3P=CH_2$
(c) acetone taken in reaction with $Ph_3P=C(CH_3)_2$
(d) benzophenone taken in reaction with $Ph_3P=CH_2$

19.23 1. 1-butanol treated with phosphorus tribromide to form 1-bromobutane; 2. treatment with NaSH

19.24

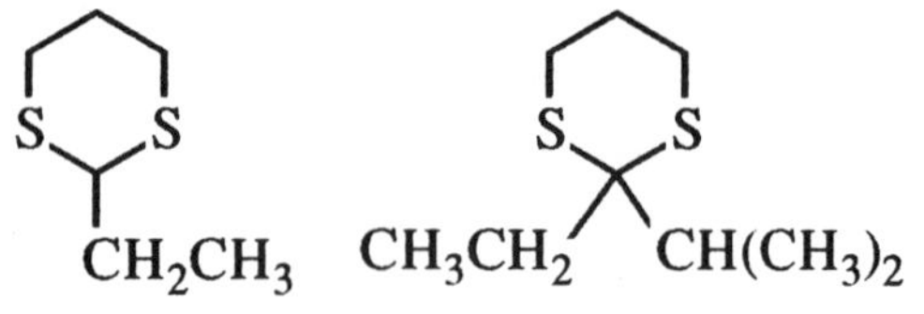

[structures: A = 1,3-dithiane with CH2CH3 substituent; B = 1,3-dithiane with CH3CH2 and CH(CH3)2 substituents; C = $(CH_3)_2CH-\underset{\underset{O}{\|}}{C}-CH_2CH_3$]

    A   B   C

19.25 10^{-3}

19.26

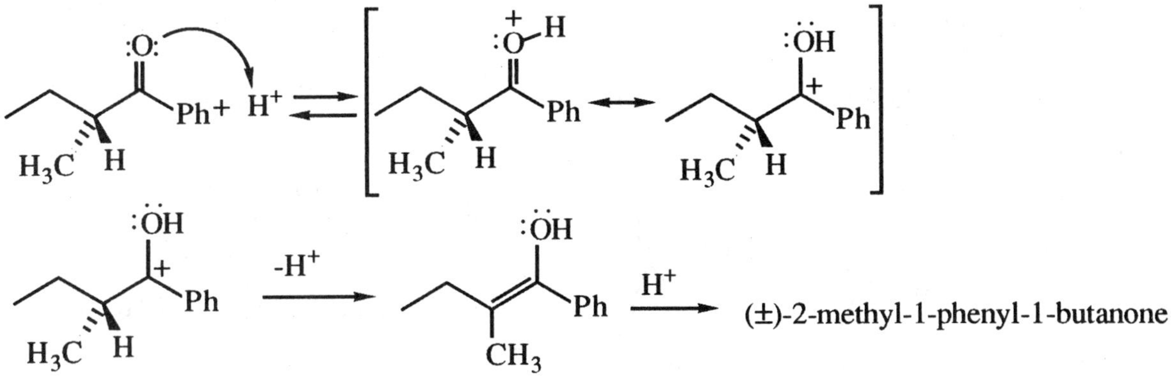

[Mechanism diagram: racemization of 2-methyl-1-phenyl-1-butanone via enol intermediate]

Starting ketone (with H₃C, H on stereocenter, C(=O)Ph) + H⁺ ⇌ [protonated ketone resonance structures with :Ö-H⁺ and :ÖH with C+] → intermediate cation → (-H⁺) → enol (with :ÖH, C=C, CH₃, Ph) → (H⁺) → (±)-2-methyl-1-phenyl-1-butanone

19.27 There are three exchangable hydrogen atoms. The molecular weight of 100 g/mole fits a formula of $C_6H_{12}O$, leading to two possible structures: $(CH_3)_2CHC(O)CH_2CH_3$ and $(CH_3)_3CC(O)CH_3$.

19.28 Three compounds, 2-pentanone, 2-pentanol, and acetophenone would give positive iodoform tests.

19.29

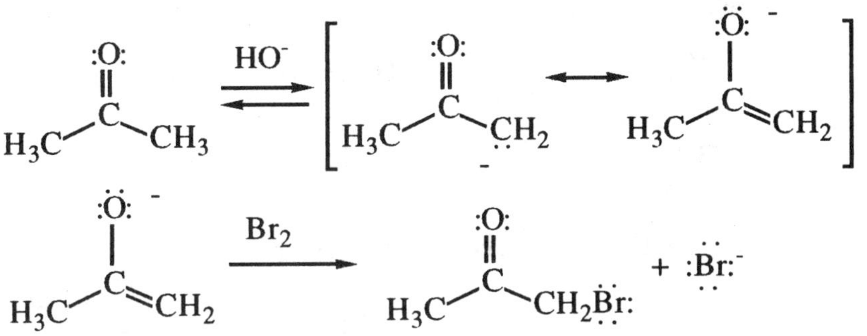

[Mechanism: acetone + HO⁻ ⇌ enolate resonance structures → + Br₂ → bromoacetone (H₃C-C(=O)-CH₂Br) + :Br⁻]

The second step is the faster, and the first step is thereby rate-determining. Iodine reacts faster than does bromine in this reaction.

19.30

[Mechanism: acetone + H₃O⁺ ⇌ protonated resonance structures → enol (via H₂O) → + Br₂ → bromoacetone + :Br⁻]

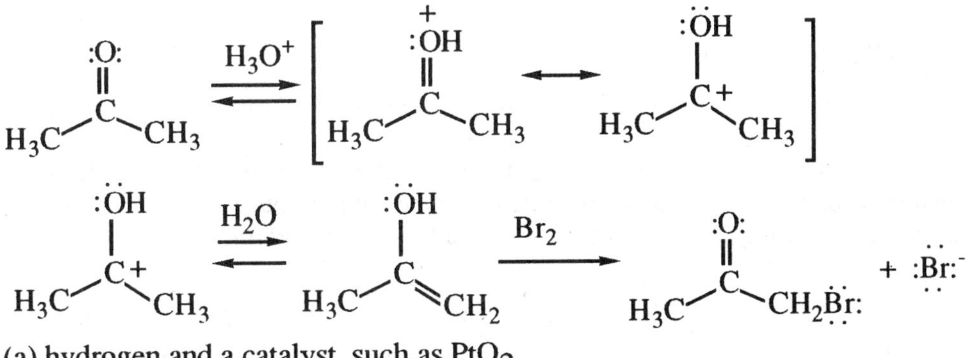

19.31 (a) hydrogen and a catalyst, such as PtO_2

(b) 1. sodium borohydride; 2. aqueous acid workup

19.32

(a) (E)-2-butenal: H₃C-CH=CH-CHO (with H's shown)

(b) PhCH₂-C(O)-Ph

(c) (CH₃)₂CHCHO

(d) 3-chloro-... CH₃CH₂CH(Cl)CH₂CHO

(e) (CH₃)₂CH-C(O)-CH₂CH₂-C≡C-CH(Br)CH₃

(f) 1-methyl-3-formylcyclobutane (H₃C and H on one carbon, CHO and H on opposite)

(g) 3-oxocyclohexanecarbaldehyde

(h) PhCH=CHCHO

(i) (CH₃)₃CCH₂-C(O)-CH₂CH(Br)CH₂CH₃

(j) 2-methyl-2-phenyl-1,3-dioxolane (Ph and CH₃ on acetal carbon)

(k) PhCH=N-NH-C₆H₃(NO₂)₂ (2,4-dinitrophenylhydrazone)

(l) CH₃CH₂C(=N-NHC(O)NH₂)CH₃ (semicarbazone of 2-butanone)

(m) cyclopentylidene=PPh₃

(n) CH₃CH(OH)C(CH₃)(H)CH₂CH=CH₂ ... with OH, H₃C, CH₂CH=CH₂

(o) CH₃C(N(CH₂CH₃)₂)=CH₂ (enamine)

(p) 1-pyrrolidinocyclopentene

(q) CH₃CH₂-C(O)-CH₂CH₃

(r) CH₃CH(Cl)CHO

(s) PhC(OH)=CH₂

(t) 3,4-dihydro-2H-pyrrole

(u) 1-methylcyclohexanol

(v) H₃C-C₆H₄-C(O)CH(CH₃)₂

(w) (CH₃)₂C=CHCH₃

(x) HOCH₂-C₆H₄-CHO

(y) Ph₂C(OH)CH₃

(z) (CH₃)₂CHC(O)CH₃

19.33 A is 2-bromobutane; B is 2-butanol; C is methyl ethyl ketone. Compounds B and C give positive iodoform tests.

19.34 (a) treatment of 2-pentanone with methylmagnesium iodide followed by aqueous acidic workup
(b) treatment of propanal with phenylmagnesium bromide followed by aqueous acidic workup
(c) treatment of propanal with ethylmagnesium iodide followed by aqueous acidic workup

19.35
(a) PhCO₂H
(b) Ph-CH(OCH₂CH₂O) (1,3-dioxolane of benzaldehyde)
(c) PhCO₂H
(d) PhCH₂OH
(e) PhCH(OH)CN
(f) PhCH(OH)SO₃H
(g) 3-nitrobenzaldehyde (O₂N-C₆H₄-CHO)
(h) PhCH₂OH
(i) Ph-CH(SCH₂CH₂CH₂S) (1,3-dithiane)
(j) PhCH(OCH₂CH₃)₂
(k) PhCH=NCH₂CH₃
(l) PhCH(OH)C≡CH
(m) PhCH₃
(n) 3-bromobenzaldehyde (Br-C₆H₄-CHO)

19.36 2-bromobutane

19.37 (a) PhCO₂H; (b) PhC(O)CD₃; (c) no reaction; (d) PhCH(OH)CH₃; (e) (CH₃CH₂)₂NC(Ph)=CH₂; (f) PhCH₂CH₃; (g) no reaction; (h) no reaction; (i) PhC(O)CH₂Br

19.38 (a) treatment of benzene with acetic anhydride in the presence of aluminum chloride with aqueous workup
(b) treatment of phenylacetylene with water, sulfuric acid, and mercuric sulfate

(c) 1. treatment of ethanal (acetaldehyde) with phenylmagnesium bromide followed by aqueous acidic workup; 2. oxidation by chromic anhydride with sulfuric acid

19.39 1. cyclopentanone treated with sodium borohydride, followed by aqueous acid workup; 2. phosphorus tribromide; 3. triphenylphosphine; 4. $CH_3CH_2CH_2CH_2Li$; 5. cyclopentanone

19.40 E is $Ph_2C(CH_3)C(O)CH_3$; F is $Ph_2C(CH_3)CO_2H$.

19.41

[Structure: CH₃-C(=O)-CH₂-CH(OH)-CH₂CH₃]   The intermediate anion   [Structure: 1,3-dithiane anion]

attacks the less hindered site of the oxirane ring, opening the ring.

19.42 (a) 1. iodomethane treated with triphenylphosphine to form a phosphonium salt; 2. butyllithium; 3. benzophenone

(b) 1. 2-butanone treated with sodium borohydride, followed by aqueous acid workup; 2. sodium hydride; 3. iodomethane

(c) 1. 1-pentene treated with borane; 2. hydrogen peroxide in aqueous basic solution; 3. chromic anhydride in pyridine

(d) 1. benzene treated with acetic anhydride in the presence of aluminum chloride, followed by water workup; 2. chlorine in the presence of iron; 3. reduction with tin and hydrochloric acid

(e) acetophenone treated with sodium borohydride, followed by aqueous acidic workup

(f) 1. 4-hydroxycyclohexanone treated with ethylene glycol in the presence of a catalytic amount of strong acid to form the ethylene ketal; 2. triphenylphosphine with carbon tetrabromide; 3. magnesium metal in ether; 4. D_2O; 5. aqueous acid

19.43

[Mechanism scheme showing acid-catalyzed ketal formation from a ketone and ethylene glycol, proceeding through protonation, addition of HOCH₂CH₂OH, proton transfers, loss of H₂O, intramolecular cyclization, and deprotonation to give the cyclic ketal.]

19.44 cyclopentanone

19.45

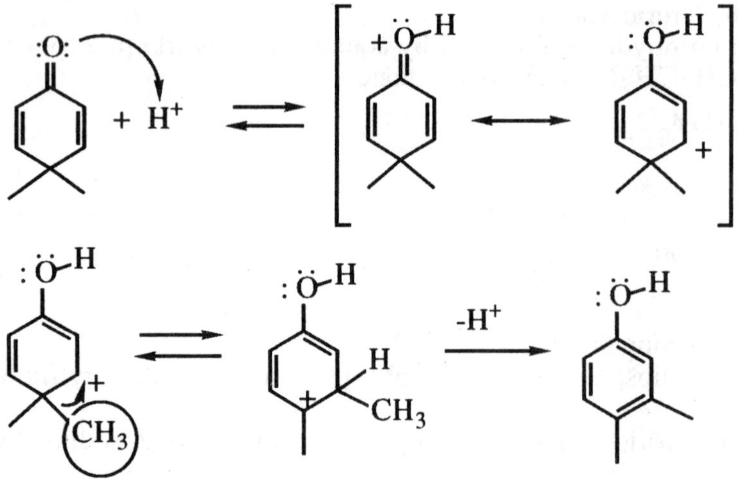

[Reaction mechanism showing protonation of 4,4-dimethylcyclohexadienone, resonance structures, methyl migration, and loss of H+ to give 3,4-dimethylphenol]

19.46

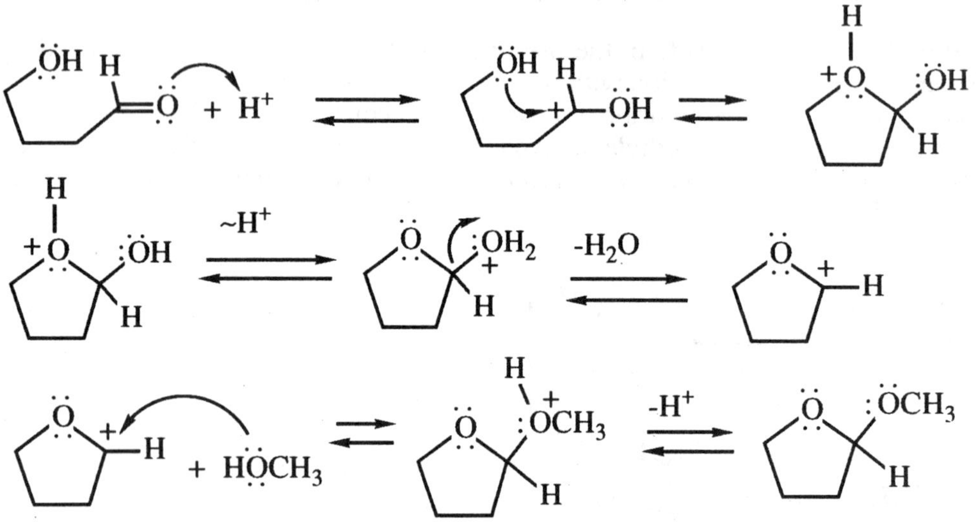

[Reaction mechanism showing acid-catalyzed cyclization of a hydroxy-aldehyde to form a cyclic hemiacetal, then acetal with methanol]

19.47

[Structure of 2-methyl-2-cyclohexenone with D labels at positions 6,6 and 4,4]

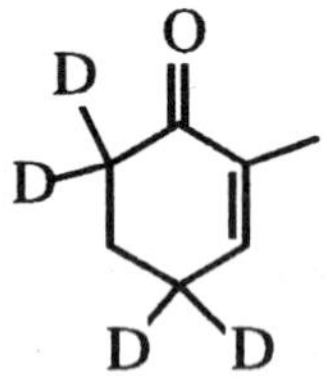

19.48 3,3-dimethyl-2,4-pentanedione
19.49 G isobutyryl chloride
     H phenyl isopropyl ketone
     I isobutylbenzene
19.50 J p-ethylacetophenone
     K 2-(p-ethylphenyl)-2-propanol
19.51 L (4-ethoxyphenyl)acetylene
     M (4-ethoxyphenyl)acetaldehyde
     N p-ethoxyethylbenzene
     O p-ethoxybenzoic acid
     P p-ethoxyacetophenone
     Q 2-(p-ethoxyphenyl)-2-propanol
     R 2-(p-ethoxyphenyl)propene

19.52

[Structure: a cyclohexanone with an OH substituent on the α-carbon, connected via CH(OH)-CH₂CH₃ chain]

Synthesis: 1. treatment of 6-bromo-2-hexanone with ethylene glycol in the presence of an acid catalyst to form the ethylene ketal; 2. triphenylphosphine; 3. butyllithium; 4. propanal (forming preferentially the *trans* alkene); 5. *anti* hydroxylation using hydrogen peroxide and formic acid in water; 6. aqueous acid

19.53

$$CH_3-C\equiv N: + H^+ \rightleftharpoons [CH_3-C\equiv \overset{+}{N}-H \leftrightarrow CH_3-\overset{+}{C}=\overset{..}{N}-H]$$

$CH_3-\overset{+}{C}=\overset{..}{N}-H$ + resorcinol → iminium intermediate

[Mechanism sequence showing attack of resorcinol on the nitrilium ion, forming an iminium Wheland intermediate, loss of H⁺ to give the aryl ketimine (HN=C(CH₃)–Ar), then hydrolysis with H₂O to give the aryl methyl ketone (2,4-dihydroxyacetophenone).]

19.54

[Structures labeled:]
- **S**: cyclopentylidene with isopropylidene (=C(CH₃)₂)
- **T**: 1-(1-hydroxycyclopentyl)-2-hydroxy-2-methyl (diol)
- **U**: 2,2-dimethylcyclohexanone
- **V**: 1,2-dimethylcyclohexan-1-ol
- **W**: 1,2-dimethylcyclohexene
- **X**: 2,3-dimethylcyclohexene (with methyl groups shifted)
- **Y**: 2-methyl-1-methylenecyclohexane; and CH₃C(O)CH₂CH₂CH₂CH₂C(O)CH₃

The formation of *U* from *T* is a pinacol rearrangement, and the formation of *W* and *X* from *V* is a carbocation rearrangement.

Solution of Study Guide Practice Problems

19.1

(a) 4-HO₂C–C₆H₄–CHO

(b) CH₃C(O)CH₂CH₂CH₂CN

(c) Ph–C(=O)–CH₂CH₃

(d) CH₃CHBrCH₂CHO

(e) 1,1-dimethylcyclohexane-3-carbaldehyde

19.2 (a) LiAlH(O-*tert*-butyl)₃ *or* Rosenmund reduction

(b) lithium dimethylcuprate *or* dimethylcadmium

(c) mercuric sulfate, sulfuric acid, water

184 Study Guide and Solutions Manual

(d) chloroform, potassium hydroxide
(e) 1. disiamyl borane; 2. hydrogen peroxide, aqueous potassium hydroxide
(f) sulfuric acid, chromic anhydride
(g) sodium periodate *or* potassium permanganate

19.3

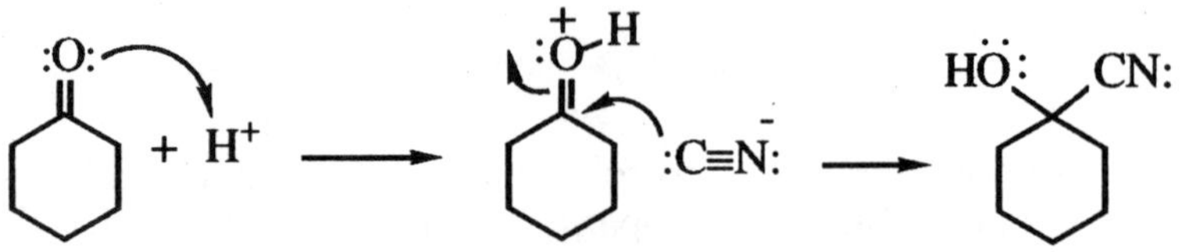

[Mechanism: cyclohexanone + H⁺ → protonated ketone → attack by :C≡N:⁻ → cyclohexanol with OH and CN groups]

19.4

[Mechanism: cyclohexanone + H⁺ ⇌ protonated ketone, attack by HOCH₂CH₂OH ⇌ tetrahedral intermediate with HO and ⁺OCH₂CH₂OH; ~H⁺ shift ⇌ intermediate with OCH₂CH₂OH and H₂O⁺ leaving group ⇌ oxocarbenium ion with :OCH₂CH₂OH; intramolecular cyclization ⇌ protonated cyclic acetal, −H⁺ ⇌ cyclic ketal (1,3-dioxolane of cyclohexanone)]

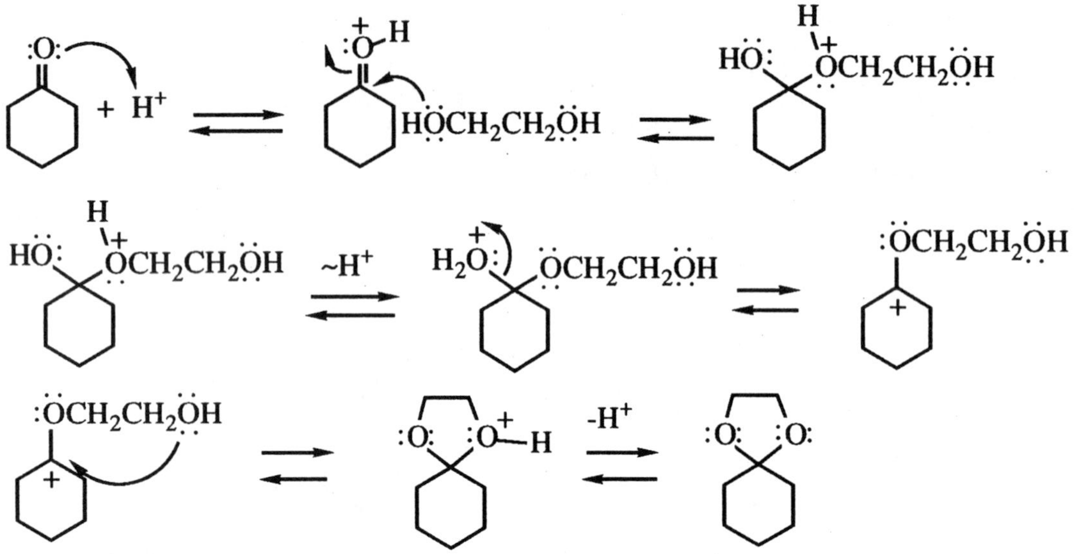

19.5 (a) 1. treatment of the bromoaldehyde with ethylene glycol in the presence of a catalytic amount of a strong acid; 2. magnesium metal, ether; 3. acetone; 4. aqueous acid
(b) 1. treatment of the ketoalkyne with ethylene glycol in the presence of a catalytic amount of a strong acid; 2. sodium amide in liquid ammonia; 3. iodomethane; 4. aqueous acid

19.6

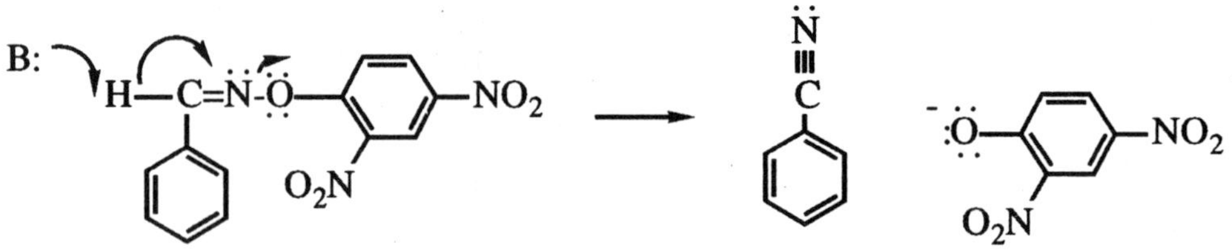

[Mechanism showing base deprotonating H—C=N—O—Ar(NO₂)₂ intermediate, giving benzonitrile (Ph—C≡N) and 2,4-dinitrophenoxide anion]

initially formed material

The two nitro groups stabilize the anion formed along with benzonitrile.

19.7
(a) [2-pyridyl-C(H)=C(OH)Ph structure]
(b) PhN=CHCH₂CH(CH₃)₂
(c) PhNCH=CHCH(CH₃)₂
          |
          CH₃

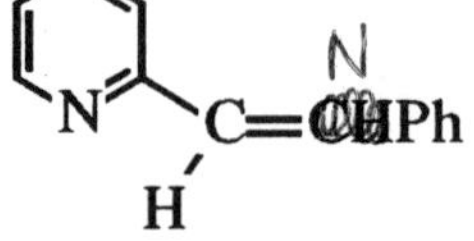

19.8 1. acetic anhydride, aluminum chloride; 2. methylmagnesium bromide, followed by aqueous workup
19.9 chloroform and *N*,*N*-dimethyl-*p*-formylaniline

19.10

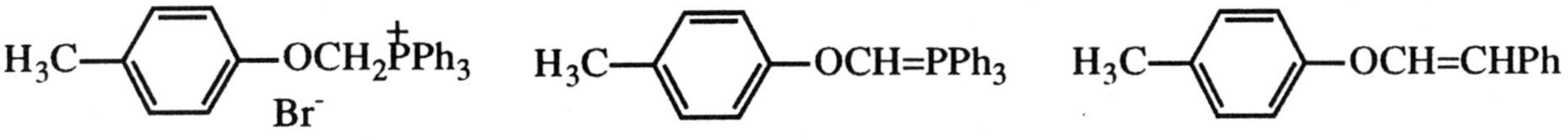

[Structures A–E shown:]

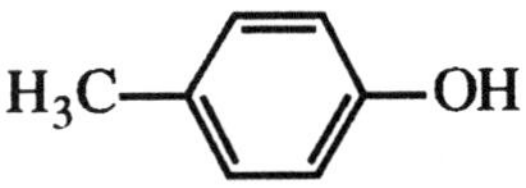

A: H₃C–C₆H₄–OCH₂⁺PPh₃ Br⁻

B: H₃C–C₆H₄–OCH=PPh₃

C: H₃C–C₆H₄–OCH=CHPh

D: H₃C–C₆H₄–OH

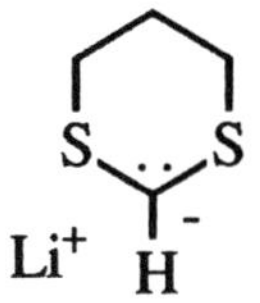

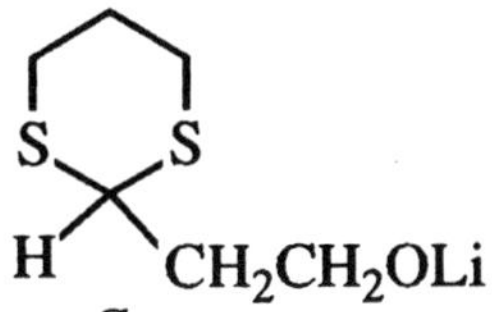

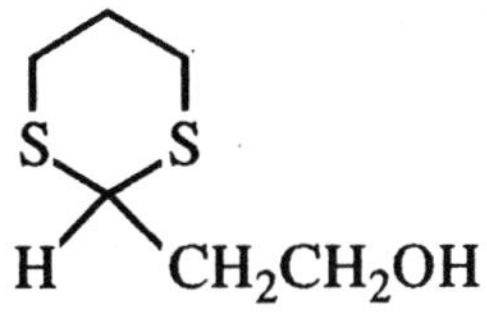

E: PhCH₂CHO

Mechanism:

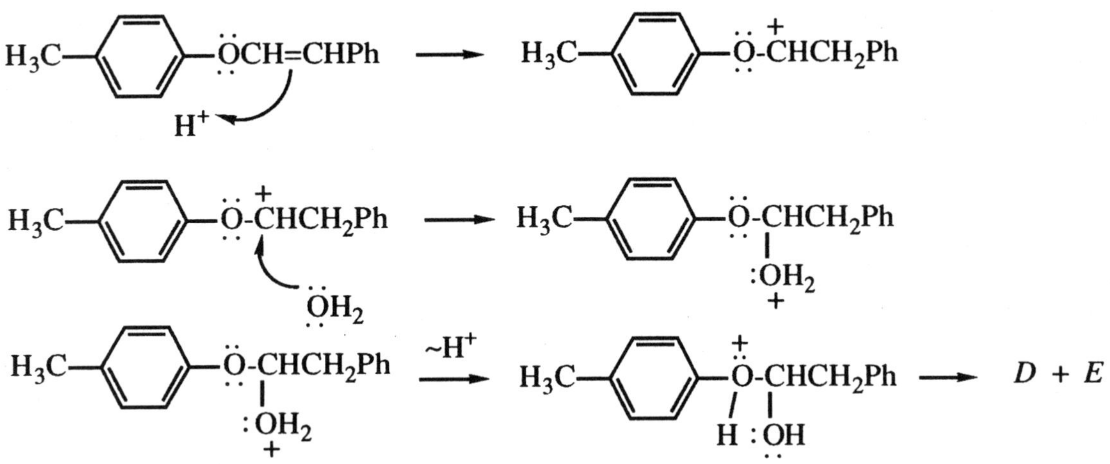

H₃C–C₆H₄–ÖCH=CHPh ⟶ H₃C–C₆H₄–Ö⁺–CHCH₂Ph
         ↑ H⁺

H₃C–C₆H₄–Ö⁺–CHCH₂Ph ⟶ H₃C–C₆H₄–Ö–CHCH₂Ph
         ↑ :ÖH₂ |
                                    :O⁺H₂

H₃C–C₆H₄–Ö–CHCH₂Ph   —~H⁺→   H₃C–C₆H₄–Ö⁺–CHCH₂Ph   ⟶   D + E
         | | |
        :O⁺H₂   H :OH

19.11

[Four 1,3-dithiane structures:]

A: 1,3-dithiane

B: 2-lithio-1,3-dithiane (Li⁺, H on C2 with negative charge)

C: 2-(2-lithiooxyethyl)-1,3-dithiane (H, CH₂CH₂OLi on C2)

D: 2-(2-hydroxyethyl)-1,3-dithiane (H, CH₂CH₂OH on C2)

19.12 (a) NaOD, D₂O
  (b) bromine, acetic acid
  (c) bromine, aqueous sodium hydroxide

19.13 Only (b) would racemize in aqueous basic solution.

19.14 2-butanol and 2-butanone

19.15 (a) sodium borohydride
  (b) Tollens' reagent
  (c) 1. 1,2-ethanedithiol; 2. Raney nickel
  (d) bromine, aqueous sodium hydroxide

CHAPTER 20
AMINES AND RELATED COMPOUNDS

Key Points

• Learn the structure and geometry of amines and quaternary ammonium ions. Also, learn to classify amines as 1°, 2°, or 3°.

Practice Problem 20.1
Consider amines of formula C_3H_9N.

(a) How many different amines of this formula are there? Classify each as 1°, 2°, or 3°.

(b) Keeping in mind that a nonbonding valence level electron pair can be regarded as a "group" for nomenclature priority purposes (see Chapter 8), complete the diagram below to represent the S enantiomer of a chiral C_3H_9N amine.

(Actually, the S and R enantiomers cannot be isolated as they rapidly interconvert - see answer.)

Practice Problem 20.2
How could you distinguish a 3° amine from a 2° or 1° amine using IR spectroscopy?

• Be able to name amines, substituted ammonium ions, and imines.

Practice Problem 20.3
Draw structures for each of the following:
  (a) 4-amino-2-methylhexane
  (b) 2-(dimethylamino)-4-methylpentane
  (c) allylethylmethylphenylammonium bromide
  (d) the imine formed by the reaction of *p*-methoxyacetophenone and aniline

• Understand how the basicity of amines is measured, and how relative basicities correlate with structure and (in solution) with solvation effects.

Practice Problem 20.4
In the *gas phase*, what will be the order of basicities for the series of amines shown below?
    $BuNH_2$, $(Bu)_2NH$, $(Bu)_3N$ $Bu = CH_3CH_2CH_2CH_2-$

Practice Problem 20.5
The pK_a values (aqueous solution, 25°C) for the *conjugate acids* of several ethylsubstituted amines are:
    Parent Amine Et_3N Et_2NH $EtNH_2$ $Et = CH_3CH_2-$
    pK_a of
    conjugate acid   10.9   11.1   10.8
  (a) Which amine is the strongest base of the group under these conditions?
  (b) Calculate the pK_b for the strongest parent amine.

Practice Problem 20.6
Of the pair of amines, *p*-cyanoaniline and *m*-cyanoaniline, which would you expect to be the more basic. Explain your choice.

• Appreciate the difficulty in synthesizing amines by the reaction of ammonia (or other amines) with haloalkanes.

Practice Problem 20.7
Which reagent, the amine or the haloalkane, should be in excess for the most efficient preparation of a tretiary amine from a secondary amine by a reaction of the type shown below? Explain your choice.

$$R_2\ddot{N}H + R'\ddot{C}l: \longrightarrow R_2NR'$$

Practice Problem 20.8
A cyclohexane derivative A, of formula $C_9H_{15}Br_3$, on treatment with ammonia yields azaadamantane (structure shown below: Suggest a structure for A.

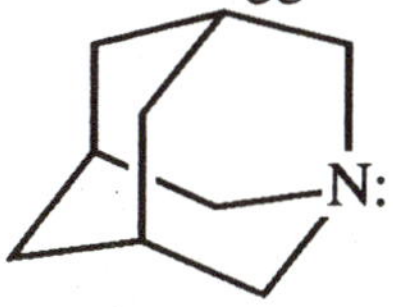

- Learn the following preparative routes for amines: the Gabriel synthesis, the reduction of nitriles, and the reduction of nitro-compounds (all covered in Section 20.6 of the text), and the reductive amination of carbonyl compounds (covered in Section 20.7 of the text).

Practice Problem 20.9
Provide the missing reagents *i-v* in the following reactions.

(a) [phthalimide NH] —*i*→ [phthalimide NK] —*ii*→ [phthalimide NR] —*iii*→ RNH_2

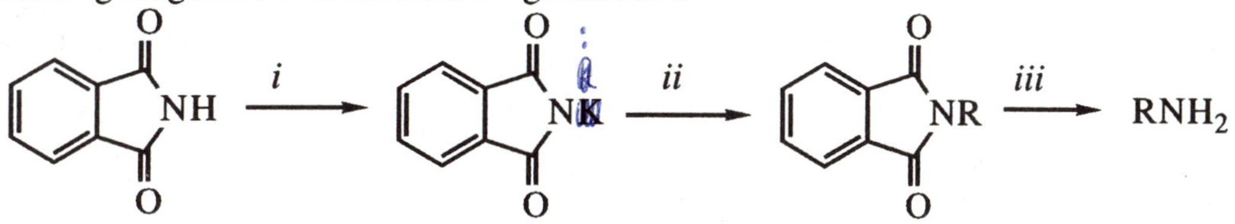

(b) $PhCH=CH-CN \xrightarrow{iv} PhCH=CH-CH_2NH_2$

(c) $p\text{-nitrotoluene} \xrightarrow{v} p\text{-methylaniline}$

Practice Problem 20.10
Give the structure of the product B in the following reaction.

$$PhCH_2NH_2 \xrightarrow[\substack{NaBH_3CN, H_2O \\ pH = 6\text{-}8}]{H_2C=O} B$$

- Learn the use of the Clarke-Eschweiler synthesis for the controlled methylation of amines.

Practice Problem 20.11
Give the mechanism for the Clarke-Eschweiler reaction of piperidine shown below.

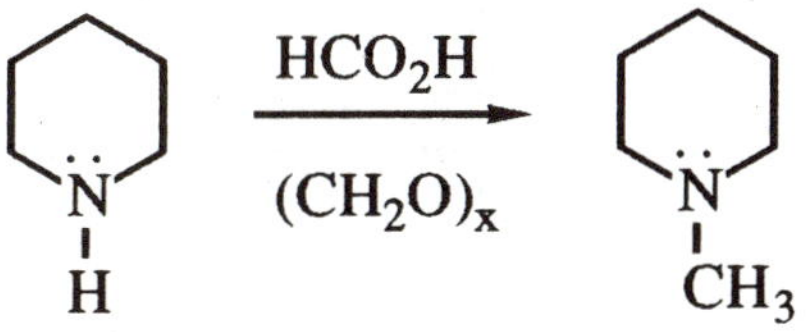

piperidine-NH $\xrightarrow[\text{(CH}_2\text{O)}_x]{HCO_2H}$ piperidine-N-CH$_3$

- Learn the chemistry of diazonium salts, including their preparations and reactions.

Practice Problem 20.12
Alkyldiazonium salts are generally quite unstable; they spontaneously lose N_2 to generate a carbocation. In light of this fact, propose a mechanism for the reaction shown below.

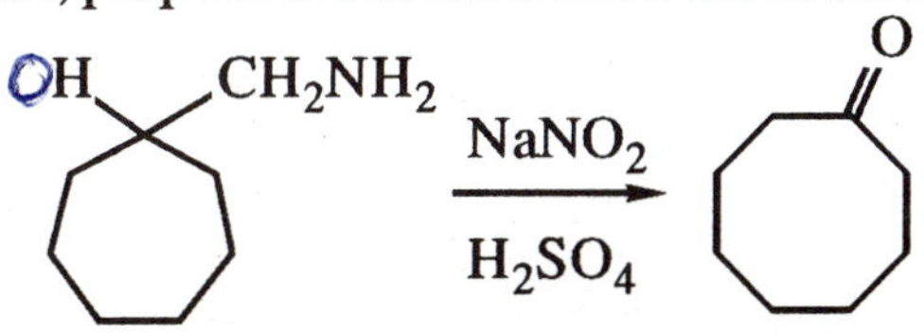

cycloheptane with OH and CH$_2$NH$_2$ $\xrightarrow[H_2SO_4]{NaNO_2}$ cyclooctanone

188 Study Guide and Solutions Manual

Practice Problem 20.13
Give the reagents for each of the following conversions:

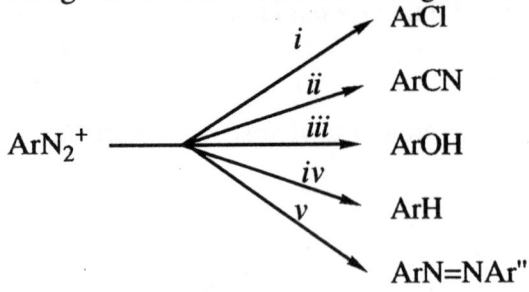

ArN₂⁺ →
  i. ArCl
  ii. ArCN
  iii. ArOH
  iv. ArH
  v. ArN=NAr"

Practice Problem 20.14
Propose methods for each of the following syntheses:
  (a) 1,3-dibromobenzene from nitrobenzene
  (b) 1,2,3-tribromobenzene from 4-nitroaniline
  (c) 4-nitroethylbenzene from 2-nitroacetophenone

• Learn the reactions of 2° and 3° amines with nitrous acid.

Practice Problem 20.15
Give the product (if any) formed in the reaction of each of the following with sodium nitrite in aqueous sulfuric acid.

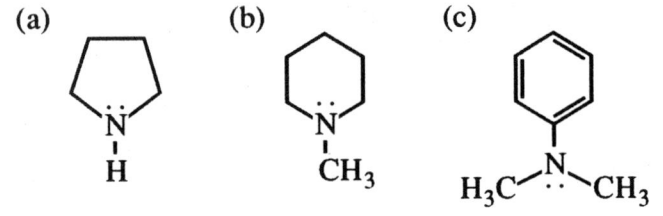

(a) pyrrolidine (N–H)
(b) N-methylpiperidine (N–CH₃)
(c) N,N-dimethylaniline (Ph–N(CH₃)₂)

• Be familiar with the Hofmann elimination procedure applied to amines. Know the regiochemistry of the reaction and its use for the degradation of more complex molecules.

Practice Problem 20.16
Which alkene product (list the major one if a mixture would result) would you expect to form when the Hofmann elimination procedure is applied to each of the following amines?
  (a) cyclohexylamine
  (b) N-ethylcyclohexylamine
  (c) 2-aminopentane

• Be aware of the application of the Cope elimination reaction, and related elimination reactions, for the generation of carbon-carbon double bonds.

Practice Problem 20.17
What major product would you expect to form upon heating N,N-dimethylcyclohexylamine-N-oxide? Give a curved-arrow depiction of the mechanism of the reaction.

Practice Problem 20.18
Which isomer, A or B, would you expect on heating to yield mainly 1-phenylcyclohexene and which you expect to yield 3-phenylcyclohexene? Explain your choice.

A: cis-2-phenylcyclohexyl-N(CH₃)₂⁺–O⁻
B: trans-2-phenylcyclohexyl-N(CH₃)₂⁺–O⁻

Practice Problem 20.19
Give structures for the compounds indicated C and D.

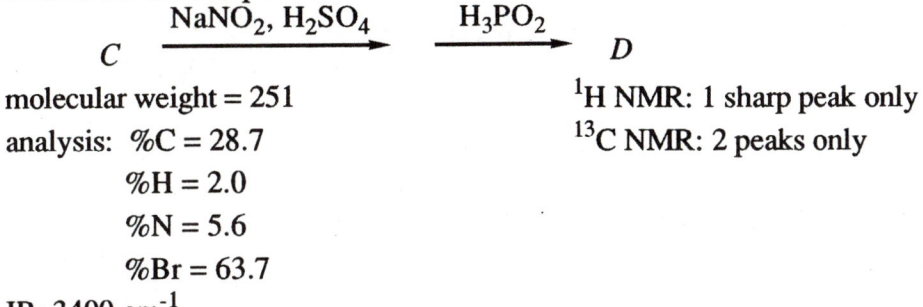

$$C \xrightarrow{\text{NaNO}_2, \text{H}_2\text{SO}_4} \xrightarrow{\text{H}_3\text{PO}_2} D$$

C: molecular weight = 251
analysis: %C = 28.7
%H = 2.0
%N = 5.6
%Br = 63.7
IR: 3400 cm^{-1}

D: ^{1}H NMR: 1 sharp peak only
^{13}C NMR: 2 peaks only

Practice Problem 20.20
Propose structures for the compounds indicated E to I.

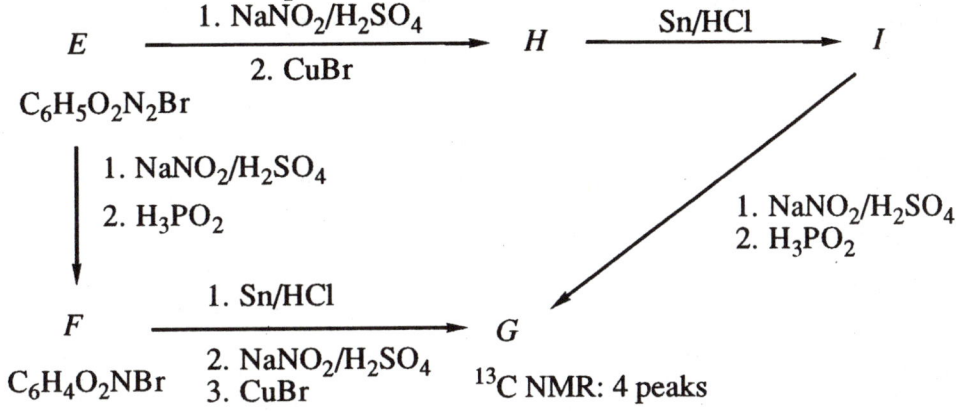

$$E \xrightarrow[\text{2. CuBr}]{\text{1. NaNO}_2/\text{H}_2\text{SO}_4} H \xrightarrow{\text{Sn/HCl}} I$$

E: C$_6$H$_5$O$_2$N$_2$Br

E $\xrightarrow[\text{2. H}_3\text{PO}_2]{\text{1. NaNO}_2/\text{H}_2\text{SO}_4}$ F

F: C$_6$H$_4$O$_2$NBr

F $\xrightarrow[\text{3. CuBr}]{\substack{\text{1. Sn/HCl} \\ \text{2. NaNO}_2/\text{H}_2\text{SO}_4}}$ G

I $\xrightarrow[\text{2. H}_3\text{PO}_2]{\text{1. NaNO}_2/\text{H}_2\text{SO}_4}$ G

G: ^{13}C NMR: 4 peaks

Solution of Text Problems

20.1 Primary: mescaline, aminoethane, aniline, *p*-nitroaniline
Secondary: *sec*-butylmethylamine
Tertiary: nicotine, cocaine, atropine, diethylpropylamine, isopropyldimethylamine, *N*,*N*-dimethylaniline

20.2
(a) CH$_3$CH$_2$CHCH$_2$CH$_3$ with NH$_2$
(b) CH$_3$CH$_2$CH$_2$CHCHCH$_2$CH$_3$ with CH$_3$ and N(CH$_3$)$_2$
(c) CH$_3$CH$_2$CHCHCH$_3$ with H–NCH$_2$CH$_3$ and OH

20.3
(a) cyclohexyldiethylamine
(b) 4-ethyl-2-methylaminoheptane
(c) *N*-ethyl-3-chloroaniline

20.4
(a) benzyl(ethyl)ammonium chloride: CH$_2$N$^+$H(CH$_2$CH$_3$)H Cl$^-$ on phenyl
(b) *p*-nitroanilinium iodide: $^+$NH$_3$ on phenyl-NO$_2$, I$^-$
(c) CH$_3$CH$_2$CH$_2$CCH(CH$_3$)$_2$ with =N–CH$_3$
(d) 3-chlorobenzaldehyde *N*-phenylimine: H–C=N–Ph on phenyl-Cl

20.5 (a) *E*-3-methylbenzaldehyde *N*-methylimine
(b) diethylpropylammonium bromide
(c) *E*-2,2-dimethylcyclohexanone *N*-1-methylbutylimine
(d) *N*,*N*-dimethyl-3-bromoanilinium chloride

Chapter 20 189

20.6

$$K_a = \frac{[H_3O^+][RNH_2]}{[RNH_3^+]}$$

$$K_b = \frac{[RNH_3^+][HO^-]}{[RNH_2]}$$

$$(K_a)(K_b) = \frac{[H_3O^+][RNH_2][RNH_3^+][HO^-]}{[RNH_3^+][RNH_2]} = [H_3O^+][HO^-] = K_w$$

20.7   The amine hydrochloride *B* has the stronger conjugate base. The pK_b values are: *A*, 5.5; *B*, 4.7.

20.8   From most basic to least basic: *N*-methyl-*m*-toluidine > *m*-toluidine > aniline > *p*-cyanoaniline > 5-amino-2-cyanoacetophenone

20.9   The compound shown to the left side is a tertiary amine with the valence level unshared electron pair localized on nitrogen. However, the compound shown to the right side has the nitrogen lone pair delocalized to each of the two oxygen atoms of the amide linkages and thereby is not basic.

20.10   The synthesis planned involves an S_N2 reaction. The proposed substrate, an aryl halide, does not undergo S_N2 reactions.

20.11   (a) 1. isobutyl bromide treated with potassium phthalimide; 2. hydrolysis with aqueous potassium hydroxide
   (b) 1. treatment of *p*-nitrobenzyl bromide with potassium phthalimide; 2. hydrolysis with aqueous potassium hydroxide
   (c) 1. 1-bromo-3-cyclohexylpropane treated with potassium phthalimide; 2. hydrolysis with aqueous potassium hydroxide

20.12   (a) 1. treatment of 1-bromobutane with potassium phthalimide; 2. hydrolysis with aqueous potassium hydroxide
   (b) 1. treatment of 1-bromobutane with sodium cyanide; 2. hydrogen, platinum oxide

20.13   (a) 1. treatment of benzene with bromine in the presence of iron; 2. nitric acid, sulfuric acid; 3. tin, hydrochloric acid
   (b) 1. treatment of benzene with sulfuric acid and nitric acid; 2. bromine in the presence of iron with heating; 3. tin, hydrochloric acid
   (c) treatment of toluene with bromine under light irradiation; 2. potassium phthalimide; 3. hydrolysis with aqueous potassium hydroxide
   (d) 1. treatment of toluene with iodomethane in the presence of aluminum chloride; 2. sulfuric acid, nitric acid; 3. tin, hydrochloric acid

20.14   Formic acid is oxidized to carbon dioxide.

20.15   A tertiary amine cannot generate any intermediate iminium ion such as would be required for further alkylation.

20.16

$:N\equiv\overset{+}{O}:$

20.17

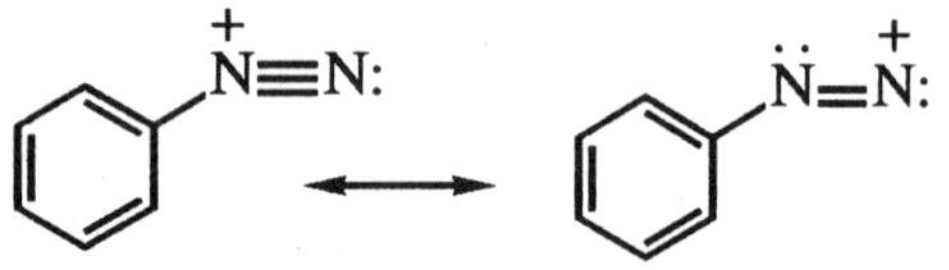

[Ph-$\overset{+}{N}\equiv N:$ ↔ Ph-$\overset{..}{N}=\overset{+}{N}:$]

20.18   Upon diazotization, 1-aminobutane yields $CH_3CH_2CH_2CH_2N_2^+$ which loses nitrogen to form the 1-butyl cation. The 1-butyl cation leads directly to 1-butene by loss of a proton, and to 1-butanol and 1-chlorobutane by combination with water and chloride ion respectively. The 1-butyl cation also undergoes hydride migration to generate the intermediate 2-butyl cation, which in turn forms 2-butanol and 2-chlorobutane by combination with water and chloride ion respectively, and the 2-butenes and 1-butene by loss of a proton.

20.19

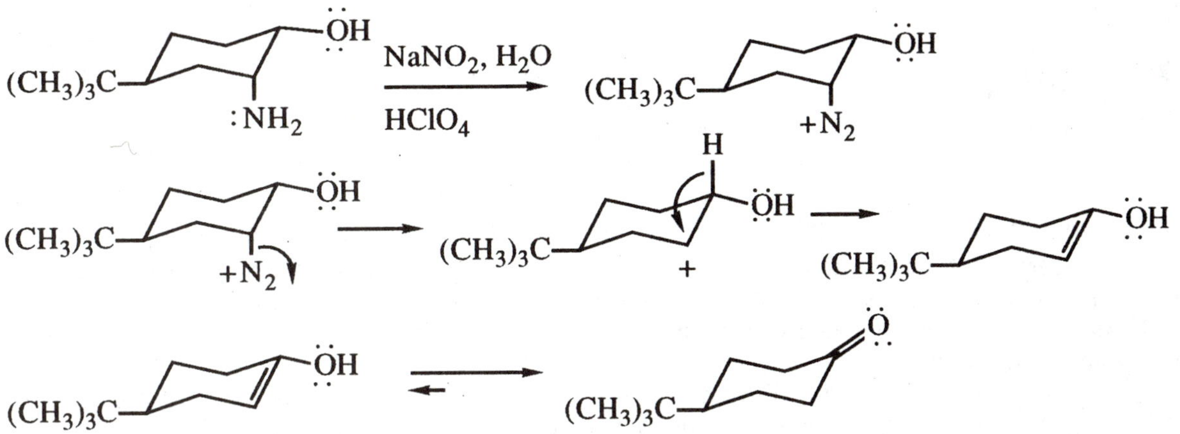

[Scheme: trans-4-tert-butyl-2-aminocyclohexanol treated with NaNO₂, H₂O, HClO₄ gives diazonium intermediate; loss of N₂ forms carbocation; hydride shift gives allylic/enol which tautomerizes to 4-tert-butylcyclohexanone.]

20.20 (a) 1. treatment of 4-methylaniline with sodium nitrtite and sulfuric acid; 2. potassium iodide
(b) 1. treatment of 4-methylaniline with sodium nitrite and sulfuric acid; 2. HBF₄; 3. heat

(c) 1. reduction of p-nitrotoluene with tin and hydrochloric acid; 2. sodium nitrite and sulfuric acid; 3. curprous bromide

20.21 N,N-dimethyl-4-nitrosobenzene

20.22

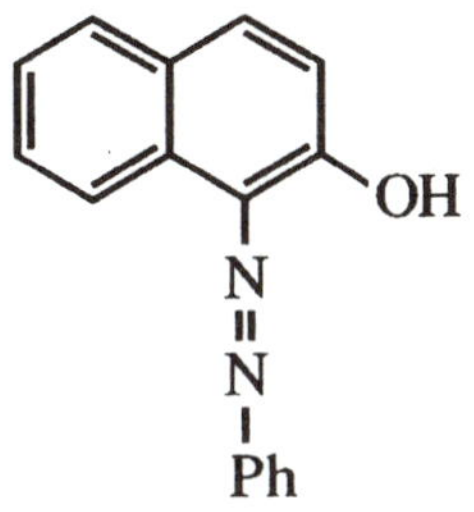

[Structure: 1-(phenylazo)-2-naphthol]

20.23

[Mechanism: electrophilic aromatic substitution of N,N-dimethylaniline with NO⁺ giving para-nitroso product after loss of H⁺]

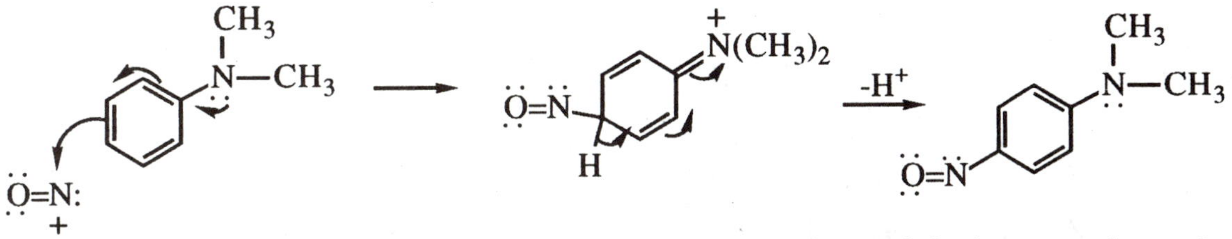

20.24 Treatment with base in the elimination reaction would simply result in deprotonation and regeneration of the free amine.

20.25

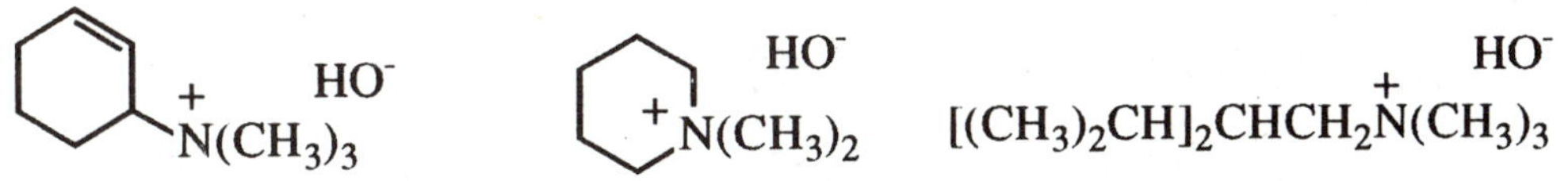

[Three structures:
- cyclohexenyl-N⁺(CH₃)₃ with HO⁻ — Eqn. 20.38
- cyclohexyl-N⁺(CH₃)₂ with HO⁻ — Eqn. 20.39
- [(CH₃)₂CH]₂CHCH₂N⁺(CH₃)₃ with HO⁻ — Eqn. 20.40]

Eqn. 20.38   Eqn. 20.39   Eqn. 20.40

20.26 An N-ethyl compound could eliminate to form ethylene; such an elimination can not occur with the N-methyl system.

20.27 In the direct dehydrohalogenation reaction an *anti* orientation of -X and -H is required. This is difficult to attain in the unsaturated ring system. With the Hofmann elimination the proton is removed first and a particular conformation is not required.

20.28

[Structures A and B: cyclohexane rings with N⁺-oxide groups shown in chair conformations]

A   B

In conformation A, although the β-hydrogen and the oxide anion site are within reach, the bonds to be broken (emphasized) are not periplanar and π bond formation cannot occur simultaneously with bond breaking. In conformation B the breaking bonds *are* periplanar, but the required atoms are not within reach.

20.29 The Clarke-Eschweiler synthesis would be used twice to generate the tertiary amine without formation of any quaternary ammonium ion.

20.30 The expected product is (Z)-4,4-dimethyl-2-phenyl-2-pentene. The (2S,3S) enantiomer gives the same product. However, the (2R,3S) diastereoisomer yields the isomeric (E)-4,4-dimethyl-2-phenyl-2-pentene.

20.31 (a) ethyldimethylamine
    (b) 1-aminobutane; 2-aminobutane; 2-amino-2-methylpropane; 1-amino-2-methylpropane
    (c) 2-amino-2-methylpropane
    (d) 1-aminobutane or 2-aminobutane
    (e) diisopropylamine

20.32 The greater electronegativity of oxygen (compared to nitrogen) allows it to remove electron density from the region of the carbon, deshielding it, and causing it to come into resonance at 48.0 ppm downfield from TMS.

20.33 There are two -NH$_2$ groups per molecule, and the structure is H$_2$NCH$_2$CH$_2$CH$_2$CH$_2$CH$_2$NH$_2$.

20.34

[Structure: 2-methyl-6-undecylpiperidine]

20.35

(a) (PhCH$_2$)$_2$NH   (b) (CH$_3$)$_3$CNH$_2$ (c) PhN(CH$_3$)$_2$ (d) PhN(CH$_3$)(N=O)   (e) (CH$_3$)$_3$N⁺H Cl⁻

(f) H$_2$N–C(CH$_3$)(H)–C(CH$_3$)(H)–NH$_2$ (with stereochemistry indicated)
(g) cyclobutyl–N⁺(CH$_3$)$_3$ I⁻
(h) cyclohexyl-CH$_2$-CH(CH$_3$)-N(H)(CH$_3$)
(i) PhN$_2$⁺ Cl⁻
(j) [(CH$_3$)$_2$CHCH$_2$]$_3$N

(k) CH$_3$N(H)CH$_2$CH$_2$CH$_2$CH$_3$
(l) (CH$_3$)$_2$C(NH$_2$)CH$_2$CH$_3$
(m) Ph-C(CH$_3$)=N-CH$_3$
(n) cyclohexanone N-benzyl imine

20.36 (a) *n*-propylamine (It has a greater surface area for a given volume.)
    (b) 4-methoxyaniline (It has an electron donating group on the ring.)
    (c) piperidine (It is an aliphatic secondary amine.)
    (d) pyridine (It retains aromatic character upon protonation.)
    (e) ethylmethylamine (A primary amine forms an imine.)
    (f) *N*-ethylaniline (A primary amine undergoes diazotization.)
    (g) phenol (It is activated for electrophilic aromatic substitution.)
    (h) ethylmethylamine (It is a secondary amine.)
    (i) pyrrole (It would lose aromatic stabilization if it formed a quaternary ammonium compound.)

20.37 Add the salt to aqueous base and extract with an organic solvent, such as ether or hexane. Dry the organic solution, filtering off the drying agent, and evaporate the solvent.

20.38 Dissolve the mixture in an organic solvent such as hexane. Add aqueous acid. The aniline goes into solution in the aqueous layer while the nitrobenzene stays in the organic layer. Evaporate the organic solvent to isolate the nitrobenzene. Add aqueous base to the aqueous layer, extract the aqueous layer with hexane, dry the organic extract and filter off the drying agent, and finally evaporate the organic solvent to isolate the aniline.

20.39

(a) 3,5-dichloroaniline (NH₂ on benzene with Cl at 3,5 positions)
(b) (CH₃)₂CHN⁺(CH₃)₃ I⁻
(c) PhN⁺H₃ Cl⁻
(d) H₂C=CHCH₂CH=CH₂
(e) 2,3-dimethylquinoxaline

20.40 (a) 4-chloroanisole
(b) anisole
(c) 4-fluoroanisole
(d) 2-hydroxy-5-methylphenyl-azo-(4-methoxyphenyl) — structure: HO-C₆H₃(CH₃)-N=N-C₆H₄-OCH₃

20.41 (a) 1. treatment of benzene with nitric acid and sulfuric acid; 2. tin, hydrochloric acid
(b) 1. treatment of aniline with sodium nitrite and sulfuric acid; 2. CuCN; 3. hydrogen, platinum oxide
(c) 1. treatment of benzene with sulfuric acid and nitric acid; 2. tin, hydrochloric acid; 3. bromine; 4. sodium nitrite, sulfuric acid; 5. H₃PO₂
(d) 1. treatment of 2-propanol with phosphorus tribromide; 2. potassium phthalimide; 3. aqueous potassium hydroxide
(e) 1. treatment of 2-nitroaniline with sodium nitrite and sulfuric acid; 2. CuCN
(f) 1. treatment of aniline with sodium nitrite and sulfuric acid; 2. addition of N,N-dimethylaniline (formed from aniline by two sequential treatment with formaldehyde and formic acid
(g) 1. treatment of 1-pentanol with phosphorus tribromide; 2. potassium phthalimide; 3. aqueous potassium hydroxide
(h) 1. treatment of acetophenone with bromine in acetic acid; 2. anhydrous ammonia
(i) 1. treatment of toluene with sulfuric acid and nitric acid; 2. bromine in the presence of iron; 3. tin, hydrochloric acid
(j) 1. treatment of toluene with sulfuric acid and nitric acid; 2. tin, hydrochloric acid; 3. bromine
(k) treatment of N,N-diethylaniline with sodium nitrite and sulfuric acid
(l) 1. treatment of benzene with bromine in the presence of iron; 2. sulfuric acid and nitric acid; 3. tin, hydrochloric acid; 4. sodium nitrite and sulfuric acid; 5. CuCN

20.42

:BrCH₂CH₂CH₂N̈(CH₃)₂ ⟶ [cyclobutyl-N⁺(CH₃)₂] :Br:⁻

20.43 H₂NCH₂CH₂CH₂NH₂

20.44

cyclopentyl-CH₂NH₂ →(H₂SO₄, NaNO₂)→ cyclopentyl-CH₂N₂⁺ →(−N₂)→ cyclopentyl-CH₂⁺ → cyclohexyl⁺ →(H₂O)→ cyclohexyl-ÖH

20.45 cyclohexanone
20.46 A is PhCH₂Cl; B is PhCH₂NH₂; C is PhCO₂H
20.47 H₂C=CHCH(NH₂)CH₃   CH₃CH₂CH(NH₂)CH₃
20.48 D is piperidine; E is N,N-dimethylpiperidinium iodide; F is 1,4-pentadiene

C=C-C-C-C-N-CH₃
              |
              CH₃

194 Study Guide and Solutions Manual

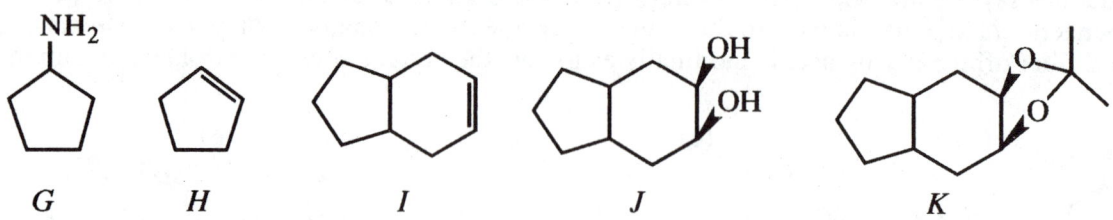

20.49

G, H, I, J, K (structures shown)

20.50 L is 3-nitrobenzyl chloride
M is 3-nitrobenzyl alcohol
N is 3-nitrobenzoic acid
O is 3-aminobenzoic acid
P is 2,4,6-tribromo-3-~~nitro~~benzoic acid (amino)

20.51 (mechanism shown)

20.52 2-amino-2,3-dimethylbutane

20.53 Synthesis: 1. treatment of 4-aminobenzenesulfonic acid with sodium nitrite and sulfuric acid; 2. addition of N,N-dimethylaniline

Reaction: Protonation is expected to occur at the nitrogen of the N=N linkage which is attached directly to the ring bearing the sulfonic acid functional group. Protonation at this site allows the positive charge to be delocalized around the ring bearing the dimethylamino group, and to the dimethylamino nitrogen itself.

20.54 (mechanism shown)

20.55
(a) (mechanism shown)
(b) (mechanism shown)

20.56 Q is 1-bromo-2-phenylethane
R is N-(2-phenylethyl)phthalimide
S is 2-phenylethylamine
T is trimethyl(2-phenylethyl)ammonium iodide
U is styrene
V is trimethylamine
W is methyl(2-phenylethyl)amine

Solution of Study Guide Practice Problems

20.1 (a) There are two 1° (2-aminopropane and 1-aminopropane), one 2° (ethylmethylamine), and one 3° (trimethylamine) amines of the given formula.

(b)

H₃C—N(:)(···H)(CH₂CH₃)

Inversion about such sites occurs quite readily, essentially being an "umbrella" type conformational flip with the orbital containing the unshared valence level electron pair ending up pointing in the opposite direction. With analogous tricoordinated phosphorus compounds this type of inversion of configuration occurs at a much slower rate and enantiomers can be isolated.

20.2 Tertiary amines have no N-H bonds and thereby exhibit no absorptions in the IR region 3300-3500 cm^{-1}.

20.3
(a) $(CH_3)_2CHCH_2CHCH_2CH_3$ with NH₂ substituent

(b) $(CH_3)_2CHCH_2CHCH_3$ with N(CH₂CH₃)₂ substituent

(c) $CH_3-\overset{+}{\underset{Ph}{N}}(CH_2CH_3)-CH_2CH=CH_2$ Br⁻

(d) CH_3O—C₆H₄—C(CH₃)=NPh

20.4 Bu₃N > Bu₂NH > BuNH₂

20.5 (a) The stronger base is the one having the weaker conjugate acid. The weaker conjugate acid is the one with the higher pK$_a$. Thus, diethylamine is the strongest base in the series.

(b) The pK$_b$ of diethylamine is 2.9.

20.6 The *m*-cyanoaniline is the more basic of the two. In the case of the *p*-cyanoaniline the basicity is diminished because of resonance delocalization of the unshared valence level electron pair from the amino nitrogen onto the cyano group. There is no resonance delocalization of this type with the *m*-cyanoaniline.

20.7 The amine should be present in excess.

20.8 *cis,cis*-1,3,5-tri(bromomethyl)cyclohexane

20.9 (a) *i* = KOH; *ii* = R-X; *iii* = reflux with aqueous KOH (or hydrazine in ethanol)

(b) *iv* = LiAlH₄

(c) *v* = tin and hydrochloric acid, or H₂/PtO₂, or NaBH₄/NiCl₂

20.10 benzylmethylamine

20.11 The mechanism of the Clarke-Exchweiler synthesis is shown in Section 20.7 of the text.

20.12

cycloheptane with HO and CH₂NH₂ —HONO→ cycloheptane with HO and CH₂N₂⁺ —−N₂→ cycloheptane with HO and ⁺CH₂ → cyclooctane with ⁺OH —−H⁺→ cyclooctanone

20.13 *i* CuCl; *ii* CuCN; *iii* H₂O; *iv* H₃PO₂; *v* Ar"H (Ar" must contain electron donating groups)

20.14 (a) 1. treatment of nitrobenzene with bromine in the presence of iron (heat); 2. tin, hydrochloric acid; 3. sodium nitrite, sulfuric acid; 4. CuBr

(b) 1. treatment of 4-nitroaniline with bromine; 2. sodium nitrite, sulfuric acid; 3. CuBr; 4. tin, hydrochloric acid; 5. sodium nitrite, sulfuric acid; 6. H₃PO₂

(c) 1. treatment of 2-nitroacetophenone with hydrazine to form the hydrazone; 2. heat with concentrated aqueous sodium hydroxide; 3. tin, hydrochloric acid; 4. sodium nitrite, sulfuric acid; 5. H₃PO₂; 6. nitric acid,

sulfuric acid
20.15 (a) N-nitrosopyrrolidine
(b) no reaction
(c) N,N-dimethyl-4-nitrosoaniline
20.16 (a) cyclohexene
(b) ethene
(c) 1-pentene
20.17

20.18 The elimination is *syn* in nature, leading to the formation of 3-phenylcyclohexene from A and 1-phenylcyclohexene from B.
20.19 C is 2,5-dibromoaniline; D is p-dibromobenzene
20.20 E is 3-bromo-5-nitroaniline; F is m-bromonitrobenzene; G is m-dibromobenzene; H is 3,5-dibromonitrobenzene; I is 3,5-dibromoaniline

PRACTICE EXAMINATION FOUR

A time limit of 90 minutes should be set for completion of this entire practice examination. Answers should be written out completely as they would be when presented for independent grading. No text or supplemental materials should be consulted during the testing period, and you should not check your answers until you have worked out the complete examination and the time limit has been reached.

1. In each case, which of the given compounds exhibits the indicated property? Give a brief justification for each choice (2 points for each part).
    (a) more basic: aniline or cyclohexylamine
    (b) more reactive toward aqueous sodium hydroxide: benzyl chloride or o-chlorotoluene
    (c) more basic: pyrrole or imidazole
    (d) more acidic: 1,3-cyclopentadiene or 1,3,7-cycloheptatriene
    (e) can be prepared readily by reduction of a ketone: 4,4-dimethylcyclohexanol or 1,4-dimethylcyclohexanol
    (f) ^{1}H NMR spectrum consists of a singlet at δ 5.0 (relative area 1) and a singlet at δ 3.9 (relative area 2):

    1,3-dioxolane   or $H_3C-C≡C-C(CH_3)_2-Cl$

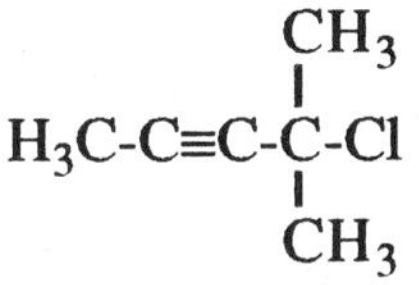

    (g) has one vertical node in its HOMO: allyl cation or allyl anion
    (h) has the lower energy HOMO: benzene or 1,3,5-hexatriene
    (i) is more reactive toward ring nitration: anisole or acetophenone
    (j) reacts with 2-methylpropanal to form an enamine: N-methylaniline or N,N-dimethylaniline

2. Consider each of the following reaction systems (6 points for each part).
    (a) Which organic compound (or compounds) do you expect to form when A (shown below) is treated with aqueous acid? Give a brief justification of your answer.

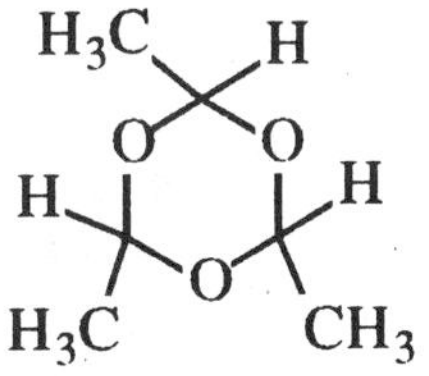

    A

    (b) In the ring bromination of p-nitroanisole there is formed an intermediate arenonium ion of formula $C_7H_7NO_3Br^+$. Draw a Lewis structure for this ion in which every atom has a noble gas electronic configuration.
    (c) In the reaction of 3,4-dinitrobromobenzene with sodium methoxide an *anionic* intermediate of formula $C_7H_6N_2O_5Br^-$ is formed. Draw a Lewis structure for this anion in which both oxygen atoms of one of the nitro groups have formal negative charges.
    (d) When benzene is allowed to react with 1,3-butadiene in the presence of acid two compounds are formed: B ($C_{10}H_{12}$) and C ($C_{16}H_{18}$). Compound B undergoes catalytic hydrogenation readily, while C does not. Suggest structures for B and C and give a brief mechanistic rationalization for their formation.
    (e) When benzene reacts with compound D (C_4H_7ClO) in the presence of aluminum chloride, a new compound E ($C_{10}H_{12}O$) is formed that has a band near 1700 cm^{-1} in its IR spectrum. When E is refluxed with zinc amalgam and hydrochloric acid, compound F ($C_{10}H_{14}$) is formed having the ^{1}H NMR spectrum as follows: δ 0.88 (doublet of relative area 6); δ 1.86 (multiplet of relative area 1); δ 2.45 (doublet of relative area 2); δ 7.12 (singlet of relative area 5). Suggest structures for compounds D, E, and F that are consistent with the above data.

3. Identify the five compounds G-K on the basis of the information given (4 points for each part).

198 Study Guide and Solutions Manual

(a) *G* IR: strong absorption at 1710 cm^{-1} and broad absorption in the region 2500-3500 cm^{-1}
^{1}NMR: δ 11.7 (singlet, relative area 1); δ 2.2 (quartet, relative area 2); δ 1.1 (triplet, relative area 3)

(b) *H* $C_5H_{12}O_2$, is hydrolyzed by aqueous sulfuric acid to yield acetone
^{1}NMR: two singlets of equal area

(c) *I* $C_2H_2Cl_2O$
IR: strong band at ~1715 cm^{-1}
^{1}NMR: two doublets of equal area, centered at δ 6.1 and δ 9.0

(d) *J* $C_6H_{12}O_2$
^{1}NMR: δ 3.5 (singlet, relative area 1); δ 1.2 (singlet, relative area 3)

(e) *K* $C_8H_{11}NO$
IR: absorption at 3300 cm^{-1}
^{1}NMR: δ 6.6 (multiplet, relative area 4); δ 3.9 (quartet, relative area 2); δ 3.4 (singlet, relative area 2); δ 1.3 (triplet, relative area 3)

4. Give structures for the products in each of the following reactions (2 points for each part).
   (a) 4-aminobutanal treated with acid to yield a compound of formula C_4H_7N
   (b) aniline treated with dilute hydrochloric acid
   (c) aniline treated with benzaldehyde in the presence of a catalytic amount of anhydrous acid
   (d) aniline treated with sodium nitrite in aqueous sulfuric acid, followed by the addition of phenol
   (e) aniline treated with a large excess of iodomethane
   (f) piperidine treated with formaldehyde followed by an excess of formic acid
   (g)

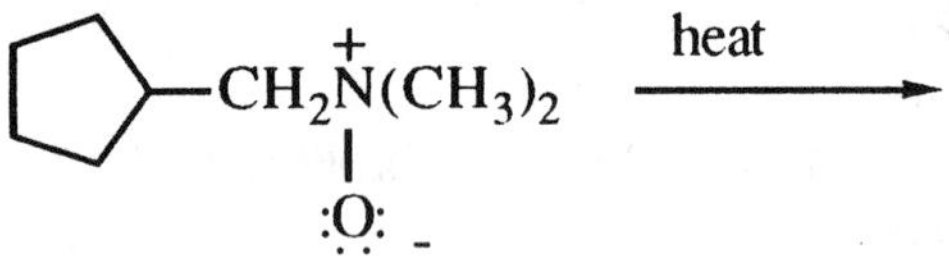

   cyclopentyl–CH$_2$N$^+$(CH$_3$)$_2$ with :O:$^-$ substituent $\xrightarrow{\text{heat}}$

5. Propose synthetic methods for each of the following conversions (4 points for each part).
   (a) 2-iodopropane to 2-methyl-1-phenylpropene (using Wittig chemistry)
   (b) benzene to 3-ethylaniline
   (c) benzene to 4-nitrostyrene
   (d) 4-nitrotoluene to *p*-methylbenzylamine

CHAPTER 21
CARBOXYLIC ACIDS

Key Points

• Be familiar with the systematic and common names of carboxylic acids.
Practice Problem 21.1
Draw structures for each of the following:
    (a) β-methylbutyric acid
    (b) 4-pentynoic acid
    (c) (3E,5E)-3,5-heptadienoic acid
    (d) hexanedioic acid
    (e) phenylacetic acid
    (f) vinylacetic acid
    (g) phthalic acid

• Understand the factors influencing the relative acidities of carboxylic acids.
Practice Problem 21.2
For each of the following compounds, tell if they are more acidic or less acidic than is acetic acid.
    (a) hydroxyacetic acid (glycolic acid)
    (b) cyanoacetic acid
Practice Problem 21.3
Consider the isomers m-methoxybenzoic acid and p-methoxybenzoic acid; one of these compounds is a stronger acid than is benzoic acid and the other is a weaker acid than is benzoic acid. Tell which is stronger and which is weaker, and give a brief explanation of the phenomenon.

• Be aware that carboxylic acids are converted to carboxylate salts by the action of bases. Know how to manipulate titration data to derive useful information about unknown carboxylic acids.
Practice Problem 21.4
Give the structures of the salts that would form upon the reaction of each of the following acids and bases:
    (a) 2-aminopropane and benzoic acid
    (b) racemic 2-amino-1-phenylpropane and (R)-2-methylbutanoic acid
Practice Problem 21.5
Compound A is a crystalline solid. A sample of A gives the following elemental analysis: 32.0% carbon; 4.0% hydrogen. A solution of A in water is acidic. Such a solution, prepared by dissolving 0.225 g of A in 25.0 mL of water requires 30.0 mL of 0.1 M aqueous sodium hydroxide solution for complete neutralization. Compound A can be resolved into enantiomers. Suggest a likely structure for A.

• Learn the important methods for the synthesis of carboxylic acids. (Study Section 21.5 of the text before attempting Practice Problem 21.6.)

200 Study Guide and Solutions Manual

Practice Problem 21.6
Provide reagents for each of the following transformations.

HO—⌬—CHO —*i*→ HO—⌬—CO₂H

Ph—CH₂CH₃ —*ii*→ Ph—CO₂H

Br—C₆H₄—CN —*iii*→ Br—C₆H₄—CO₂H

cyclopentene —*iv*→ HO₂C-(CH₂)₃-CO₂H

(2-methyl-3-pentanone type ketone) —*v*→ (2-methylbutanoic acid)

PhMgBr —*vi*→ PhCO₂H

• Be familiar with the reagents needed to convert carboxylic acids into acid halides, acid anhydrides, esters, amides, and nitriles.

Practice Problem 21.7
What reagents are needed for each of the following conversions?

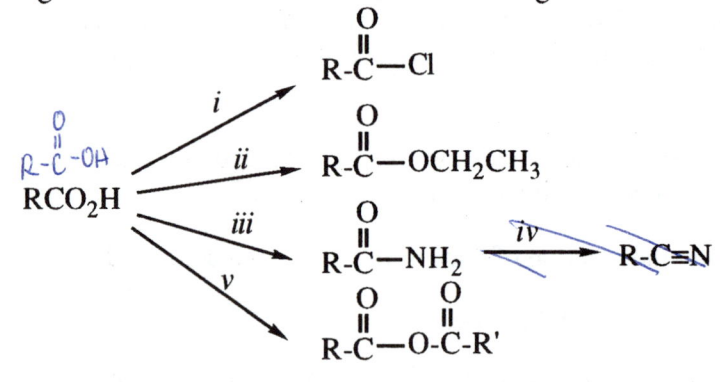

RCO₂H (R-C(=O)-OH)
- *i* → R-C(=O)-Cl
- *ii* → R-C(=O)-OCH₂CH₃
- *iii* → R-C(=O)-NH₂ —*iv*→ R-C≡N
- *v* → R-C(=O)-O-C(=O)-R'

• Make sure that you know the mechanism for acid catalyzed ester formation starting with carboxylic acids and alcohols.

Practice Problem 21.8
From memory, write out the mechanism for the reaction:

CH₃C(=O)-OH + CH₃OH —H⁺→ CH₃C(=O)-OCH₃ + H₂O

• Be familiar with the Hell-Volhard-Zelinski reaction, used to prepare α-halocarboxylic acids.

Practice Problem 21.9
Suggest a synthesis of malonic acid from acetic acid that utilizes the Hell-Volhard-Zelinski reaction.

• Know the reagents used to reduce carboxylic acids.

Practice Problem 21.10
Give the reagents requried for each of the following conversions.

- Know the various types of decarboxylation reactions undergone by carboxylic acids.

Practice Problem 21.11
In each of the following sets, choose the compound that will undergo decarboxylation most readily
  (a) acetic acid, malonic acid, succinic acid
  (b) $CH_3CH_2C(O)CH_2CO_2H$, $CH_3C(O)CH_2CH_2CO_2H$, $CH_3CH_2CH_2C(O)CO_2H$

- Be able to use spectroscopic data, along with other analytical data and reactivity data, to deduce structural information.

Practice Problem 21.12
A compound of formula $C_5H_8O_4$, whose ^1H NMR spectrum consists of two singlets and loses carbon dioxide readily upon heating. Suggest a structure for this compound.

Practice Problem 21.13
Suggest a structure for a compound of formula $C_5H_9ClO_2$ based on the following spectroscopic characteristics:

^1H NMR: δ 1.4 (singlet, relative area 6); δ 3.6 (singlet, relative area 2); δ 11.8 (singlet, relative area 1)

IR: strong absorptions at 2600-3400 cm^{-1} and 1730 cm^{-1}

Solution of Text Problems

21.1 (a) 6-chloro-3-ethylheptanoic acid
  (b) 4-methyl-4-phenylpentanoic acid
  (c) (S)-3-hydroxy-3-methylpentanoic acid
  (d) 3-ethoxypropanoic acid
  (e) m-(3-hydroxy-1-propyl)benzoic acid

21.2

21.3 1. heating cyclohexanol with sulfuric acid to form cyclohexene; 2. treatment with aqueous basic potassium permanganate; 3. acidification with aqueous acid

21.4 Oxalic will have the larger ratio K_{a1}/K_{a2}. Once the first acidic event has occurred, the anionic site of oxalic acid is closer to the remaining acidic site than is the case with malonic acid. Thereby, the effect (decreasing acidity) on the second acidic event is greater with oxalic acid than with malonic acid. The result is that K_{a1}/K_{a2} is greater with oxalic acid than with malonic acid.

21.5 For a diacid HRH, K_{a1}/K_{a2} = [HR$^-$]^2/[HRH][R^{2-}]. The formation of HR$^-$ from HRH is statistically twice as probable as the formation of either HRH or of R^{2-} from HR$^-$. Thus, we find that in the limiting case, $K_{a1}/K_{a2} = (2)^2/(1)(1) = 4$

21.6

An aqueous solution of 3-chloropropanoic acid is titrated with one equivalent amount of potassium hydroxide in aqueous solution. The solvent is then evaporated under reduced pressure to isolate the solid salt.

21.7 10.25 mL

21.8

$R =$ [3-methylphenyl-1-methylcyclohexyl group]

$HO:^- + R-C\equiv N: \rightleftharpoons$ [tetrahedral intermediate with OH, N⁻, R] $\rightleftharpoons$ [protonation steps leading to imine-diol] $\rightleftharpoons$ [further steps to R-C(OH)(NH)(OH)]

[continues through mechanism to form] $R-C(=O)-OH + {}^-NH_2$

$R-C(=O)-OH + {}^-NH_2 \longrightarrow R-C(=O)-O^- + NH_3$

21.9 1. heating of cyclopentene with aqueous basic potassium permanganate, and worked up with aqueous acid to form $HO_2CCH_2CH_2CH_2CO_2H$; 2. lithium aluminum hydride, acidic workup to form $HO(CH_2)_5OH$; 3. phosphorus trichloride to form $Cl(CH_2)_5Cl$; 4. sodium cyanide; 5. heat with aqueous acid

21.10 (a) 1. treatment of 1-pentene with borane; 2. aqueous hydrogen peroxide; 3. aqueous basic potassium permanganate; 4. aqueous acid

(b) 1. heating of 1-pentene with aqueous basic potassium permanganate; 2. aqueous acid

(c) 1. treatment of 1-pentene with borane; 2. aqueous hydrogen peroxide; 3. phosphorus trichloride; 4. sodium cyanide; 5. heat with aqueous acid

(d) 1. treatment of 1-butene with aqueous acidic mercuric sulfate; 2. sodium borohydride; 3. phosphorus trichloride; 4. magnesium metal, ether solution; 5. carbon dioxide; 6. aqueous acid workup

(e) 1. treatment of benzene with bromine in the presence of iron; 2. magnesium metal in ether; 3. carbon dioxide; 4. aqueous acid workup

(f) 1. treatment of benzene with iodomethane in the presence of aluminum chloride; 2. sulfuric acid, nitric acid; 3. heating with aqueous basic potassium permanganate; 4. aqueous acidic workup

21.11 anhydrides > esters > amides > carboxylic acids

21.12 Acetic benzoic anhydride could be prepared either by the reaction of sodium benzoate with acetyl chloride, or by the reaction of sodium acetate with benzoyl chloride

21.13 (a) methyl acetate
(b) acetamide

21.14

$H_2C=C=\ddot{O}$ + $H-\ddot{O}-C(=O)-CH_3$ $\longrightarrow$ $H_2\bar{C}-C(-\overset{+}{O}H-)-O-C(=O)CH_3$ $\longrightarrow$ $H_3C-C(=O)-O-C(=O)CH_3$

21.15 (a) acetyl chloride
(b) acetamide
(c) ethyl acetate

21.16

PhCH2-O-C(=O)-CH3 (benzyl acetate: benzyl–O–C(=O)CH3)

21.17 (a) phenyl propanoate
(b) isopropyl benzoate

21.18 Acid is consumed in steps 1 and 4, and is formed in steps 3 and 6. There is equivalent consumption and generation of acid leaving the reaction catalytic in acid.

21.19

(a) Ph–C(=^{16}O)–^{18}OCH$_2$CH$_3$ All product molecules contain 1 ^{18}O.

(b) Ph–C(=^{18}O)–^{16}OCH$_2$CH$_3$ (50%)   and   Ph–C(=^{16}O)–^{16}OCH$_2$CH$_3$ (50%)

(c) Ph–C(=^{16}O)–^{16}OCH$_2$CH$_3$ Ultimately, all ^{18}O will be lost to the aqueous solution.

21.20 There was formed polymeric ester (polyester) product with the repeating unit -C(O)[O(CH$_2$)$_5$C(O)]$_n$O-. A better yield of the lactone will result using a more dilute solution of the starting hydroxy acid.

21.21

$$R\text{-}\ddot{N}=C(H)\text{-}\ddot{N}\text{-}R \; (+) \longleftrightarrow \; R\text{-}\ddot{N}=\overset{+}{C}(H)\text{-}\ddot{N}\text{-}R \longleftrightarrow \; R\text{-}\overset{+}{N}\equiv C(H)\text{-}\ddot{N}\text{-}R$$

21.22

(a) CH$_3$-C(=O)-O-C$_6$H$_{11}$ (cyclohexyl)

(b) H-C(=O)-O-CH(CH$_3$)CH$_2$CH$_3$

(c) Ph-C(=O)-O-CH$_2$-C$_6$H$_4$-NO$_2$

(d) PhCH$_2$CH$_2$-C(=O)-O-CH$_2$CH(CH$_3$)$_2$

(e) CH$_3$CH(Br)-C(=O)-O-CH$_2$CH$_3$

(f) CH$_3$CH$_2$-O-C(=O)-CH$_2$CH$_2$CH$_2$CH$_2$-C(=O)-O-CH$_2$CH$_3$

(g) CH$_3$CH$_2$CH(CH$_3$)-C(=O)-O-CH$_3$

21.23

H$_3$C-substituted 1,4-dioxane-2,5-dione (lactide-type ring with H$_3$C and CH$_3$ substituents)

21.24 (a) 1. treatment of 3,3-dimethyl-2-butanone with iodine in aqueous sodium hydroxide (haloform reaction); 2. aqueous acid; 3. methanol with an anhydrous acid catalyst

(b) 1. treatment of 1-hexanol with aqueous basic potassium permanganate; 2. aqueous acid; 3. ethanol with an anhydrous acid catalyst

(c) 1. treatment of 1-bromobutane with sodium cyanide to form 1-cyanobutane; 2. heat with aqueous acid to form pentanoic acid; 3. methanol in the presence of an anhydrous acid catalyst

21.25

(a) CH₃-C(=O)-NHPh

(b) CH₃CH₂CH₂-C(=O)-N(CH₃)₂

(c) PhCH₂-C(=O)-NH₂

21.26 The added acid reacts with the base (ammonia) and is removed from serving as a catalyst for the amide formation.

21.27

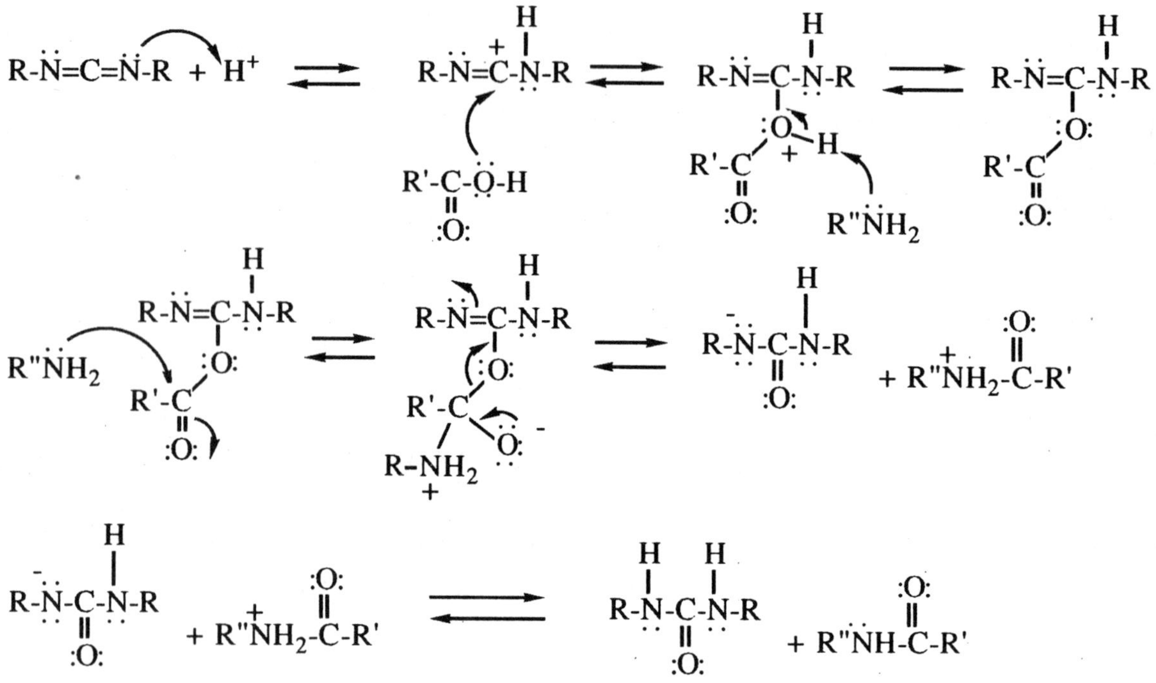

[Mechanism scheme showing protonation of carbodiimide R-N=C=N-R by H⁺, addition of carboxylic acid R'-C(=O)-O-H, attack by amine R"NH₂, proton transfers, and final products: R-NH-C(=O)-NH-R (urea) + R"NH-C(=O)-R' (amide).]

21.28

Ammonium carbamate: H₂N-C(=O)-O⁻ ⁺NH₄

Urea: H₂N-C(=O)-NH₂

21.29 A portion (62.5%) of the lithium aluminum hydride is consumed by reaction with the acidic hydrogen of the benzoic acid.

21.30

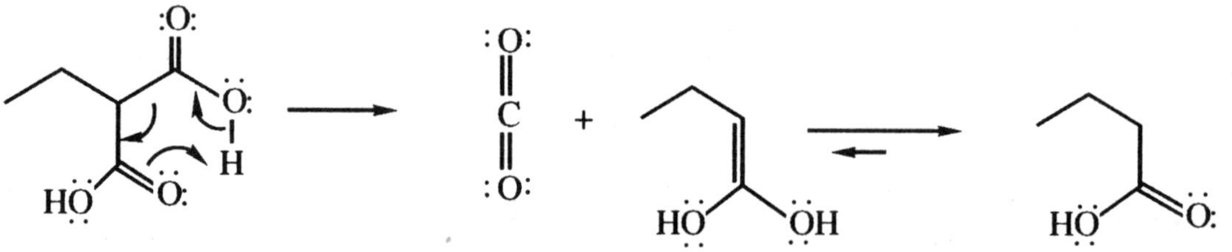

[Mechanism showing β-keto acid decarboxylation via cyclic transition state, loss of CO₂, formation of enol intermediate HO-C(R)=CH-R, and tautomerization to ketone HO-C(=O)-CH₂-R.]

21.31 Note that with the molecular models of the enol intermediate (if it can be made without breaking the models) that the bonds to the (planar) bridgehead carbon are extremely strained and the bonds are twisted making π bond formation quite unfavorable.

21.32 As *n* becomes large, the molecule has sufficient flexibility to allow the molecule to become planar about the bridgehead site and for *p* orbital interaction and π bond stabilization to occur.

21.33

21.34 The normal position of the C=O stretching band in the IR spectrum, as is the case with all IR absorptions, is related to the strength of that bond. In a carboxylate salt the π bonding electrons are delocalized over a larger region than in a simple C=O compound, significantly changing the force constant associated with the stretching of the linkage.

21.35

(a) CH$_3$CH$_2$CHCO$_2$H with Ph substituent

(b) PhCHCO$_2$H with OH substituent

(c) H$_2$C=CHCO$_2$H

(d) HO$_2$C–(long chain with one internal C=C)–

(e) Ph-C(=O)-O-C(=O)-CH$_2$CH$_2$CH$_2$CH$_2$CH$_2$CH$_3$

(f) succinic anhydride (γ-butyrolactone-like cyclic anhydride)

(g) (CH$_3$)$_2$CHCH$_2$C(=O)-NH$_2$

(h) Ph-C(=O)-N-CH(CH$_3$)$_2$ with CH$_3$

(i) (CH$_3$)$_2$CHCN

(j) (CH$_3$)$_2$CHCN

(k) 3,5-dinitrobenzoate, O$_2$N–C$_6$H$_3$(NO$_2$)–C(=O)–OCHCH$_2$CH$_3$ with CH$_3$

(l) 4-bromo-phenyl-NHC(=O)CH$_3$

(m) CH$_3$OCH$_2$CO$_2$H

(n) BrCH$_2$CHCHCO$_2$H with Br Br

(o) CH$_3$CH(CH$_2$CO$_2$H)$_2$

(p) CH$_3$CH$_2$CO$_2$K

(q) (CH$_3$)$_2$CHCCl(=O)

(r) (CH$_3$)$_3$CC(=O)CH$_3$

(s) NCCH$_2$CH$_2$CN

21.36 (a) 4-pentenoic acid
(b) 3-hydroxy-2-methylpropanoic acid
(c) 2,2-diethylbutanoic acid
(d) sec-butyl ~~propenoate~~ 3-butenoate

21.37 (a) fluoroacetic acid - The stronger electron withdrawing group increases the acidity more.
(b) methoxyacetic acid - The methoxy group is inductively electron withdrawing whereas the alkyl groups are electron donating.
(c) pK_a = 4 - (K_a = 10^{-4})
(d) HO$_2$CCH$_2$CO$_2$H - For pK_{a1}, the electron withdrawing effect is greater with malonic acid.
(e) cis-cyclopenatnedicarboxylic acid - The trans isomer has its carboxyl groups pointing away from each other such that they cannot reach to form a ring.

(f) formic acid - The formate anion is resonance stabilized, and the two carbon-oxygen linkages are equivalent (each has a bond order of 1.5).

(g) calcium carbonate - The carbonate anion has greater single bond character in each carbon-oxygen linkage. Each C-O bond is single in two resonance structures and double in one resonance structure.

(h) 1,1-cyclobutanedicarboxylic acid - The decarboxylation proceeds through the more favorable six-membered activated complex.

(i) cyclohexanecarboxylic acid - This material has an α-hydrogen available for substitution.

(j) propionyl chloride - The chloride ion is a better leaving group than is a carboxylate ion.

(k) *cis*-3-hydroxycyclohexanecarboxylic acid - The *trans* isomer has the carboxyl and hydroxyl groups pointing away from each other.

21.38 (a) PhCH$_2$CH$_2$CO$_2$Na

(b) PhCH$_2$CH$_2$CO$_2$Na

(c) PhCH$_2$CH$_2$CO$_2$NH$_4$

(d) PhCH$_2$CH$_2$CH$_2$OH

(e) PhCH$_2$CH$_2$C(O)Cl

(f) PhCH$_2$CH(Br)CO$_2$H

(g) PhCH$_2$CH$_2$C(O)OCH$_2$CH$_2$CH$_2$CH$_3$

(h) PhCH$_2$CH$_2$C(O)OCH$_2$CH$_2$OH plus Ph-C-C-C(=O)-O-C-C-O-C(=O)-C-C-Ph

(i) PhCH$_2$CH$_2$C(O)Ph

(j) PhCH$_2$CH$_2$CO$_2$MgBr

(k) PhCO$_2$H

21.39

$$\begin{array}{c} \text{H} \\ :\ddot{O}\text{---}H\text{—}\ddot{N}: \\ R\text{—}C \diagup \diagdown C\text{—}R \\ :\ddot{N}\text{—}H\text{---}\ddot{O}: \\ \text{H} \end{array}$$

21.40 1. Dissolve the mixture in a water insoluble organic solvent. 2. Wash the solution with aqueous base. 3. Separate the organic layer and dry it (use a water absorbing solid), filter to remove the drying agent, and evaporate the solvent to isolate the alcohol. 4. Make the aqueous layer acidic by the addition of aqueous acid. 5. Extract the acidified aqueous layer with an organic solvent. 6. Dry the organic solution (use a water absorbing solid), filter to remove the drying agent, and evaporate the solvent to isolate the carboxylic acid.

21.41 (a) aqueous sodium hydroxide

(b) 1. aqueous basic potassium permanganate; 2. aqueous acid

(c) 1. thionyl chloride; 2. ammonia

(d) lithium aluminum hydride

(e) 1. lithium aluminum hydride; 2. chromic anhydride in pyridine

21.42 (a) sulfuric acid, nitric acid

(b) bromine in the presence of iron

(c) 1. magnesium metal in ether; 2. carbon dioxide; 3. aqueous acid

(d) 1. chlorine with aqueous sodium hydroxide; 2. aqueous acid

(e) 1. aqueous basic potassium permanganate; 2. sulfuric acid, nitric acid

(f) 1. bromine in the presence of iron; 2. magnesium in ether solution; 3. carbon dioxide; 4. aqueous acid

(g) 1. phosphorus tribromide, bromine; 2. sodium cyanide; 3. heat with aqueous acid

(h) 1. iodoethane in the presence of aluminum chloride; 2. chlorine in aqueous base; 3. aqueous acid

(i) 1. sodium nitrite, sulfuric acid; 2. CuCN; 3. heat with aqueous acid

(j) 1. chromic anhydride, sulfuric acid; 2. sodium cyanide in aqueous acid; 3. heat with aqueous acid

(k) 1. sulfuric acid, nitric acid; 2. aqueous basic potassium permanganate; 3. thionyl chloride; 4. methylamine

(l) 1. mercuric sulfate, sulfuric acid, water; 2. sodium borohydride; 3. chromic anhydride, sulfuric acid

(m) 1. bromine with light irradiation; 2. magnesium in ether; 3. carbon dioxide; 4. aqueous acid

(n) 1. HBr; 2. magnesium in ether; 3. ethylene oxide; 4. phosphorus trichloride; 5. NaCN; 6. heat with

aqueous acid
    (o) 1. phosphorus trichloride, chlorine; 2. ethanol with a trace of anhydrous acid
    (p) 1. bromine under light irradiation; 2. NaCN; 3. heat with aqueous acid
    (q) 1. bromine, phosphorus tribromide; 2. NaCN; 3. heat with aqueous acid
    (r) 1. four equivalents of methyllithium; 2. sodium borohydride; 3. heat with aqueous acid
    (s) 1. chromic anhydride, sulfuric acid; 2. chlorine, phosphorus trichloride; 3. NaCN; 4. heat with aqueous acid

21.43

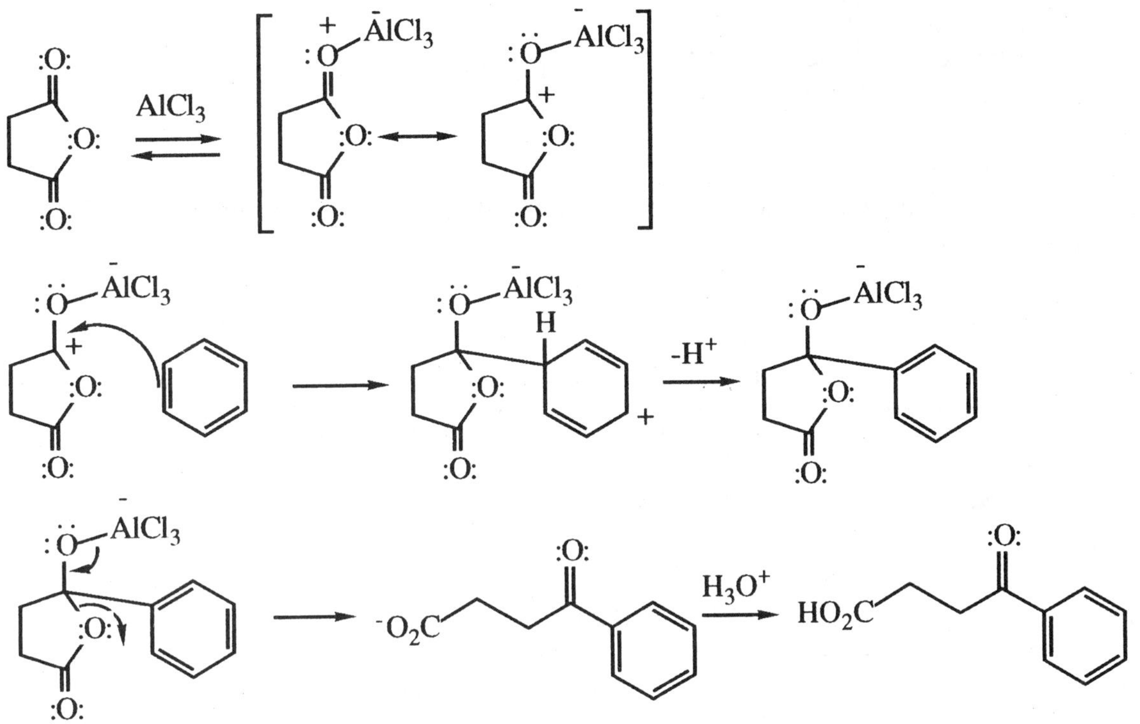

[mechanism scheme showing succinic anhydride + AlCl$_3$ reacting with benzene via Friedel-Crafts acylation to give 3-benzoylpropanoic acid (HO$_2$C-CH$_2$CH$_2$-C(=O)-Ph)]

21.44

A: CH$_3$C(Br)(CO$_2$H)$_2$
B: CH$_3$CH(Br)CO$_2$H
C: (CH$_3$)$_2$CHP$^+$Ph$_3$ Br$^-$
D: (CH$_3$)$_2$C=PPh$_3$
E: [β-lactone with PPh$_3$ group, (CH$_3$)$_2$C—C(=O)—O ring with PPh$_3$]

F: (CH$_3$)$_2$CHCO$_2$H
G: HO$_2$CCH$_2$CH$_2$CH$_2$C(=O)CH$_3$
H: (CH$_3$)$_2$C(CO$_2$H)$_2$
I: (CH$_3$)$_2$CHCO$_2$H

J: cyclopentenyl-CH$_2$CH$_2$CH$_2$CH$_2$CO$_2$H
K: cyclopentyl-CH$_2$CH$_2$CH$_2$CH$_2$CO$_2$H
L: cyclopentyl-CH$_2$CH$_2$CH$_2$CH$_2$Br
M: cyclopentyl-CH$_2$CH$_2$CH=CH$_2$
N: cyclopentyl-CH$_2$CH(Br)CH=CH$_2$
O: cyclopentyl-CH=CHCH=CH$_2$

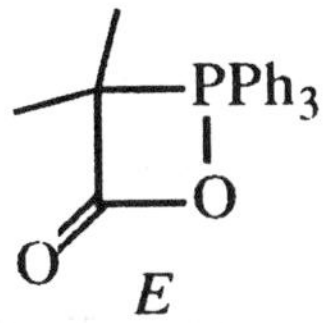

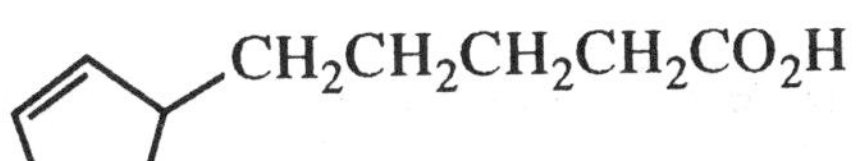

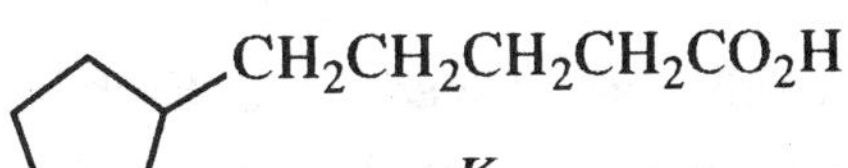

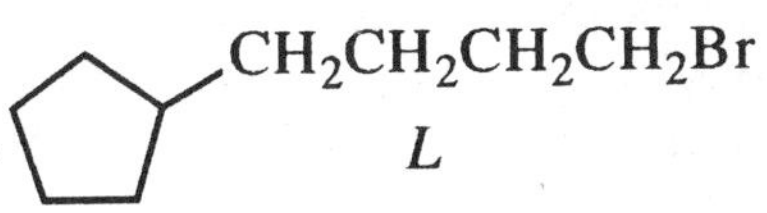

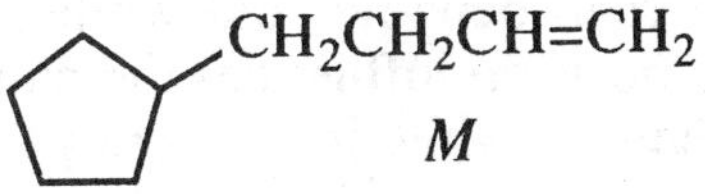

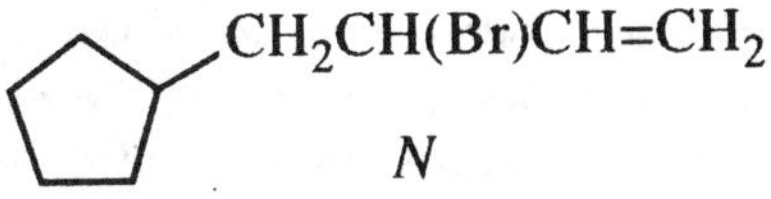

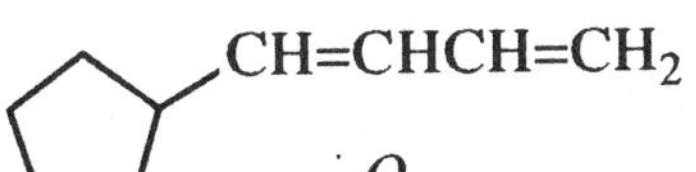

21.45 P is (CH$_3$)$_3$CCO$_2$H; Q is (CH$_3$)$_3$CCH$_2$OH; R is (CH$_3$)$_2$C(OH)CH$_2$CH$_3$

21.46 S is HOCH$_2$CH$_2$CO$_2$H; T is HO$_2$CCH$_2$CO$_2$H; U is CH$_3$CO$_2$H

21.47

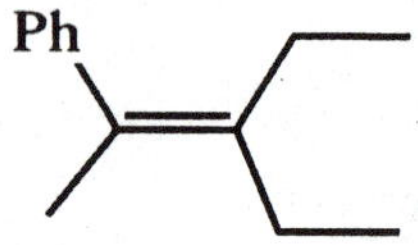

V: (Ph)(CH₃)C=C(CH₂CH₃)₂ (2-phenyl-3-ethyl-2-pentene structure)

W: PhC(O)CH₃

X: CH₃CH₂C(O)CH₂CH₃

Y: PhCO₂H

21.48

CH₃-C(=O)-O:⁻ →(-e⁻)→ CH₃-C(=O)-O• → •CH₃ + CO₂

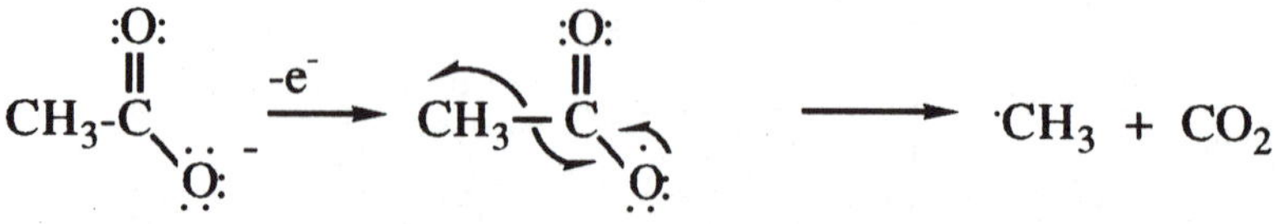

2 •CH₃ → H₃C-CH₃

21.49 Z is PhCCl₃. Hydroxide ion displacement of the three chlorides leads to formation of PhC(OH)₃, which loses water to form PhCO₂H.

21.50

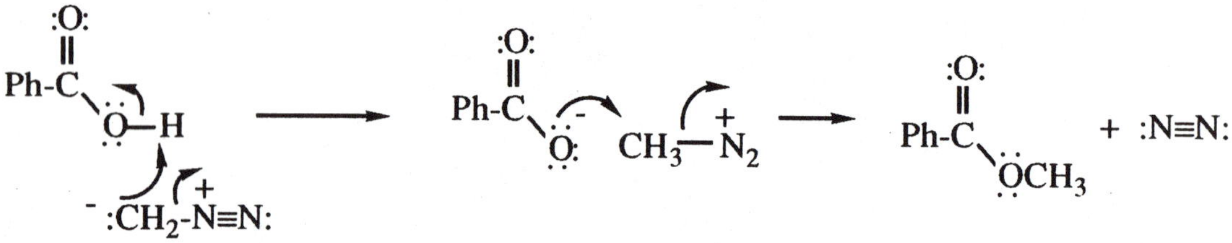

Ph-C(=O)-O-H + ⁻:CH₂-N⁺≡N: → Ph-C(=O)-O:⁻ + CH₃-N₂⁺ → Ph-C(=O)-OCH₃ + :N≡N:

21.51

R-C≡N: + R'-MgX → R-C(=N:⁻ MgX⁺)-R' →(H₂O)→ R-C(=O)-R'

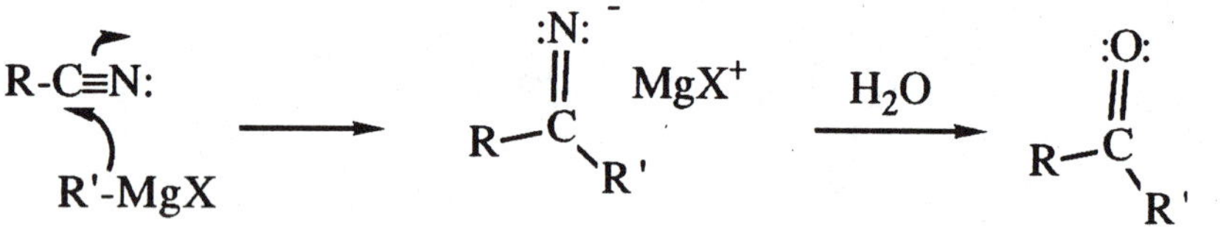

21.52 AA is (CH₃)₂CHC(O)NH₂; BB is (CH₃)₂CHCN; CC is (CH₃)₂CHC(O)Ph

Solution of Study Guide Practice Problems

21.1
(a) CH₃CH(CH₃)CH₂CO₂H
(b) H-C≡C-CH₂CH₂CO₂H
(c)  CH₃CH=CHCH=CHCH₂CO₂H (hexa-2,4-dienoic acid)
(d) HO₂CCH₂CH₂CH₂CH₂CO₂H
(e) PhCH₂CO₂H
(f) H₂C=CHCH₂CO₂H
(g) 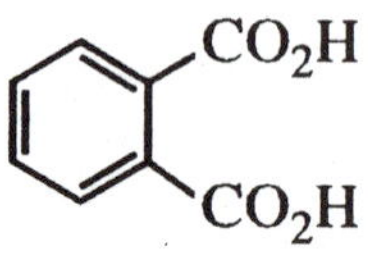 phthalic acid (1,2-benzenedicarboxylic acid)

21.2 Both compounds are more acidic than is acetic acid. Each has an electron-withdrawing group substituted at the α-carbon site that helps to stabilize the conjugate base formed on deprotonation of the parent acid.

21.3 *m*-Methoxybenzoic acid is stronger than benzoic acid while *p*-methoxybenzoic acid is weaker than is benzoic acid. A methoxy group in the *meta* position influences the carboxyl group only by way of its inductive effect (electron-withdrawing, acid-strengthening), while in the *para* position it exerts an electron releasing (acid-weakening) effect that outweighs the inductive effect.

21.4 (a) Ph-C(=O)-O:⁻   H₃N⁺CH(CH₃)₂

(b) Fischer projection: CH₃ top, H—|—CO₂⁻ with CH₂CH₃ bottom; H₃N⁺CH(CH₃)CH₂Ph

mixture of *R,R* and *R,S* diastereoisomers

(HO₂C)₂C(OH)CH₂OH

21.5 Consider first the elemental composition data. If we assume that the "missing" material is oxygen, we arrive at an empirical formula of C₂H₃O₃. If this is a correct assumption, the titration data should (and does) reveal that the molecular weight of *A* is 85 g/mole, or a multiple of 85 g/mole. Because no structure can be drawn for C₂H₃O₃ in which all of the normal valences are used, we should first consider the formula C₄H₆O₆. In this case we would need to consider a dicarboxylic acid. As we are told that the compound can be resolved, a likely structure is that of racemic tartaric acid, HO₂CCH(OH)CH(OH)CO₂H.

21.6 *i* silver ion in aqueous ammonia solution
 ii heating with aqueous basic potassium permanganate
 iii heating with dilute aqueous sulfuric acid
 iv ozone with workup using hydrogen peroxide
 v iodine in aqueous sodium hydroxide solution
 vi addition of carbon dioxide and workup with aqueous acid

21.7 *i* thionyl chloride
 ii ethanol with DCC or ethanol in the presence of a trace amount of anhydrous acid
 iii heating with anhydrous ammonia (see Chapter 22 for an alternative, better method)
 iv heat with phosphorus pentoxide
 v R'C(O)Cl

21.8 The complete mechanism for the Fischer esterification is given in the text in Section 21.6.

21.9 1. treatment of acetic acid with phosphorus tribromide and bromine; 2. cold aqueous sodium hydroxide solution; 3. NaCN; 4. heat with aqueous acid

21.10 *i* BH₃, THF
 ii lithium aluminum hydride

21.11 (a) malonic acid
 (b) CH₃CH₂C(O)CH₂CO₂H

21.12 (a) (CH₃)₂C(CO₂H)₂
 ~~(b) (CH₃)₂CH(Cl)CH₂CO₂H~~

13. (CH₃)₂C(Cl)CH₂CO₂H

CHAPTER 22
DERIVATIVES OF CARBOXYLIC ACIDS

Key Points

• Know the general structures and the nomenclature of carboxylic acid esters, amides, anhydrides, halides, and nitriles.

Practice Problem 22.1
Give an acceptable name for each of the following:

(a) CH$_3$-C(=O)-O-Ph
(b) CH$_3$-O-C(=O)-Ph
(c) Br-C$_6$H$_4$-C(=O)-NHCH$_3$
(d) CH$_3$CH$_2$-C(=O)-O-C(=O)-CH$_2$CH$_3$
(e) O$_2$N-C$_6$H$_4$-C(=O)-Cl

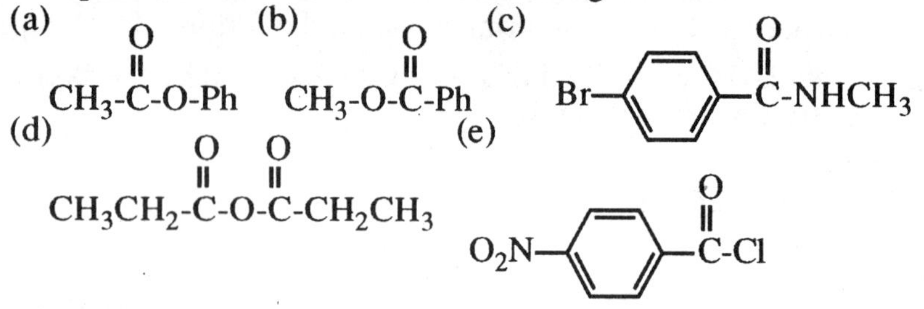

Practice Problem 22.2
Draw structures for each of the following:
 (a) trifluoroacetonitrile
 (b) trifluoroacetic anhydride
 (c) 4-bromobenzamide

• Know the general mechanism for the reaction of nucelophiles with carboxylic acid halides, anhydrides, esters, and amides.

Practice Problem 22.3
Give all steps in the mechanism for the reaction of ammonia with benzoyl chloride to yield benzamide.

• Know the relative reactivities of carboxylic acid derivatives toward nucleophiles.

Practice Problem 22.4
List the following in order of increasing reactivity toward hydrolysis under neutral conditions:
    PhC(O)OPh; PhC(O)Cl; PhC(O)NH$_2$; PhC(O)OC(O)Ph

• Be able to predict the products of reactions of carboxylic acid derivatives and nucleophiles.

Practice Problem 22.5
Give the organic products for each of the following reactions.
    (a) acetic anhydride reacting with ammonia
    (b) benzoyl chloride reacting with ethylamine
    (c) sodium acetate reacting with acetyl chloride
    (d) phthalic anhydride reacting with ammonia

• Learn the reaction conditions for transesterification reactions, ester to amide conversions, and amide to nitrile conversions.

Practice Problem 22.6
How would you convert methyl acetate to ethyl acetate? (Give a description of the laboratory procedures you would use.)

Practice Problem 22.7
Give the structure of the organic product in the reaction shown below.

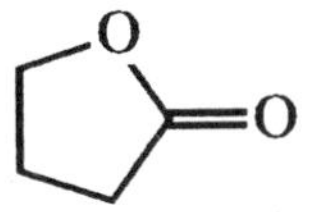

γ-butyrolactone + excess CH₃OH, H⁺ → C₅H₁₀O₃

Practice Problem 22.8
Propose a synthetic sequence that will convert PhCH₂CO₂CH₂CH₃ to PhCH₂CN.

- Know the mechanisms and the reaction conditions (and products) for the hydrolysis of esters, amides, and nitriles.

Practice Problem 22.9
Suppose that benzyl acetate is hydrolyzed in H₂^{18}O, in the presence of an acid catalyst. Which of the products would you expect to contain ^{18}O?

Practice Problem 22.10
A neutral compound, A, of formula C₁₀H₁₃NO, is refluxed for one hour with aqueous sodium hydroxide solution, and then a distillation is performed. The distillate contains an oily basic liquid, B, whose broad-band decoupled ^{13}C NMR spectrum consists of a single peak. A crystalline substance, C, is obtained by acidification of the alkaline solution remaining behind in the distilling flask. Compound C proves to be acidic, and 0.272 g of it are exactly neutralized in titration with 20.0 mL of 0.01 M sodium hydroxide solution. On refluxing C with alkaline potassium permanganate solution and worked up by acidification, a new acid, D, is formed that exhibits four signals in its ^{13}C NMR spectrum, and forms phthalic anhydride on heating. Propose structures for compounds A-D.

Practice Problem 22.11
A neutral compound, E, exhibits an IR band near 2200 cm^{-1}, but no band in the 3300-3500 cm^{-1} region. On reflux of E with aqueous sodium hydroxide solution ammonia gas is liberated. The remaining solution on evaporation yields a salt, F, that exhibits only a triplet and a quartet in its ^{1}H NMR spectrum measured in D₂O solution. What are the structures of E and F?

- Know the various methods used to reduce carboxylic acid derivatives. Know the reagents used, the products formed, and be able to incorporate this knowledge into your arsenal of reactions for syntheses.

Practice Problem 22.12
Give the missing reagents or organic products in each of the following reactions.

(a) ClCH₂C(=O)Cl —LiAlH₄, ether→ —H₂O→ ?

(b) PhC(=O)-O-CH₂CH₃ —LiAlH₄, ether→ —H₂O→ ?

(c) CH₃CH₂-C(=O)-NHCH₃ —BH₃-THF→ ?

(d) PhCH₂C(=O)-Cl —?→ PhCH₂CHO

(e) PhCH₂CH₂CN —Na / CH₃CH₂OH→ ?

- Learn the reactions of carboxylic acid derivatives with organometallic reagents.

Practice Problem 22.13
Supply the missing products or reagents in each of the following reactions.

(a)
$$PhC(=O)\text{-}O\text{-}CH_2CH_3 \xrightarrow[\text{ether}]{\text{excess } CH_3CH_2MgI} \xrightarrow{H_2O} ?$$

(b)
$$PhCH_2CN \xrightarrow[\text{ether}]{CH_3MgI} \xrightarrow{H_2O} ?$$

(c)
$$PhC(=O)\text{-}Cl \xrightarrow{?} PhC(=O)\text{-}CH_2CH_2CH_3$$

- Learn the following special reactions: Hofmann rearrangement of amides; Baeyer-Villiger oxidation of esters; pyrolysis of acetate and xanthate esters.

Practice Problem 22.14
Give the expected products for each of the following reactions.
  (a) treatment of *p*-nitrobenzamide with bromine in aqueous basic solution, followed by heating
  (b) treatment of cyclohexanone with peroxybenzoic acid in the presence of an acid catalyst
  (c) pyrolysis of the acetate ester of (*R*,*R*)-2-deuterio-1,2-diphenylethanol, assuming that the preferred conformation is one in which the phenyl groups are as far removed from each other as is possible
  (d) cyclohexene with CH$_2$OAc and CH$_2$OAc substituents, heated to 500° in an inert atmosphere
  (e) treating 3,3-dimethyl-1-butanol with carbon disulfide in the presence of potassium hydroxide, followed by addition of iodomethane, and finally pyrolyzing the resultant material

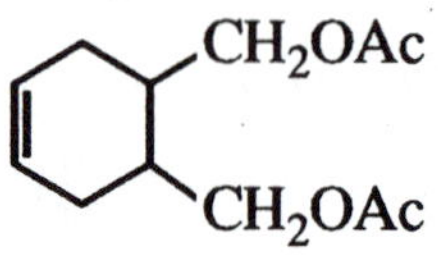

Practice Problem 22.15
Propose series of reactions to accomplish each of the following conversions.
  (a) pentanoic acid to 1-aminobutane
  (b) phenyl(4-methoxyphenyl)methanol to 4-methoxyphenyl benzoate
  (c) methyl 4-methylphenyl ketone to *p*-methylphenol

- Review the main spectroscopic properties of carboxylic acid derivatives as presented in Section 22.8 of the text.

Solution of Text Problems

22.1

Ar = biphenyl-4-yl

Ar-O-H attacks CH$_3$-C(=O)-O-C(=O)-CH$_3$ → Ar-O(+)(H)-C(CH$_3$)(=O)...O-C(=O)-CH$_3$ → Ar-O-C(=O)-CH$_3$ + HO-C(=O)-CH$_3$

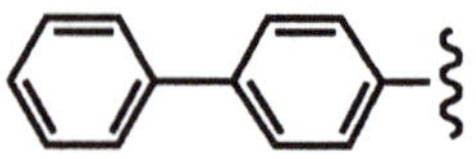

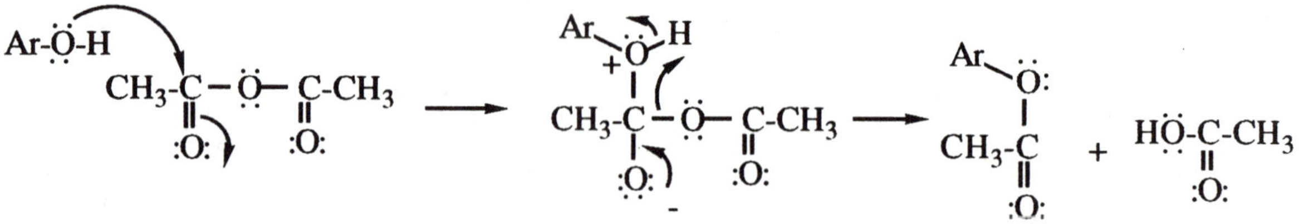

22.2

from NH$_3$	from PhNH$_2$	from (CH$_3$CH$_2$)$_2$NH
PhC(O)NH$_2$	PhC(O)NHPh	PhC(O)N(CH$_2$CH$_3$)$_2$

With (CH$_3$CH$_2$)$_3$N there is formed as an intermediate PhC(O)N$^+$(CH$_2$CH$_3$)$_3$, which on the addition of water is cleaved to benzoic acid and triethylamine

Chapter 22   213

22.3

R = H₂C=CHCH₂–

[mechanism showing R-O-H attacking CH₃-C(=O)-O-C(=O)-CH₃ acetic anhydride, tetrahedral intermediate with R-O⁺H, then products: CH₃-C(=O)-OR + HO-C(=O)-CH₃]

The formyl carbon is attacked rather than the acetyl carbon because the formyl carbon is the more electrophilic of the two. The attached methyl group directs electron density toward the acetyl carbon site making it less electrophilic.

22.4 (a) ethyl 2,2-dimethylpropanoate
(b) *N*-(*p*-tolyl)benzamide
(c) 2,2-dimethylpropanamide
(d) benzoic acid (on workup with water)   N,N-diethylbenzamide

22.5 (a) treatment of cyclohexanol with propanoic anhydride
(b) treatment of ethanol with 4-nitrobenzoyl chloride in the presence of triethylamine
(c) treatment of *tert*-butyl alcohol with acetyl chloride in the presence of triethylamine at 0°
(d) treatment of 1-hexanol with formic acetic anhydride

22.6 (a) 1. treatment of benzoic acid with thionyl chloride; 2. diethylamine
(b) 1. treatment of propanoic acid with thionyl chloride; 2. 4-methoxyaniline
(c) 1. treatment of phenylacetic acid with thionyl chloride; 2. ammonia

22.7

R = H₂C=CH–   R' = *n*-C₁₀H₂₁

[mechanism sequence for acid-catalyzed transesterification showing protonation of ester R-C(=O)-OCH₃, addition of HOR', proton transfer steps, and ultimately loss of CH₃OH to give R-C(=O)-OR']

22.8 HOCH₂CH(OH)CH₂OH

22.9 The *tert*-butyl ester reacts with acid in the following way:

[mechanism: R-C(=O)-O-C(CH₃)₃ + H⁺ ⇌ protonated ester ⇌ RCO₂H + (CH₃)₃C⁺]

(CH₃)₃C⁺ —−H⁺→ (CH₃)₂C=CH₂

Esters derived from primary and secondary alcohols do not react in this way as they would lead to 1° and 2° carbocations, both of which are less stable than the 3° carbocation from a *tert*-butyl ester. These esters hydrolyze by way of a different mechanism (see Section 22.3 of text).

22.10

(a) $F_3\overset{-}{B}-\overset{+}{N}H_3$

(b) [Mechanism showing Ph-C(=O)-NH$_2$ reacting with BF$_3$, then with HOCH$_3$, proton transfer, and subsequent steps giving Ph-C(=O)-OCH$_3$ + F$_3\overset{-}{B}\overset{+}{N}H_3$]

21.11

[Mechanism showing Cl$_3$CC≡N: protonated by H$^+$ to Cl$_3$C-C=N-H, then attacked by H$_2$PO$_4^-$ via O-PO$_3$H$_2$, then HOR addition, proton transfers, giving Cl$_3$C-C(=O)-NH$_2$ + H$_2$O$_3$P-O-R]

22.12 The reaction proceeds reversibly through the intermediate:

$$H_3C-\underset{\underset{:OCH_2CH_3}{|}}{\overset{\overset{^{18}:OH}{|}}{C}}-\ddot{O}H$$

In continuing reaction, this species can lose either of the two HO oxygen atoms as water with equal likelihood as ethyl acetate is regenerated.

22.13

[Mechanism: R-C(=O)-NH$_2$ + HO⁻ ⇌ R-C(OH)(O⁻)-NH$_2$ ⇌ R-C(=O)-OH + ⁻NH$_2$ ⇌ RCO$_2^-$ + :NH$_3$]

Esters hydrolyze more readily under alkaline conditions as the negative charge of the anion produced in the rate determining step is stabilized better by an -OR group than by -NH$_2$.

22.14 [HO$_2$C(CH$_2$)$_5$NH$_3$]$^+$ Cl$^-$

22.15

[Mechanism: R-C≡N: + :OH⁻ ⇌ R-C(OH)=N:⁻ ⇌ (H$_2$O) R-C(OH)=NH ⇌ R-C(O⁻)-NH ⇌ (H$_2$O) R-C(=O)-NH$_2$ → continued further hydrolysis to carboxylate anion and ammonia]

22.16 ethyl phenylacetate

Chapter 22 215

22.17 (HO$_2$CCH$_2$CH$_2$CH$_2$)$_2$C=O
22.18 For each of the illustrated Equations, the organic by-product is ethanol.
22.19 (a) 1. treatment of benzoyl chloride with ammonia; 2. lithium aluminum hydride
       (b) 1. treatment of acetic anhydride with cyclohexylamine; 2. lithium aluminum hydride
       (c) 1. treatment of propanoyl chloride with dipropylamine; 2. lithium aluminum hydride
22.20 (a) 1. treatment of benzaldehyde with ammonia in the presence of a catalytic amount of anhydrous acid; 2. hydrogen, platinum oxide; 3. benzoyl chloride; 4. lithium aluminum hydride
       (b) 1. treatment of cyclohexanone with ammonia in the presence of a catalytic amount of anhydrous acid; 2. hydrogen, platinum oxide; 3. acetyl chloride; 4. lithium aluminum hydride
22.21

[mechanism scheme showing hemiaminal/imine hydrolysis steps]

22.22 (a) treatment of *p*-bromobenzonitrile with ethylmagnesium bromide in ether, followed by workup with aqueous acid
       (b) treatment of propionitrile with isobutylmagnesium bromide in ether, followed by workup with aqueous acid
       (c) treatment of isobutyronitrile with isopentylmagnesium bromide in ether, followed by workup with aqueous acid
22.23 1. treatment of benzene with bromine in the presence of iron; 2. sulfuric acid, nitric acid; 3. tin, hydrochloric acid; 4. sodium nitrite, sulfuric acid; 5. CuCN
22.24 Two hydrogens are required on the amide nitrogen for the reaction to go to completion. The product formed is: PhC(O)N(Br)CH$_3$
22.25

[mechanism scheme for Hofmann rearrangement of isocyanate to amine]

22.26 The statistical result would be: 3:1 vinylcyclopentane:ethylidenecyclopentane. For the formation of the internal alkene there is necessarily an alkyl group in an axial-type position about the six-membered cyclic activated complex which makes it disfavored.
22.27 (a) *trans*-2-octene, *cis*-2-octene, *trans*-3-octene, and *cis*-3-octene
       (b) styrene
       (c) 3-methylcyclohexene
       Only the reactions in parts (b) and (c) would be useful synthetically.
22.28 We use an alkyl iodide with no β-hydrogens which could take part in an elimination reaction.
22.29 At low temperature the molecule remains in a conformation in which all six atoms lie in a plane (to allow resonance delocalization of the nitrogen valence level unshared electron pair) and which keeps one of the N-H linkages *cis* relative to the C=O and the other N-H linkage *trans* relative to the C=O. At higher temperatures there is rapid rotation about the C-N bond yielding the two hydrogens equivalent on the NMR time scale (the observed signal correlates with their average environment). Both N-H by N-CH$_3$
22.30 4-phenylbutanoyl chloride

22.31 (a) methyl propanoate
(b) ethyl acetate
The position of the methyl singlet and the methylene quartet show which is bonded to oxygen in each case.

22.32 B is *p*-methoxyphenylacetic acid
C is 2-(*p*-methoxyphenyl)ethanol
D is *p*-methoxybenzoic acid
E is *p*-hydroxyphenylacetic acid

22.33 (a) phenyl propanoate
(b) methyl benzoate
(c) benzyl benzoate
(d) *N*,*N*-dimethylbenzamide
(e) ketene
(f) *N*,*N*-diethylethanamide
(g) 4-bromobenzonitrile
(h) propionitrile
(i) *N*,*N'*-dicyclohexylcarbodiimide
(j) sodium butanoate
(k) δ-valerolactone
(l) *N*-methylpyrrolidone

22.34 (a) formamide
(b) isopropyl acetate
(c) butanoyl chloride
(d) acetic propionic anhydride
(e) propanal
(f) ethylene glycol diacetate
(g) *N*-2-butylbenzamide
(h) 2-butyl formate
(i) methyl acetate
(j) benzonitrile
(k) methyl 4-hydroxybutanoate
(l) 1-methylaminobutane
(m) 1-decanol
(n) 1-aminobutane
(o) hexanal
(p) 2-phenyl-2-propanol
(q) benzophenone
(r) aniline
(s) ethylisobutylamine
(t) *N*-methylethanamide
(u) isopropyl phthalate

22.35 (a) 1. lithium aluminum hydride, ether; 2. aqueous acid
(b) 1. lithium aluminum tri *tert*-butoxy hydride; 2. aqueous acid
(c) 1. lithium aluminum hydride, ether; 2. aqueous acid
(d) 1. lithium aluminum hydride, ether; 2. aqueous acid
(e) bromine, aqueous sodium hydroxide
(f) aqueous acid
(g) peroxybenzoic acid
(h) tertiary alcohol, triethylamine, cool to 0° (If the reaction is not cooled, decomposition of the ester occurs immediately.)
(i) 1. ammonia with a catalytic amount of anhydrous acid; 2. hydrogen, platinum oxide

22.36 (a) 1. treatment of benzamide with aqueous acid; 2. ethanol with a catalytic amount of anhydrous acid
(b) 1. treatment of benzamide with lithium aluminum hydride in ether; 2. aqueous acid
(c) 1. treatment of benzamide with phosphorus pentoxide; 2. phenylmagnesium bromide; 3. aqueous acid
(d) treatment of benzamide with bromine in aqueous potassium hydroxide solution

22.37 (a) formic acetic anhydride - The formyl carbon is more electrophilic than the acetyl carbon.
(b) ethyl *p*-nitrobenzoate - The acyl carbon is more electrophilic.
(c) methylamine - The electron pair is localized on nitrogen.

(d) acetamide - Hydrogen is more easily removed from the electron-deficient carbon.
(e) phthalimide - The hydrogen is activated by two acyl sites.
(f) phenyl acetate - The group is an *ortho,para* director
(g) $CH_3(CH_2)_{13}CO_2CH_3$ - There are an odd number of carbons in the acid.
(h) benzamide - The molecule has the weaker C=O linkage.
(i) methyl *p*-aminobenzoate - There is greater single bond character to the C=O linkage.
(j) *cis*-1,2-ethenedicarboxylic acid - It can easily bring the two carboxyl groups together to form the anhydride linkage.

22.38

F: 2-(acetyloxy)benzoic acid structure with OC(O)CH₃ and CO₂H groups
G: γ-butyrolactone structure
H: PhCH₂CH₂CH₂C(O)Cl
I: HOCH₂CH₂CH₂CH(CH₃)CO₂CH₂CH₃

J: coumarin structure
K: (CH₃)₃CC(O)NH₂
L: (CH₃)₃CNH₂
M: PhCH₂CH₂CH(OH)CH₃

N: CH₃CH₂CH(Ph)OC(O)CH₂OCH(CH₃)₂
O: CH₃CH₂CH(Ph)OH
P: (CH₃)₂CHOCH₂CO₂H

Q: CH₃CH₂CO₂CH₂CH₂CH₃
R: CH₃CH₂CH₂OH
S: PhCH₂CO₂CH₂CH₂Ph
T: PhCH₂CH₂OH

U: 4-ethyl-N-propanoyl-aniline (CH₃CH₂-C₆H₄-NHC(O)CH₂CH₃)
V: 4-ethylaniline (CH₃CH₂-C₆H₄-NH₂)
W: 2,6-dibromo-4-ethylaniline
X: 4-ethylbenzonitrile (CH₃CH₂-C₆H₄-CN)

Y: terephthalic acid (HO₂C-C₆H₄-CO₂H)
Z: CH₃NHC(O)CH(CH₃)₂ or CH₃NHC(O)CH₂CH₂CH
AA: (CH₃)₂CHCO₂H or CH₃CH₂CH₂CO₂H

22.39

BB: 1-cyanocyclopentanol (cyclopentane with OH and CN)
CC: 1-hydroxycyclopentanecarboxylic acid (cyclopentane with OH and CO₂H)
DD: bis-spiro anhydride structure

22.40 $HC(OCH_2CH_3)_3$

22.41 δ 0.8, 3H, singlet; δ 1.0, 2H, triplet; δ 2.1, 3H, singlet; δ 3.2, 2H, triplet

22.42 Ethane gas is evolved. The ethylmagnesium bromide abstracts a hydrogen from the nitrogen of the benzamide.

22.43 The ethylmagnesium bromide attacks the carboxyl carbon of the *N,N*-diethylbenzamide to form an intermediate in which the original carboxyl carbon is tetracoordinated. Upon workup this intermediate loses the elements of diethylamine to form the propiophenone.

218 Study Guide and Solutions Manual

22.44

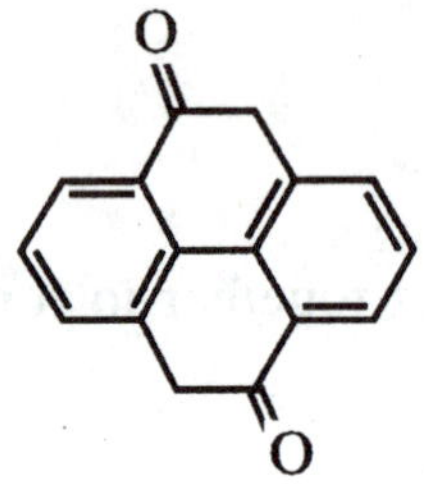

[structure: a tetracyclic diketone]

22.45

[mechanism scheme showing benzamide with ortho-nitroso group + HO⁻ proceeding through cyclic intermediates, loss of H₂O, formation of benzotriazine-type intermediate, loss of N₂, to give benzoate anion, then ~H⁺ to benzoate]

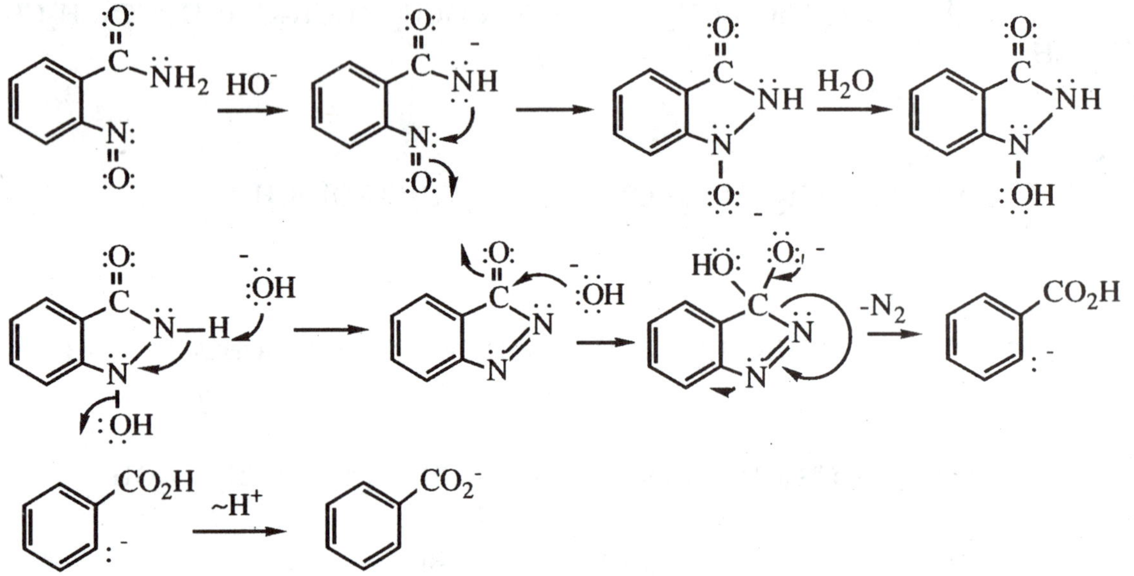

22.46 The cyclopentyl group would migrate.
22.47 4-methoxyphenyl benzoate
22.48

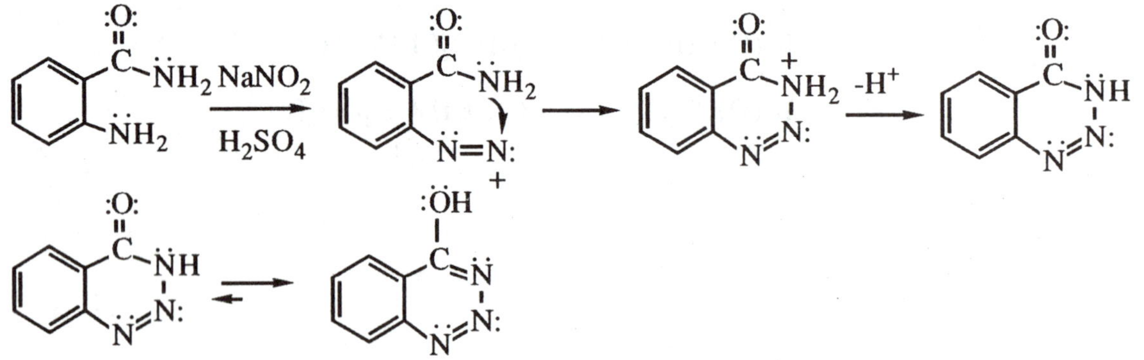

[mechanism: anthranilamide + NaNO₂/H₂SO₄ → diazonium intermediate → cyclization with loss of H⁺ → benzotriazinone tautomers]

22.49

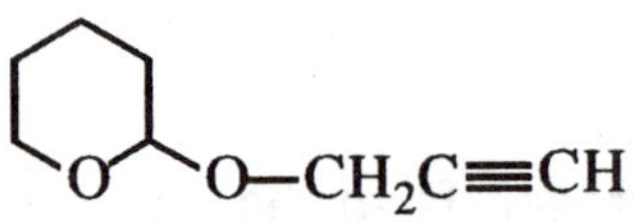

[structures: THP-O-CH₂C≡CH (GG); THP-O-CH₂C≡C:⁻ MgBr⁺ (IIII); HOCH₂C≡C-CO₂H (II)]

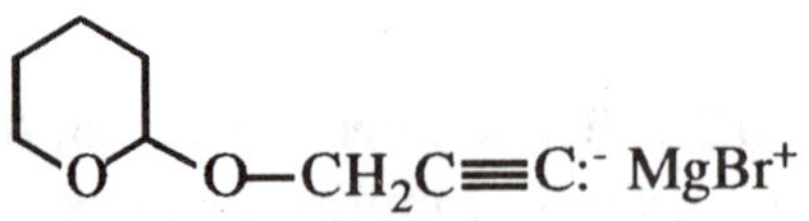

22.50 The Grignard reagent would react preferentially with the more acidic hydroxyl hydrogen.
22.51 (a) 1. treatment of toluene with potassium permanganate in aqueous basic solution; 2. aqueous acid; 3. iodomethane in the presence of aluminum chloride; 4. thionyl chloride; 5. diethylamine
   (b) 1. treatment of toluene with nitric acid and sulfuric acid; 2. potassium permanganate in aqueous basic solution; 3. aqueous acid; 4. ethanol in the presence of a catalytic amount of anhydrous acid; 5. hydrogen with platinum oxide catalyst
   (c) 1. treatment of toluene with bromine under light irradiation; 2. NaCN; 3. heat with aqueous acid; 4.

d) 1. thionyl chloride 2. N-N diethylamide

ethanol in the presence of a catalytic amount of anhydrous acid

(d) 1. treatment of toluene with potassium permanganate in aqueous basic solution; 2. aqueous acid; 3. thionyl chloride; 4. benzylamine

(e) treatment of benzyl cyanide [as formed in part (c)] with lithium aluminum hydride in ether

(f) 1. treatment of benzoic acid [as formed in part (d)] with ethanol in the presence of a catalytic amount of anhydrous acid; 2. two equivalents of phenylmagnesium bromide, with water workup

(g) 1. treatment of toluene with nitric acid and sulfuric acid; 2. potassium permanganate in aqueous basic solution; 3. aqueous acid; 4. thionyl chloride

(h) 1. treatment of benzoic acid [as formed in part (d)] with nitric acid and sulfuric acid; 2. thionyl chloride

22.52

Solution of Study Guide Practice Problems

22.1 (a) phenyl acetate
(b) methyl benzoate
(c) N-methyl-p-bromobenzamide
(d) propanoic anhydride
(e) p-nitrobenzoyl chloride

22.2 (a) F$_3$C-C≡N (b) F$_3$C-C(O)-O-C(O)-CF$_3$ (c) Br-C$_6$H$_4$-C(O)-NH$_2$

22.3 [mechanism: Ph-C(O)-Cl + :NH$_3$ → tetrahedral intermediate → Ph-C(O)-NH$_3^+$ → Ph-C(O)-NH$_2$]

22.4 PhC(O)Cl > PhC(O)OC(O)Ph > PhC(O)OPh > PhC(O)NH$_2$

22.5 (a) acetamide and ammonium acetate
(b) benzamide N-ethyl
(c) acetic anhydride
(d) o-C$_6$H$_4$(C(O)NH$_2$)(CO$_2$H)

22.6 Reflux the methyl acetate with a large excess of ethanol in the presence of an anhydrous acidic catalyst.

22.7 HOCH$_2$CH$_2$CH$_2$CO$_2$CH$_3$

22.8 1. heating ethyl phenylacetate with an excess of anhydrous ammonia; 2. phosphorus pentoxide

22.9 The ^{18}O will be contained in the phenylacetic acid.

22.10 A is N,N,2-trimethylbenzamide; B is dimethylamine; C is 2-methylbenzoic acid; D is phthalic acid

22.11 E is propionitrile; F is sodium propanoate

22.12 (a) 2-chloroethanol
(b) benzyl alcohol and ethanol
(c) methylpropylamine
(d) lithium aluminum tri(tert-butoxy) hydride

(e) 1-amino-3-phenylpropane
22.13 (a) 3-phenyl-3-pentanol
(b) benzyl methyl ketone
(c) di(propyl)cadmium

22.14

(a) $O_2N-\underset{}{\bigcirc}-NH_2$

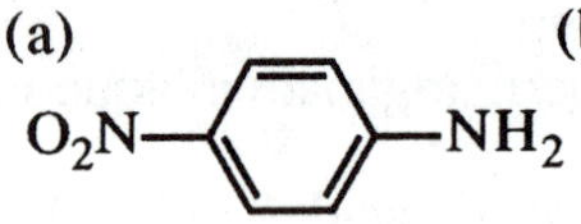

(b) 7-membered lactone ring (oxepan-2-one)

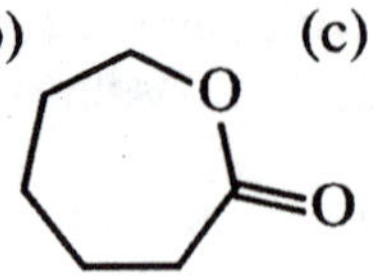

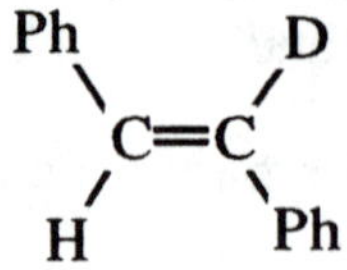

(c) $\underset{H}{\overset{Ph}{>}}C=C\underset{Ph}{\overset{D}{<}}$

(d) 1,2-bis(methylene)cyclohexane

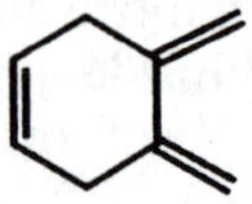

(e) $(CH_3)_3CCH=CH_2$

22.15 (a) 1. treatment of pentanoic acid with thionyl chloride; 2. ammonia; 3. bromine in aqueous sodium hydroxide solution
(b) 1. oxidation of phenyl(4-methoxyphenyl)methanol with a suitable oxidizing agent, such as chromic anhydride in sulfuric acid; 2. hydrogen peroxide, peroxybenzoic acid
(c) 1. treatment of methyl 4-methylphenyl ketone with hydrogen peroxide and peroxybenzoic acid; 2. heating with aqueous acid

CHAPTER 23
ENAMINES, ENOLATES, AND α,β-UNSATURATED CARBONYL COMPOUNDS

Key Points

• Know how enamines are prepared and how they are used in synthesis.

Practice Problem 23.1
Propose a series of reactions using enamine chemistry that will accomplish the synthetic conversion shown below.

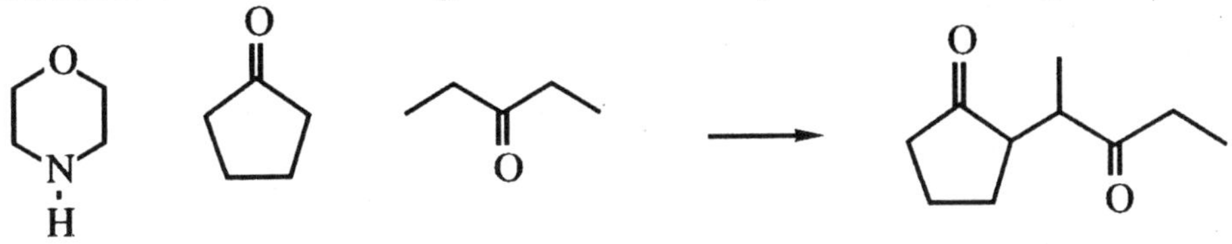

organic starting materials → target molecule

Practice Problem 23.2
Propose structures for the substances A-E.

$$A \xrightarrow{CH_3I} B \xrightarrow{HCl, reflux} C + D$$

A: $C_{11}H_{19}N$, IR: no bands 3300-3500 cm^{-1}
B: $C_{12}H_{22}NI$
C: $C_7H_{12}O$, IR: 1715 cm^{-1}
D: $C_5H_{12}NCl$

$$D \xrightarrow{NaOH, H_2O} E$$

E: $C_5H_{11}N$, IR: 3400 cm^{-1}, ^{13}C NMR: 3 peaks

• Understand enolate ion formation and reactivity.

Practice Problem 23.3
Suggest structures for substances A and B in the following sequence.

cyclohexyl-CO$_2$CH$_3$ $\xrightarrow{A}$ $\xrightarrow{B}$ cyclohexyl with CO$_2$CH$_3$ and CH$_2$CH$_2$CH$_2$CH$_2$CH$_3$

Practice Problem 23.4
Propose a structure for the product (formula $C_8H_{14}O$; IR: 1715 cm^{-1}) formed by the reaction of 6-bromo-3,3-dimethyl-2-hexanone with lithium diisopropylamide in ether.

• Recognize the types of compounds that have particularly acidic α-hydrogen atoms and know how to use them in synthesis. Be particularly familiar with the malonic ester synthesis and the acetoacetic ester synthesis.

222 Study Guide and Solutions Manual

Practice Problem 23.5
Suggest a synthesis of the compound shown below starting with cyclopentene and diethyl malonate.

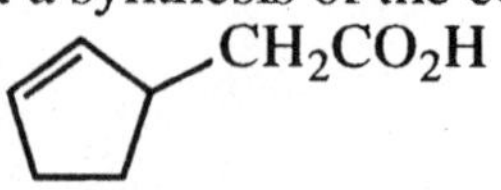

[cyclopentene with CH$_2$CO$_2$H substituent]

Practice Problem 23.6
Give structures for the compounds *A-F* indicated in the sequence below. (Product *F* is a spirocyclic compound.)

1,3-propanediol $\xrightarrow{\text{HBr}}$ *A* $\xrightarrow[\text{two equivalents of NaOCH}_2\text{CH}_3]{\text{CH}_2(\text{CO}_2\text{CH}_2\text{CH}_3)_2}$ *B* $\xrightarrow{\text{LiAlH}_4}$ *C* $\xrightarrow{\text{HBr}}$ *D*

$\quad\quad\quad\quad\quad\quad\quad$ C$_3$H$_6$Br$_2$ $\quad\quad\quad\quad\quad$ C$_{10}$H$_{16}$O$_4$ $\quad\quad$ C$_6$H$_{12}$O$_2$ $\quad\quad$ C$_6$H$_{10}$Br$_2$

D $\xrightarrow[\text{two equivalents of NaOCH}_2\text{CH}_3]{\text{CH}_2(\text{CO}_2\text{CH}_2\text{CH}_3)_2}$ *E* $\xrightarrow[\text{3. heat}]{\text{1. aqueous base; 2. aqueous acid}}$ *F*

$\quad\quad\quad\quad\quad\quad\quad\quad\quad\quad$ C$_{13}$H$_{20}$O$_4$ $\quad\quad\quad\quad$ C$_8$H$_{12}$O$_2$

Practice Problem 23.7
Propose structures for substances *G-I* indicated below.

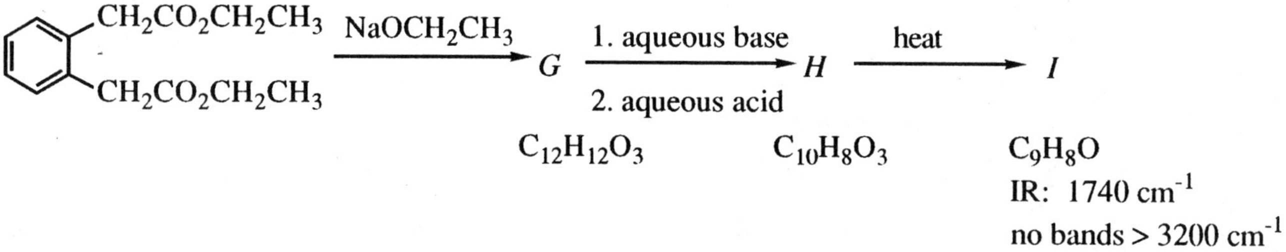

[o-disubstituted benzene with two -CH$_2$CO$_2$CH$_2$CH$_3$ groups] $\xrightarrow{\text{NaOCH}_2\text{CH}_3}$ *G* $\xrightarrow[\text{2. aqueous acid}]{\text{1. aqueous base}}$ *H* $\xrightarrow{\text{heat}}$ *I*

$\quad\quad\quad\quad\quad\quad\quad\quad\quad\quad\quad\quad\quad\quad$ C$_{12}$H$_{12}$O$_3$ $\quad\quad$ C$_{10}$H$_8$O$_3$ $\quad\quad$ C$_9$H$_8$O
$\quad\quad\quad\quad\quad\quad\quad\quad\quad\quad\quad\quad\quad\quad\quad\quad\quad\quad\quad\quad\quad\quad\quad\quad\quad\quad\quad\quad\quad\quad$ IR: 1740 cm^{-1}
$\quad\quad\quad\quad\quad\quad\quad\quad\quad\quad\quad\quad\quad\quad\quad\quad\quad\quad\quad\quad\quad\quad\quad\quad\quad\quad\quad\quad\quad\quad$ no bands > 3200 cm^{-1}

• Know the Mannich reaction and its use in synthesis.
Practice Problem 23.8
Predict the product of reaction of acetone with formaldehyde and diethylamine in aqueous solution at pH = 2.
Practice Problem 23.9
Give the structure of the highly electrophilic intermediate iminium ion that is formed in the reaction shown in Practice Problem 23.8.

• Learn all aspects of aldol addition and condensation reactions that are presented in the text. Topics to study thoroughly are: simple aldol additions and condensations, mechanistic aspects, crossed aldol condensations, intramolecular aldol condensations, and retroaldol reactions.
Practice Problem 23.10
For each of the following, determine what starting compound or compounds you would use to prepare the indicated products by an aldol type reaction.
$\quad$(a) CH$_3$CH$_2$CH$_2$CH(OH)CH(CH$_2$CH$_3$)CHO
$\quad$(b) CH$_3$CH$_2$CH$_2$CH$_2$CH=C(CH$_2$CH$_2$CH$_3$)CHO
$\quad$(c) (CH$_3$)$_3$C-C(O)CH=CHPh
$\quad$(d)

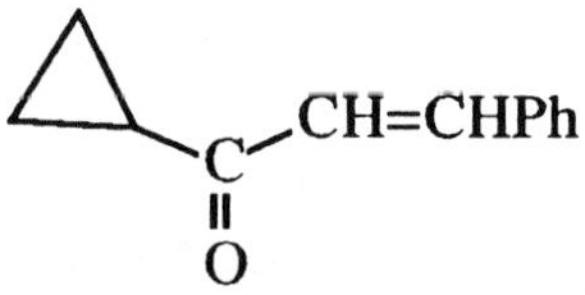

[cyclopropyl-C(=O)-CH=CHPh]

$\quad$(e)

[bicyclic enone structure]

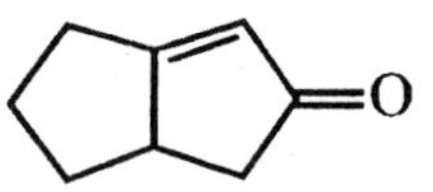

Chapter 23 223

Practice Problem 23.11
What is the structure of a compound (of formula $C_{10}H_{16}O$) that undergoes a retroaldol reaction when treated with aqueous sodium hydroxide to yield 3-methylcyclohexanone and acetone?

• Acquire a thorough knowledge of the Claisen and related condensations, and recognize the synthetic utility of the products formed in such reactions.

Practice Problem 23.12
Which of the compounds, $(CH_3CH_2)_2CHCO_2CH_2CH_3$ or $(CH_3)_2CHCH_2CO_2CH_2CH_3$, does *not* yield a β-ketoester product on reaction with sodium ethoxide? Explain your choice.

Practice Problem 23.13
Give structures for the compounds indicated *J-M*.

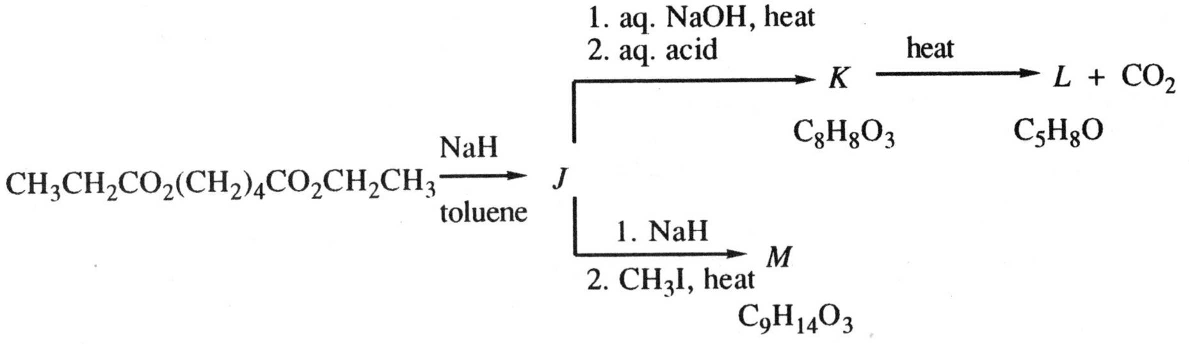

$$CH_3CH_2CO_2(CH_2)_4CO_2CH_2CH_3 \xrightarrow[\text{toluene}]{\text{NaH}} J$$

1. aq. NaOH, heat
2. aq. acid → K ($C_8H_8O_3$) → heat → L + CO_2 (C_5H_8O)

1. NaH
2. CH_3I, heat → M ($C_9H_{14}O_3$)

Practice Problem 23.14
Suggest a Claisen condensation reaction to prepare ethyl 2-isopropyl-5-methyl-3-oxohexanoate.

Practice Problem 23.15
Suggest a structure for the starting material which would lead to the compound shown below upon performance of a Dieckmann condensation using NaH.

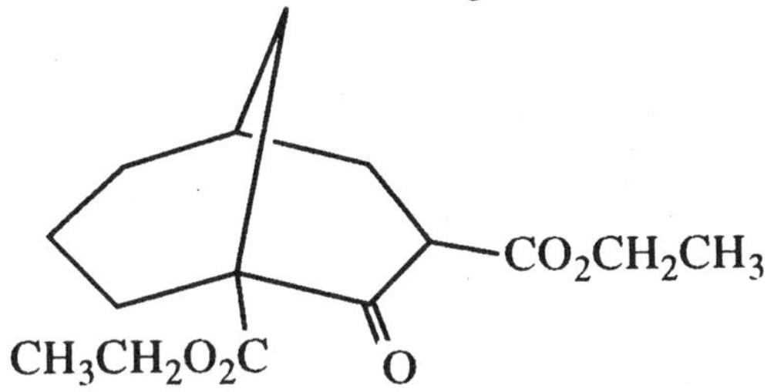

[Structure: bicyclic compound with $CO_2CH_2CH_3$ and $CH_3CH_2O_2C$ substituents and a ketone (O)]

• Recognize that a number of reactions (Perkin, Knoevenagel, *etc.*) are mechanistically related to the aldol and Claisen condensations.

Practice Problem 23.16
Which reagents would you use to prepare the compound shown below by:
   (a) the Perkin condensation
   (b) the Knoevenagel condensation

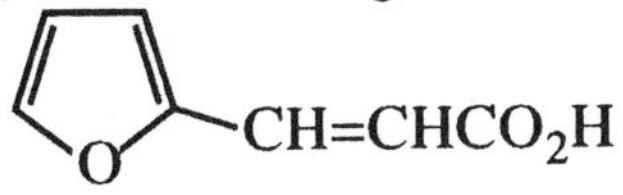

[Structure: furan-CH=CHCO$_2$H]

Practice Problem 23.17
Propose a structure for the product formed (of formula C_7H_5NO) upon heating furfural with cyanoacetic acid in pyridine solution in the presence of ammonium acetate. (Carbon dioxide and water are also formed as by-products in this reaction.)

• Understand what is meant by the term *conjugate addition*, particularly in regard to the Michael reaction.

Practice Problem 23.18
Give the products of conjugate addition in each of the following reactions.
   (a) HCN with PhCH=CHC(O)Ph

(b) ethylamine with $(CH_3)_2C=CHC(O)CH_3$

Practice Problem 23.19
When $PhCH=CHC(O)Ph$ is treated with hydrazine, H_2NNH_2, an initial conjugate addition occurs and is followed by an intramolecular reaction leading to a compound of formula $C_{15}H_{14}N_2$. Suggest a structure for this product.

Practice Problem 23.20
Give the expected products in each of the following reactions.
(a) methyl vinyl ketone treated with nitroethane in the presence of base
(b) methyl acrylate treated with diethyl malonate in the presence of base

Practice Problem 23.21
Cyclic compounds can at times be prepared by double Michael addition reactions. Propose a structure for the product (of formula $C_{24}H_{24}O_5$) of the reaction of $PhCH=CHC(O)CH=CHPh$ with diethyl malonate in the presence of base.

• Appreciate the utility of the Robinson annulation reaction.

Practice Problem 23.22
Complete the following statement: "The Robinson annulation involves the use of a ..?.. addition followed by an intramolecular ..?..."

Practice Problem 23.23
Suggest a structure for the compound of formula $C_{14}H_{20}O$ formed by the Robinson annulation reaction shown below.

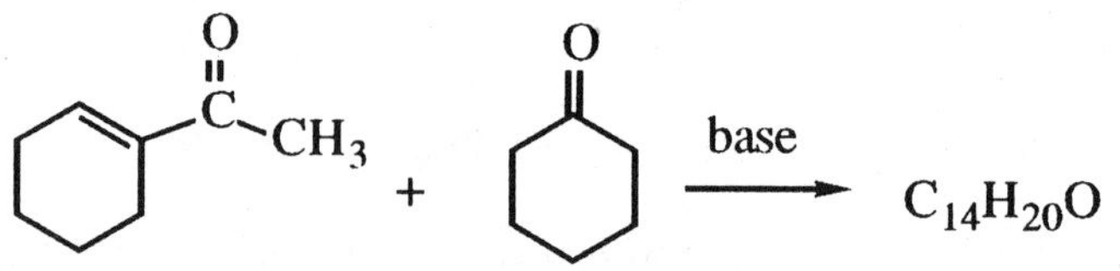

Solution of Text Problems

23.1 1. treatment of propanoic acid with bromine and phosphorus tribromide; 2. heating with ethanol in the presence of a catalytic amount of an anhydrous acid

23.2

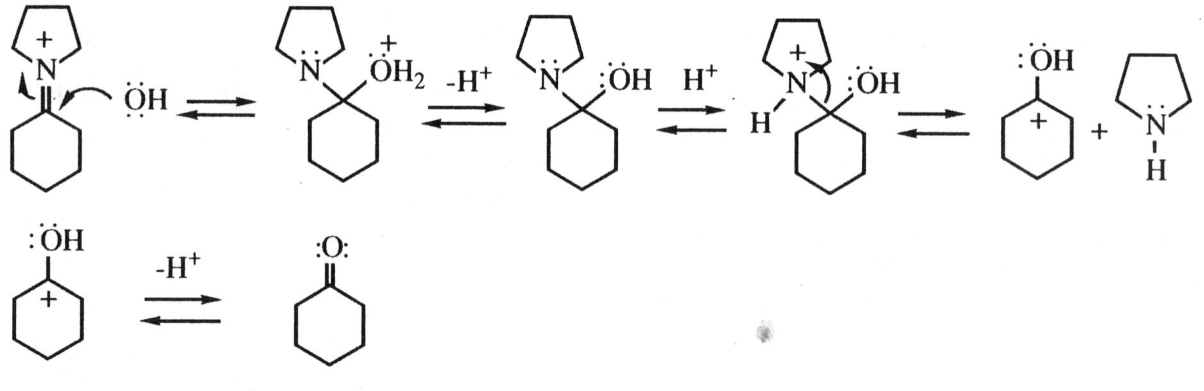

23.3

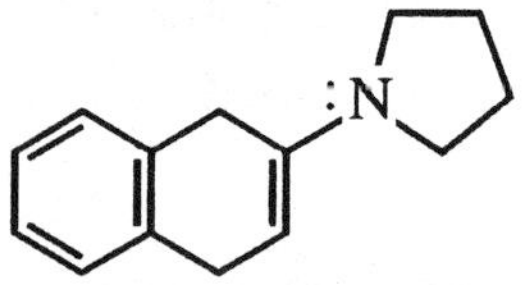

This species is *not* formed as the alternative product has the enamine C=C bond in conjugation with the aromatic ring.

23.4 (a) 1. treatment of hexanal with pyrrolidine in the presence of a catalytic amount of anhydrous acid; 2. iodomethane; 3. aqueous acid
(b) 1. treatment of cyclopentanone with pyrrolidine in the presence of a catalytic amount of anhydrous

acid; 2. iodoethane; 3. aqueous acid

23.5 (a) 1. treatment of cyclohexanone with pyrrolidine in the presence of a catalytic amount of anhydrous acid; 2. propionyl chloride; 3. aqueous acid

(b) 1. treatment of cyclohexanone with pyrrolidine in the presence of a catalytic amount of anhydrous acid; 2. iodoethane; 3. aqueous acid; 4. sodium borohydride in isopropyl alcohol with aqueous workup

23.6 2-phenyl-2-propanol

23.7

A: bicyclic lactone with CH$_2$OH substituent
B: cyclohexenyl acrylic acid (vinyl-CO$_2$H on cyclohexene)

23.8 The carboxylate anion would be generated immediately, retarding virtually completely (under these conditions) the generation of an anionic site at the α-position.

23.9 In an aprotic solvent, once the initial acid-base reaction had occurred there would be no way for reprotonation of the enolate ion to occur to establish an equilibrium. The distribution of products would reflect the kinetic nature of the acid-base reaction rather than the thermodynamically determined position of the equilibrium. An excess of the starting ketone allows equilibration to occur, even in the absence of an added protic solvent.

23.10

Z enolate: H$_3$C and C(CH$_3$)$_3$ on same side; E enolate: H and C(CH$_3$)$_3$ arrangement with CH$_3$ opposite.

We expect the enolate with the favored conformation in the ketone prior to proton loss to be the one which is formed preferentially. For the given system, the Z enolate comes from a lower energy conformation than does the E enolate.

23.11 1. treatment of 1,3-cyclohexanediol with chromic anhydride in sulfuric acid; 2. potassium carbonate; 3. allyl bromide (formed by reaction of propene with N-bromosuccinimide)

23.12 1. treatment of diethyl malonate with sodium ethoxide; 2. benzyl bromide; 3. aqueous acid; 4. heat

In order to prepare phenylacetic acid we would need to perform an S$_N$2 reaction on a halobenzene, a reaction which does not occur under normal circumstances.

23.13 In a malonic ester synthesis of 2-ethyl-3-methylbutanoic acid we would need to add an isopropyl group and an ethyl group to the fundamental malonic ester skeleton. Thus, the overall route is: 1. treatment of diethyl malonate with sodium ethoxide; 2. 2-iodopropane; 3. sodium ethoxide; 4. iodoethane; 5. aqueous acid; 6. heat

23.14

C: cyclobutane with two CO$_2$CH$_2$CH$_3$ groups on same carbon
D: cyclobutane with CO$_2$H and H

23.15 1. treatment of diethyl malonate with sodium ethoxide; 2. addition of the resultant solution to an excess of 1,4-dibromobutane (this procedure minimizes formation of diadduct); 3. sodium ethoxide, heat; 4. aqueous acid; 5. heat

23.16 (a) 1-bromohexane
(b) 1-bromohexane and iodoethane
(c) iodocyclohexane and iodomethane

23.17

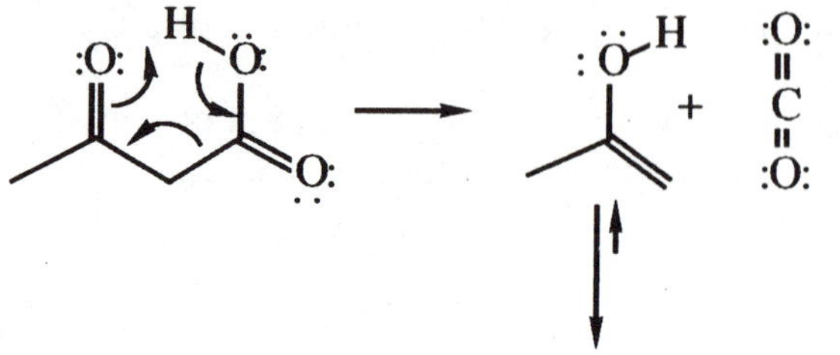

[Mechanism showing decarboxylation of acetoacetic acid via cyclic transition state, producing an enol of acetone plus CO₂, which tautomerizes to CH₃C(O)CH₃]

23.18 (a) 1. treatment of ethyl acetoacetate with sodium ethoxide; 2. benzyl bromide; 3. aqueous acid; 4. heat

(b) 1. treatment of ethyl acetoacetate with sodium ethoxide; 2. iodoethane; 3. aqueous acid; 4. heat

(c) 1. treatment of ethyl acetoacetate with sodium ethoxide; 2. ethylene ketal of 1-chloroacetone (use of the ketal is required that side-reactions do not occur); 3. aqueous acid (cleaves both the ester and ketal linkages); 4. heat

(d) 1. treatment of ethyl acetoacetate with sodium ethoxide; 2. iodomethane; 3. sodium ethoxide; 4. 1-iodopropane; 5. aqueous acid; 6. heat

23.19 For the synthesis of methyl *tert*-butyl ketone by an acetoacetic ester route it would be necessary that *three* groups (methyl groups) be substituted at the α-position. The acetoacetic ester route allows the introduction of only *two* groups at the α-position, with a hydrogen remaining after decarboxylation.

23.20

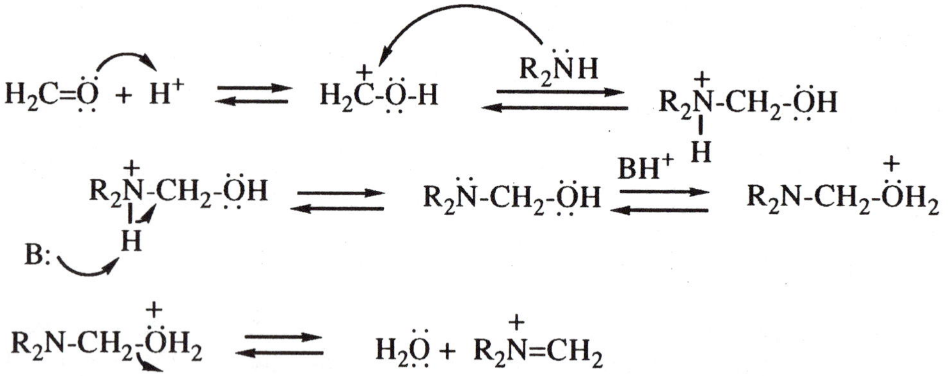

[Mechanism for acid-catalyzed formation of iminium ion from formaldehyde and secondary amine R₂NH, proceeding through protonated formaldehyde, carbinolamine, protonated carbinolamine, and loss of water to give R₂N=CH₂⁺]

23.21 Acidic conditions lead to protonation of the oxygen of formaldehyde and thereby leave the carbon more electrophilic (electron deficient) and subject to attack by nucleophiles.

23.22 CH₃N[CH₂C(CH₃)₂CHO]₂

23.23

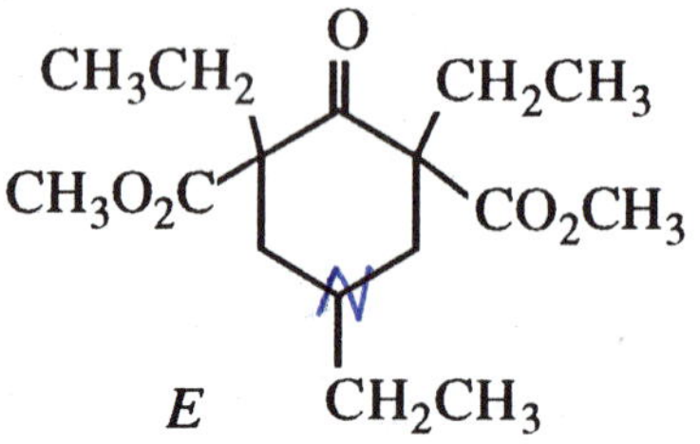

[Structure E: a piperidine ring with N–CH₂CH₃, a ketone at the 4-position, and at the 3,5-positions two CH₂CH₃ and two CO₂CH₃ substituents]

E

This reaction proceeds through the initial formation of the structure shown below by a Mannich condensation, which continues in a second Mannich condensation to yield product E.

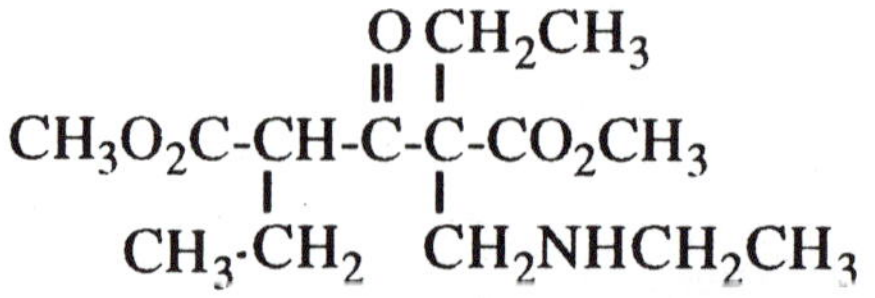

$$\text{CH}_3\text{O}_2\text{C-CH-C-C-CO}_2\text{CH}_3$$

with substituents: CH₃-CH₂ and CH₂NHCH₂CH₃ on one carbon, O and CH₂CH₃ on the other.

23.24 (a) reaction of acetophenone with formaldehyde and dimethylamine in the presence of an acid catalyst

(b) reaction of 3-pentanone with formaldehyde and diethylamine in the presence of an acid catalyst

(c) reaction of cyclohexanone with formaldehyde and piperidine in the presence of an acid catalyst

23.25

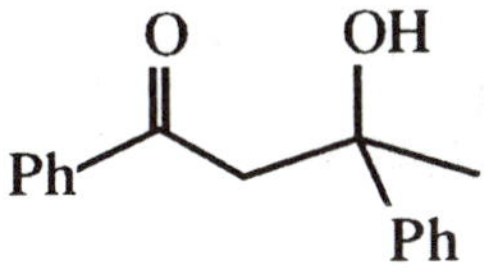

[Structure: Ph-C(=O)-CH₂-C(OH)(CH₃)-Ph]

23.26

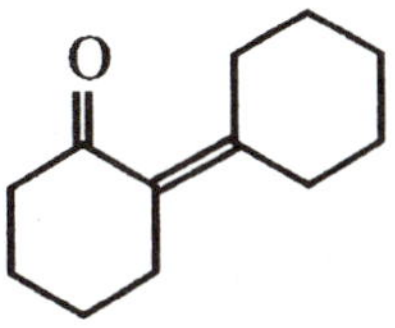

[Structure: 2-cyclohexylidenecyclohexan-1-one]

23.27 Yes.

[Mechanism showing acid-catalyzed aldol of acetone: protonation of carbonyl, enol formation via base removing α-H, enol attacking protonated acetone, proton transfers, yielding 4-hydroxy-4-methylpentan-2-one]

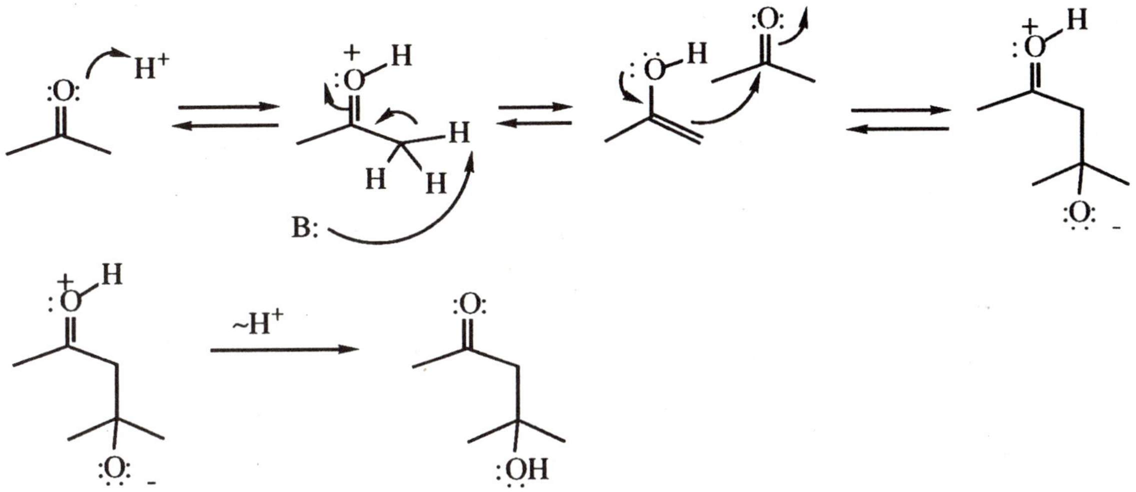

23.28 The reaction involves acetal formation.

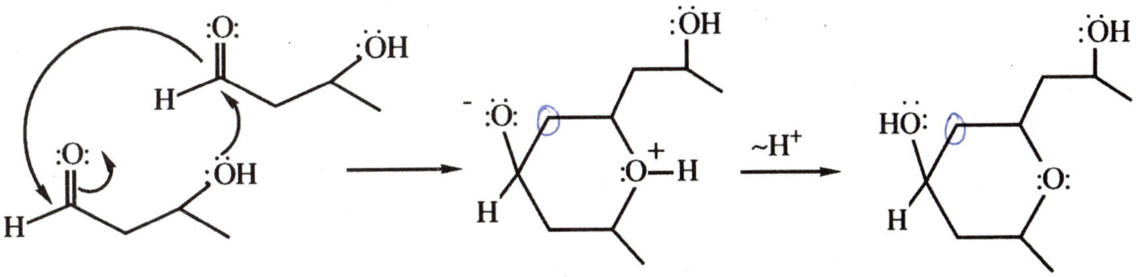

[Mechanism: intramolecular hemiacetal/acetal formation from a hydroxy-dialdehyde, forming a six-membered cyclic acetal with pendant –CH(OH)CH₃]

23.29 The four aldol addition products are as follows: 3-hydroxy-2-methylpentanal; 3-hydroxy-2-methylhexanal; 2-ethyl-3-hydroxypentanal; 2-ethyl-3-hydroxyhexanal.

23.30 There are eight possible products. They are: 5-hydroxy-5-methyl-3-heptanone; 4-hydroxy-3,4-dimethyl-2-hexanone; 5-hydroxy-5-methyl-3-octanone; 4-hydroxy-3,4-dimethyl-2-heptanone; 6-hydroxy-6-methyl-4-nonanone; 3-ethyl-4-hydroxy-4-methyl-2-heptanone; 6-hydroxy-6-methyl-4-octanone; 3-ethyl-4-hydroxy-4-methyl-2-hexanone

23.31 2-ethyl-3-hydroxy-2-methylpropanal

23.32

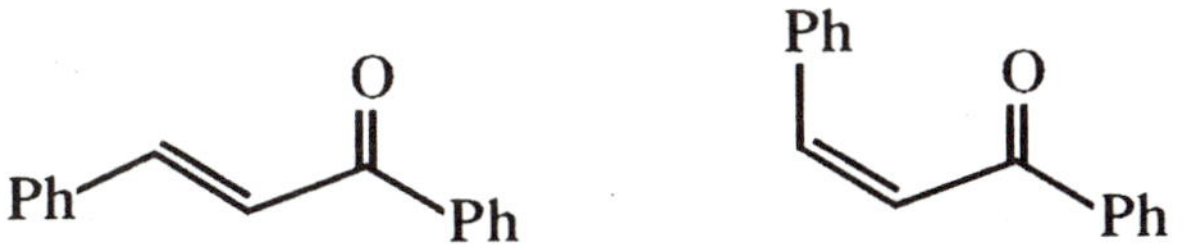

[Structures: major = (E)-PhCH=CH–C(=O)–Ph (trans-chalcone); minor = (Z)-isomer]

major   minor

23.33 (a) furfural and acetophenone (A minor amount of the *cis* product would also form in this reaction.)

(b) (CH₃)₃CCHO and acetone (A minor amount of the *cis* compound also forms in this reaction.)

(c) cyclohexanone and benzaldehyde (A minor amount of the product with the phenyl group *cis* to the carbonyl linkage also forms in this reaction.)

23.34 4-Methoxybenzaldehyde and acetone would be treated with NaOH in aqueous methanol.

23.35 With an intramolecular aldol condensation, leading to the formation of an unstrained or relatively unstrained ring, the electrophilic site to be attacked by the carbanionic site is held in the proper position for that reaction to occur. With an intermolecular aldol reaction the two reactive sites must find each other in the reaction mixture.

23.36

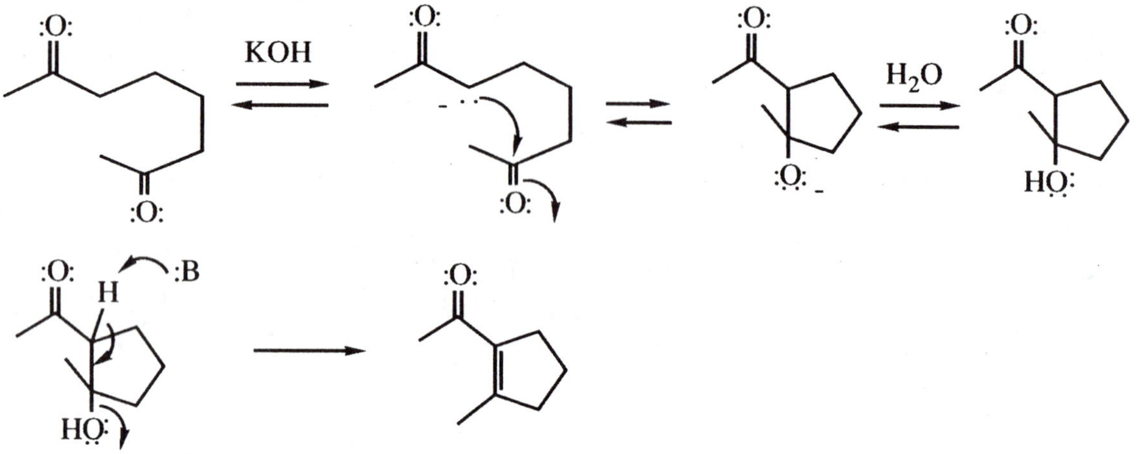

23.37

(a)

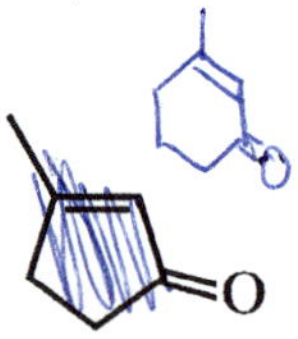

(b)

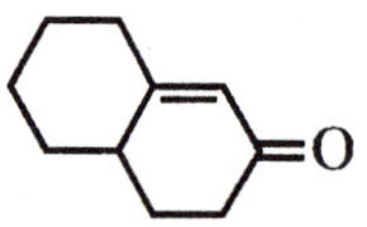

23.38 1. heating the given starting material (bicyclo[4.4.0]dec-1-ene) with aqueous basic potassium permanganate; 2. aqueous potassium hydroxide to effect intramolecular aldol condensation

23.39

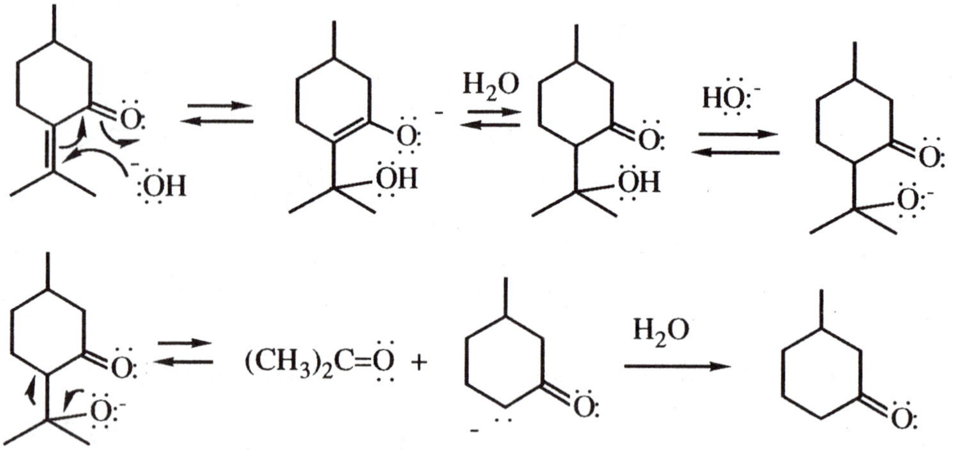

23.40 $CH_3C(O)CH_2CH_2CH(CH_3)CHO$

23.41

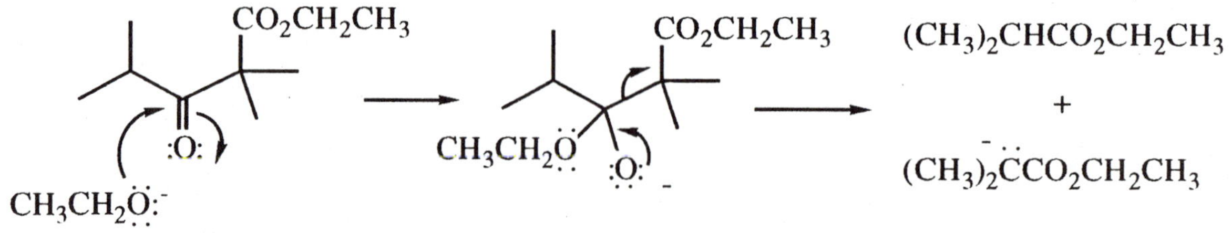

$$(CH_3)_2\ddot{C}^-CO_2CH_2CH_3 \xrightarrow{CH_3CH_2OH} (CH_3)_2CHCO_2CH_2CH_3$$

23.42 Partial hydrolysis of the esters will occur resulting in the formation of a mixture of products, as well as the inactivation of reaction to a certain extent caused by the formation of carboxylate salts.

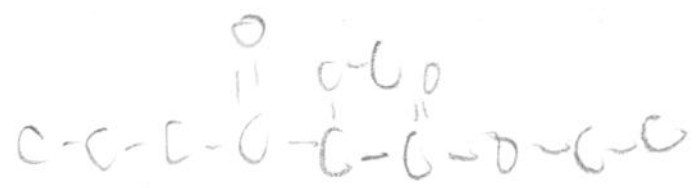

23.43 (a) CH$_3$CH$_2$CH$_2$C(O)CH(CH$_2$CH$_3$)CO$_2$CH$_2$CH$_3$
(b) PhCH$_2$C(O)CH(Ph)CO$_2$CH$_2$CH$_3$

23.44 (a) treatment of cyclohexanone and ethyl formate with sodium ethoxide
(b) treatment of acetone and ethyl hexanoate with sodium ethoxide
(c) treatment of benzyl phenyl ketone and diethyl oxalate with sodium ethoxide

23.45

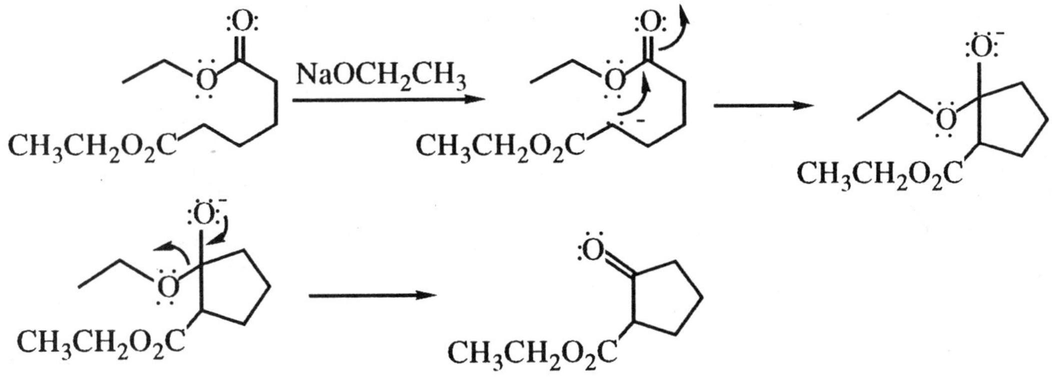

23.46

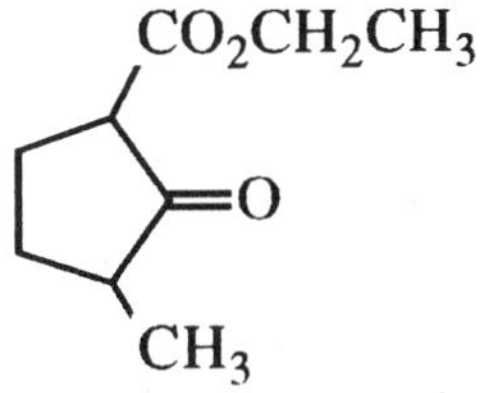

23.47 1. treatment of the designated starting material with potassium permanganate in aqueous base; 2. ethanol with a catalytic amount of an anhydrous acid; 3. sodium ethoxide to effect condensation

23.48

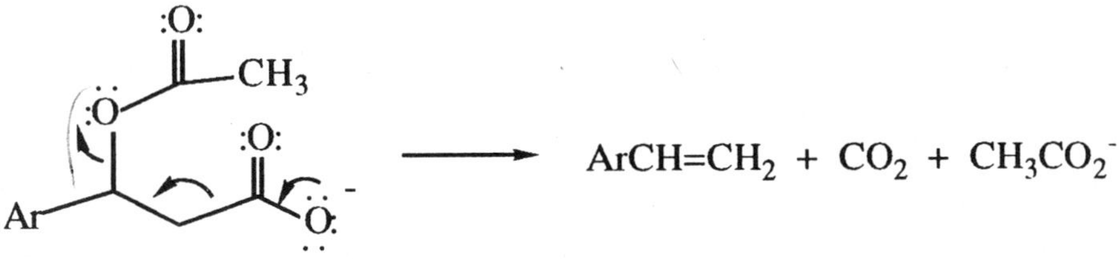

ArCH=CH$_2$ + CO$_2$ + CH$_3$CO$_2^-$

23.49

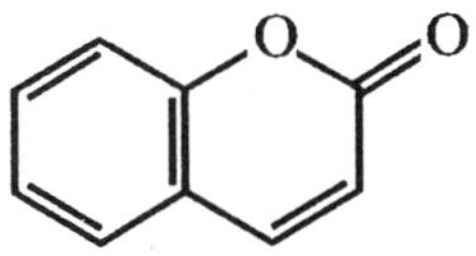

23.50 The resultant anion would have its charge delocalized only to *carbon* sites; attack the the β-position, which allows delocalization of charge to the *oxygen* site (more able to accommodate such a charge) is more favorable.

23.51 The more stable (resonance delocalized) cation involves initial protonation at oxygen, yielding specifically [CH$_2$=CH-CH=OH]$^+$. The product is ClCH$_2$CH$_2$CHO.

23.52

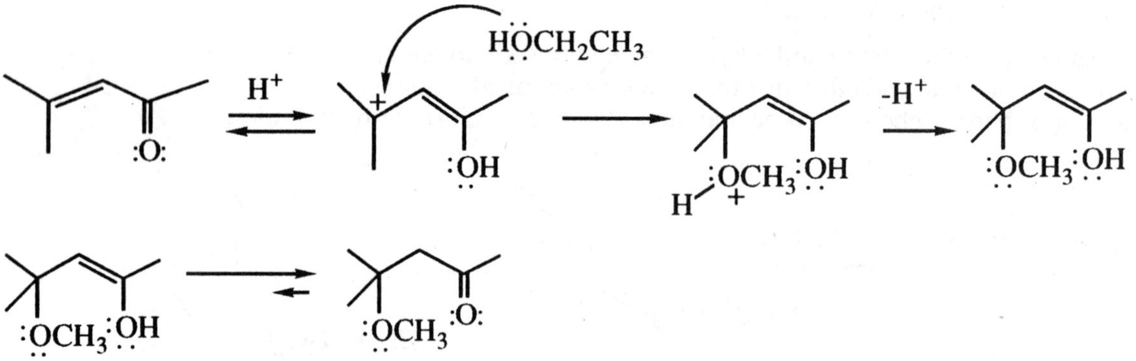

23.53 (a) treatment of PhC(O)CH=CHPh and diethyl malonate with sodium ethoxide in ethanol

(b) 1. treatment of the diketone shown below and ethyl acetoacetate with sodium ethoxide in ethanol; 2. aqueous acid; 3. heat

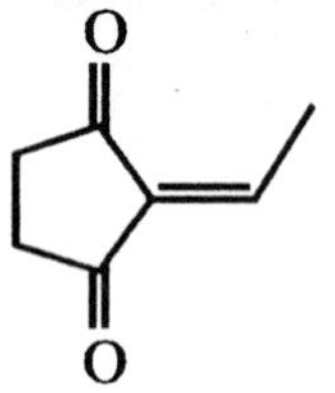

(c) 1. treatment of 2-cyclopentenone and diethyl malonate with sodium ethoxide in ethanol; 2. aqueous acid; 3. heat; 4. methanol with a catalytic amount of anhydrous acid

23.54

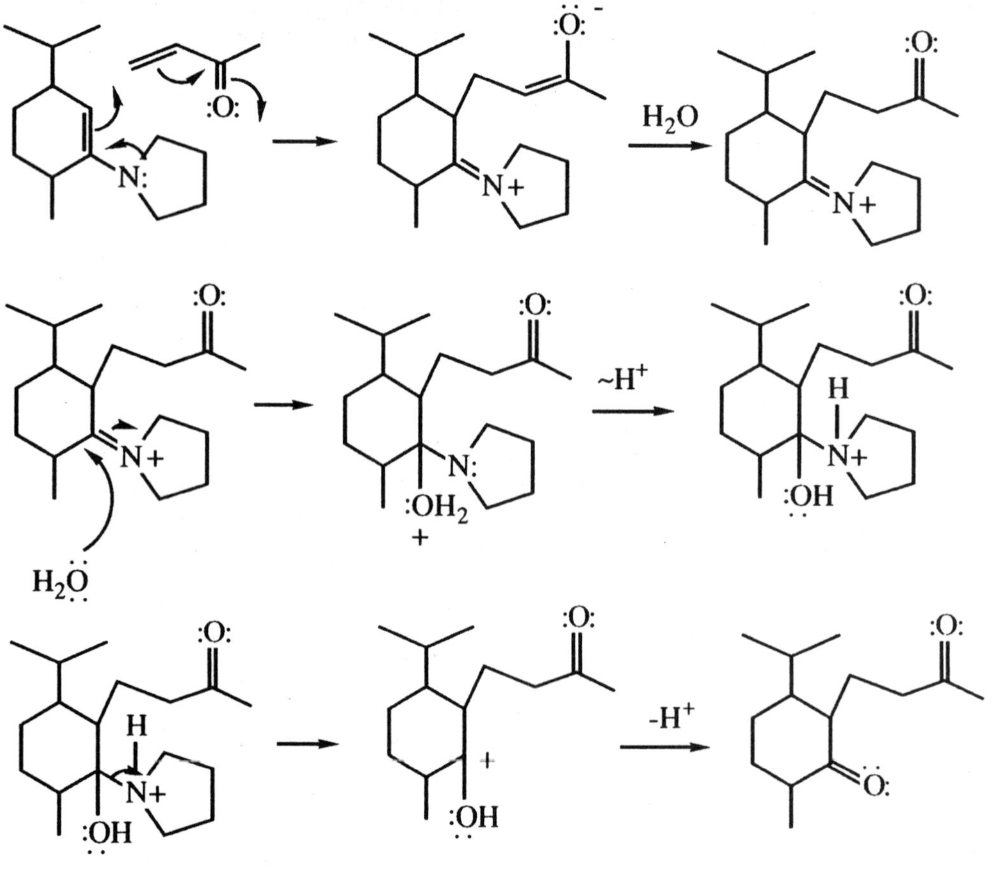

23.55

[Structure: bicyclic enone with Ph substituent]

23.56

[Structure: cyclohexane-1,3-dione with gem-dimethyl, methyl, and CO₂CH₂CH₃ substituents]

23.57 treatment of 2-methyl-1,3-cyclohexanedione and methyl vinyl ketone with sodium ethoxide in ethanol

23.58 (a) (R)-3-methyl-2-hexanone
(b) diethylamine
(c) cyclohexanone
(d) p-nitroacetophenone
(e) nitromethane
(f) 2,4-pentanedione
(g) cyclopentanone
(h) 5-phenyl-2-hexenal

23.59

(a) CH₃CH₂CH(CH₃)CH₂C(O)CH₃

(b) [cyclohexane with CH₂C(O)CH₃ and CHO substituents]

(c) [cyclopentenone with HO, Ph, Ph, and methyl substituents]

23.60

(a) [cyclohexane=CHC(O)CH₃ and cyclohexanone with =C(CH₃)₂] and

(b) [structure with CHO and OH on quaternary carbon]

(c) [cyclohexene with methyl and C(O)CH₃]

(d) [CH₃-CH(CO₂CH₂CH₃) with C(O)CH₂CH₃]

(e) CH₃C(O)CH₂CHO

(f) [2-oxocyclohexane with CO₂CH₂CH₃ substituent]

(g) CH₃CH₂C(O)CH(CH₃)CH₂CH₂CO₂CH₃

(h) PhCH₂C(O)CH(CO₂CH₂CH₃)₂

(i) [4-O₂N-C₆H₄-CH=CH-CO₂H]

(j) [cyclohexane=CH-NO₂]

23.61 (a) treatment of acetone with formaldehyde and sodium hydroxide
(b) 1. treatment of ethyl acetoacetate with sodium ethoxide; 2. 2-bromopropane; 3. aqueous acid; 4. heat
(c) 1. treatment of acetone and benzaldehyde with sodium ethoxide; 2. hydrogen, platinum oxide
(d) 1. treatment of cyclopentanol with chromic anhydride and sulfuric acid; 2. pyrrolidine in the presence of a catalytic amount of anhydrous acid; 3. iodomethane; 4. aqueous acid
(e) 1. treatment of ethyl acetoacetate with sodium ethoxide; 2. ethylene oxide; 3. aqueous acid; 4. heat
(f) 1. treatment of diethyl malonate with sodium ethoxide; 2. iodomethane; 3. sodium ethoxide; 4. iodomethane; 5. aqueous acid; 6. heat
(g) 1. treatment of diethyl malonate with sodium ethoxide; 2. ethylene oxide; 3. aqueous acid; 4. heat
(h) 1. treatment of 2-cyclohexenone with pyrrolidine in the presence of a catalytic amount of anhydrous

acid; 2. iodomethane; 3. aqueous acid
   (i) 1. treatment of ethyl acetoacetate with sodium ethoxide; 2. chloromethyl benzyl ketone; 3. aqueous acid; 4. heat

23.62

$(CH_3)_2C=O$ $Ph_2C=O$ $CH_3C(O)CH=CPh_2$ $CH_3C(O)CH=C(CH_3)_2$ $(CH_3)_2C=CHCO_2H$
    A   B   C   D   E

$(CH_3)_2CHCH_2CO_2H$ $CH_3C(O)CH_2CH(CO_2CH_2CH_3)C(O)CH_3$ $CH_3C(O)CH_2CH_2C(O)CH_3$
          F   G   H

I (3-methylcyclopent-2-enone)   J (1-(pyrrolidin-1-yl)-1-butene-like enamine)   K (branched diketone)   L (3-ethyl-4-methylcyclohex-2-enone)

$CH_3N(CH_2CH_2CO_2CH_3)_2$ N (methyl 1-methyl-4-oxopiperidine-3-carboxylate) $CH_3C(O)CH(CO_2CH_2CH_3)CH_2CH_2C(O)CH_3$
          M   N   O

$CH_2[CH_2C(O)CH_3]_2$ $CH_2[CH_2CH(OH)CH_3]_2$ $CH_3CH_2C(CH_3)_2Br$ $CH_3CH_2C(CH_3)_2CN$
        P   Q   R   S

$CH_3CH_2C(CH_3)_2CO_2H$ $CH_3CH_2C(CH_3)_2CO_2CH_2CH_3$ $HOCH_2C(CH_3)_2CHO$
          T   U   V

W: $HOCH_2$ and CO_2H on a C=C with H_3C and CH_3

23.63 When the α-hydrogen is removed to generate an enolate anion, delocalization of the charge to the terminal position (the γ-carbon) also occurs. When a proton is regained, attachment at the γ-position yields 2-butenal. In light of this mechanism, we predict that X will isomerize to Z, but Y will not so isomerize.

23.64 (a) Sodium borohydride removes an α-hydrogen to generate a resonance stabilized anion which then performs an nucleophilic attack on the -CN carbon of another molecule of the nitrile. Workup with aqueous acid generates the final product.

(b) A hydrogen at the α-position relative to the keto-function is removed to form an enolate which undergoes intramolecular attack on the esteric site to generate a new ring. The carbonyl group remaining in this ring is strained and subject to attack by ethoxide ion which opens the ring leading to the final product.

(c) Ethyl acetoacetate undergoes a condensation reaction with benzaldehyde to yield an α,β-unsaturated keto-ester which reacts in a Michael manner with another equivalent of the anion of ethyl acetoacetate. Transfer of a proton through solution allows an intramolecular aldol condensation to yield the final product.

(d) The enolate anion from benzyl phenyl ketone performs a Michael addition on the α,β-unsaturated ketone followed by an intramolecular aldol condensation to yield the final product.

(e) The free -NH$_2$ group of the phenylhydrazine performs a Michael addition on the ester followed by an intramolecular ester-amide conversion by the PhNH nitrogen.

23.65

CH₃O₂CCH₂CH₂CO₂CH₃   HO₂CCH₂CH₂CO₂H

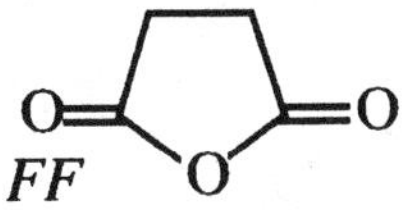

DD   EE   FF (succinic anhydride)

CH₃O₂CCH(CHO)CH₂CO₂CH₃   HO₂CCH₂CH₂CHO   HOCH₂CH₂CH₂CO₂H

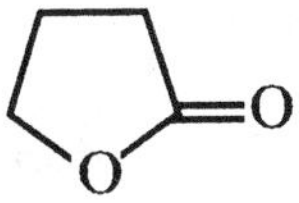

GG   HH   II   JJ (γ-butyrolactone)

23.66

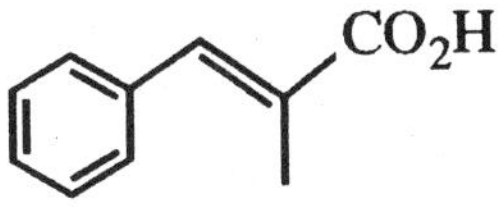

KK: PhCH=C(CH₃)CO₂H

LL: PhCH₂CH(CH₃)CO₂H

MM: PhCH₂CH(CH₃)C(O)Cl

NN: 2-methyl-1-indanone

23.67

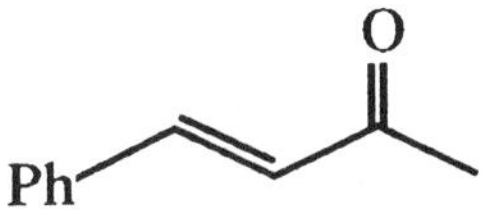

OO: PhCH=CHC(O)CH₃   PP: NaBH₄ (or DIBAL)   QQ: CH₂=CH–CH=CH–Ph   RR: H₂C=CHCHO

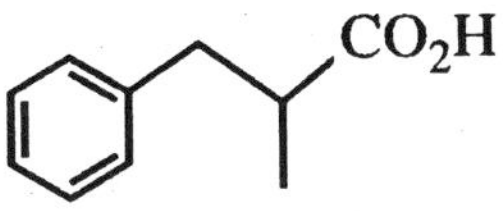

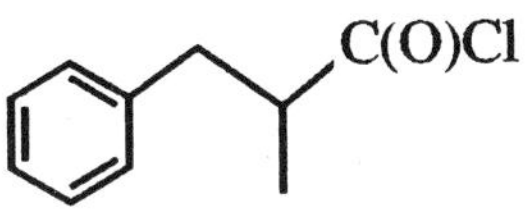

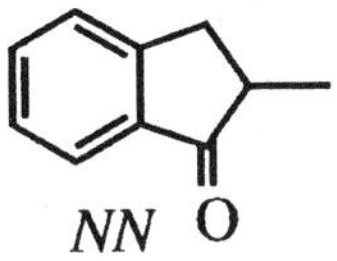

23.68

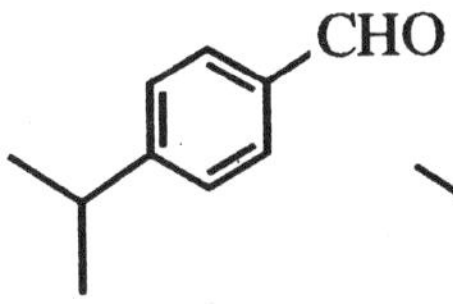

SS: 4-isopropylbenzaldehyde (p-iPr-C₆H₄-CHO)

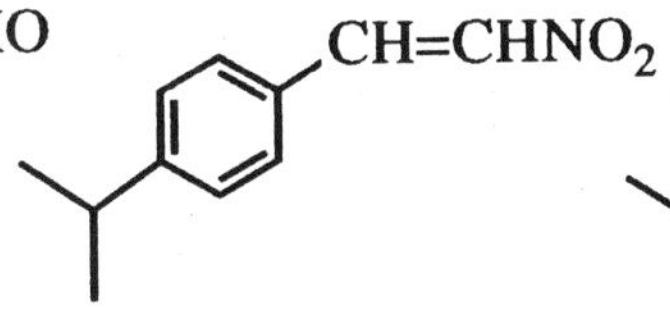

TT: p-iPr-C₆H₄-CH=CHNO₂

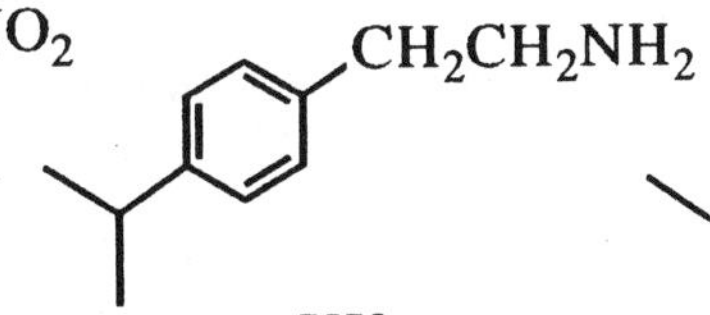

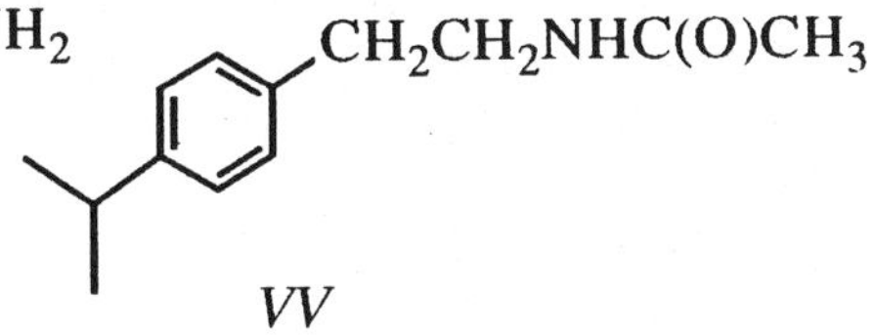

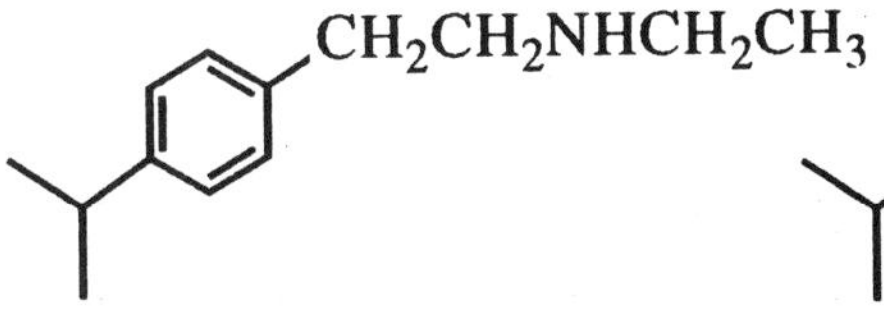

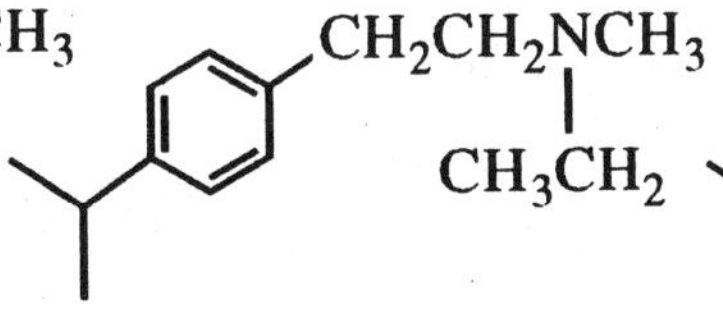

UU: p-iPr-C₆H₄-CH₂CH₂NH₂

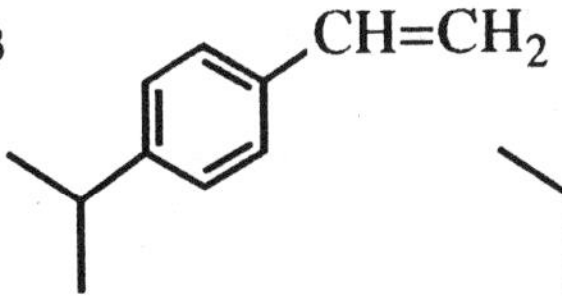

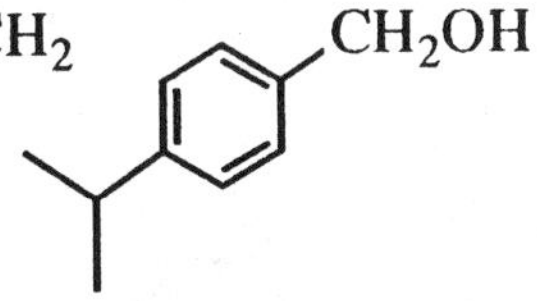

VV: p-iPr-C₆H₄-CH₂CH₂NHC(O)CH₃

WW: p-iPr-C₆H₄-CH₂CH₂NHCH₂CH₃

XX: p-iPr-C₆H₄-CH₂CH₂N(CH₃)(CH₂CH₃)

YY: p-iPr-C₆H₄-CH=CH₂

ZZ: p-iPr-C₆H₄-CH₂OH

Solution of Study Guide Practice Problems

23.1 1. formation of 2-bromo-3-pentanone by treatment of 3-pentanone with bromine in acetic acid; 2. addition of the product from step 1 to the enamine formed from morpholine and cyclopentanone using a catalytic amount of anhydrous acid; 3. aqueous acid

23.2

A: 1-(cyclohex-1-en-1-yl)piperidine (enamine)

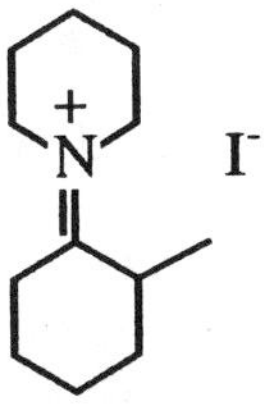

B: N-(2-methylcyclohexylidene)piperidinium iodide

C: 2-methylcyclohexanone

D: piperidinium chloride

E: piperidine

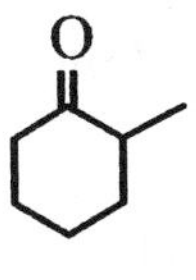

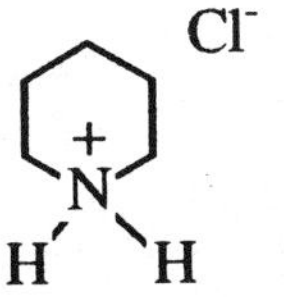

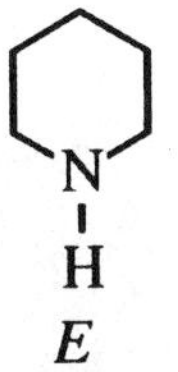

23.3   A is [(CH$_3$)$_2$CH]$_2$NLi (LDA); B is 1-bromopentane

23.4   2,2-dimethylcyclohexanone

23.5   1. treatment of cyclopentene with N-bromosuccinimide; 2. addition of the sodium salt of diethyl malonate formed from sodium ethoxide and diethyl malonate; 3. aqueous acid; 4. heat

23.6

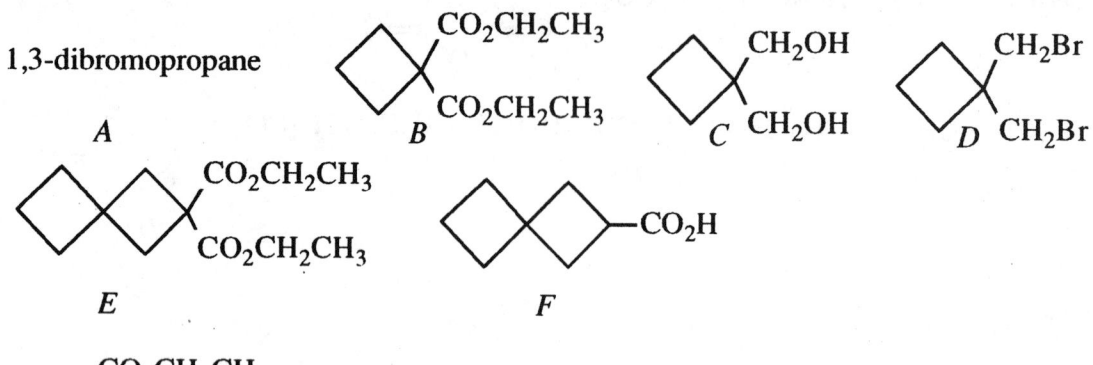

1,3-dibromopropane

A

B: cyclobutane with two CO$_2$CH$_2$CH$_3$ groups on one carbon

C: cyclobutane with two CH$_2$OH groups on one carbon

D: cyclobutane with two CH$_2$Br groups on one carbon

E: spiro bicyclobutane with two CO$_2$CH$_2$CH$_3$ groups

F: spiro bicyclobutane with CO$_2$H group

23.7

G: indanone with CO$_2$CH$_2$CH$_3$ substituent

H: indanone with CO$_2$H substituent

I: indanone

23.8   CH$_3$C(O)CH$_2$CH$_2$N(CH$_3$)$_2$

23.9   H$_2$C=$\overset{+}{N}$(CH$_3$)$_2$

23.10 (a) treatment of butanal with KOH
       (b) treatment of pentanal with KOH
       (c) treatment of tert-butyl methyl ketone and benzaldehyde with KOH
       (d) treatment of cyclopropyl methyl ketone and benzaldehyde with KOH
       (e) 2-(2-oxopropyl)cyclopentanone (treated with KOH)

23.11 2-isopropylidene-5-methylcyclohexanone

23.12 Ethyl 2-ethylbutanoate will *not* participate in a Claisen reaction with NaOR as it has only one α-hydrogen atom (see Section 23.8 of text).

23.13

J: 2-(ethoxycarbonyl)cyclopentanone

K: 2-(carboxy)cyclopentanone

L: cyclopentanone

M: 2-methyl-2-(ethoxycarbonyl)cyclopentanone

23.14 treatment of ethyl 3-methylbutanoate with NaH in toluene followed by aqueous acid workup

23.15 CH$_3$CH$_2$O$_2$CCH$_2$CH$_2$— cyclohexane with two CO$_2$CH$_2$CH$_3$ groups

23.16 (a) furfural (*i.e.* furan-2-carbaldehyde) with acetic anhydride and sodium acetate

(b) furfural with diethyl succinate treated first with sodium ethoxide and then with aqueous acid, and finally heated

23.17

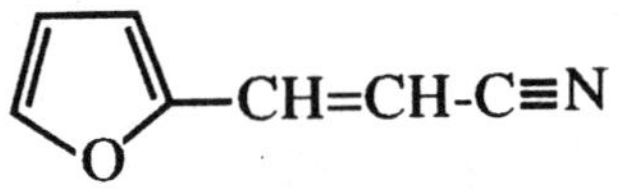

23.18 (a) $PhCH(CN)CH_2C(O)Ph$

(b) $(CH_3)_2C(NHCH_2CH_3)CH_2C(O)CH_3$

23.19

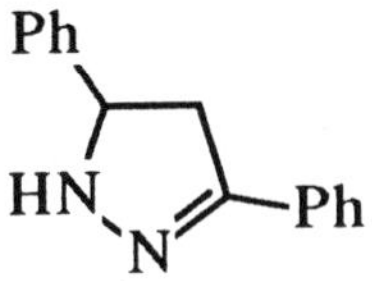

23.20 (a) $CH_3CH(NO_2)CHCH_2CH_2C(O)CH_3$

(b) $(CH_3CH_2O_2C)_2CHCH_2CH_2CO_2CH_3$

23.21

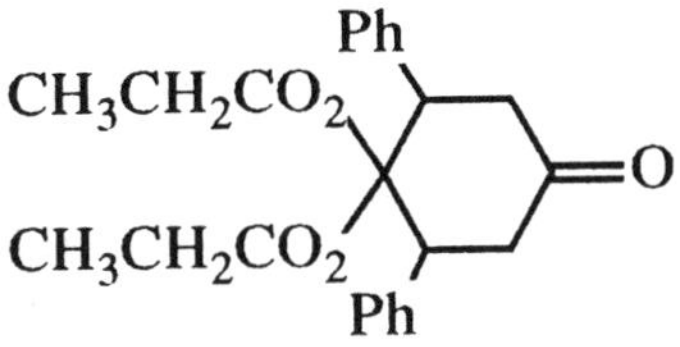

23.22 Michael; aldol condensation

23.23

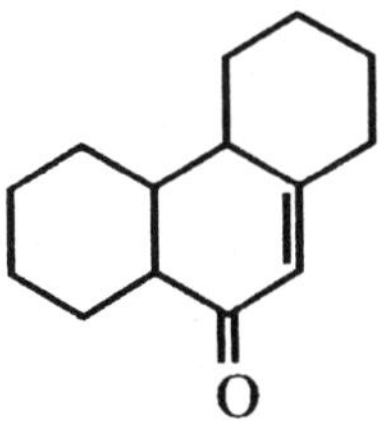

CHAPTER 24
CARBOHYDRATES

Key Points

• Know the fundamentals of nomenclature and the classification of simple carbohydrates.

Practice Problem 24.1
Consider the following structural designations of carbohydrates:
    (a) What are the general formulas ($C_xH_yO_z$) for :
        i. ketotetroses
        ii. aldopentoses
    (b) Classify each of the following (aldopentose, ketohexose, *etc.*).

i. 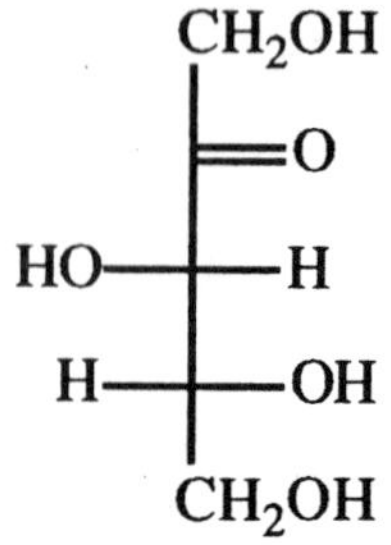
```
      CH₂OH
       |
       C=O
       |
   HO--C--H
       |
    H--C--OH
       |
      CH₂OH
```

ii. 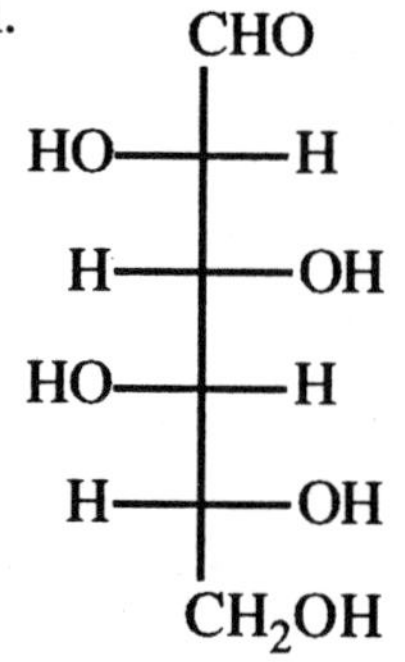
```
       CHO
        |
    HO--C--H
        |
     H--C--OH
        |
    HO--C--H
        |
     H--C--OH
        |
       CH₂OH
```

• Understand the D- and L- stereochemical designations.

Practice Problem 24.2
The structure shown on the left side is D-erythrose. Which of the structures noted (I) and (II) is L-erythrose?

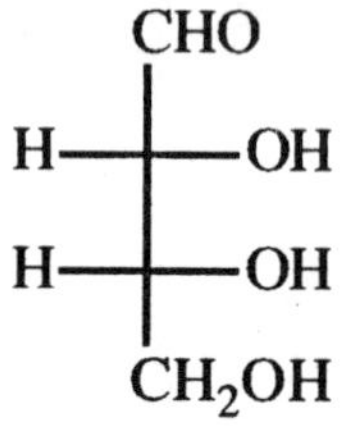

```
     CHO              CHO             CHO
      |                |               |
   H--C--OH        HO--C--H         H--C--OH
      |                |               |
   H--C--OH        HO--C--H        HO--C--H
      |                |               |
     CH₂OH            CH₂OH           CH₂OH
  D-erythrose          (I)             (II)
```

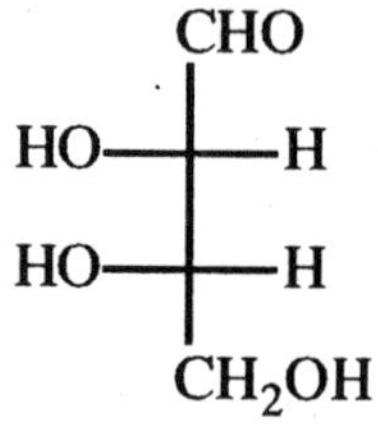

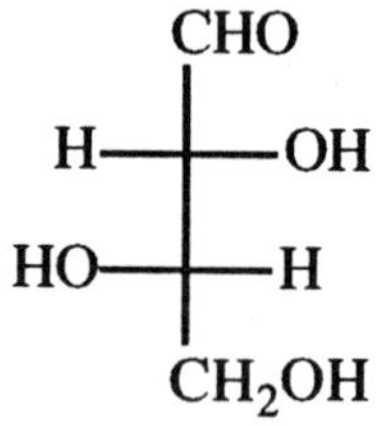

Practice Problem 24.3
If D-erythrose is dextrorotatory at the wavelength of the sodium D-line, does it follow that L-erythrose is *necessarily* levorotatory at the same wavelength? Explain your conclusion.

Practice Problem 24.4
What compound is produced if an inversion of configuration is performed at C-5 in D-glucose?

Practice Problem 24.5

D-Arabinose is oxidized by Tollens' reagent to compound *A*, which readily forms a γ-lactone, *B*. Draw structures for *A* and *B*. Can we tell for certain whether *B* is of the D- or L-family, or would you need more information to decide?

• Be familiar with the cyclic hemiacetal and hemiketal forms of carbohydrates, and learn about the formation and reactivity of glycosides.

Practice Problem 24.6
Using the open-chain structures shown below, draw the Haworth projections of the indicated compounds.

(a) 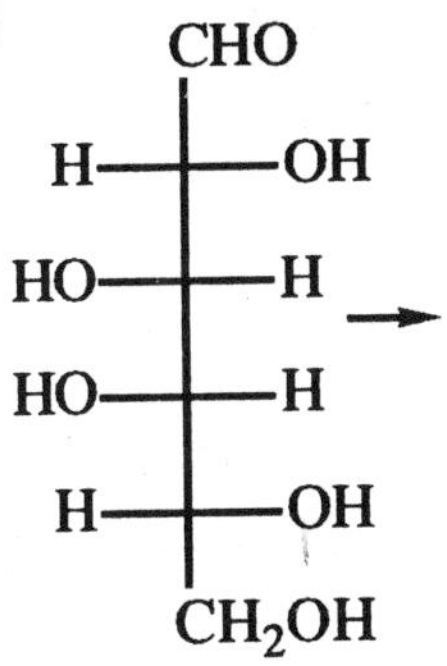
```
        CHO
    H───┼───OH
   HO───┼───H      → α-D-galactopyranose
   HO───┼───H
    H───┼───OH
        CH₂OH
      D-galactose
```

(b) 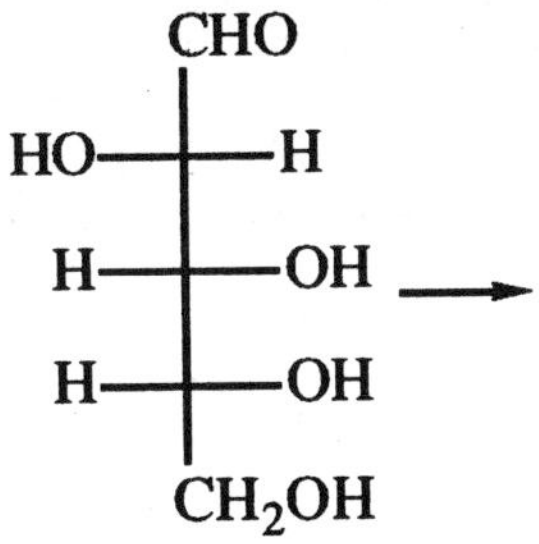
```
        CHO
   HO───┼───H
    H───┼───OH     → β-D-arabinopyranose
    H───┼───OH
        CH₂OH
      D-arabinose
```

Practice Problem 24.7
For the following three compounds, state which are enantiomers relative to each other and which are diastereoisomers relative to each other: α-D-glucopyranose; β-D-glucopyranose; α-L-glucopyranose.

Practice Problem 24.8
Tell if each of the following statements is necessarily true or false.

(a) In aqueous solution α- and β-D-galactopyranose are present in equal amounts at equilibrium.

(b) The two compounds α- and β-D-glucopyranose have equal but opposite specific rotations at the sodium D-line.

Practice Problem 24.9
Draw the structure of α-D-mannopyranose in its favored chair conformation.

Practice Problem 24.10
Consider each of the following aspects of glycoside chemistry.

(a) Describe the laboratory procedure used to prepare methyl glycosides.

(b) Why is only one hydroxyl group of a carbohydrate (cyclic form) converted to a methoxy group under the conditions noted in part (a)?

(c) Describe the laboratory procedure used to convert a methyl glycoside to its parent carbohydrate.

• Know the classical methods for ascending and descending the carbohydrate series - the Kiliani-Fischer synthesis, and the Wohl and Ruff degradations.

Practice Problem 24.11
Suppose that you wished to synthesize D-xylose by way of the Kiliani-Fischer synthesis.

(a) With which carbohydrate would you start?

(b) Would you produce any other carbohydrate along with D-xylose? If you would, name that carbohydrate.

Practice Problem 24.12
Identify the compounds A-C indicated in the following reaction scheme.

$$\text{D-ribose} \xrightarrow[\text{H}_2\text{O}]{\text{Br}_2} A \xrightarrow[\text{H}_2\text{O}_2]{\text{CaCO}_3} \xrightarrow{\text{FeCl}_3} B \xrightarrow[\text{heat}]{\text{aq. acid}} C$$

• Learn the important reactions of carbohydrates. Know how to effect these reactions and how to use their results to make structural deductions. Important reactions include: ester formation, ether formation, osazone formation, and oxidations using nitric acid, bromine/water, and periodate.

238 Study Guide and Solutions Manual

Practice Problem 24.13
Give the products for each of the following reactions.

(a) [structure of a pyranose sugar with HO, CH₂OH, OH groups] $\xrightarrow[\text{acid}]{\text{CH}_3\text{OH}}$ $\xrightarrow{\text{HIO}_4}$

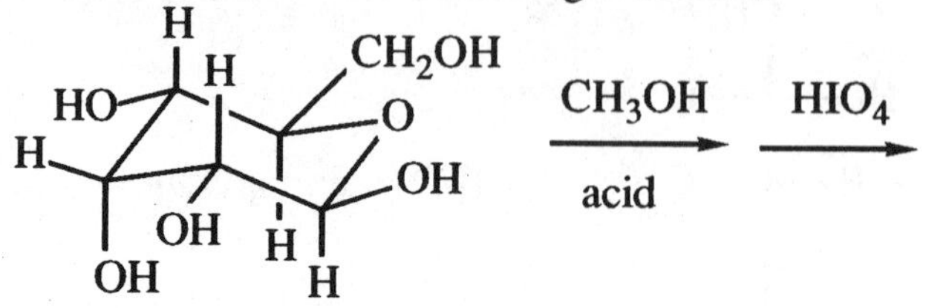

(b)
```
    CHO
H───OH
H───OH         excess PhNHNH₂
H───OH         ─────────────→
    CH₂OH
```

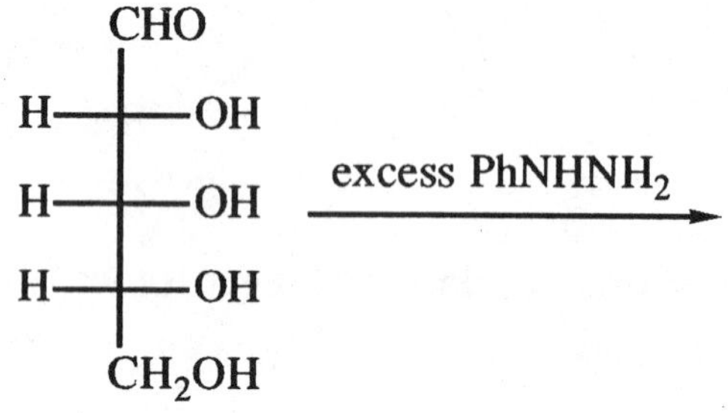

Practice Problem 24.14
How would you prepare the material having the structure shown below, and what would be the product(s) of its reaction with hydrochloric acid?

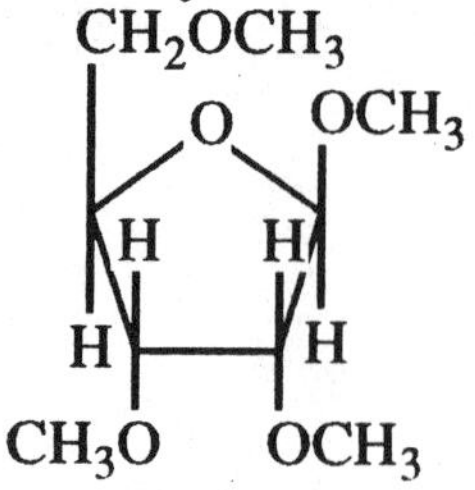

[furanose structure with CH₂OCH₃, OCH₃, CH₃O, OCH₃ substituents]

Practice Problem 24.15
What products would you obtain by nitric acid oxidation of D-glucose and of D-mannose? The product from *one* of these oxidations can also be prepared by the nitric acid oxidation of an L-series aldohexose. Which product can be obtained in this manner and which L-aldohexose would be used? (HINT: Carboxyl groups are obtained by the nitric acid oxidation of either -CHO or -CH₂OH groups.)

• Know the basic structural and nomenclature aspects of di- and polysaccharides.

Practice Problem 24.16
Draw the chair conformation of the β-anomer of lactose, which is a 1,4'-β-glycoside of β-D-galactose and β-D-glucose.

Solution of Text Problems

24.1 (a) There are a total of four stereoisomers, two pairs of enantiomers, each pair diastereoisomeric with regard to the other pair.
     (b) There are a total of eight stereoisomers, four pairs of enantiomers, each pair diastereoisomeric with regard to the other pairs.

24.2 HOCH₂C(O)CH₂OH (dihydroxyacetone)

24.3 No. The direction of rotation of plane polarized light at a particular wavelength bears no fundamental relationship to the designation of absolute configuration.

24.4

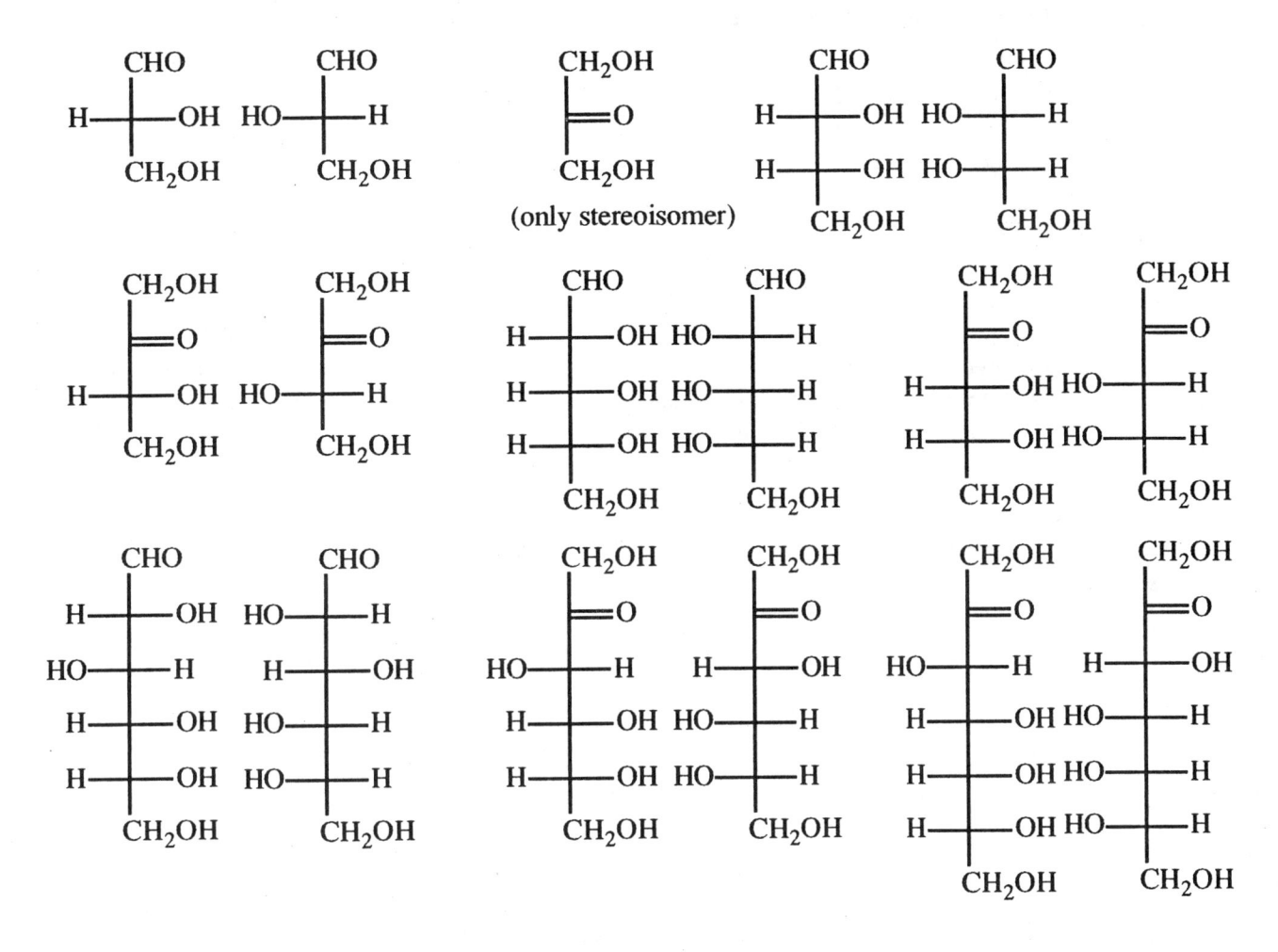

$$\begin{array}{c}\text{CHO}\\\text{H}\!\!-\!\!\!-\!\!\text{OH}\\\text{CH}_2\text{OH}\end{array}\quad\begin{array}{c}\text{CHO}\\\text{HO}\!\!-\!\!\!-\!\!\text{H}\\\text{CH}_2\text{OH}\end{array}\quad\begin{array}{c}\text{CH}_2\text{OH}\\=\!\!\text{O}\\\text{CH}_2\text{OH}\end{array}\text{ (only stereoisomer)}\quad\begin{array}{c}\text{CHO}\\\text{H}\!\!-\!\!\!-\!\!\text{OH}\\\text{H}\!\!-\!\!\!-\!\!\text{OH}\\\text{CH}_2\text{OH}\end{array}\quad\begin{array}{c}\text{CHO}\\\text{HO}\!\!-\!\!\!-\!\!\text{H}\\\text{HO}\!\!-\!\!\!-\!\!\text{H}\\\text{CH}_2\text{OH}\end{array}$$

$$\begin{array}{c}\text{CH}_2\text{OH}\\=\!\!\text{O}\\\text{H}\!\!-\!\!\!-\!\!\text{OH}\\\text{CH}_2\text{OH}\end{array}\quad\begin{array}{c}\text{CH}_2\text{OH}\\=\!\!\text{O}\\\text{HO}\!\!-\!\!\!-\!\!\text{H}\\\text{CH}_2\text{OH}\end{array}\quad\begin{array}{c}\text{CHO}\\\text{H}\!\!-\!\!\!-\!\!\text{OH}\\\text{H}\!\!-\!\!\!-\!\!\text{OH}\\\text{H}\!\!-\!\!\!-\!\!\text{OH}\\\text{CH}_2\text{OH}\end{array}\quad\begin{array}{c}\text{CHO}\\\text{HO}\!\!-\!\!\!-\!\!\text{H}\\\text{HO}\!\!-\!\!\!-\!\!\text{H}\\\text{HO}\!\!-\!\!\!-\!\!\text{H}\\\text{CH}_2\text{OH}\end{array}\quad\begin{array}{c}\text{CH}_2\text{OH}\\=\!\!\text{O}\\\text{H}\!\!-\!\!\!-\!\!\text{OH}\\\text{H}\!\!-\!\!\!-\!\!\text{OH}\\\text{CH}_2\text{OH}\end{array}\quad\begin{array}{c}\text{CH}_2\text{OH}\\=\!\!\text{O}\\\text{HO}\!\!-\!\!\!-\!\!\text{H}\\\text{HO}\!\!-\!\!\!-\!\!\text{H}\\\text{CH}_2\text{OH}\end{array}$$

$$\begin{array}{c}\text{CHO}\\\text{H}\!\!-\!\!\!-\!\!\text{OH}\\\text{HO}\!\!-\!\!\!-\!\!\text{H}\\\text{H}\!\!-\!\!\!-\!\!\text{OH}\\\text{H}\!\!-\!\!\!-\!\!\text{OH}\\\text{CH}_2\text{OH}\end{array}\quad\begin{array}{c}\text{CHO}\\\text{HO}\!\!-\!\!\!-\!\!\text{H}\\\text{H}\!\!-\!\!\!-\!\!\text{OH}\\\text{HO}\!\!-\!\!\!-\!\!\text{H}\\\text{HO}\!\!-\!\!\!-\!\!\text{H}\\\text{CH}_2\text{OH}\end{array}\quad\begin{array}{c}\text{CH}_2\text{OH}\\=\!\!\text{O}\\\text{HO}\!\!-\!\!\!-\!\!\text{H}\\\text{H}\!\!-\!\!\!-\!\!\text{OH}\\\text{H}\!\!-\!\!\!-\!\!\text{OH}\\\text{CH}_2\text{OH}\end{array}\quad\begin{array}{c}\text{CH}_2\text{OH}\\=\!\!\text{O}\\\text{H}\!\!-\!\!\!-\!\!\text{OH}\\\text{HO}\!\!-\!\!\!-\!\!\text{H}\\\text{HO}\!\!-\!\!\!-\!\!\text{H}\\\text{CH}_2\text{OH}\end{array}\quad\begin{array}{c}\text{CH}_2\text{OH}\\=\!\!\text{O}\\\text{HO}\!\!-\!\!\!-\!\!\text{H}\\\text{H}\!\!-\!\!\!-\!\!\text{OH}\\\text{H}\!\!-\!\!\!-\!\!\text{OH}\\\text{H}\!\!-\!\!\!-\!\!\text{OH}\\\text{CH}_2\text{OH}\end{array}\quad\begin{array}{c}\text{CH}_2\text{OH}\\=\!\!\text{O}\\\text{H}\!\!-\!\!\!-\!\!\text{OH}\\\text{HO}\!\!-\!\!\!-\!\!\text{H}\\\text{HO}\!\!-\!\!\!-\!\!\text{H}\\\text{HO}\!\!-\!\!\!-\!\!\text{H}\\\text{CH}_2\text{OH}\end{array}$$

24.5

(a) It shows that an aldehyde group is present.

$$\begin{array}{c}\text{CHO}\\\text{H}\!\!-\!\!\!-\!\!\text{OH}\\\text{HO}\!\!-\!\!\!-\!\!\text{H}\\\text{H}\!\!-\!\!\!-\!\!\text{OH}\\\text{H}\!\!-\!\!\!-\!\!\text{OH}\\\text{CH}_2\text{OH}\end{array}\longrightarrow\begin{array}{c}\text{CO}_2\text{H}\\\text{H}\!\!-\!\!\!-\!\!\text{OH}\\\text{HO}\!\!-\!\!\!-\!\!\text{H}\\\text{H}\!\!-\!\!\!-\!\!\text{OH}\\\text{H}\!\!-\!\!\!-\!\!\text{OH}\\\text{CH}_2\text{OH}\end{array}$$

(b) This also shows that a carbonyl group is present.

$$\begin{array}{c}\text{CH=NNHPh}\\\text{H}\!\!-\!\!\!-\!\!\text{OH}\\\text{HO}\!\!-\!\!\!-\!\!\text{H}\\\text{H}\!\!-\!\!\!-\!\!\text{OH}\\\text{H}\!\!-\!\!\!-\!\!\text{OH}\\\text{CH}_2\text{OH}\end{array}$$

(c) The reaction consumes five moles of acetic anhydride.

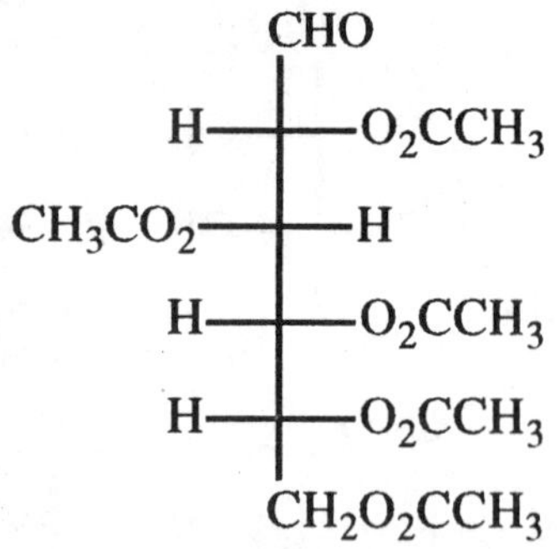

$$\begin{array}{c} CHO \\ H \underline{} O_2CCH_3 \\ CH_3CO_2 \underline{} H \\ H \underline{} O_2CCH_3 \\ H \underline{} O_2CCH_3 \\ CH_2O_2CCH_3 \end{array}$$

(d) Two hydroxyl groups on a single carbon atom constitute a hydrated carbonyl group (a geminal diol), which readily loses water to give the carbonyl compound (see discussion in Chapter 19).

24.6 The two structures differ simply by rotation about the C5-C4 bond.

24.7

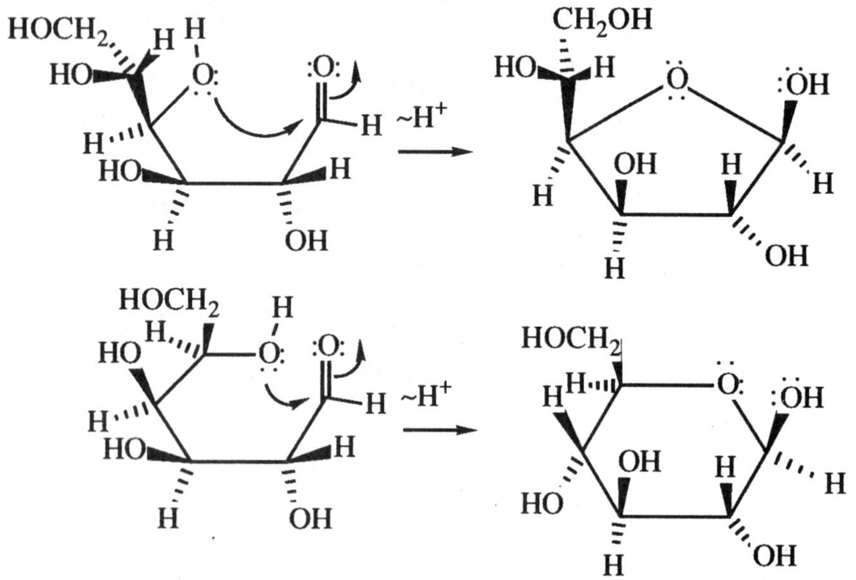

24.8 The α anomer is S and the β anomer is R. The designation does not vary among the carbohydrates in the D-series nor with the size of the ring.

24.9 For the α anomer: $1R, 2R, 3S, 4R$. For the β anomer: $1S, 2R, 3S, 4R$. They are diastereoisomers.

24.10 (a) β-D-arabinofuranose

(b) α-D-lyxofuranose

(c) β-D-allopyranose

24.11 (a) [β-D-fructofuranose and α-D-fructofuranose structures]

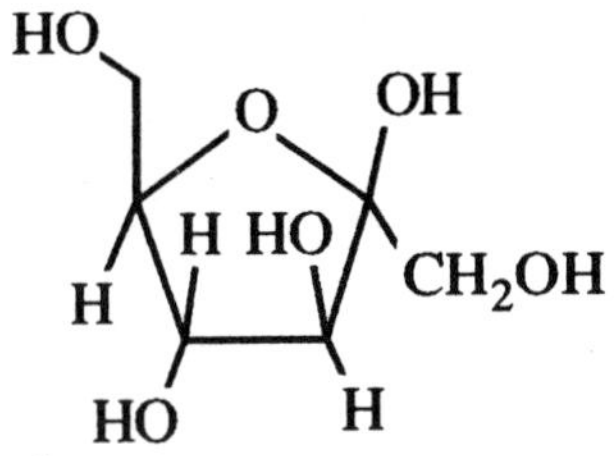

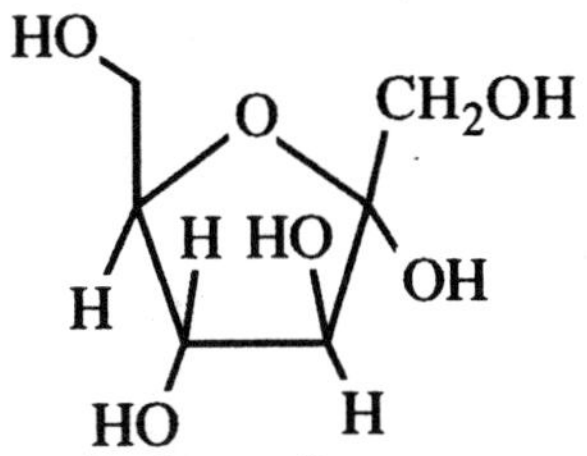

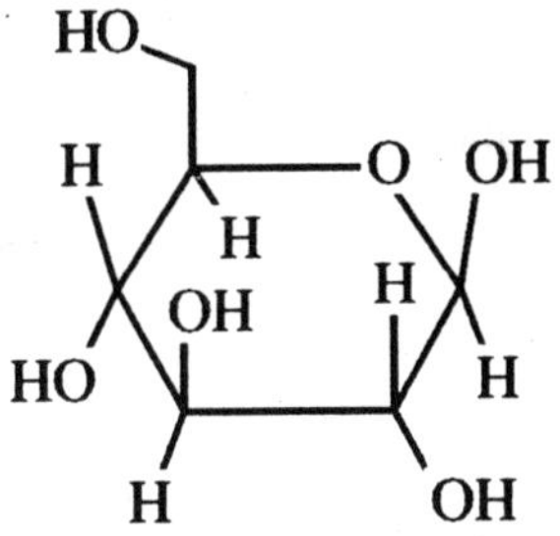

(b) [β-D-glucopyranose and α-D-glucopyranose structures]

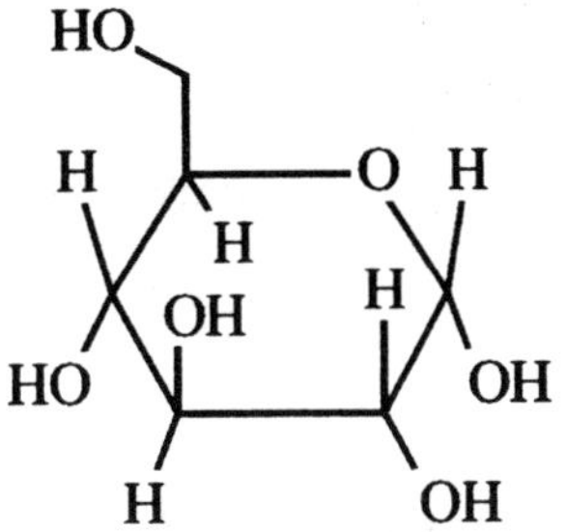

24.12 (a) No. They are diastereoisomers and will have different physical properties.
(b) They will have some other value. One stereogenic center only is changed from that in the +112° rotating form. It would be coincidence for either +112° or -112° to be its rotation.

24.13 (a) [structures shown]
(b) [structures shown]

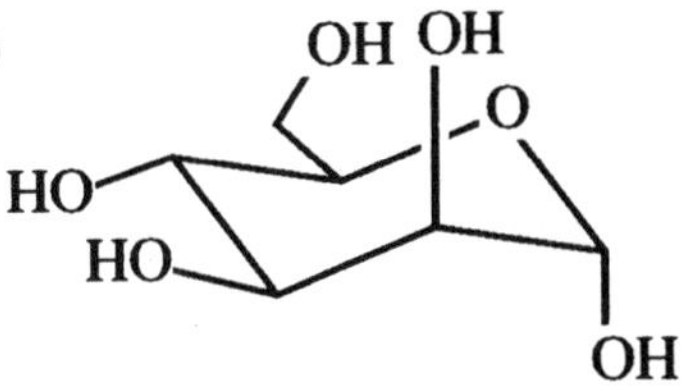

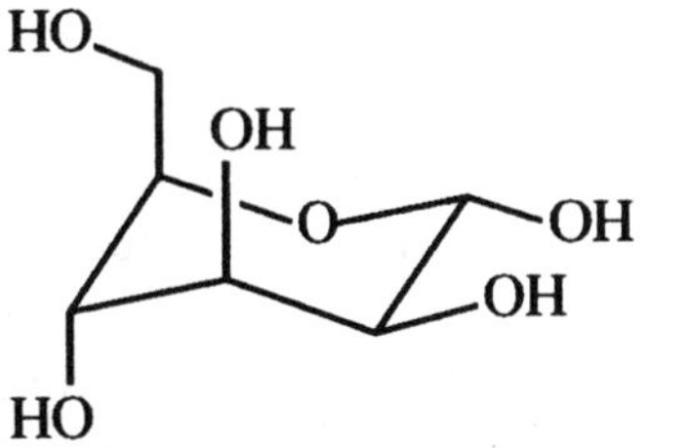

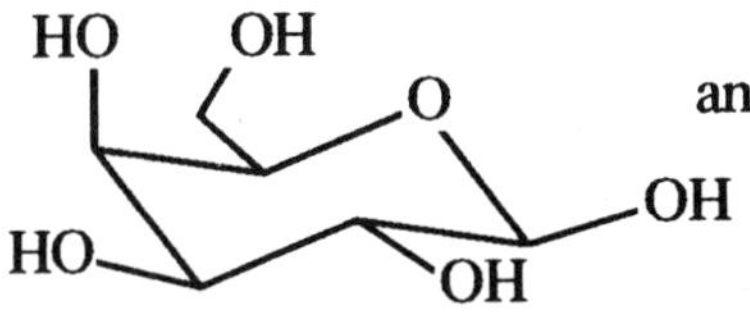

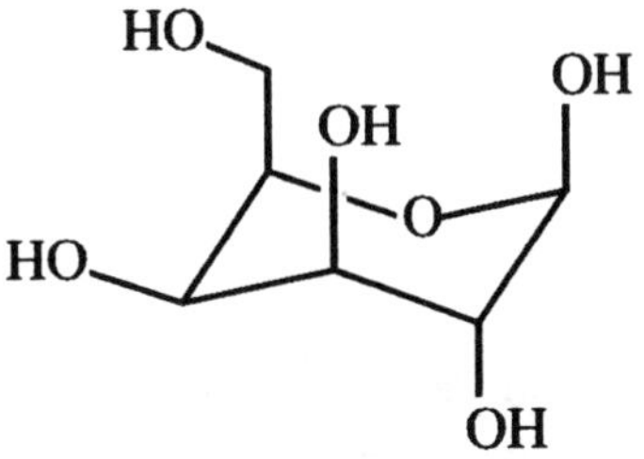

24.14 (a) methyl α-D-arabinofuranoside
(b) ethyl β-D-glucopyranoside
(c) methyl β-D-fructofuranoside

24.15 (a), (b), 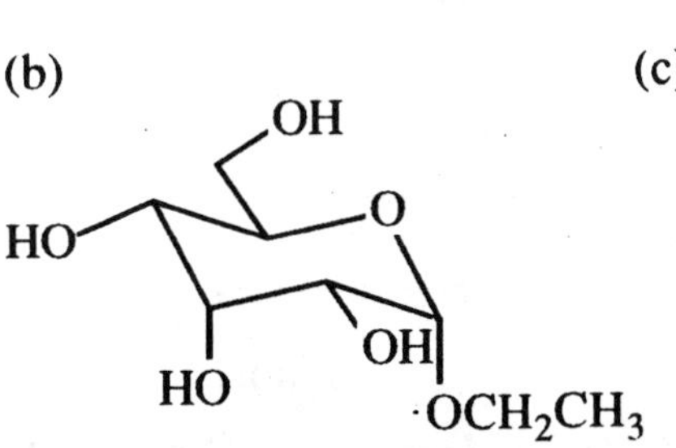 (c) [structures shown]

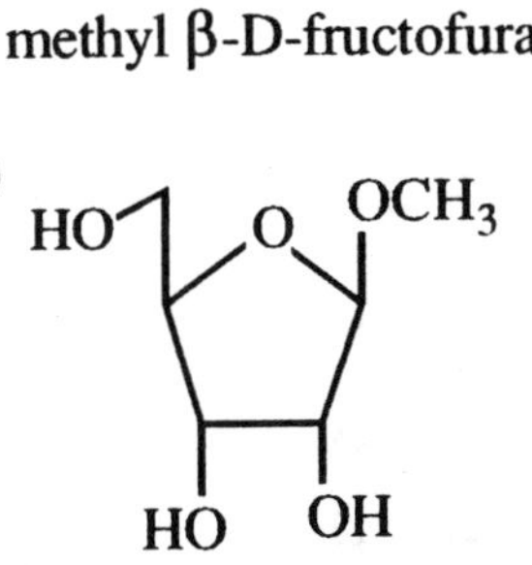

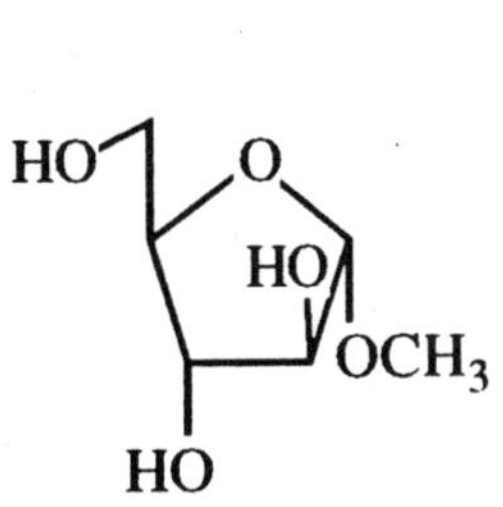

24.16 They would be expected to undergo mutarotation under aqueous acidic conditions. The rings would not open under nuetral or basic conditions.

24.17 Hindrance to approach at the α and β sides are not the same. Thereby, we expect different amounts of the

242 Study Guide and Solutions Manual

two diastereoisomers to form.

24.18 D-threose
24.19 D-ribose and D-arabinose
24.20

24.21 At the higher temperature the starting anomer opens its ring, reclosing with a portion of it in the α anomeric form.
24.22 Both D-tetroses give the same osazone. For the aldopentoses and aldohexoses, the following pairs give the same osazones:
  D-ribose and D-arabinose
  D-xylose and D-lyxose
  D-allose and D-altrose
  D-glucose and D-mannose
  D-gulose and D-idose
  D-galactose and D-talose
24.23 D-threose; D-arabinose; D-lyxose; D-altrose; D-glucose; D-mannose; D-gulose; D-idose; D-talose
24.24 Yes. One of the two anomeric sites is a hemiacetal.
24.25

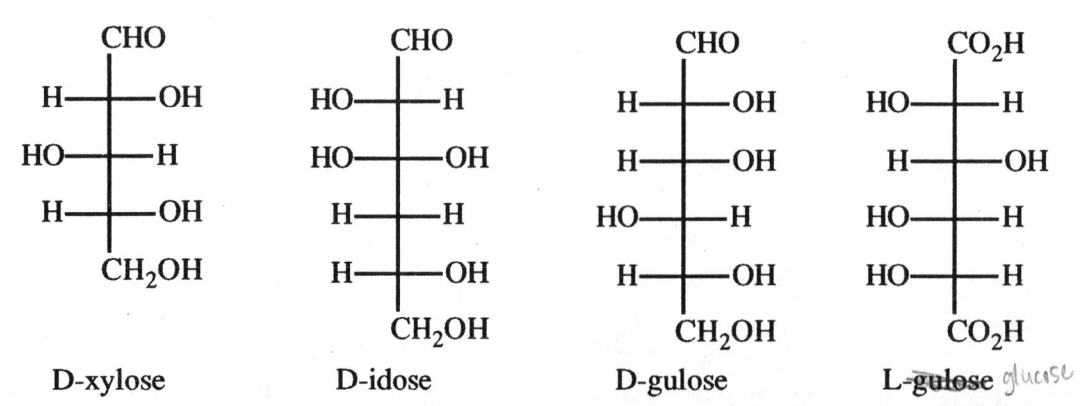

D-xylose   D-idose   D-gulose   L-gulose (glucose)

24.26 Although both would consume two equivalent amounts of periodate, the products of oxidation would be different for the two compounds. The methyl β-D-glucopyranoside would produce one equivalent amount each of

D-glyceraldehyde, formic acid, glyoxal (O=CH-CH=O), and methanol, while the methyl β-D-glucofuranoside would produce one equivalent each of formaldehyde, methanol, glyoxal and HOCH(CHO)$_2$.

24.27 (a) Two equivalent amounts of periodate are consumed and there is produced one equivalent amount of each of the following: methanol, formic acid, D-glyceraldehyde, and glyoxal.

(b) Two equivalent amounts of periodate are consumed and there is produced one equivalent amount of each of the following: methanol, glyoxal, formaldehyde, and HOCH(CHO)$_2$.

(c) One equivalent amount of periodate is consumed and there is produced one equivalent amount of each of the following: methanol, glyoxal, and dihydroxyacetone.

24.28

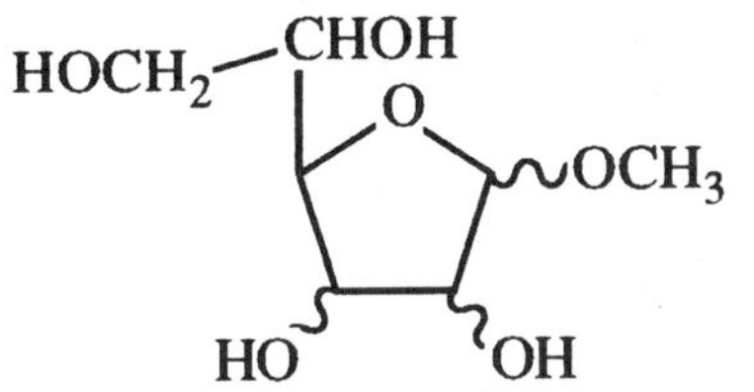

We cannot deduce the stereochemistry about the sites indicated with indefinite orientation of substituents.

24.29 The conversion might have been accomplished using sodium borohydride in alcohol or lithium aluminum hydride in ether.

24.30 The following experiments would be performed: (i) Tollens' test for a reducing sugar; (ii) acidic hydrolysis to determine which monosaccharides were involved in the compound and the relative amounts of each; (iii) test for cleavage with an α-glycosidase and a β-glycosidase to determine the nature of the glycosidic linkage; (iv) methylation followed by acidic hydrolysis to determine the points of union of the monosaccharide units.

24.31

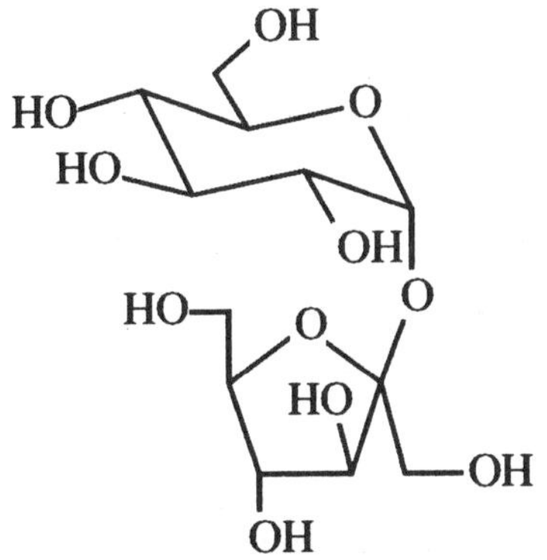

24.32

(a)   (b)

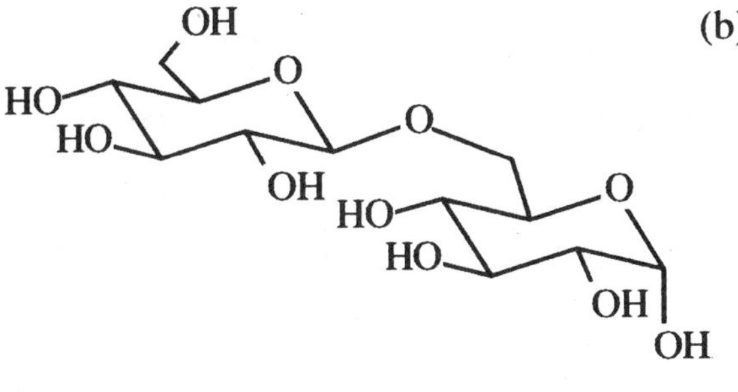

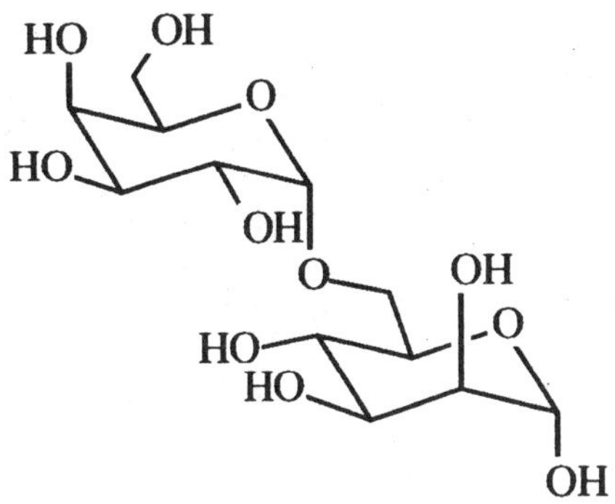

24.33 (a) 6-*O*-(α-D-galactopyranosyl)-β-D-fructofuranose

(b) β-D-galactopyranosyl-β-D-galactopyranoside

244 Study Guide and Solutions Manual

24.34

24.35 For both L-fucose and L-rahmnose the 6-position of D-glucose must be reduced. With L-fucose, the stereochemistry at positions -5, -3, and -2 must be inverted. For L-rahmnose, the stereochemistry at positions -5, -4, and -3 must be inverted.

24.36

α-D-fucopyranose   α-D-glucopyranose   A change in configuration at position -4 is required.

24.37 (a) (b) (c) (d) (e)

24.38

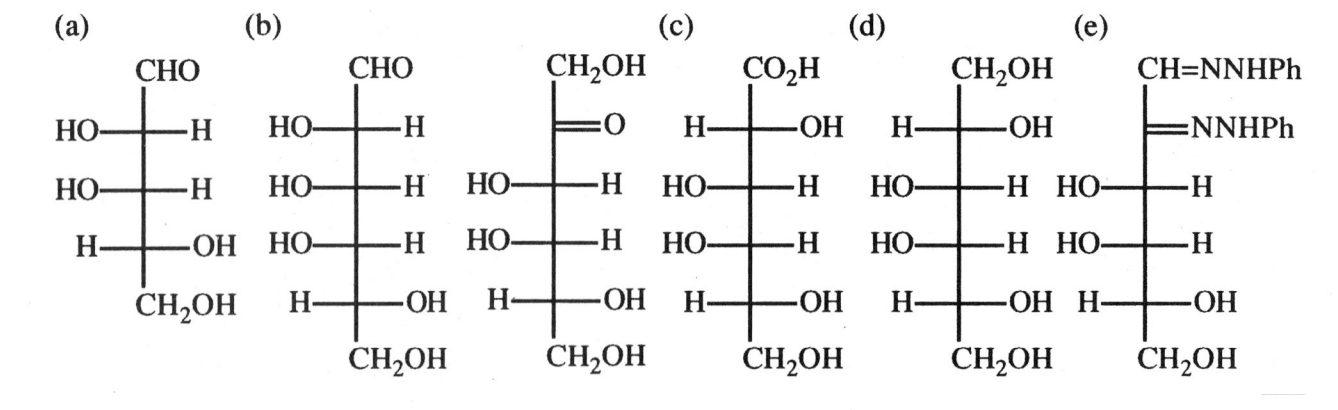

(a)	(b)	(c)	(d)	(e)
CHO	CHO	CH₂OH	CO₂H	CH₂OH	CH=NNHPh
HO—H	HO—H	=O	H—OH	H—OH	=NNHPh
HO—H	HO—H	HO—H	HO—H	HO—H	HO—H
H—OH	HO—H	HO—H	HO—H	HO—H	HO—H
CH₂OH	H—OH	H—OH	H—OH	H—OH	H—OH
	CH₂OH	CH₂OH	CH₂OH	CH₂OH	CH₂OH

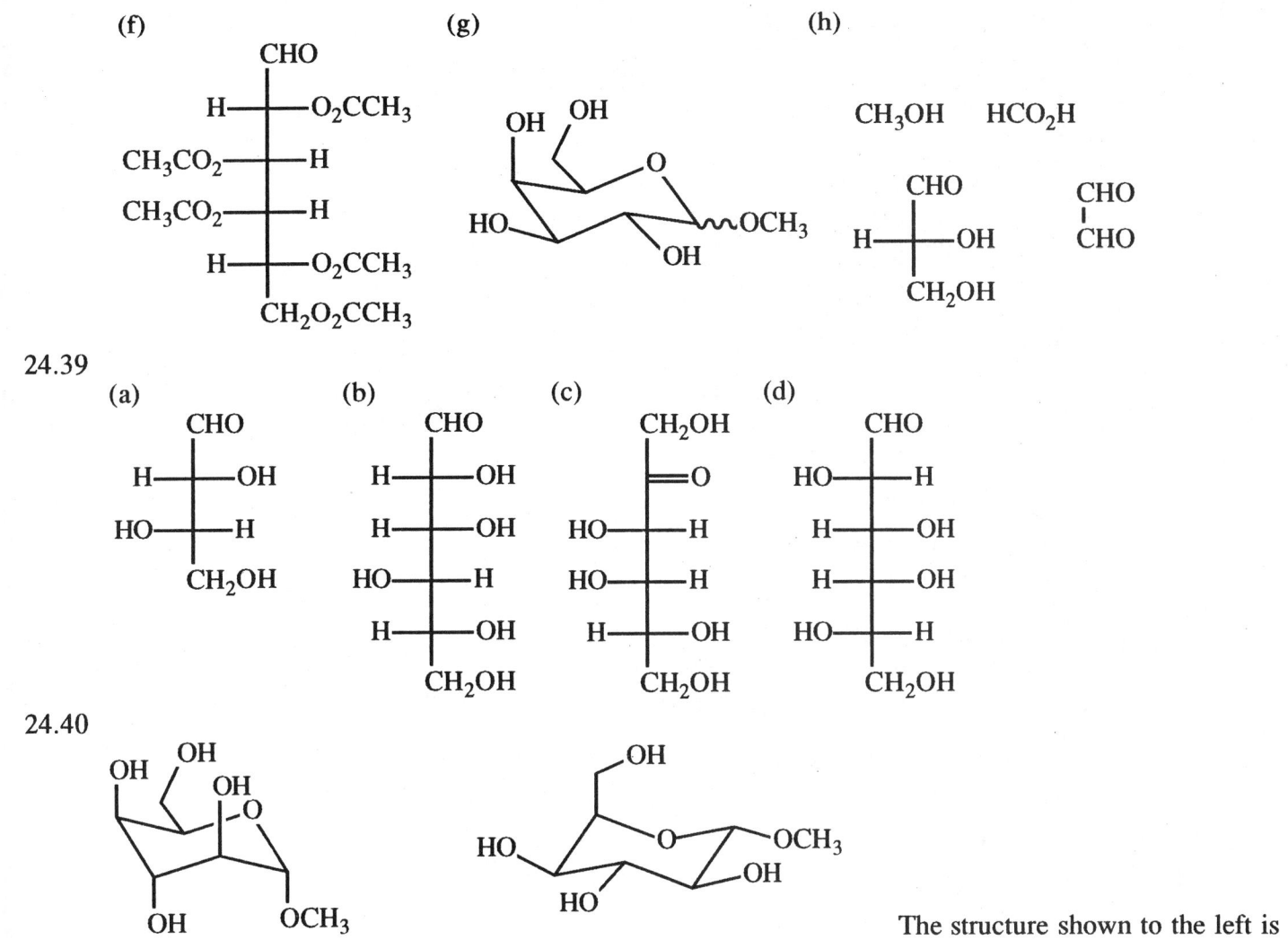

(f)

```
        CHO
H   ——  O₂CCH₃
CH₃CO₂ ——  H
CH₃CO₂ ——  H
H   ——  O₂CCH₃
       CH₂O₂CCH₃
```

(g) pyranose ring with OH, OH, HO, OH substituents and anomeric OCH₃ (mixed configuration)

(h) CH₃OH   HCO₂H

```
    CHO                CHO
H —— OH                CHO
    CH₂OH
```

24.39

(a)
```
    CHO
H —— OH
HO —— H
    CH₂OH
```

(b)
```
    CHO
H —— OH
H —— OH
HO —— H
H —— OH
    CH₂OH
```

(c)
```
    CH₂OH
     =O
HO —— H
HO —— H
H —— OH
    CH₂OH
```

(d)
```
    CHO
HO —— H
H —— OH
H —— OH
HO —— H
    CH₂OH
```

24.40

[pyranose chair with OH, OH, OH, OH, OCH₃] [pyranose chair with OH, HO, HO, OCH₃, OH] The structure shown to the left is dominant as it is stabilized by cross-ring hydrogen bonding.

24.41

(a) pyranose chair with OH, OH, HO, OH, OCH₃

(b) pyranose chair with OH, OCH₃, OH, OH

(c) pyranose chair with OH, HO, OH, OCH₃

24.42 *cis*-1,2-cyclohexanediol

24.43 1,2,3,4-pentanetetraol

24.44 There is formed one equivalent amount each of methanol, glyoxal, formic acid, and D-glyceraldehyde.

24.45

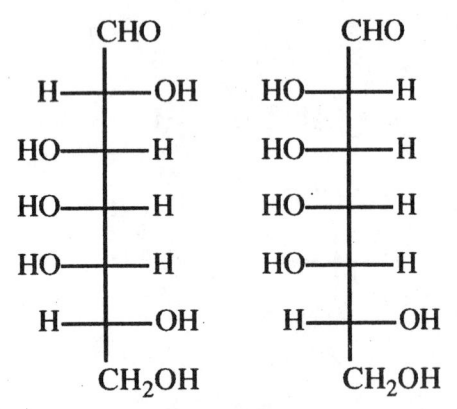

```
      CHO              CHO
  H──┼──OH         HO──┼──H
 HO──┼──H          HO──┼──H
 HO──┼──H          HO──┼──H
 HO──┼──H          HO──┼──H
  H──┼──OH          H──┼──OH
     CH₂OH             CH₂OH
       A                 B
```

A reduction of each would be performed using sodium borohydride in alcohol. Upon performing this reaction, *A* would yield an optically inactive product while *B* would yield an optically active product. Thus we could distinguish the two aldoheptoses.

24.46 I: ethanol, HCl
  II: sulfuric acid, acetone
  III: chromic anhydride, pyridine
  IV: methylenetriphenylphosphorane (Ph₃P=CH₂)
  V: sodium periodate or osmium tetroxide

24.47 (a) non-reducing - All anomeric sites are bound as full acetals or full ketals.
  (b) two equivalent amounts of D-glucose and one equivalent amount of D-fructose
  (c) two equivalent amounts each of glyoxal, formic acid, and D-glyceraldehyde, and one equivalent amount of D-fructose

24.48 *A* is D-ribose; *B* is ribonic acid; *C* is D-altrose; *D* is D-allose; *E* is D-mannonic acid; *F* is allonic acid

24.49

24.50

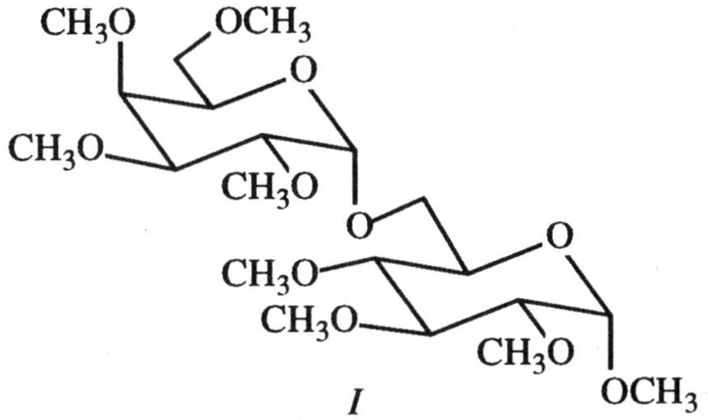

24.51

[Structure I: permethylated disaccharide with OCH₃ groups]

[Structure: hydrolyzed trisaccharide with OH groups]

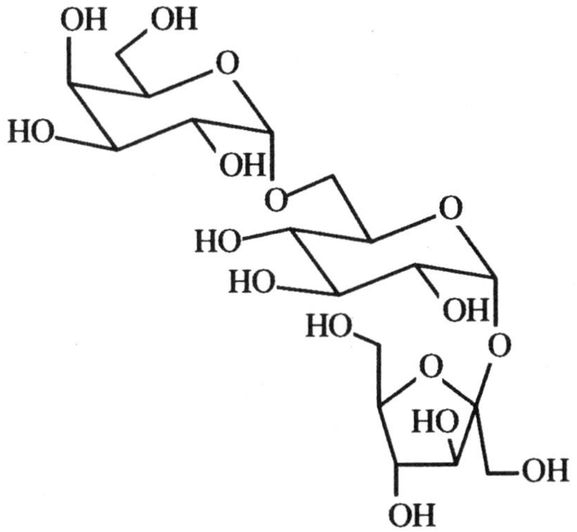

24.52 1. treatment of D-mannose with sodium borohydride in ethanol to form D-mannitol; 2. treatment with acetone in the presence of an anhydrous acid to form the 1,2:5,6-di-*O*-isopropylidene-D-mannitol; 3. sodium periodate (to yield two equivalent amounts of D-glyceraldehyde acetonide); 4. Ph₃P=CHCH₂CH₃; 5. hydrogen, platinum oxide; 6. aqueous acid.

24.53 *J* is D-ribose; *K* is D-arabinose; *L* is *meso*-1,2,3,4-butanetetraol; *M* is (2*S*,3*S*)-1,2,3,4-butanetetraol; *N* is *meso*-3,4-tetrahydrofurandiol; *O* is (3*S*,4*S*)-tetrahydrofurandiol

24.54

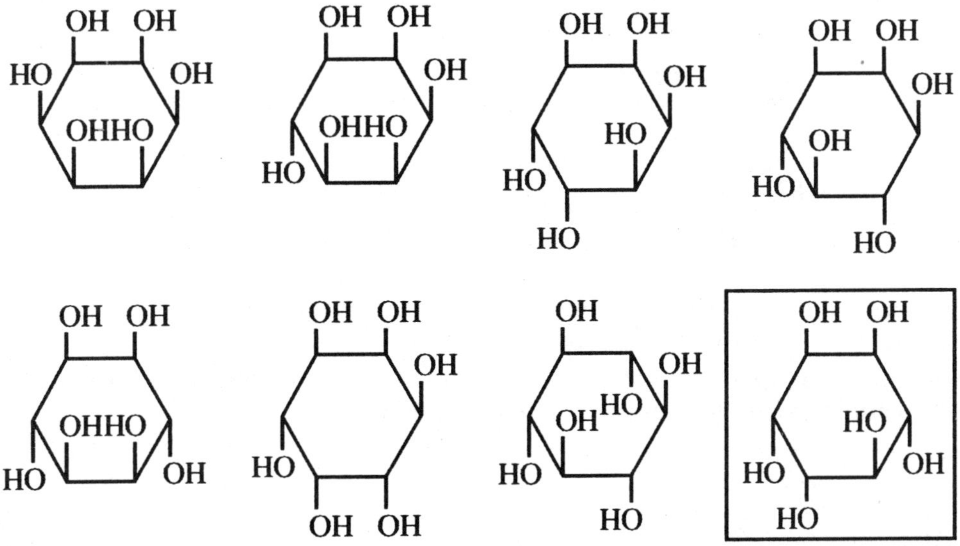

[Eight inositol stereoisomer structures shown, with the last one boxed]

There are a total of nine stereoisomers. Only the structure shown enclosed is optically active (its enantiomer is not shown).

24.55

[Structures of two furanose rings: first with HO-CH2, O, OH, HO, OH, OH substituents; second similar but with OCH3 instead of one OH]

24.56 1. treatment of D-ribose with methanol and HCl to form the methyl glycoside; 2. acetone with an anhydrous acid to form the acetonide of the 2- and 3-position hydroxyl groups; 3. triphenylphosphine, tetrabromomethane to replace the free hydroxyl group by a bromine; 4. excess ammonia to displace the bromide; 5. aqueous acid, followed by aqueous base to give the target compound

Solution of Study Guide Practice Problems

24.1 (a) i) $C_4H_8O_4$ ii) $C_5H_{10}O_5$
      (b) i) ketopentose   ii) aldohexose

24.2 (I) is L-erythrose, the non-superimposable mirror image of D-erythrose.

24.3 Yes. Enantiomers always have equal and opposite optical rotations.

24.4 L-gulose (*not* L-glucose)

24.5

[Fischer projections:]

A:
- CO₂H
- HO—H
- H—OH
- H—OH
- CH₂OH

B:
- C(O)O— (cyclic, connected to C5)
- HO—H
- H—OH
- H—
- CH₂OH

Compound B is clearly of the D-family of carbohydrate derivatives as no change in the stereochemistry about the stereogenic center most distant from the "carbonyl" site has been made.

24.6

(a) [pyranose ring with CH₂OH, HO, O, OH, OH, OH]

(b) [pyranose ring with O, OH, HO, HO, OH]

24.7 The α-D-glucopyranose and α-L-glucopyranose are enantiomers; the β-D-glucopyranose is diastereoisomeric with regard to both of the other compounds.

24.8 (a) False - The two compounds are diastereoisomers with different properties and stabilities in solution.
      (b) False - The two compounds are diastereoisomers, not enantiomers.

24.9

[Chair conformation of a pyranose with OH, OH, O, HO, HO, OH substituents]

24.10 (a) Treat the parent carbohydrate with excess methanol in the presence of a catalytic amount of anhydrous acid.

(b) Only one hydroxyl group is part of a hemiacetal linkage. Only it can be converted to a methoxy group under the conditions of reaction. All other hydroxyl groups are of the "normal" alcohol type.

(c) Treat the glycoside with aqueous acid.

24.11 (a) D-threose
(b) D-lyxose

24.12

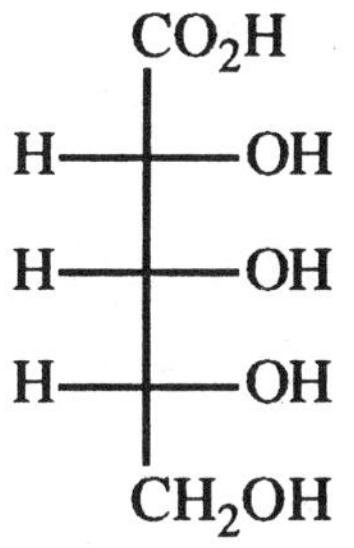

```
     CO₂H              ⎡ CO₂⁻ ⎤           CO₂H
H ──┼── OH             │ ══O  │ Ca²⁺  H ──┼── OH
H ──┼── OH        H ──┼── OH           H ──┼── OH
H ──┼── OH        H ──┼── OH              CH₂OH
    CH₂OH             ⎣ CH₂OH⎦₂
      A                   B                  C
```

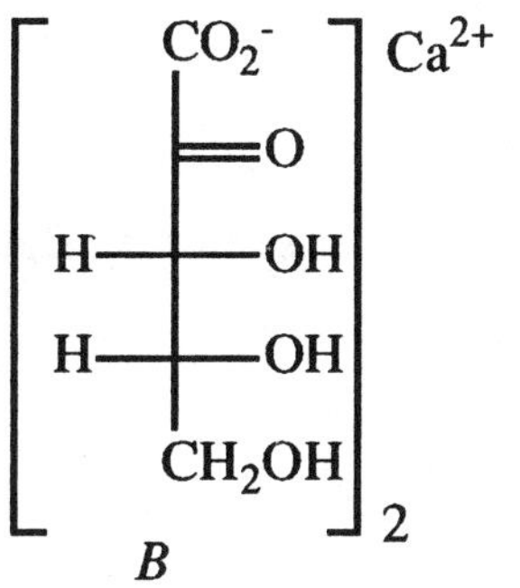

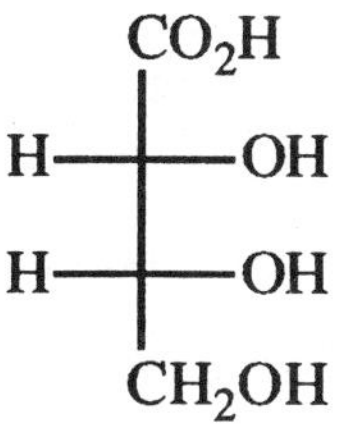

24.13

(a) HCO₂H + [structure with CH₂OH, O, OH]

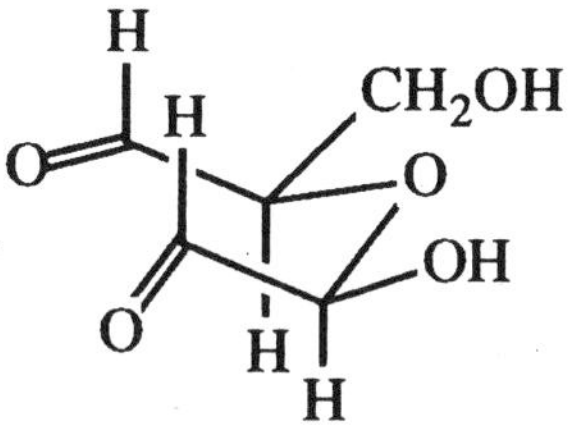

(b)
```
    CH=NNHPh
    ══NNHPh
H ──┼── OH
H ──┼── OH
    CH₂OH
```

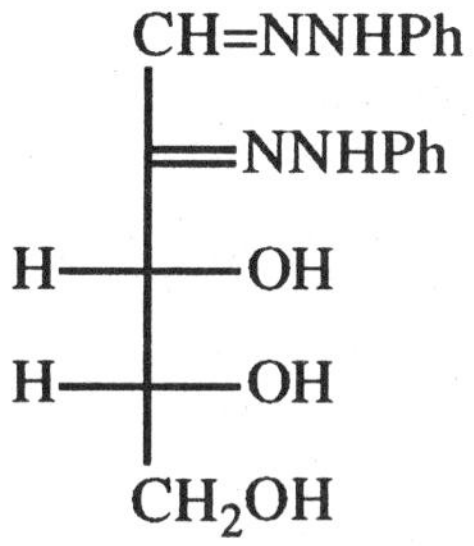

24.14 We could prepare the indicated compound (along with the α-anomer) by first treating D-ribose with methanol in the presence of a catalytic amount of anhydrous acid, followed by treatment with dimethyl sulfate with sodium hydroxide. Upon treatment of the methyl glycoside with hydrochloric acid, only the glycosidic methyl group would be removed to generate 2,3,5-tri-O-methyl-D-ribofuranose, as a mixture of anomeric forms.

24.15 From D-glucose we would produce D-gluconic acid, and from D-mannose we would produce D-mannonic acid. Using L-gulose we would generate the same dicarboxylic acid (aldonic acid) as we produced using D-mannose.

24.16

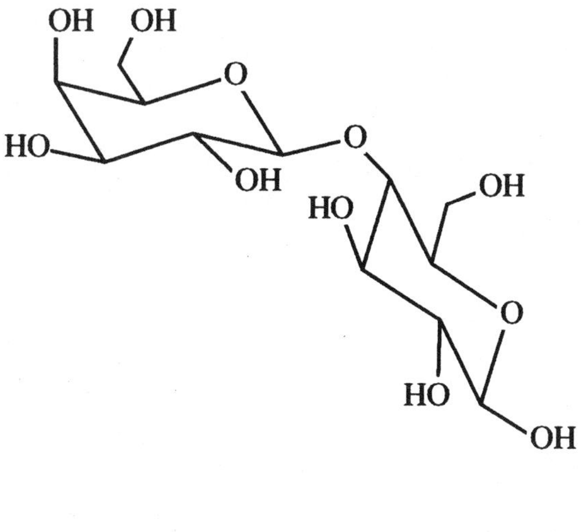

CHAPTER 25
ULTRAVIOLET/VISIBLE AND MASS SPECTROMETRY

Key Points

- Know the basic terminology used in UV/VIS spectroscopy, *e.g.* absorbance, molar absorptivity, λ_{max}, and learn the shorthand notation for electronic transitions (n→π*, π→π*, *etc.*).

Practice Problem 25.1
From memory, write equations showing each of the following relationships:
  (a) absorbance, A to I and I_o
  (b) molar absorptivity, ε, to A

Practice Problem 25.2
Could a band in a UV/VIS spectrum possibly be due to a σ→π transition? Explain your answer.

Practice Problem 25.3
Which has more energy per photon, UV light or visible light; blue light or red light?

- Appreciate the general realtionship between conjugation and the λ_{max} for π→π* transitions.

Practice Problem 25.4
One of the compounds shown below exhibits a UV band at ~260 nm and the other exhibits a band at ~220 nm. Assign the proper absorption to each structure.
  $H_2C=CH-CH=CH_2$ $H_2C=CH-CH=CH-CH=CH_2$

- Know how to use Woodward's rules to estimate λ_{max} for π→π* transitions in polyenes and enones.

Practice Problem 25.5
Estimate the λ_{max} as measured using pure ethanol as the solvent for each of the steroid molecules shown below. In each case, R = 1,5-dimethylpentyl.

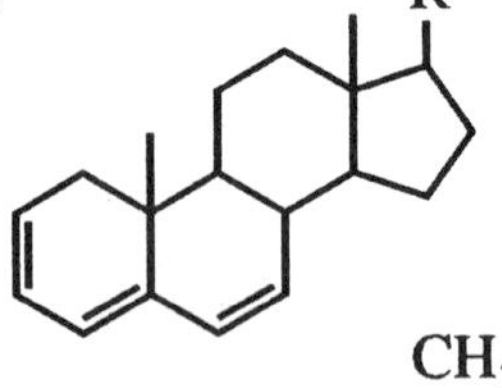

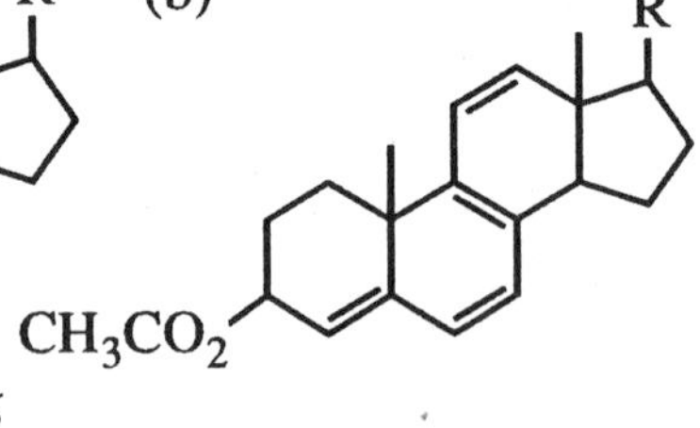

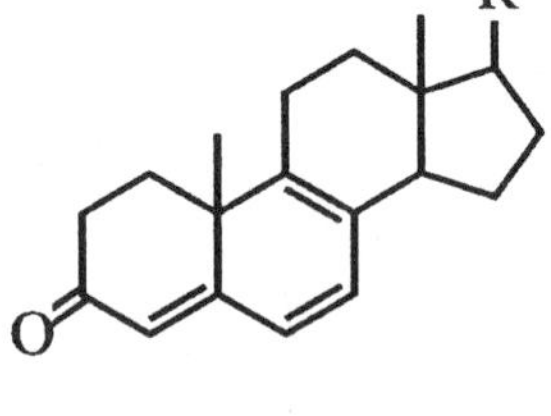

Practice Problem 25.6
Of the structures shown below, which would exhibit a UV absorption band at 250 nm?

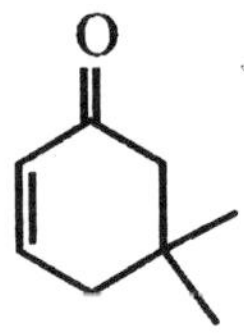

A 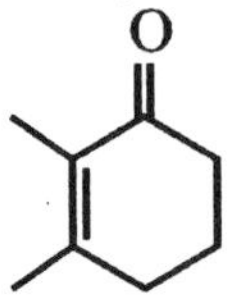   B

- Learn the basic principles of mass spectrometry. Understand how to use fragment ion peaks to infer structural information, and also learn how mass spectra are influenced by the presence of naturally occurring isotopes.

Practice Problem 25.7
A straight-chain ketone exhibits prominent peaks at m/e 43, 57, 71, and 128, the last being the molecular ion peak. Suggest a structure for this material.

Practice Problem 25.8
The mass spectrum of ethanol exhibits prominent peaks at m/e 29, 31, 45, and 46. Account for each of these peaks. There is also a minor peak at m/e 47. Estimate its approximate intensity as a percentage of the intensity of the m/e 46 peak.

Practice Problem 25.9
Compounds A and B are isomeric alcohols of formula $C_5H_{12}O$. Compound A exhibits a major peak in its mass spectrum at m/e 59, while B exhibits a major peak in its mass spectrum at m/e 45 (but *no* peak at m/e 59). Suggest structures for A and B.

Practice Problem 25.10
A gas gives a parent (molecular ion) peak at m/e 42, and M+1 and M+2 peaks of relative intensity ~3.3% and ~0.04% respectively. Of the following possibilities, which is most likely this gas?
    (a) H_2CN_2 (diazomethane); (b) cyclopropane; (c) $H_2C=C=O$ (ketene); (d) propyne

Practice Problem 25.11
Methyl esters of the type $R(CH_2)_3CO_2CH_3$ give rise to a fragment ion of m/e 74 when studied using mass spectrometry. The fragment ion arises by a McLafferty rearrangement. Suggest a structure for this ion.

Solution of Text Problems

25.1 $\varepsilon = 90$

25.2 $1.90 \times 10^{-4} M$

25.3   (a) 1,3,5,7-octatetraene
       (b) 1,3-pentadiene
       (c)

25.4   A is 1,3-cyclohexadiene; B is cyclohexane

25.5   (a) 240 nm
       (b) 225 nm
       (c) 263 nm
       (d) 260 nm

25.6   (a) 249 nm
       (b) 237 nm
       (c) 239 nm
       (d) 272 nm

25.7

25.8   Changing a ^1H to a ^2H will have the greater effect on the IR stretching frequency. It constitutes a greater percentage change in the mass of the atom at one end of the bond, as shown in the formula:

$$\nu = \frac{1}{2\pi c}\sqrt{\frac{f(m_1 + m_2)}{m_1 m_2}}$$

For ^{12}C-^1H with $f = 4.96 \times 10^5$ g sec.$^{-2}$, $\nu = 3000$ cm^{-1}.

For ^{13}C-^1H, $\nu = 2995$ cm^{-1}, and for ^{12}C-^2H, $\nu = 2210$ cm^{-1}.

25.9   A is 4-methylundecane; B is *n*-dodecane; C is 2,2,4,4,6-pentamethylheptane

25.10 There are 6 carbon atoms in each molecule of D.

25.11 $C_{12}H_{10}O$

25.12 C_6H_6O

25.13 $[^{13}CH_3^{79}Br]^+$. The intensity of this peak is ~1.1% that of the peak at m/e = 94.

252 Study Guide and Solutions Manual

25.14 Peaks are indicated with their approximate intensities relative to each other
(a) M   1
     M+2   3
     M+4   3
     M+6   1
(b) M   1
     M+2   4
     M+4   6
     M+6   4
     M+8   1

25.15 Peaks are indicated with their m/e value and relative intensities.
(a) 50   100.0
     51   1.1
     52   33.3
     53   0.4
(b) 98   100.0
     99   2.2
     100   133.3
     101   2.9
     102   33.3
     103   0.7
(c) 146   100.0
     147   3.3
     148   233.3
     149   7.7
     150   166.6
     151   5.5
     152   33.3
     153   1.1

25.16 (a) 2.93%
      (b) 2.2%

25.17 We would also expect the fragmentation to occur on the C2-C3 side of C3, producing ions with m/e = 99, as well as cleaving methyl radicals, which would produce ions of m/e = 113.

25.18 Loss of an electron from the C3-C4 σ bond produces the molecular ion shown in Figure 25.12. For the ion of m/e = 99, loss of an electron from the C2-C3 σ bond is required, and for the ion of m/e = 133, loss of an electron from the C3-CH$_3$ σ bond is required.

25.19 (a) 57
      (b) 71

25.20

$$CH_3CH_2CH_2CH_2-\overset{+}{\underset{..}{O}}-CH_2 \longleftrightarrow CH_3CH_2CH_2CH_2-\overset{+}{O}=CH_2$$

25.21

from isobutylamine: $H_2C=\overset{+}{N}H_2 \longleftrightarrow H_2\overset{+}{C}-\overset{..}{N}H_2$

from *tert*-butylamine: $(CH_3)_2C=\overset{+}{N}H_2 \longleftrightarrow (CH_3)_2\overset{+}{C}-\overset{..}{N}H_2$

25.22

(a)

```
      +  H
      :O
      ‖
      C      H
  H₃C   \C/
          |
          H
```
m/e = 58

(b)

```
      +  H
      :O
      ‖
      C      H
   H   \C/
          |
          CH₃
```
m/e = 58

(c)

```
      +  H
      :O
      ‖
      C      H
   H   \C/
          |
          H
```
m/e = 44

25.23

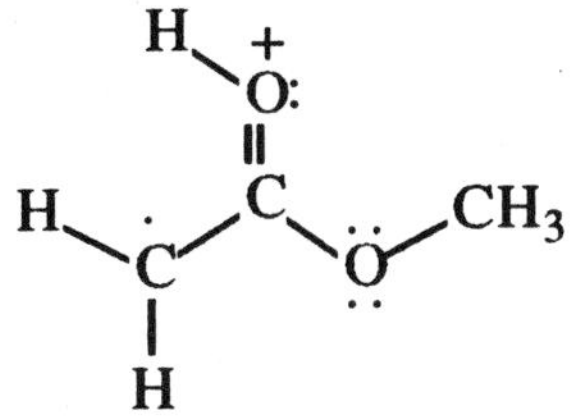

[Structure: protonated methyl acetate — CH with two H's, bonded to C(=O⁺H)(OCH₃)]

25.24 ε = 2.63 × 10³

25.25 (a) UV light (shorter wavelength, higher energy)
(b) low energy electron bombardment (less energy for fragmentation)
(c) cyclohexane (no absorption in UV)
(d) p-dibromobenzene (approximately equal amounts of ⁷⁹Br and ⁸¹Br)
(e) trimethylamine (non-bonding electrons are no longer available for excitation)

25.26
(a) 1,3-pentadiene
(b) [bicyclic structure: tetrahydronaphthalene with one double bond in one ring]
(c) [CH₃CH₂-CH=CH-C(=O)-CH₃, pent-3-en-2-one]
(d) [bicyclic ketone: octahydronaphthalenone with a double bond]

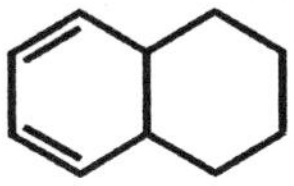

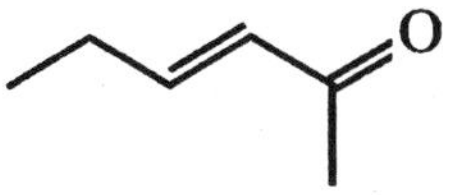

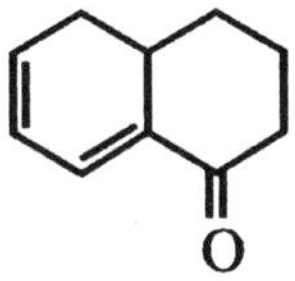

25.27 (a) For λ_max = 320 nm, n→π*; for λ_max = 218 nm, π→π*.
(c) H₂C=CH-C(O)-CH₃

25.28

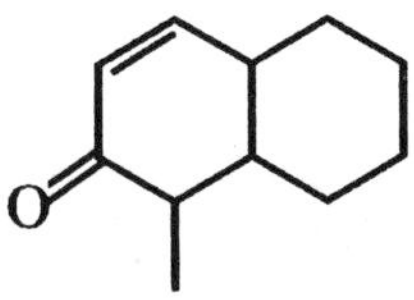

[Structure: bicyclic enone with methyl substituent]

The UV band would disappear upon catalytic hydrogenation.

25.29 (a) C₁₁H₁₆
(b) C₆H₁₂O₄
(c) C₈H₈N₂O

25.30 C₆H₅Cl

25.31 chlorobenzene

25.32 (m/e = 394)/(m/e = 396) = 1.5

25.33 BrCH₂CH(OCH₃)₂

25.34 (a) C₇H₁₈N₂
(b) C₆H₁₂

Solution of Study Guide Practice Problems

25.1 (a) $A = \log(I/I_o)$

(b) $\varepsilon = A/cl$

25.2 No. An electron must be promoted to a *vacant* orbital.
25.3 UV; blue.
25.4 The longer wavelength absorption occurs with $H_2C=CH-CH=CH-CH=CH_2$.
25.5 (a) 303 nm
 (b) 343 nm
 (c) 385 nm
25.6 Structure *B* would exhibit the UV absorption band at 250 nm.
25.7 4-octanone
25.8 The m/e = 46 peak corresponds to the molecular ion, $(C_2H_5OH)^{+\cdot}$.

The m/e = 45 peak corresponds to the ion $(C_2H_5O)^+$.

The m/e = 31 peak corresponds to the ion $(CH_2OH)^+$.

The m/e = 29 peak corresponds to the ion $(C_2H_5)^+$.

25.9 Compound *A* is 2-methyl-2-butanol. The m/e = 59 peak corresponds to the ion $[(CH_3)_2C=OH]^{+\cdot}$.

Compound *B* is 3-methyl-2-butanol. The m/e = 41 peak corresponds to $[CH_3CH=OH]^{+\cdot}$, that is, an isopropyl group has been lost from the original molecule by cleavage of the C-C bond α- to the oxygen atom.

25.10 cyclopropane

25.11

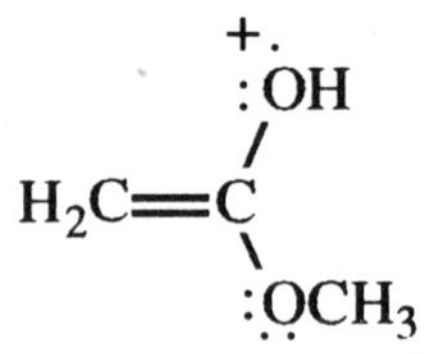

$$H_2C=C\begin{matrix} \overset{+\cdot}{:}OH \\ \\ :\underset{\cdot\cdot}{O}CH_3 \end{matrix}$$

PRACTICE EXAMINATION FIVE

A time limit of 90 minutes should be set for completion of this entire practice examination. Answers should be written out completely as they would be when presented for independent grading. No text or supplemental materials should be consulted during the testing period, and you should not check your answers until you have worked out the complete examination and the time limit has been reached.

1. Give structures matching the descriptions given below (4 points for each part).

   (a) the enamine formed by the reaction of cyclohexanone with diethylamine

   (b) a substance containing no oxygen atoms that, on hydrolysis with aqueous acid yields *p*-bromobenzoic acid, and on reduction with lithium aluminum hydride yields *p*-bromobenzylamine

   (c) a substance of formula $C_8H_{12}O_3$ that undergoes decarboxylation readily on heating to form 2-methylcyclohexanone

   (d) the compound formed by the reaction of *p*-bromoaniline and propanoic anhydride

   (e) the product obtained from the reaction of diethyl malonate anion $[:CH(CO_2CH_2CH_3)_2]^-$ with isopentyl chloride, followed by treatment with aqueous acid and final heating

   (f) β-D-glucofuranose, shown as a Haworth projection

   (g) the product obtained by the base induced Michael addition of nitromethane to methyl vinyl ketone

   (h) the reactant needed to prepare 2-carboethoxycyclopentanone by way of a Dieckmann condensation (the carboethoxy group is $-CO_2CH_2CH_3$)

   (i) a compound of formula $C_4H_8O_3$ that exhibits the following spectroscopic properties: IR, strong absorptions at 2500-3300 cm^{-1} and ~1715 cm^{-1}; ^1H NMR, two singlets (δ 4.1, two hydrogens, and δ 11.0, one hydrogen), a triplet (δ 1.3, three hydrogens), and a quartet (δ 3.7, two hydrogens)

   (j) an acyclic compound of formula C_5H_8 that has a UV band showing approximately the same λ_{max} as the band in the corresponding spectrum of 1-pentene, but with a molar absorptivity that is approximately twice as large

   (k) the substance having the following spectroscopic properties: Mass Spectrum, molecular ion at m/e = 106 with M+1 and M+2 peaks of intensity 4.4% and 33% respectively of the molecular ion; IR, strong absorption at ~1800 cm^{-1}; ^1H NMR, δ 1.0 (triplet, 3H), δ 1.8 (triplet of quartets, 2H), δ 2.9 (triplet, 2H)

   (l) a compound of formula $C_{12}H_{10}O_4$ formed by the reaction of deithyl phthalate with ethyl acetate in the presence of sodium ethoxide in ethanol

   (m) the D-aldohexose that gives the same dicarboxylic acid as that given by its enantiomer when oxidized with dilute nitric acid

   (n) the product formed when γ-butyrolactone is heated with excess methanol and a catalytic amount of anhydrous acid

   (o) the product formed when γ-butyrolactone is reduced with lithium aluminum hydride

2. When D-glyceraldehyde is added to 1,3-dihydroxy-2-propanone (dihydroxyacetone) in basic solution a crossed aldol condensation occurs producing a mixture of ketohexoses. Propose structures for these ketohexoses and predict if they will be formed in equal or unequal amounts (10 points).

3. When phenylacetonitrile ($PhCH_2CN$) is treated with methylmagnesium iodide in ether followed by workup with aqueous acid there are isolated two products, *A* and *B*. Compound *A*, of formula $C_9H_{12}O$, is formed in ~8% yield and compound *B*, of formula $C_{16}H_{13}NO$, is formed in ~70% yield. Both *A* and *B* exhibit IR absorptions at ~1715 cm^{-1}, and *B* (but not *A*) also exhibits an absorption at ~2200 cm^{-1}. Suggest structures for *A* and *B* and explain how they are formed (10 points).

4. Show how to accomplish each of the following synthetic transformations. Use any common laboratory reagents deemed necessary (5 points for each part).

  (a) D-glucose converted to 2,3,4,6-tetra-O-methyl-β-D-glucopyranoside
  (b) ethyl acetoacetate converted to 5-methyl-2-hexanone
  (c) acrylonitrile ($H_2C=CH-CN$) converted to $(CH_3)_2NCH_2CH_2CH_2NHC(O)CH_3$
  (d) benzaldehyde and acetone converted to

CHAPTER 26
AMINO ACIDS, PEPTIDES, AND PROTEINS

Key Points

• Know the structures and three-letter abbreviations of the twenty amino acids commonly found as components of proteins. [*Special Note:* One *letter abbreviations are also used for amino acids, as listed in Practice Problem 26.1 below.*]

Practice Problem 26.1
From memory, draw the structures and give the three-letter abbreviations of the following amino acids. (*One* letter abbreviations for these same amino acids are also shown.)
alanine (A), valine (V), leucine (L), isoleucine (I), proline (P), phenylalanine (F), tryptophan (W), methionine (M), cysteine (C), glycine (G), asparagine (N), glutamine (Q), serine (S), threonine (T), tyrosine (Y), lysine (K), arginine (R), histidine (H), aspartic acid (D), glutamic acid (E)

• Understand the stereochemistry and the acid-base properties of amino acids.

Practice Problem 26.2
Some D-amino acids exist free in nature, *e.g.* D-alanine is present in some larvae. Complete the structure shown below to represent D-alanine and give it a label as to R or S.

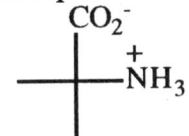

$$\begin{array}{c} CO_2^- \\ | \\ \underline{\qquad\qquad}-\overset{+}{N}H_3 \\ | \end{array}$$

Practice Problem 26.3
Give the R,S designation for each of the following:
  (a) L-cysteine
  (b) L-methionine

Practice Problem 26.4
Identify any amino acids from Table 26.1 of the text that have more than one stereogenic center.

Practice Problem 26.5
Calculate the ratio

$$\frac{[H_3\overset{+}{N}CH_2CO_2^-]}{[H_3\overset{+}{N}CH_2CO_2H]} \quad \text{for glycine at pH} = 7.34$$

Practice Problem 26.6
Histidine has a basic side chain (R). On which atom of the R group of histidine would you expect protonation to be most likely to occur? Explain your choice.

Practice Problem 26.7
Which of the following will exist largely in a *cationic* form at pH = 6?
    alanine, lysine, aspartic acid

Practice Problem 26.8
The side chain in arginine is -(CH$_2$)$_3$NHC(NH$_2$)=NH. Without consulting a tabulation of values, predict whether the pI value of arginine is > 7 or < 7.

• Know how to synthesize α-amino acids from various precursors.

Practice Problem 26.9
How would you prepare racemic leucine starting with a suitable carboxylic acid and using the Hell-Volhard-Zelinsky reaction for one of the steps?

Practice Problem 26.10
Draw structures for the following important intermediates in the various types of amino acid syntheses.
   (a) N-phthalimidomalonic ester
   (b) acetamidomalonic ester

Practice Problem 26.11
Show structures for the compounds A and B below, and identify the named synthetic route to amino acids which is being represented.

PhCH₂CHO →[NH₄Cl/KCN, H₂O] A (C₉H₁₀N₂) →[aq. acid] B (C₉H₁₁NO₂)

Practice Problem 26.12
Draw the structure of the benzyl ester of proline. How would you prepare this ester from proline?

Practice Problem 26.13
Draw the structure of N-acetyl histidine and indicate how to prepare this compound from histidine.

• Know how amino acids are separated and analyzed.

Practice Problem 26.14
Describe how an automated amino-acid analyzer operates.

Practice Problem 26.15
Consider the electrophoresis of a mixture of phenylalanine, arginine, glutamic acid, and leucine at pH 7.5. For each amino acid, what is the direction of migration (toward the anode or toward the cathode) that we would observe?

Practice Problem 26.16
The first cycle of an Edman degradation of a polypeptide results in the formation of the compound shown below. What is the N-terminal amino acid of the polypeptide?

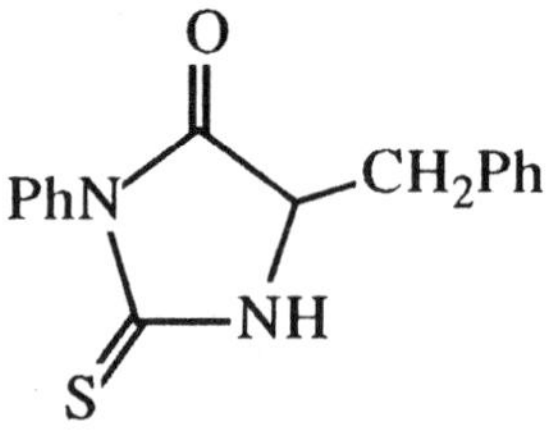

[Structure: 5-membered ring with PhN, C=O, CH(CH₂Ph), NH, C=S]

Practice Problem 26.17
A peptide has the following composition: Arg, Asp₂, Glu₂, Gly₃, Leu, Phe, Val₃. After partial hydrolysis the following sequenced peptides were identified:
   (a) Asp-Glu-Val-Gly-Gly-Glu-Phe
   (b) Val-Asp-Val-Asp-Glu
   (c) Val-Asp-Val
   (d) Glu-Phe-Leu-Gly-Arg
   (e) Val-Gly-Gly-Glu-Phe-Leu-Gly-Arg
   (f) Leu-Gly-Arg
What is the amino acid sequence in the original polypeptide?

Practice Problem 26.18
Vasopressin is a nonapeptide that plays an important role in the control of water balance in the body. Synthetic vasopressin is used to treat diabetes insipidus. Partial hydrolysis of vasopressin yields the mixture of peptides shown below.
   (a) Tyr-Phe-Glu-Asp
   (b) Asp-Cys-Pro-Arg
   (c) Cys-Tyr-Phe
   (d) Pro-Arg-Gly
   (e) Phe-Glu-Asp
Suggest a structure for vasopressin, recognizing that there is present a disulfide bridge.

• Appreciate the strategies used to synthesize polypeptides.

Practice Problem 26.19
Propose a synthesis of Ala-Phe from BOC-Ala. Use protecting groups so that a high yield of the target material

Chapter 26 259

will be obtained.

• Understand the methods used to classify proteins and know what is meant by secondary, tertiary, and quaternary structure.

Practice Problem 26.20
For each of the following, tell if it is an aspect of secondary, tertiary, or quaternary structure.

(a) a pleated sheet arrangement
(b) the aggregation of individual proteins into a cluster
(c) the folding of a protein chain into a spherical shape
(d) the adoption of an α-helical arrangement by a portion of a protein molecule
(e) the description of a protein as *fibrous*

Solution of Text Problems

26.1 (a) S
(b) S
(c) S

26.2 No. The absolute configuration tells us nothing about the magnitude or direction of optical rotation at a particular wavelength.

26.3

(a)
$$HO_2CCHCH_2CO_2H + H_2O \xrightleftharpoons{K_1} HO_2CCHCH_2CO_2^- + H_3O^+$$
$$\quad\quad |\quad\quad\quad\quad\quad\quad\quad |$$
$$\quad +NH_3 \quad\quad\quad\quad\quad\quad\quad\quad\quad\quad +NH_3$$

$$HO_2CCHCH_2CO_2^- + H_2O \xrightleftharpoons{K_2} {}^-O_2CCHCH_2CO_2^- + H_3O^+$$
$$\quad\quad |\quad\quad\quad\quad\quad\quad\quad\quad\quad\quad\quad\quad\quad |$$
$$\quad +NH_3 \quad\quad\quad\quad\quad\quad\quad\quad\quad\quad +NH_3$$

$${}^-O_2CCHCH_2CO_2^- + H_2O \xrightleftharpoons{K_3} {}^-O_2CCHCH_2CO_2^- + H_3O^+$$
$$\quad\quad |\quad\quad\quad\quad\quad\quad\quad\quad\quad\quad\quad\quad\quad |$$
$$\quad +NH_3 \quad\quad\quad\quad\quad\quad\quad\quad\quad\quad NH_2$$

(b)
$$H_3\overset{+}{N}CH_2CH_2CH_2CHCO_2H + H_2O \xrightleftharpoons{K_1} H_3\overset{+}{N}CH_2CH_2CH_2CHCO_2^- + H_3O^+$$
$$\quad\quad\quad\quad\quad\quad\quad\quad |\quad\quad\quad\quad\quad\quad\quad\quad\quad\quad\quad\quad\quad\quad\quad |$$
$$\quad\quad\quad\quad\quad\quad +NH_3 \quad\quad\quad\quad\quad\quad\quad\quad\quad\quad\quad\quad\quad +NH_3$$

$$H_3\overset{+}{N}CH_2CH_2CH_2CHCO_2^- + H_2O \xrightleftharpoons{K_2} H_2NCH_2CH_2CH_2CHCO_2^- + H_3O^+$$
$$\quad\quad\quad\quad\quad\quad\quad\quad |\quad\quad\quad\quad\quad\quad\quad\quad\quad\quad\quad\quad\quad\quad\quad |$$
$$\quad\quad\quad\quad\quad\quad +NH_3 \quad\quad\quad\quad\quad\quad\quad\quad\quad\quad\quad\quad\quad +NH_3$$

$$H_2NCH_2CH_2CH_2CHCO_2^- + H_2O \xrightleftharpoons{K_3} H_2NCH_2CH_2CH_2CHCO_2^- + H_3O^+$$
$$\quad\quad\quad\quad\quad\quad\quad\quad |\quad\quad\quad\quad\quad\quad\quad\quad\quad\quad\quad\quad\quad\quad\quad |$$
$$\quad\quad\quad\quad\quad\quad +NH_3 \quad\quad\quad\quad\quad\quad\quad\quad\quad\quad\quad\quad\quad NH_2$$

26.4 (a) pK_{a1} and pK_{a2}
(b) pK_{a2} and pK_{a3}

26.5
$$Na^+ \; {}^-O_2CCH_2CH_2CHCO_2^-$$
$$\quad\quad\quad\quad\quad\quad\quad |$$
$$\quad\quad\quad\quad\quad +NH_3$$

26.6 The anionic form would be present in the higher concentration

26.7 1. treatment of 3-methylpentanoic acid with bromine and phosphorus tribromide; 2. treatment with an excess of anhydrous ammonia

26.8 The required starting carboxylic acid, $HOCH_2CH_2CO_2H$, would react at the hydroxyl group as well as the carboxyl and α-carbon sites. We would need to protect the hydroxyl group prior to the performance of the Hell-Volhard-Zelinsky reaction.

26.9 RC(NH$_2$)(CO$_2$H)$_2$

26.10 1. treatment of the diethyl phthalimidomalonate with sodium hydride; 2. addition of 2-bromopropane to the resultant diethyl phthalimidomalonate anion; 3. heat with aqueous sodium hydroxide to hydrolyze the ester and amide linkages; 4. aqueous acid; 5. heat to accomplish decarboxylation and formation of the target material.

26.11

[Structure: phthalimide-N–C(CO$_2$CH$_2$CH$_3$)(CO$_2$CH$_2$CH$_3$)(CH$_2$CH$_2$CO$_2$CH$_2$CH$_3$)]   HO$_2$CCH$_2$CH$_2$CH(NH$_2$)CO$_2$H

26.12

HOCH$_2$C(CO$_2$CH$_2$CH$_3$)$_2$ HOCH$_2$CH(NH$_2$)CO$_2$H
       |
    NHC(O)CH$_3$

26.13 1. treatment of phenylacetaldehyde (PhCH$_2$CHO) with ammonium cyanide; 2. heat with aqueous acid to hydrolyze the nitrile linkage; 3. neutralize with aqueous sodium hydroxide

26.14 Intramolecular (and intermolecular) condensation of the aldehyde and distant amino sites would occur. The amino group would need to be protected before the aldehyde site were generated for cyanohydrin formation.

26.15

[Mechanism scheme showing acetaldehyde + NH$_3$ addition, proton transfers, loss of water, and addition of cyanide to give CH$_3$CH(NH$_2$)CN]

26.16 Hippuric acid - Hippuric acid is a true acid having a free carboxylic acid group, whereas glycine is a zwitterion of reduced acidity.

26.17 CH$_3$C(O)NHCH(CH$_3$)CO$_2$H   Synthesis: 1. treatment of alanine with sodium hydroxide; 2. addition of acetyl chloride; 3. add aqueous acid to generate the free carboxylic acid form of the target material

26.18 With ninhydrin the hydrated carbonyl group is favored since the dipole moment is reduced by hydration; the unfavorable effect of having three aligned dipole moments within the same molecule is reduced by hydration.

26.19 Yes. A primary amine (RNH$_2$) is required for the ninhydrin reaction to go to completion as shown in the mechanism of the ninhydrin test.

26.20 The species eluting with the less acidic buffer are basic amino acids.

26.21

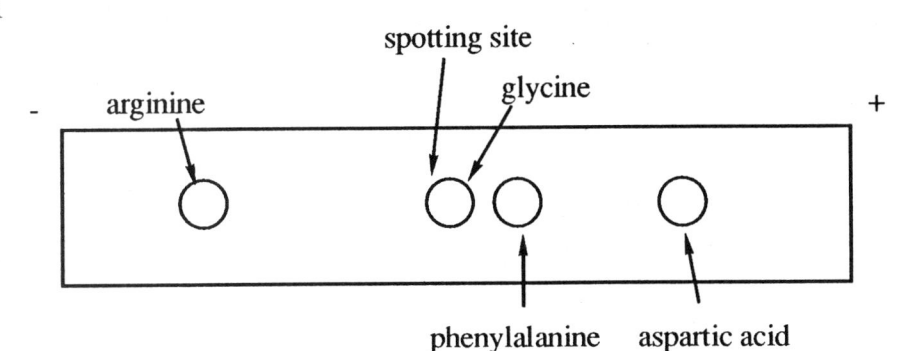

- arginine ... spotting site / glycine ... phenylalanine ... aspartic acid ... +

26.22 Phe-Ala-Gly; Ala-Phe-Gly; Gly-Ala-Phe; Gly-Phe-Ala; Ala-Gly-Phe; Phe-Gly-Ala

26.23

26.24 Phe-Gly-Val-Tyr-Cys-Ala-Leu-Ile

26.25 Ala-Phe-Gly-Glu
Gly-Phe-Ala-Glu

26.26

(a) and (b)

(c) Gly-Tyr-Ala-Leu

26.27 Normal hemoglobin would migrate faster toward the positive electrode. There is an additional glutamic acid unit in the normal hemoglobin which causes this effect relative to sickle-cell anemia hemoglobin.

26.28 The free amino function could react rapidly to perform a displacement at the benzylic chloride site and become attached to the polymer.

26.29 1. reaction of the sodium salt of BOC-Val with the benzylic chloride resin; 2. treatment with trifluoroacetic acid to remove the BOC group; 3. addition of the isobutylchloroformate anhydride derivative of BOC-Ala to form a peptide linkage with the Val attached to the resin; 4. treatment with trifluoroacetic acid to remove the BOC group; 5. addition of the isobutylchloroformate anhydride derivative of BOC-Phe to form a peptide linkage with the units

262 Study Guide and Solutions Manual

already attached to the resin; 6. treatment with HBr and trifluoroacetic acid to liberate the free Phe-Ala-Val.

26.30 (a) basic amino acid
 (b) α-amino acid
 (c) S
 (d) threonine
 (e) aspartic acid
 (f) high pI
 (g) Gly-Ala-Val
 (h) Ala-Lys-Gly
 (i) Phe
 (j) Pro
 (k) glycine in the cationic form

26.31 Gly-Ser-Thr-Lys   Ala-Arg   Ser-Gly

26.32 (a) ~6.8
 (b) toward the anode (+)
 (c) (i) Ser-Phe and Ala-Glu
   (ii) Ser-Phe-Ala and Glu
   (iii)
    Phe-Ala-Glu and [structure: Ph-N, C=O, C with H and CH$_2$OH, NH, C=S ring]

26.33
 (a) CH$_3$C(O)NHCHCO$_2$H with CH$_3$ side chain
 (b) H$_2$NCHC(O)NHCHCO$_2$H with CH$_2$Ph and CH$_2$CH(CH$_3$)$_2$ side chains
 (c) H$_2$NCH$_2$CO$_2$CH$_2$CH$_3$
 (d) PhCH$_2$CHCO$_2$H with $^+$NH$_3$ Cl$^-$
 (e) [polymer structure with N-H, CH$_2$, C=O, repeating n]
 (f) [diketopiperazine structure with NH and HN, two C=O]
 (g) PhCH$_2$O$_2$CNHCHCO$_2$H with CH$_3$ side chain
 (h) H$_2$NCH$_2$CO$_2^-$ K$^+$
 (i) H$_2$NCH$_2$CO$_2$CH$_2$CH$_3$

26.34 (a) 2-aminopropanoic acid
 (b) 2-amino-3-phenylpropanoic acid
 (c) 2-amino-3-(4-hydroxyphenyl)propanoic acid

26.35 1. treatment of methyl nitroacetate with sodium hydride to remove a proton from the α-carbon site, generating the anion; 2. addition of the appropriate alkyl halide (RI) for the side chain; 3. reduction of the nitro group using hydrogen and platinum oxide catalyst; 4. hydrolysis of the ester linkage using aqueous acid

26.36 1. addition of ^{14}CO$_2$ to an ether solution of methylmagnesium iodide; 2. treatment of the resultant labeled acetic acid with bromine and phosphorus tribromide; 3. heat with anhydrous ammonia

26.37 Ser-Leu-Ala-Pro-Phe-Ala-Tyr

26.38 1. substitution of the benzylic chloride site of the Merrifield resin by the addition of the sodium salt of BOC-Gly; 2. treatment with trifluoroacetic acid to remove the BOC group; 3. addition of the isobutyl chloroformate anhydride derivative of BOC-Phe to form a peptide linkage; 4. treatment with trifluoroacetic acid to remove the BOC group; 5. addition of the isobutyl chloroformate anhydride derivative of BOC-Ala to form another peptide linkage; 6. treatment with HBr and trifluoroacetic acid to liberate the Ala-Phe-Gly from the resin

26.39 (a) 1. treatment of α-bromoacetaldehyde with ammonium cyanide to form a cyanohydrin and displace the bromide; 2. heat with aqueous acid to hydrolyze both cyano groups; 3. neutralize with aqueous base

(b) 1. treatment of 2-methylbutanal with ammonium cyanide to form the cyanohydrin; 2. heat with aqueous acid to hydrolyze the cyano group; 3. neutralize with aqueous base

26.40 A - N-bromosuccinimide or bromine under irradiation with light
B - Na$^+$ $^-$:CH(CO$_2$CH$_2$CH$_3$)$_2$
C - 3-(4-methoxyphenyl)propanoic acid
D - phosphorus tribromide and bromine
E - aqueous HI

26.41

F is phthalimide-N-CH(CO$_2$CH$_2$CH$_3$)$_2$

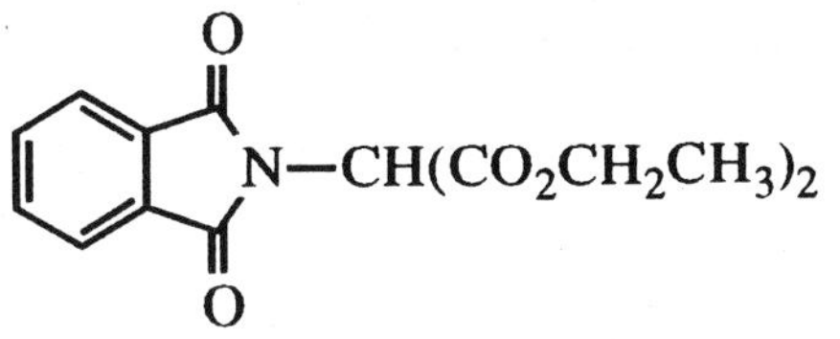

G is phthalimide-N-C(CO$_2$CH$_2$CH$_3$)$_2$-CH$_2$CH$_2$CH$_2$Br

H is phthalimide-N-C(CO$_2$CH$_2$CH$_3$)$_2$-CH$_2$CH$_2$CH$_2$O$_2$CH$_3$

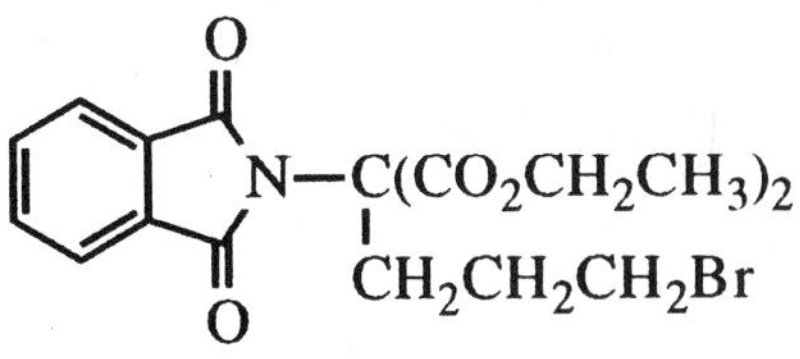

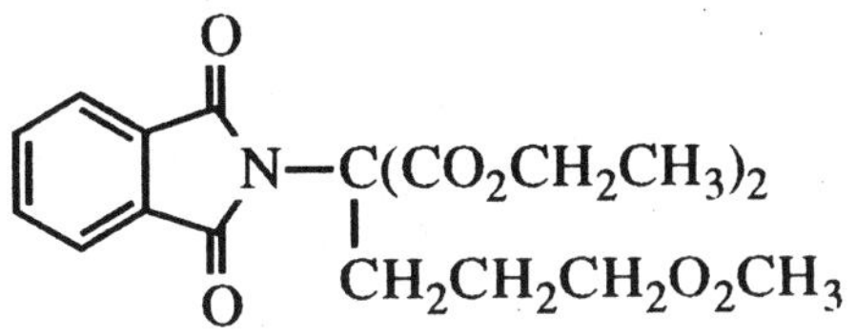

I is HOCH$_2$CH$_2$CH$_2$CHCO$_2$H with NH$_2$

J is ClCH$_2$CH$_2$CH$_2$CHCO$_2$H with NH$_2$

K is proline (pyrrolidine-CO$_2$H with NH)

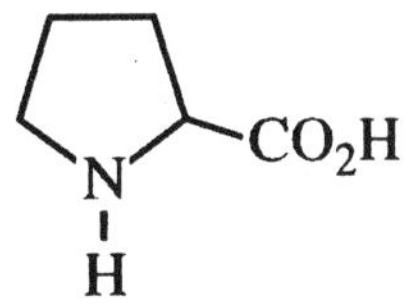

26.42 Val-Gly-Lys-Ala-Ser-Phe-Gly-Lys-Asp-Glu-Tyr-Ala-Arg-Tyr-Gly-Leu

Solution of Study Guide Practice Problems

26.1 Check your answers with Table 26.1 in the text.

26.2

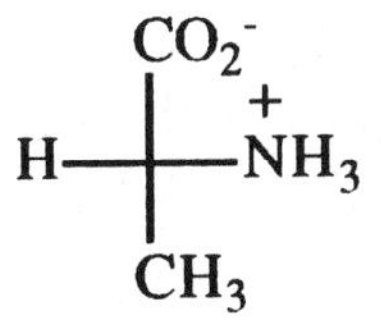

Fischer projection:
CO$_2^-$ at top
H on left, $^+$NH$_3$ on right
CH$_3$ at bottom

The designation of the stereogenic site is R.

26.3 (a) R (Cysteine is unique among the common naturally occurring α-amino acids in that the *designation* of its stereogenic center is not S.)
(b) S

26.4 Ile and Thr

26.5 10^5:1

26.6 Protonation occurs to generate the structure shown below such that no loss of aromatic stabilization is involved.

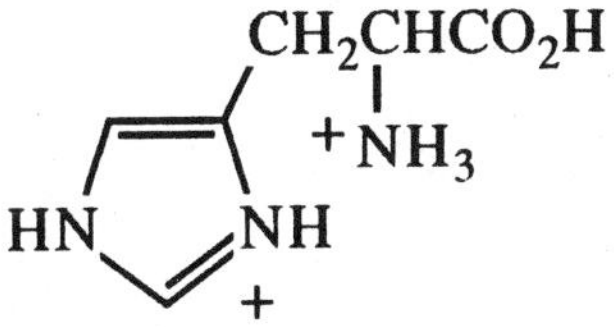

Imidazole ring with CH$_2$CHCO$_2$H side chain, $^+$NH$_3$ on α-carbon; ring nitrogens shown as HN and NH with + charge.

26.7 lysine

26.8 > 7

26.9 1. treatment of 4-methylpentanoic acid with phosphorus tribromide and bromine; 2. treatment with an excess of anhydrous ammonia

26.10

(a) [phthalimide]N–CH(CO$_2$CH$_2$CH$_3$)$_2$

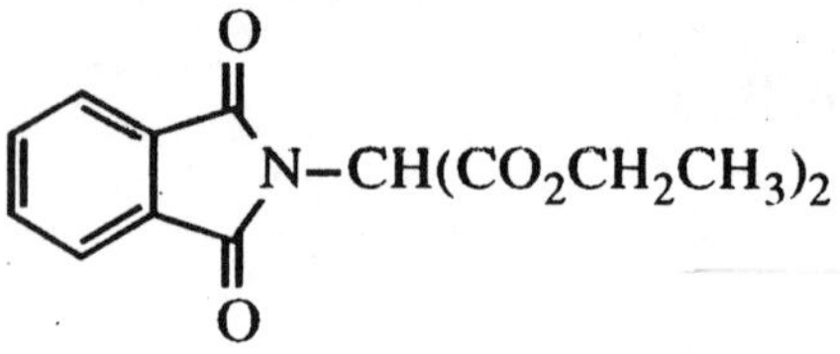

(b) CH$_3$C(O)NHCH(CO$_2$CH$_2$CH$_3$)$_2$

26.11 A is PhCH$_2$CH(NH$_2$)CN;   B is PhCH$_2$CH(NH$_2$)CO$_2$H
This is the Strecker synthesis.

26.12

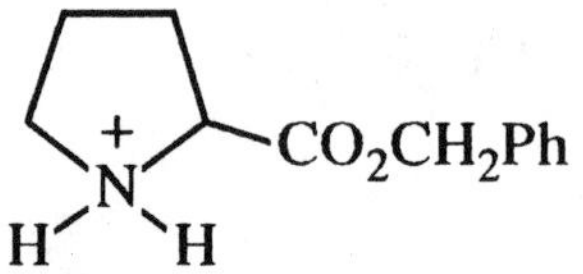

[pyrrolidinium]–CO$_2$CH$_2$Ph   (protonated form)

This material could be prepared by the treatment of proline with benzyl alcohol in the presence of anhydrous HCl.

26.13

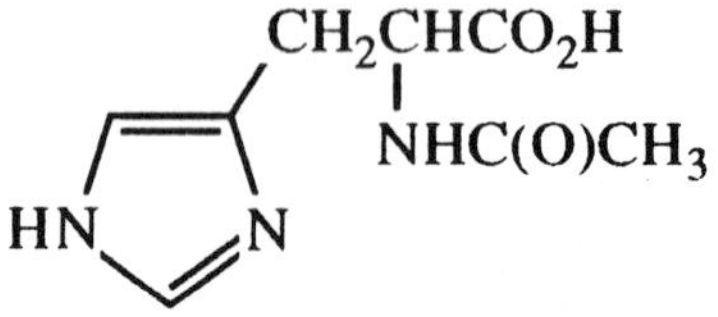

[imidazole ring with]–CH$_2$CHCO$_2$H | NHC(O)CH$_3$

This material can be prepared by the reaction of histidine with acetic anhydride.

26.14 See the discussion in Section 26.5 of the text.

26.15 Phenylalanine, leucine, and glutamic acid would migrate toward the anode. Arginine would migrate toward the cathode.

26.16 Phenylalanine

26.17 Val-Asp-Val-Asp-Glu-Val-Gly-Gly-Glu-Phe-Leu-Gly-Arg

26.18

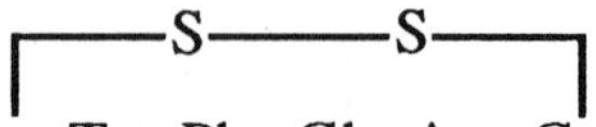

⌐—S———S—⌐
Cys-Tyr-Phe-Glu-Asp-Cys-Pro-Arg-Gly

26.19 1. treatment of BOC-Ala with isobutyl chloroformate to activate the carboxyl group for amide formation; 2. addition of the *tert*-butyl ester of phenylalanine (prepared from phenylalanine upon treatment with isobutene in the presence of anhydrous HCl); 3. treatment with HBr and trifluoroacetic acid to liberate the dipeptide

26.20 (a) secondary
(b) tertiary
(c) quaternary
(d) secondary
(e) tertiary

CHAPTER 27
HETEROCYCLIC COMPOUNDS

Key Points

• Recognize that intramolecular reactions of compounds containing two functional groups that can be induced to react with each other is often a useful approach to the preparation of many heterocycles.

Practice Problem 27.1
Draw structures for the heterocyclic compounds formed in each of the following reactions, and give a brief explanation for the formation of compounds C and D.
  (a) 1,4-dibromobutane + PhNH$_2$ ⟶ A (C$_{10}$H$_{13}$N)
  (b) 2,6-dibromohexanoic acid + Na$_2$S ⟶ B (C$_6$H$_{10}$O$_2$S)
  (c) 4-oxopentanoic acid + NaCN (acidic medium) ⟶ C (C$_6$H$_7$NO$_2$)
  (d) 3-chloropropyl acetate + KOH ⟶ D (C$_3$H$_6$O)

• Know the relative tendencies for rings of different sizes to form.

Practice Problem 27.2
The relative rates shown below refer to the ring closure reactions of compounds of the structural type Br(CH$_2$)$_n$NH$_2$, where n = 2-5, but *are not listed in order*. Associate each relative rate with a value of "n" for the chain length.
            100; 1; 0.07; 0.001

• Recognize that the reactions of many non-aromatic heterocyclic compounds are exactly anelogous to the reactions of the corresponding open-chain compounds.

Practice Problem 27.3
Predict the organic products in each of the following reactions.

(a) 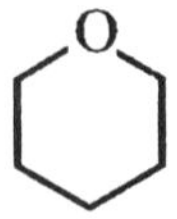 tetrahydropyran + excess hot conc. HI ⟶

(b) 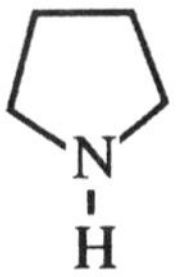 pyrrolidine (N-H) + CH$_3$C(O)Cl ⟶

(c) 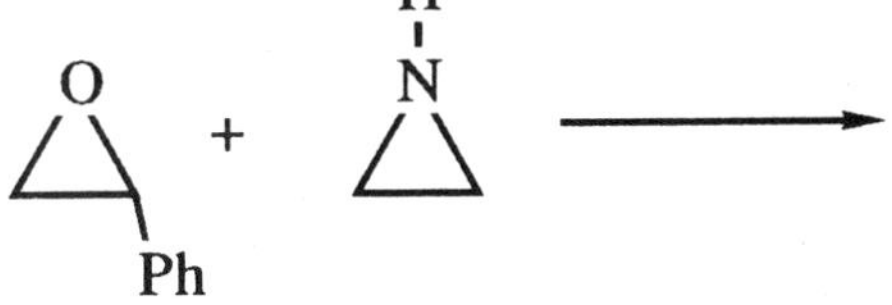 2-phenyloxirane + aziridine (N-H) ⟶

• Learn the relative reactivities of various aromatic heterocycles toward electrophilic aromatic substitution reactions, and the regioselectivity of these reactions. Also, review the reagents used, appropriate mechanisms, and products obtained in electrophilic aromatic substitution reactions (Chapter 18).

Practice Problem 27.4
Arrange the following list of compounds in increasing order of facility of ring bromination:
    benzene; pyrrole; pyridine; pyrimidine

Practice Problem 27.5
Predict the products in each of the following reactions.
    (a) treatment of pyridine with nitric acid and sulfuric acid to generate a product of formula

$C_5H_4N_2O_2$

(b) 2-furoic acid undergoing ring chlorination to form a product of formula $C_5H_3O_3Cl$

- Be able to predict and/or explain other types of reactions for substituted heteroaromatic compounds.

Practice Problem 27.6
How many halogen atoms do you expect to be displaced by the treatment of 3,4,5-tribromopyridine with sodium ethoxide in ethanol?

Practice Problem 27.7
Consider the following aspects of nitrogen heteroaromatic compounds.
    (a) Which compound, pyridine or pyrrole, reacts readily with sodium amide in liquid ammonia to form an amino-substituted compound?
    (b) On which ring nitrogen would you expect the compound shown below to undergo methylation using iodomethane? Give an explanation of your choice.

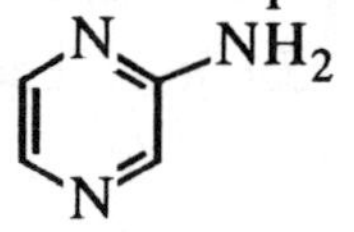

Practice Problem 27.8
One of the diazonium ions shown below is extremely reactive and reacts with water in the solution used for its preparation. Tell which is the particularly reactive species and give an explanation for its reactivity.

Practice Problem 27.9
Treatment of either 3-chloropyridine or 4-chloropyridine with sodium amide in liquid ammonia produces a mixture of 44% 4-aminopyridine and 22% 3-aminopyridine. Offer an explanation for this result.

- Learn that furan derivatives undergo useful ring opening reactions with dilute aqueous sulfuric acid.

Practice Problem 27.10
Give the structure of the product E (of formula $C_{16}H_{14}O_2$, and which exhibits an IR absorption at ~1700 cm^{-1}) formed in the reaction of 2,5-diphenylfuran with aqueous sulfuric acid, and give a mechanism for the reaction.

Practice Problem 27.11
Predict the structures of compounds F and G in the following reaction scheme.

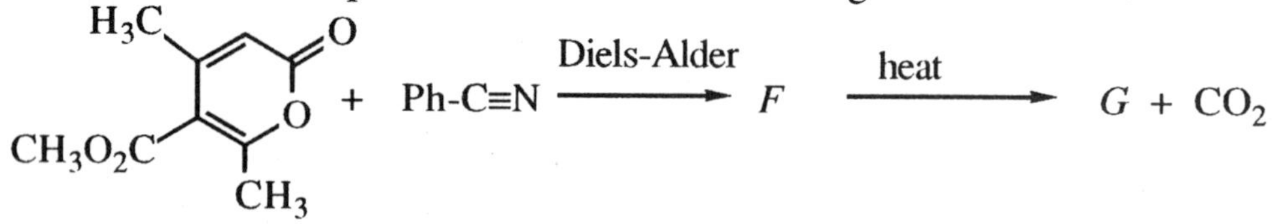

- Appreciate the use of pyridine N-oxides in syntheses.

Practice Problem 27.12
Give reagents and conditions for accomplishing the conversion shown below.

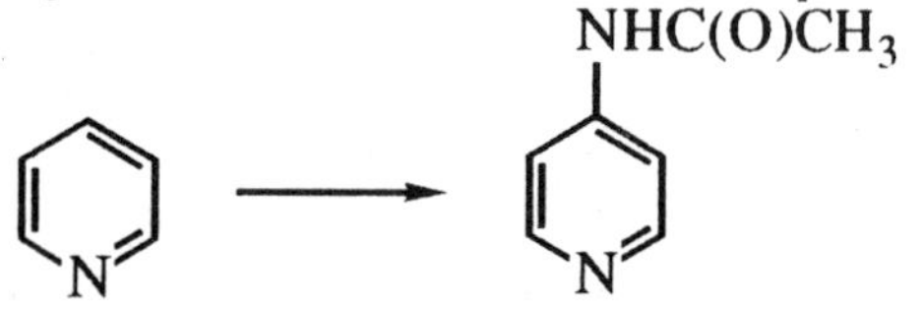

- Know the chief synthetic routes for the preparation of quinolines, isoquinolines, and indoles.

Practice Problem 27.13
A *double*-Skraup synthesis can be used to prepare 4,7-phenanthroline (structure shown below). Which aromatic compound of formula $C_6H_8N_2$ would you use as the starting material in this synthesis, and what other reagents would be required?

Practice Problem 27.14
Give routes for the preparation of each of the following compounds.

(a) by the Bischler-Napieralski approach

(b) by the Fischer approach

Solution of Text Problems
27.1
(a) (b) (c)

27.2

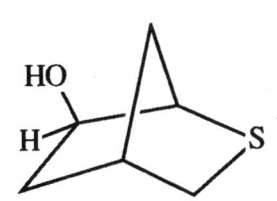

27.3 (a). treatment of furfural with sodium borohydride in 2-propanol, followed by aqueous workup
(b) 1. treatment of furfural with sodium borohydride in 2-propanol; 2. addition of *p*-toluenesulfonyl chloride in pyridine to form a tosylate ester; 3. treatment with lithium aluminum hydride in ether, with aqueous workup
(c) treatment of furfural with silver nitrate in aqueous ammonia followed by workup with aqueous acid
(d) 1. treatment of furfural with silver nitrate in aqueous ammonia followed by aqueous acid workup; 2. addition of thionyl chloride to generate the carboxylic acid chloride; 3. addition of an excess of ammonia

27.4 (a) heat tetrahydrofuran with sodium bromide in sulfuric acid
(b) 1. heat tetrahydrofuran with sodium bromide in sulfuric acid; 2. treatment of the dibromide with sodium cyanide in DMSO
(c) 1. formation of the 1,4-dicyanobutane as shown in part (b); 2. heat with aqueous acid

27.5 With 2-aminopyridine, the 5-position is activated for electrophilic aromatic substitution, as illustrated with the resonance stabilized intermediate shown below:

We would not expect 2-picolinic acid to be so activated as there is no valence level unshared electron pair to be donated onto the ring.

27.6 Aniline - The nitrogen directly in the ring of pyrrole stabilizes the cationic intermediates, as does aniline;

27.7 (a)

most reactive site for electrophilic aromatic substitution ⟶ [pyrimidine structure]

(b) We would predict pyrimidine to be less reactive than pyridine in electrophilic aromatic substitution reactions. There are *two* sp^2 nitrogen atoms in the ring in pyrimidine providing unfavorable interactions for electrophilic aromatic substitution.

27.8 3-chloro-4-ethoxypyridine

27.9 (a) The material would be dissolved in water and the pH would be raised to ~11. The resultant solution would be extracted with ether. The ether layer would be separated, dried over a solid drying agent such as anhydrous calcium chloride, filtered, and evaporated to give the 4-chloropyridine.

(b) The nitrogen of one 4-chloropyridine molecule would attack the 4-position of another 4-chloropyridine molecule, by addition-elimination displacing the chloride there, and leaving the attacking nitrogen as a pyridinium site. The remaining trivalent nitrogen in the molecule, the one from the molecule of 4-chloropyridine which was *attacked*, would remain capable of performing the same type of reaction on yet another molecule of 4-chloropyridine, ultimately leading to polymeric material.

27.10 The electron withdrawing effect of the additional nitrogen in pyrimidine decreases the basicity of each site in pyrimidine relative to pyridine, which has only the single nitrogen. In a consideration of pyrrole *vs*. imidazole, the sp^2 hybridized nitrogen of imidazole with the unshared valence level electron pair *in* an sp^2 orbital is the basic site. It is much more basic than the "N-H" nitrogen in either pyrrole or in imidazole, each of which has its unshared valence level electron pair in a *p* orbital which is part of the aromatic system of the ring.

27.11 The basicity of a compound says nothing directly about its acidity, only about the acidity of its *conjugate acid*. Imidazole should be more acidic than pyrrole as well as more basic. There is greater stabilization of the conjugate base of imidazole owing to charge delocalization onto the more electronegative nitrogen atom, while in the conjugate acid of pyrrole the negative charge can be delocalized only to carbon.

27.12 There is a greater degree of charge delocalization (more and better resonance structures may be written) available for α-protonation of pyrrole than for β-protonation.

27.13

27.14

Any other resonance structures would have positive charges on adjacent atoms.

27.15 The 2-dimethylaminopyridine *N*-oxide is nitrated chiefly at the 3-position. The intrmediate for attack at the 3-position is favored by resonance stabilization of the intermediate. Such stabilization is not posssible in the 2-acetamidopyridine *N*-oxide owing to electron delocalization to the oxygen.

27.16

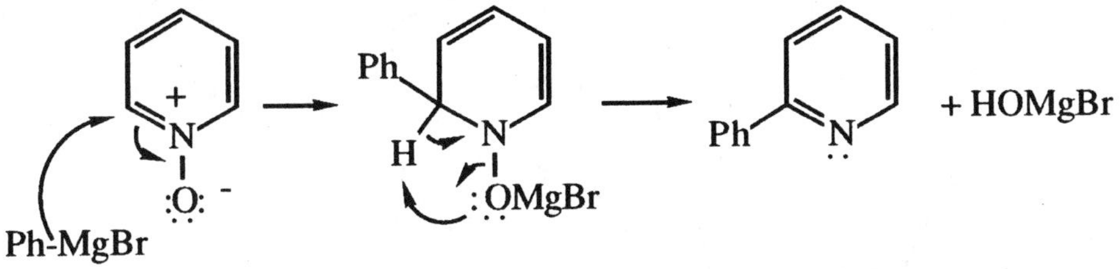

[Mechanism: pyridine N-oxide + Ph-MgBr → addition intermediate → 2-phenylpyridine + HOMgBr]

27.17

[Mechanism: 2-methylpyridine + CH₃CO₂K ⇌ 2-pyridylmethyl anion + CH₃CO₂H; anion adds to PhCHO → alkoxide → (with CH₃CO₂H) alcohol; E2-type elimination with CH₃CO₂⁻ → 2-(2-phenylethenyl)pyridine + CH₃CO₂⁻ + H₂O]

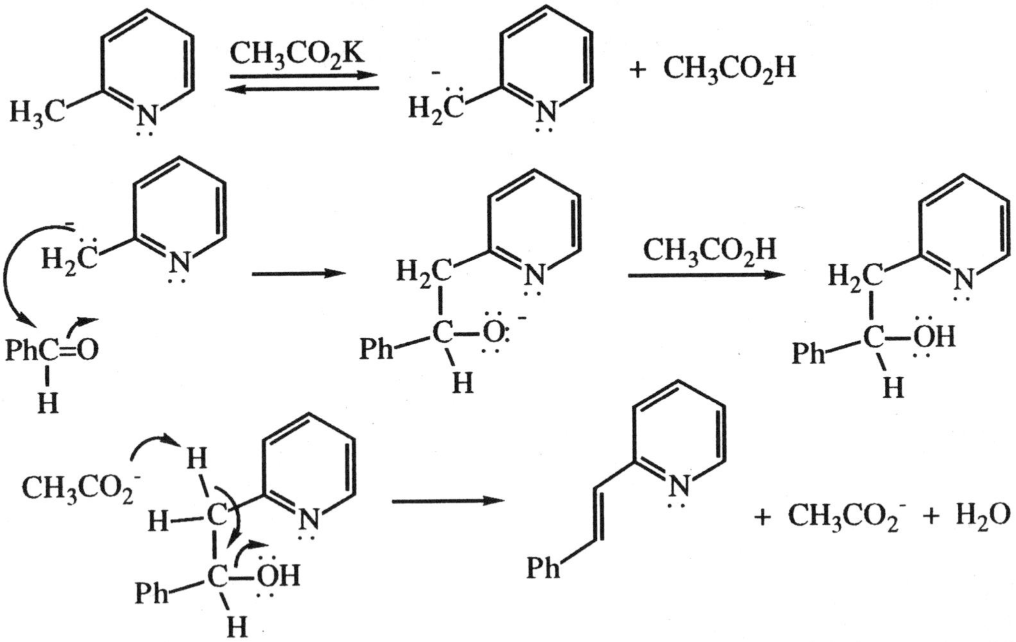

27.18 1. treatment of toluene with bromine under light irradiation; 2. oxidation to the aldehyde by treatment with DMSO; 3. treatment with 4-methylpyridine and potassium acetate; 4. reduction with hydrogen over platinum oxide to generate the target material.

27.19 (a) 3,5-dimethylaniline
(b) 4-methoxy-2-nitroaniline

27.20 For Eqn. 27.12: treatment of acetophenone with phenylhydrazine hydrochloride
For Eqn. 27.13: treatment of 2-pentanone with phenylhydrazine hydrochloride

27.21 Indole is more reactive than is quinoline.

27.22 The "benzene" ring portion of indole provides greater resonance stabilization for the intermediate derived from attack at the 3-position than for that derived from attack at the 2-position. No associated "benzene" ring portion is present in pyrrole to provide such stabilization.

27.23

(a)–(k) [structures]

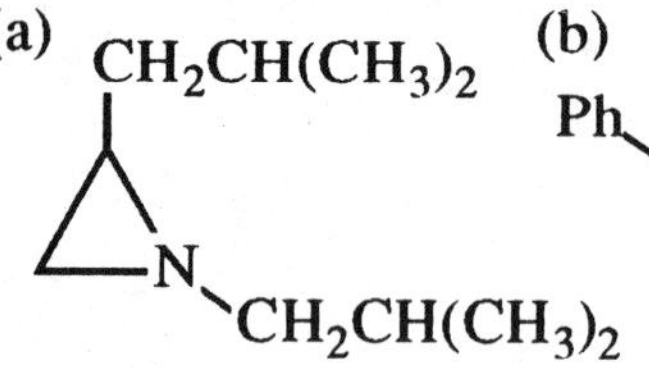

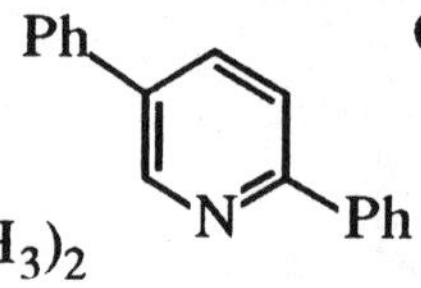

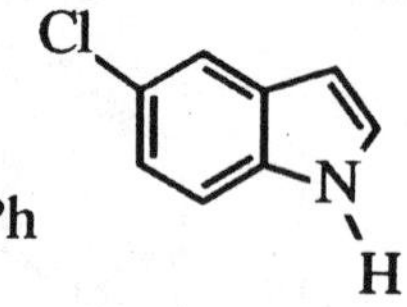

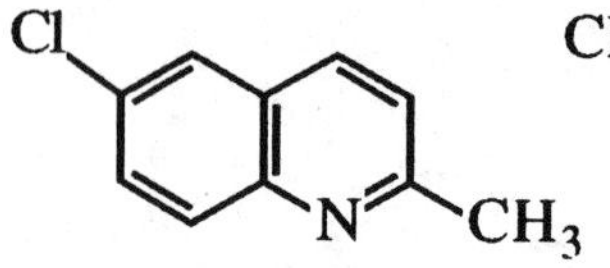

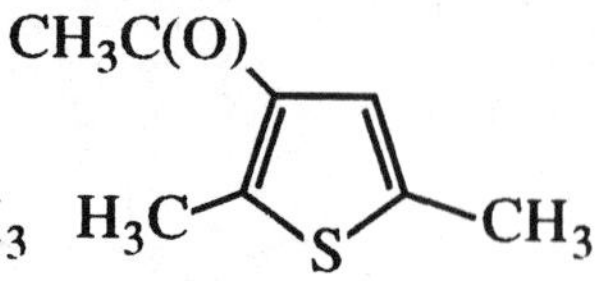

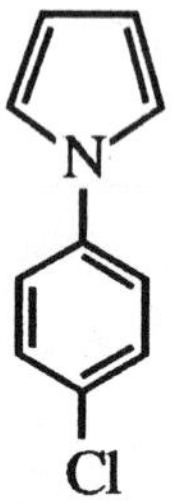

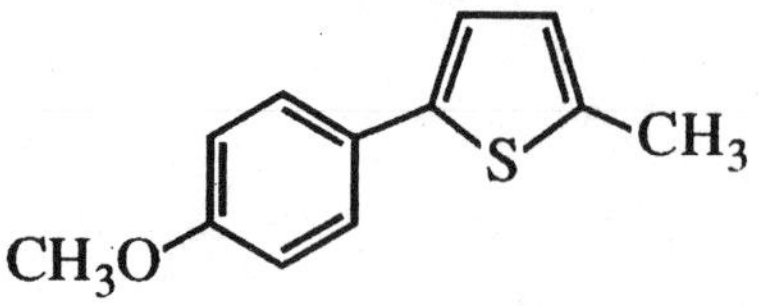

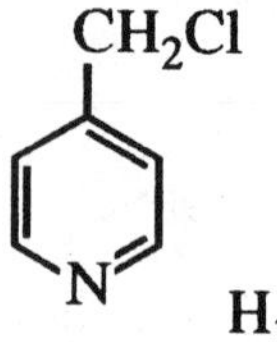

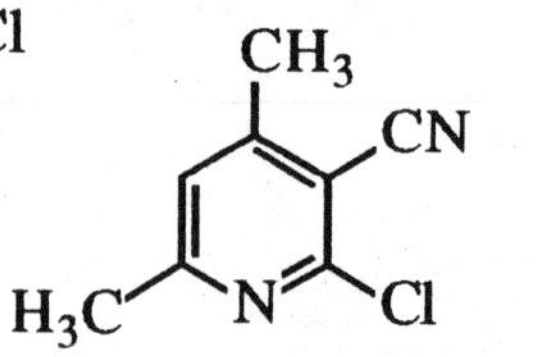

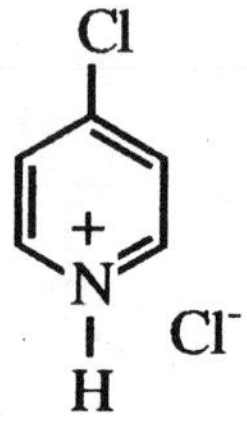

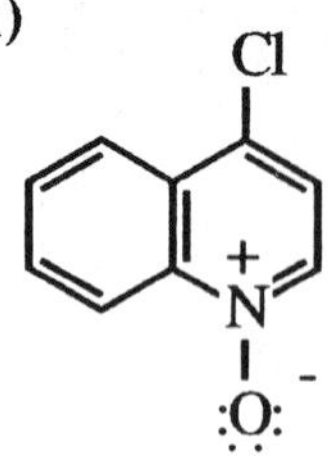

27.24
(a) 2-aminopyridine
(b) 2-*sec*-butylthiophene
(c) 2-chloro-3,5-dinitropyridine
(d) 1-methylpyrrole
(e) 8-methoxyquinoline
(f) 3-isobutylindole
(g) 2,2-dimethyloxirane
(h) 1-ethylpyrrolidine
(i) 3-bromo-4-chlorofuran

27.25 (a) pyrrole - Pyrrole is more reactive in electrophilic aromatic substitution.
(b) pyridine - The pyrrole nitrogen is not a basic site; aromatic character is lost from pyrrole upon protonation.
(c) imidazole - There is a greater stabilization of the anion with imidazole.
(d) pyridine - The nitro group is electron withdrawing.
(e) quinoline - There is greater stabilization in indole for reaction on the "pyrrole" ring position.
(f) furan - 2-Furoic acid has an electron withdrawing group on the ring.
(g) thiophene - Thiophene has a lower degree of aromatic stabilization.
(h) 2-chloropyridine - With 2-chloropyridine there is a greater stabilization of the intermediate.
(i) 4-chloropyridine - With 4-chloropyridine there is a greater stabilization of the intermediate.
(j) 2-picoline - With 2-picoline there is a greater stabilization of the intermediate anion.
(k) pyridine - With pyridine there is a greater stabilization of the intermediate.
(l) The ion with the oxygen in the ring is the stronger acid as it has a weaker conjugate base.
(m) pyrrole - The nitrogen of pyrrole is not a nucleophilic site.

27.26

(a) 2-bromothiophene
(b) thiophene-2-carboxylic acid
(c) 3-nitropyridine
(d) pyridine N-oxide
(e) 3,5-dinitropyridine
(f) pyridine
(g) 1-methylpyridinium iodide
(h) isonicotinic acid (pyridine-4-carboxylic acid)
(i) 2-methyltetrahydrofuran
(j) 2-phenylpyridine
(k) 1-acetylpyrrolidine
(l) 1-(furan-2-yl)-3-buten-2-one [(E)-4-(furan-2-yl)but-3-en-2-one]
(m) 1-benzylpyridinium chloride
(n) $CH_3CH_2C(O)CH_2CH_2C(O)CH_2CH_3$
(o) 1,7-naphthyridine

27.27
2,4,5-trimethylpyridine

27.28
We would not expect such an equilibrium for 3-hydroxypyridine as aromatic stabilization would be lost.

27.29
(a) 1. treatment of 2-phenylethylamine with phenylacetyl chloride to form an amide; 2. ring closure using phosphorus pentoxide; 3. aromatization (dehydrogenation) by passing nitrogen over the compound in the presence of palladium on charcoal.

(b) 1. treatment of 4-methylaniline with glycerol in the presence of sulfuric acid; 2. aromatization by dehydrogenation using nitrobenzene; 3. side-chain oxidation using potassium permanganate in aqueous base.

(c) treatment of 4-amino-2-methoxyaniline with two equivalent amounts of glycerol in the presence of phosphorous acid

27.30
(a) 1. treatment of 2-methoxyaniline with glycerol in the presence of phosphorous acid; 2. reaction with aqueous HI

(b) treatment of 1,2-diaminobenzene with two equivalent amounts of glycerol in the presence of phosphorous acid

(c) 1. treatment of 2-aminopyridine with sodium nitrite in aqueous sulfuric acid at 0°; 2. adjustment of pH to ~9 with the addition of 3-hydroxyphenol

27.31
(a) 1. treatment with bromine under light irradiation; 2. aqueous potassium hydroxide

(b) 1. treatment with sulfuric acid and nitric acid; 2. tin, hydrochloric acid; 3. acetic anhydride

(c) 1. treatment with aqueous basic potassium permanganate; 2. sulfuric acid, nitric acid, heat

(d) 1. treatment with Tollens' reagent; 2. thionyl chloride; 3. ethanol in the presence of pyridine

(e) treatment with acetone in the presence of potassium hydroxide

(f) 1. treatment with thionyl chloride; 2. ammonia; 3. bromine, aqueous potassium hydroxide; 4. two sequential treatments with formaldehyde and formic acid

27.32

A is $Li^+\ ^-:C\equiv C-CO_2^-\ Li^+$

B is $CH_3CH-C\equiv C-CO_2H$ with $CH(OH)CH_3$ substituent

C is (Z/E)-alkene: $CH_3CH(OH)CH(CH_3)-CH=CH-CO_2H$

272 Study Guide and Solutions Manual

27.33

(a) [reaction scheme: epoxide + H⁺ → protonated epoxide → carbocation with :OH → ketone, −H⁺]

(b) H₂C=Ö + HCl: ⟶ H₂C⁺-ÖH + :Cl:⁻ ⟶ [thiophene intermediate with CH₂OH] ⟶ (−H⁺) [2-(hydroxymethyl)thiophene]

(c) The base in the solution generates the carboxylate anion of the starting molecule, which performs an intramolecular displacement of the chloride ion to form the lactone.

(d) Each of the amino groups of the ethylenediamine (H₂NCH₂CH₂NH₂) performs an imine formation reaction with a carbonyl groups of the starting molecule.

(e) On of the two methoxy groups is protonated, followed by loss of a molecule of methanol. The resultant carbocation undergoes cleavage of a C-O ring bond to generate one carbonyl group with a carbocation site remaining at the carbon from which the oxygen departed. Water adds to the carbocation site, and after loss of a proton, yields a hemiacetal at that site. The hemiacetal then loses methanol to generate the dicarbonyl intermediate. This species then undergoes two imine forming reactions with the ethylenediamine to generate the final product.

(f) [Diels-Alder addition scheme showing oxazole + CH₃O₂C-C≡C-CO₂CH₃ → bicyclic adduct → CH₃CN + substituted furan]

(g) The aqueous base converts the amine hydrochloride to the free amine. The nitrogen of the free amine performs a nucleophilic substitution reaction to displace chloride ion and form a three-membered ring with a quaternary nitrogen atom as one member of the ring. Chloride ion then attacks the more highly substituted carbon of the three membered ring to displace the free tertiary amine site of the final product.

(h) The double bond of the enamine attacks a carbon of the three membered ring breaking a C-N bond and leaving a negative charge on nitrogen. The electron rich nitrogen attacks the electron-deficient carbon of the C=N linkage (formed with the enamine nitrogen in the first step) to generate a new ring with a C-N bond and free the original enamine nitrogen as a tertiary amine site.

Solution of Study Guide Practice Problems

27.1

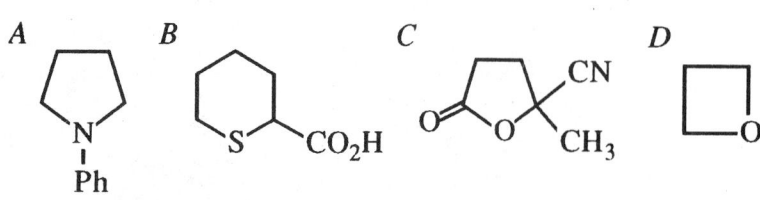

A: N-phenylpyrrolidine
B: tetrahydrothiopyran-2-carboxylic acid
C: γ-butyrolactone with CN and CH₃ substituents
D: oxetane

27.2
n	2	3	4	5
Rel Rate	0.07	0.001	100	1

The general order of reactivity in reactions of this kind is governed by the size of the ring being formed, as discussed in Section 27.2.

27.3
   (a) I-CH$_2$CH$_2$CH$_2$CH$_2$CH$_2$-I

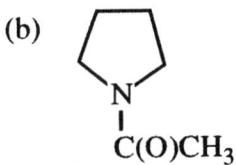

   (b) pyrrolidine with N-C(O)CH$_3$

   (c) PhCH(OH)CH$_2$-N (aziridine)

27.4 pyrrole > benzene > pyridine > pyrimidine

27.5 (a) 3-nitropyridine
     (b) 4-chlorofuroic acid

27.6 Only one will be displaced, the bromine in the 4-position.

27.7 (a) pyridine
     (b) Methylation will occur preferentially at the ring nitrogen most distant from the amino substituent on the ring. There is better charge delocalization with this product, and the activating complex leading to it.

27.8 The 2-substituted diazonium ion is very reactive. Remember, nucleophiles readily displace suitable leaving groups from the 2- and 4-positions of pyridine rings. The -N$_2^+$ group is one of the best leaving groups of all.

27.9 Both reactions proceed through a common benzyne intermediate, as shown below.

27.10 PhC(O)-(CH$_2$)$_3$C(O)Ph   The mechanism for the formation of this material is entirely analogous to that shown in the text in Section 27.5 for the acid-catalyzed ring-opening of 2,5-dimethylfuran.

27.11 F is (bicyclic adduct with N, Ph, CH$_3$, CO$_2$CH$_3$, CH$_3$); G is pyridine with CH$_3$O$_2$C, CH$_3$, H$_3$C, Ph substituents.

27.12 1. Treatment of pyridine with hydrogen peroxide and acetic acid to form the N-oxide; 2. nitric acid, sulfuric acid; 3. phosphorus trichloride; 4. tin, hydrochloric acid; 5. acetic anhydride

27.13 1,4-diaminobenzene

CHAPTER 28
NUCLEOSIDES, NUCLEOTIDES, AND NUCLEIC ACIDS

Key Points

• Know the structural components of nucleosides and nucleotides, and learn the relevant nomenclature.

Practice Problem 28.1
Examine structures *A* and *B* shown below and answer the questions posed.

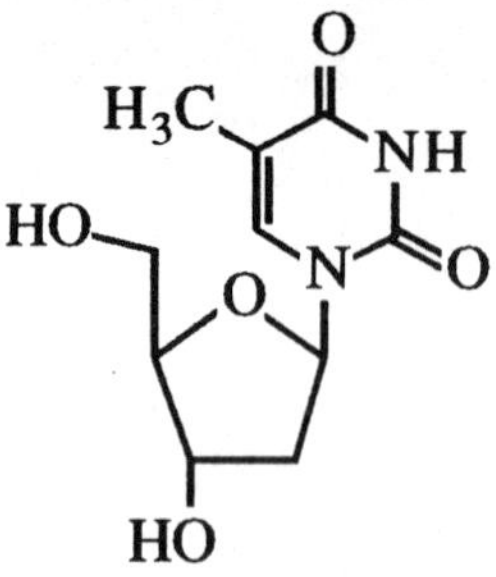

(a) Which is a nucleotide and which is a nucleoside?
(b) What is the nitrogenous heterocyclic base in *A*?
(c) What is the nitrogenous heterocyclic base in *B*?

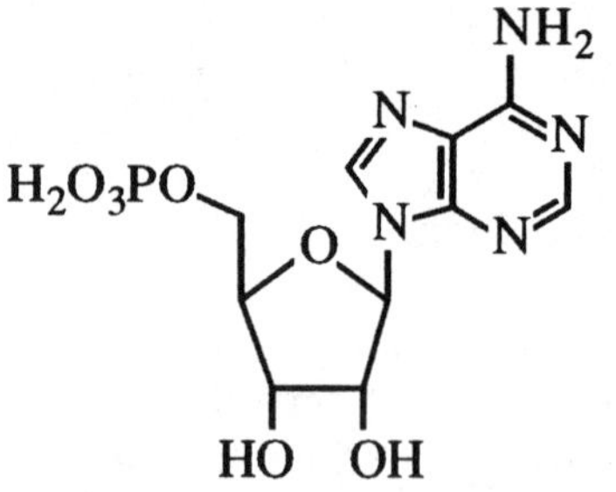

Practice Problem 28.2
From memory, draw the structures for each of the following compounds:

(a) 2-deoxy-β-D-ribose-5-phosphate
(b) thymine
(c) deoxythymidine-5'-phosphate

Practice Problem 28.3
Give the systematic name for the dinucleoside phosphate shown below.

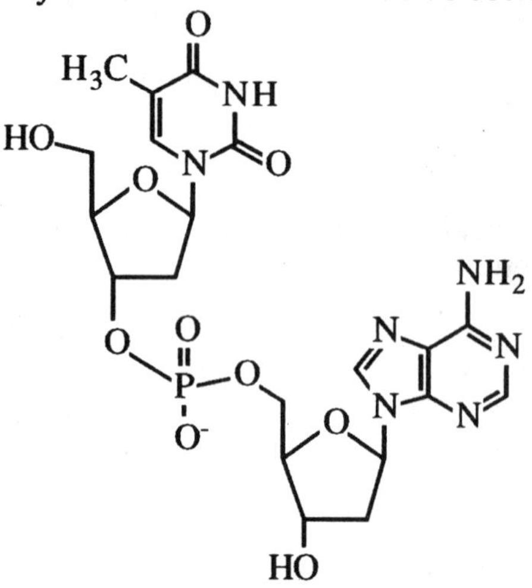

[Commonly abbreviated as:
T   A
    —OH
  P
HO—

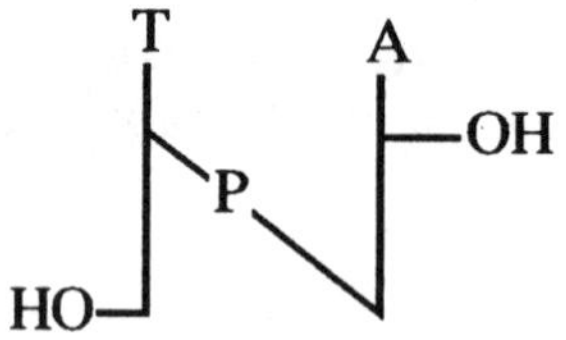

where T = thymine and A = adenine]

Practice Problem 28.4
Draw the full structure for the oligonucleotide represented by the abbreviated notation shown below. (See the previous problem for an explanation of this type of abbreviation.)

Practice Problem 28.5
The principal drug currently used in the treatment of AIDS is a structural analogue of a nucleoside. The compound, commonly known as AZT, is 3'-azidodeoxythymidine, in which the 3'-OH group of the sugar unit has been replaced by an azide group (-N₃). Draw the structure of AZT.

• Understand the importance of hydrogen bonding in nucleic acids.
Practice Problem 28.6
For which of the following pairs of bases, when present in complementary positions of the double-stranded α-helical form of DNA, will there be more than one hydrogen bond?
   (a) thymine and guanine
   (b) thymine and adenine

• Understand the process of transcription, the role of *t*RNA and *m*RNA in polypeptide synthesis, and know what is meant by the term "codon".
Practice Problem 28.7
Comparing *m*RNA and *t*RNA:
   (a) In the first step of protein synthesis, which species interacts with the DNA strand to receive information regarding the specific amino acids to be introduced?
   (b) Which has the higher molecular weight?
   (c) Which is present in the cytoplasm?
   (d) Which transports individual amino acids to the ribosome so that they can be assembled into proteins?
Practice Problem 28.8
What amino acid sequence is indicated for synthesis by the following codons of *m*RNA?
   (a) -GAU-AUC-UAC-GGU-
   (b) -UUU-GGU-UUA-GAA-

Solution of Text Problems

28.1   These are β-anomeric forms.
28.2

28.3

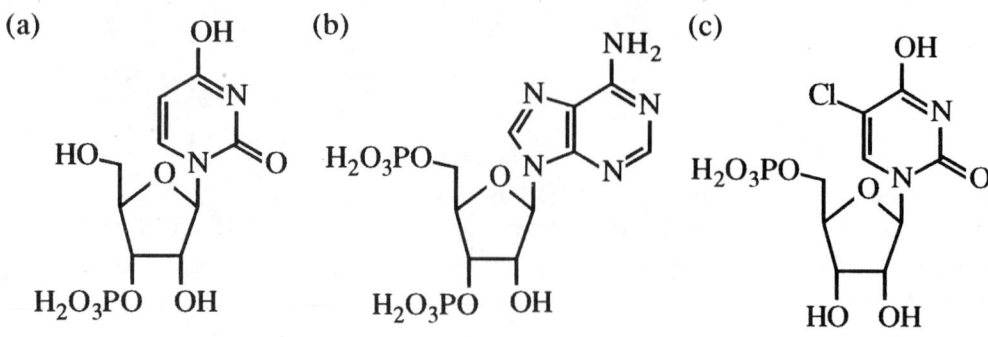

(a) 2'-deoxycytidine-3'-phosphate (structure shown with cytosine base, deoxyribose, and 3'-phosphate)

(b) adenosine-5'-phosphate (AMP) with 3'-OH

(c) 5-chloro-2'-deoxyuridine-5'-phosphate analog (structure shown with 5-chlorouracil base and ribose with 5'-phosphate)

28.4

(a) Trinucleotide: A–C–U (adenosine linked 3'→5' to cytidine linked 3'→5' to uridine, via phosphodiester bonds)

(b) Trinucleotide: C–G–A (cytidine linked 3'→5' to guanosine linked 3'→5' to adenosine, via phosphodiester bonds)

(c) [structure]

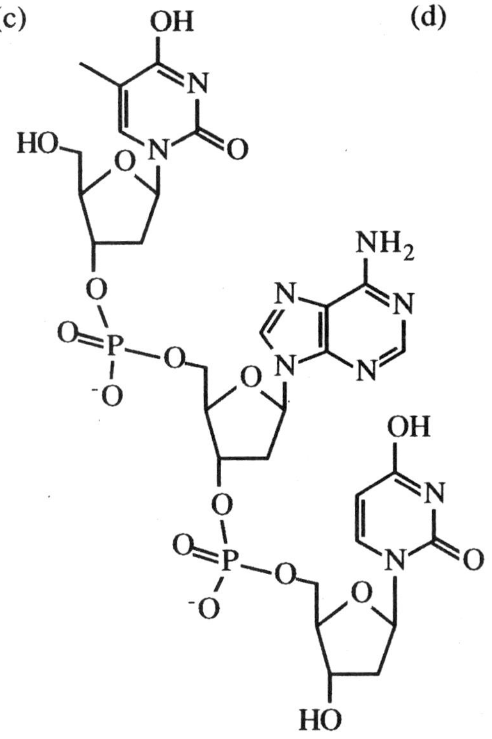

(d) [structure]

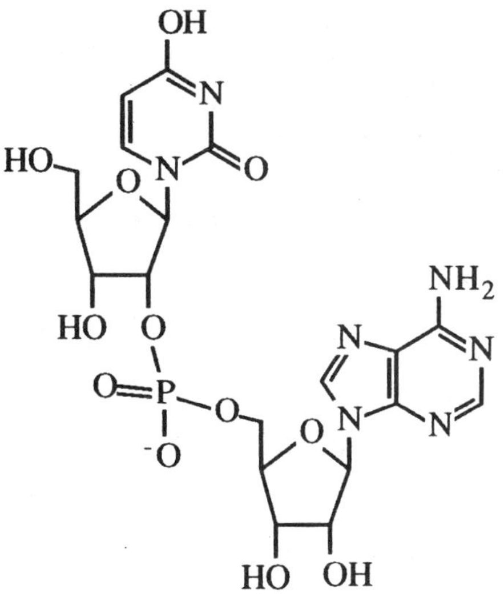

28.5 (a) guanidyl(3'→5')uridyl(3'→5')adenosine

(b) deoxyadenylyl(3'→5')deoxyguanidyl(3'→5')deoxythymidine

28.6 (a) CGTTACG
(b) GCAATATC

28.7 (a) GCUCAAAUGUUU
    (GCCCAG   UUC)
    (GCA   )
    (GCG   )

(b) UGUGGUCAUCCU
    (UGCGGCCACCCC)
    (GGA   CCA)
    (GGG   CCG)

(c) AAAAUUGAU
    (AAGAUCGAC)
    (AUA   )

28.8 (a) 16
(b) 64
(c) 12

28.9 (a) Gln-Thr-Asn
(b) Asp-Leu-Trp-Leu
(c) Cys-Ile-Val-Tyr
(d) Met-Lys-Arg-Ser-Cys-Cys

Solution of Study Guide Practice Problems

28.1 (a) Compound A is a nucleoside, and B is a nucleotide.
(b) thymine
(c) adenine

278 Study Guide and Solutions Manual

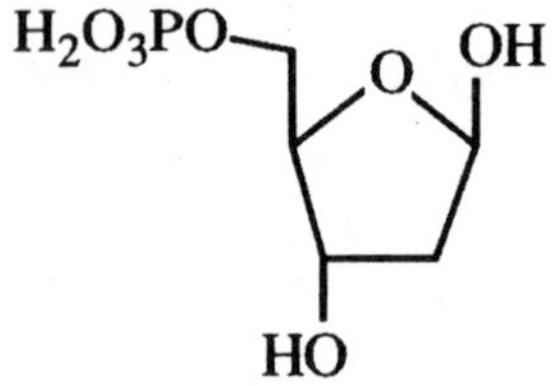

28.2 (a), (b), 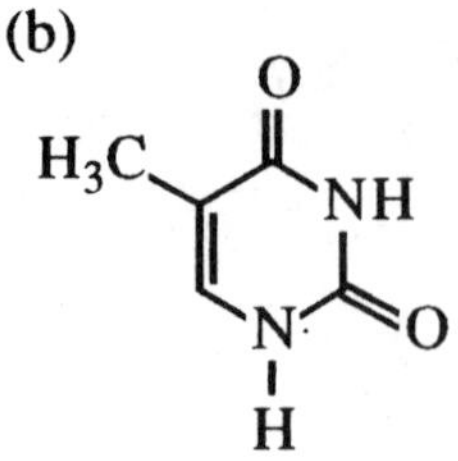 (c) [structures shown]

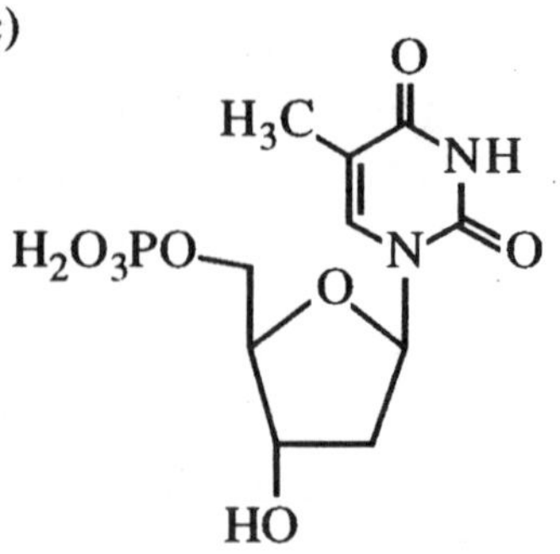

28.3 deoxycytidinyl(3'→5')deoxuguanosine

28.4 [structure shown]

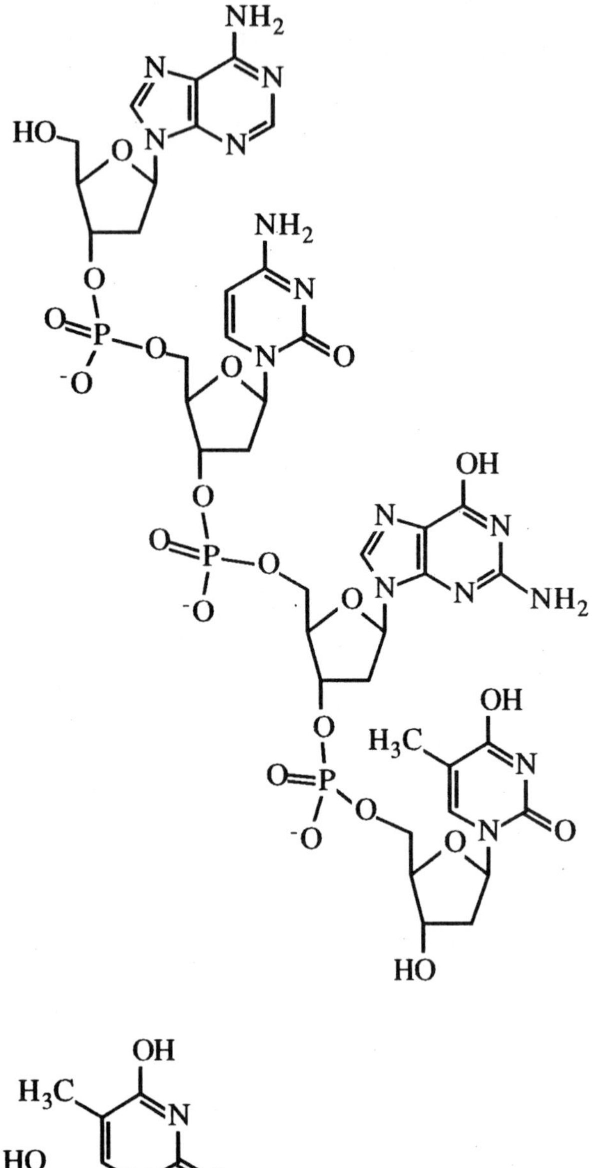

28.5 [structure shown]

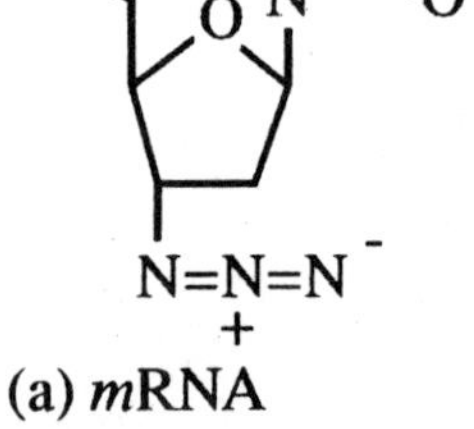

28.6 (a) *m*RNA

(b) *m*RNA
(c) *t*RNA
(d) *t*RNA
28.8 (a) Asp-Ile-Tyr-Gly
(b) Phe-Gly-Leu-Glu

CHAPTER 29
SYNTHETIC POLYMERS

Key Points

• Know how addition polymers are prepared. Understand the polymerization mechanism, and be able to identify the monomer needed to prepare a given addition polymer. Be able to predict the structure of an addition polymer, given the structure of the starting material.

Practice Problem 29.1
Acrylonitrile ($H_2C=CH-CN$) can be polymerized in aqueous solution. If the polymer is then dissolved in DMF it can be spun into orlon. Suggest a structure for orlon.

Practice Problem 29.2
Lucite is an addition polymer of the structure shown below. What monomer would be used to synthesize lucite, and what conditions of polymerization would most likely be used.

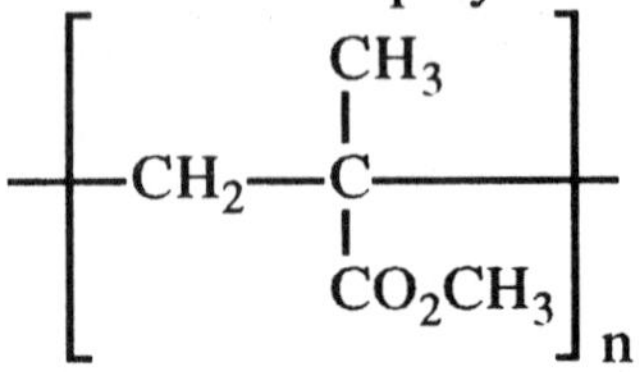

$$\left[-CH_2-\underset{\underset{CO_2CH_3}{|}}{\overset{\overset{CH_3}{|}}{C}}- \right]_n$$

Practice Problem 29.3
Give the immediate product of the reaction shown below, which is the first chain-propagating step in the growth of a polystyrene chain, the initial radical being generated from styrene using di-*tert*-butylperoxide as initiator.

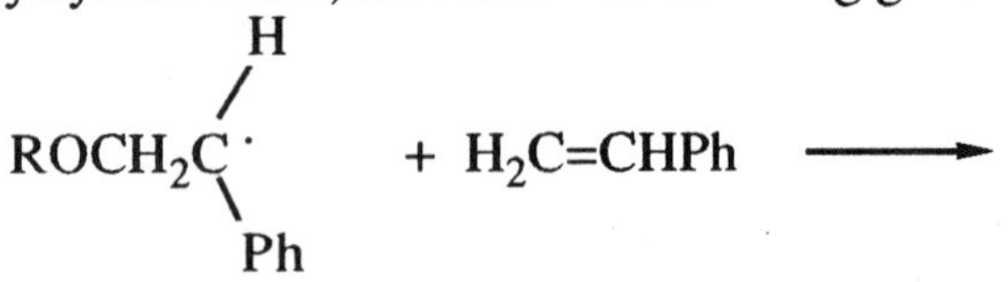

$$ROCH_2\overset{H}{\underset{Ph}{C\cdot}} \quad + \quad H_2C=CHPh \longrightarrow$$

Practice Problem 29.4
Explain the difference between atactic and isotactic addition polymers. What type of catalyst should be used to produce an isotactic polymer? Which type of polymer is more "crystalline"?

Practice Problem 29.5
Which of the polymers listed below could *not* be prepared in atactic, isotactic, and syndiotactic forms? Explain your answer.
    polypropylene; polyisobutylene; polystyrene

• Understand how condensation polymers are prepared. Be able to identify the precursors to a condensation polymer and to predict the structures of condensation polymers given the monomer precursors.

Practice Problem 29.6
Give the repeating unit of the condensation polymer formed by the reaction of the two components shown below.

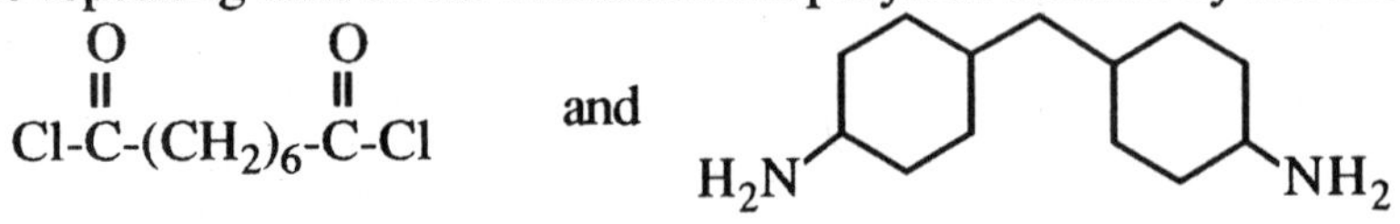

$$Cl-\overset{\overset{O}{\|}}{C}-(CH_2)_6-\overset{\overset{O}{\|}}{C}-Cl \quad \text{and} \quad H_2N-\text{[cyclohexyl-cyclohexyl]}-NH_2$$

Practice Problem 29.7
Formaldehyde and primary amines, RNH_2, react to form water and compounds of the general structure shown below. Give an explanation of how this occurs.

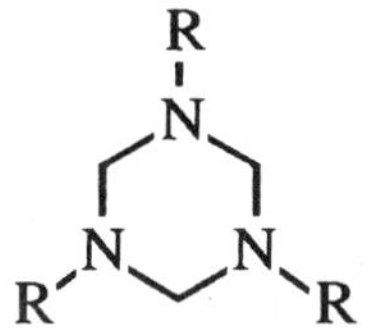

A polymer with the structure shown below can be made using a variation of this reaction. What precursors are required for the formation of this type of polymer?

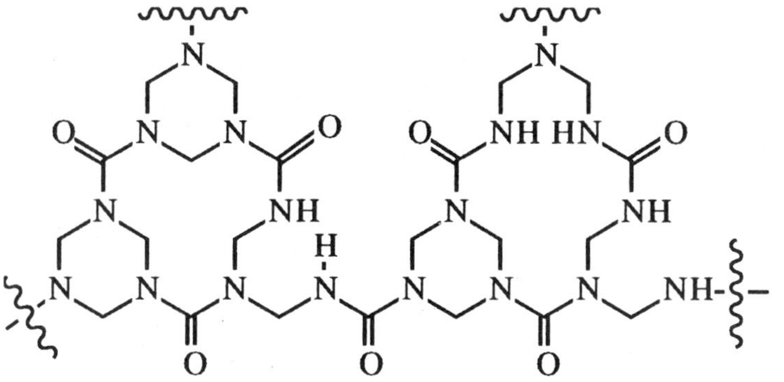

Practice Problem 29.8
One example of a "ladder polymer" is that made by the reaction of the component molecules A and B shown below. Ladder polymers are highly rigid structures. Propose a structure for the polymer formd by the reaction of A and B.

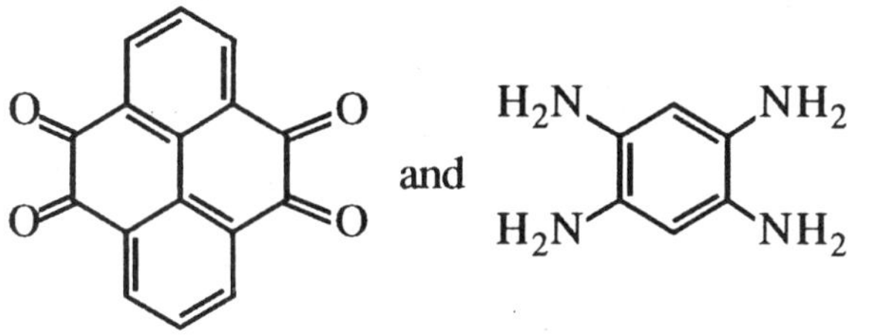

Solution of Text Problems

29.1 H$_2$C=CH-CO$_2$CH$_2$CH$_2$OH

29.2 When polymerization occurs in this manner each step of the polymerization process provides the more stable of the possible intermediate radicals. Each intermediate radical is a resonance stabilized benzylic radical.

29.3 Each stage of the polymerization process provides the more stable of the possible carbocation intermediates. Each intermediate is a tertiary carbocation.

29.4

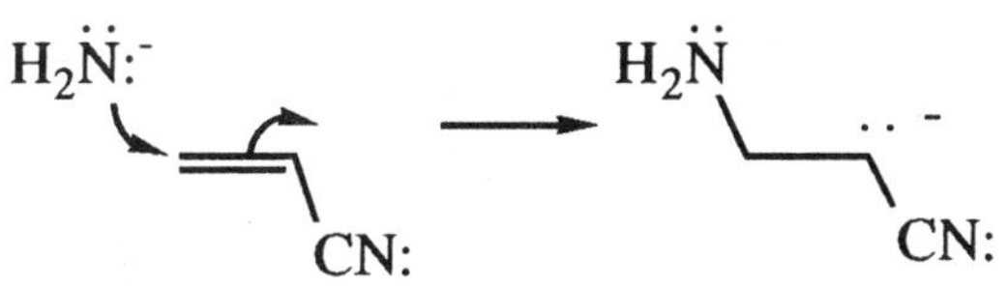

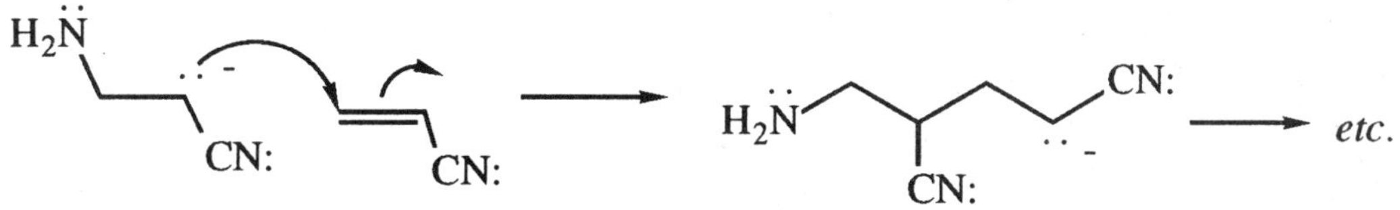

29.5 No. Starting with *only* achiral materials the products are optically inactive. For any product species which contains a stereogenic center, there is produced an equivalent amount of each enantiomer, that is, a racemic mixture.

29.6 The H$_2$C=CCl$_2$ does not give distinct isotactic, syndiotactic, and atactic polymers. It is a symmetrical alkene in the sense that there are no *E* or *Z* forms. There is only one type of polymer possible with regiospecific addition reactions occurring.

29.7 There are three benzene dicarboxylic acids, the *ortho*, *meta*, and *para* species. The *para* isomer would lead

282 Study Guide and Solutions Manual

to a more linear polymer than would the others, which would have a great tendency to form coils. The product from the *para* isomer would thereby be a more crystalline polymer.

29.8 For the production of the particular polyester in question we would start with the dimethyl ester of benzene-1,4-dicarboxylic acid (terephthalic acid) and 1,4-di(hydroxymethyl)cyclohexane. The transesterification would be initiated by heating the mixture with a catalytic amount of acid, the by-product methanol being distilled from the reaction system as it formed.

29.9

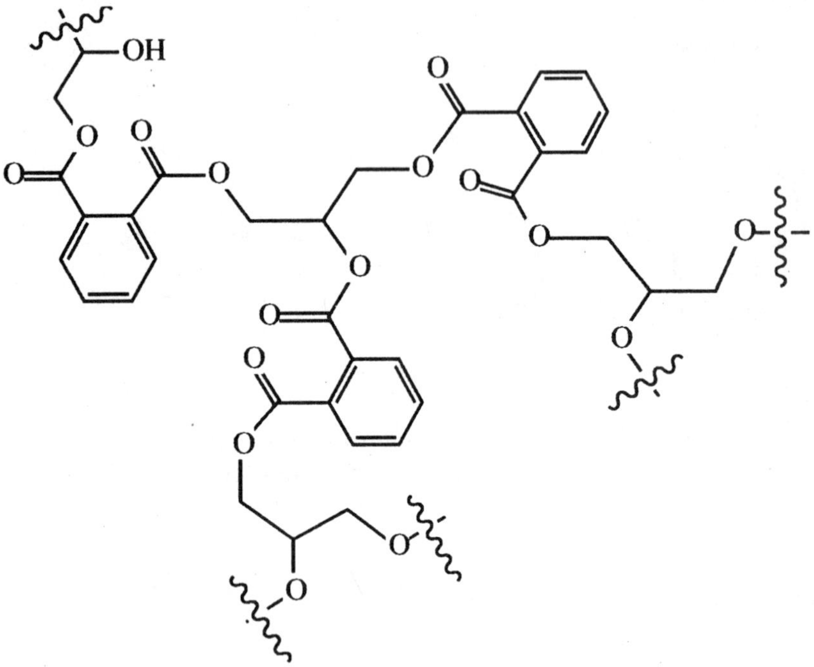

29.10

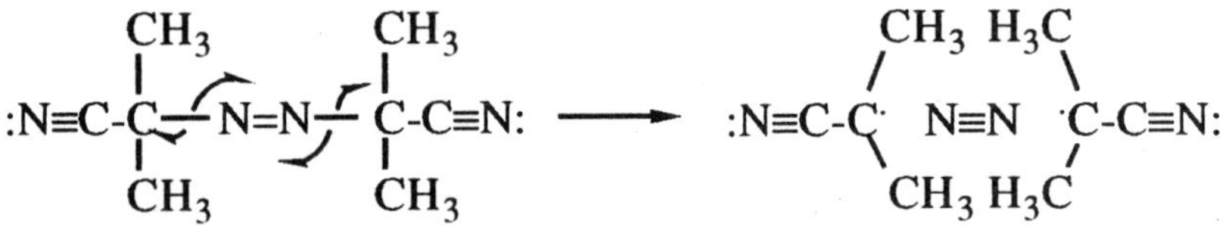

The end-group would be:

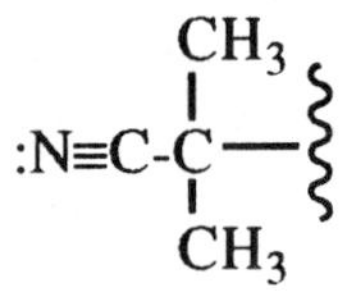

29.11

(a)

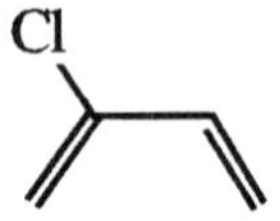

(b)

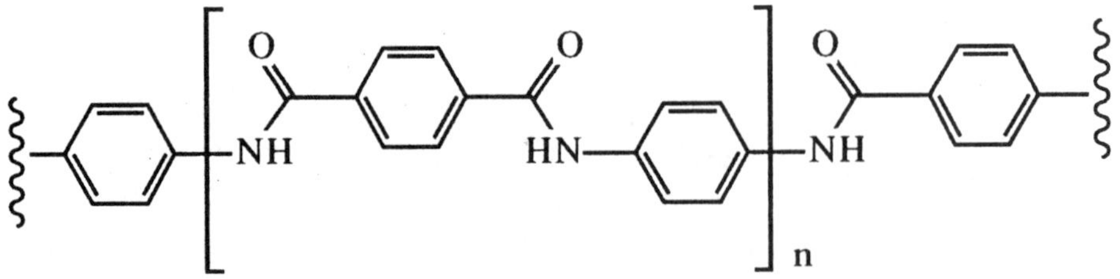

(c)

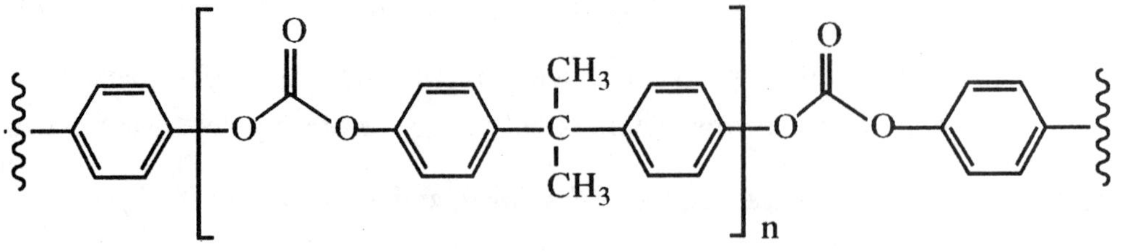

Chapter 29 283

29.12 A deficiency in the amount of initiator would be used to generate a higher molecular weight polymer. If a fewer number of chains are initiated, those which are initiated have the chance to become longer.

29.13

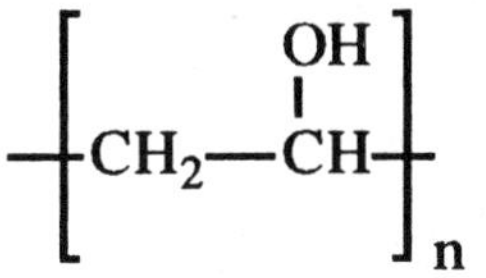

$$\left[-CH_2-CH(OH)- \right]_n$$

Vinyl alcohol itself ($H_2C=CHOH$) is an unfavorable enol form of acetaldehyde. If vinyl acetate is first polymerized and then hydrolyzed, poly(vinyl alcohol) is produced.

29.14

(a) [β-propiolactone opening by hydroxide to give $^-O\text{-CH}_2\text{CH}_2\text{-CO}_2H$, then further ring-openings (two repetitions) to give the polyester $HO^- \rightarrow [^-O\text{-CH}_2\text{CH}_2\text{-C(O)-O-CH}_2\text{CH}_2\text{-C(O)-O-}]_n\text{-CH}_2\text{CH}_2\text{-CO}_2^-$]

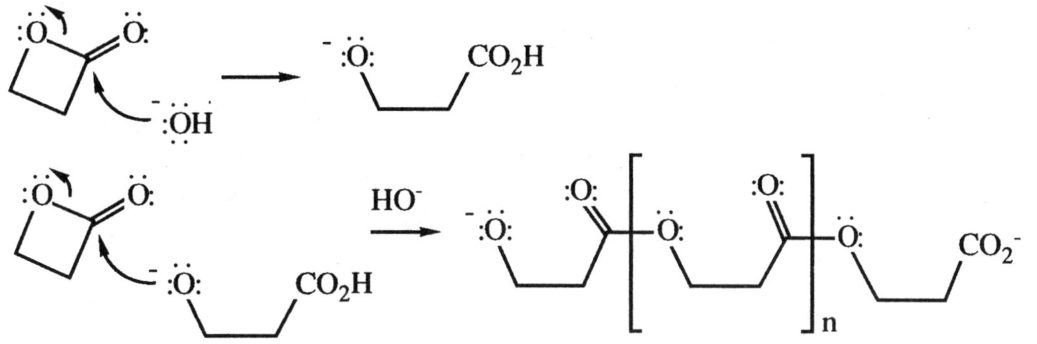

(b) [caprolactam ring-opening by hydroxide: attack of ^-OH on the C=O, cleaving C–N, giving $HO_2C\text{-(CH}_2)_5\text{-NH}^-$; this amide anion attacks another caprolactam, etc. to give polyamide $^-O_2C\text{-[(CH}_2)_5\text{-NH-C(O)-]}_n\text{-(CH}_2)_5\text{-NH-C(O)-}\sim$]

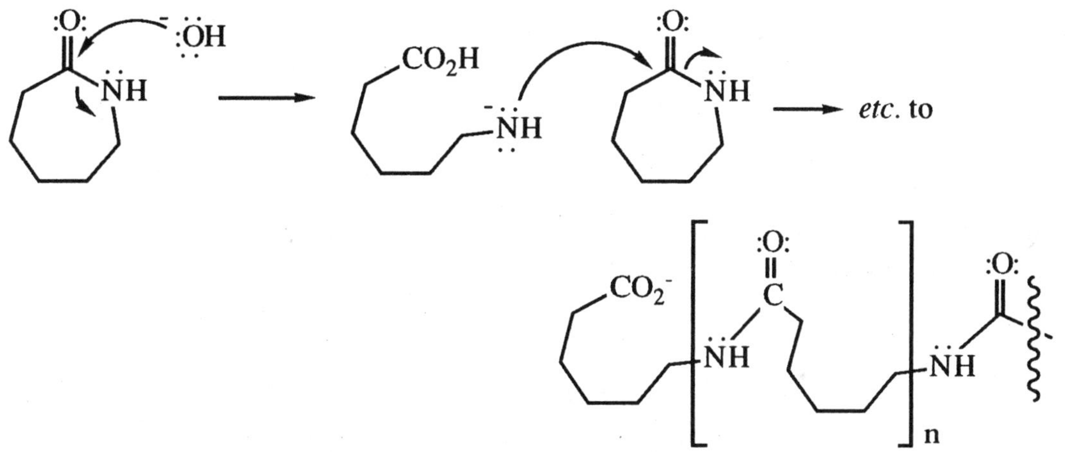

(c) transesterification leading to:

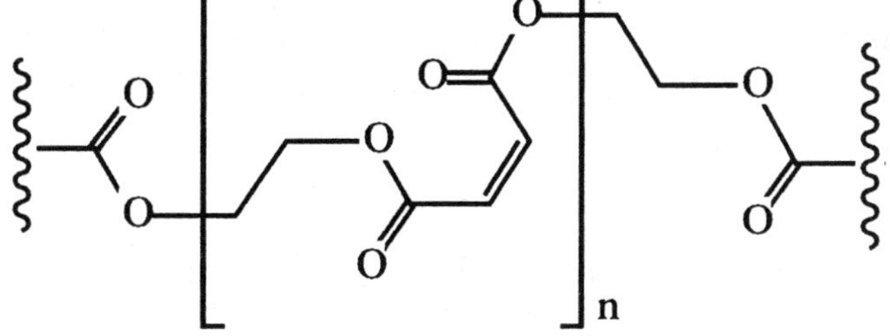

$$\left[\sim\text{-O-C(O)-CH}_2\text{-CH(O-C(O)-CH=CH-C(O)-O-)-CH}_2\text{-O-C(O)-}\sim \right]_n$$

29.15

$$R\text{-N=C=O} + R'\text{-OH} \longrightarrow \underset{R\text{-N-C=O}}{\overset{H}{\underset{|}{\overset{|}{\text{:O-R'}^+}}}} \xrightarrow{\sim H^+} R\text{-N(H)-C(=O)-O-R'}$$

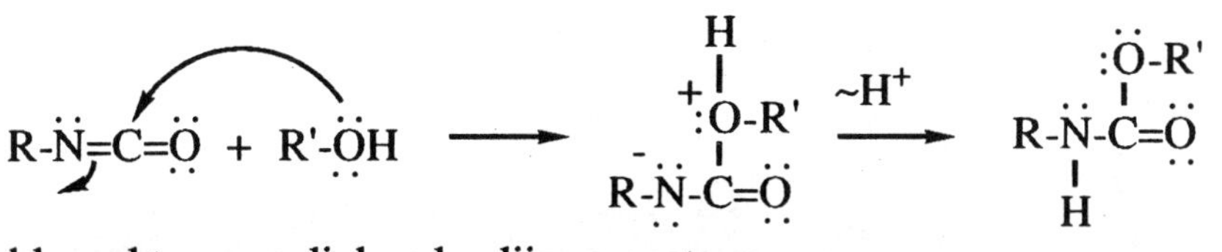

For the formation of a polyurethane we would need to use a diol and a diisocyanate.

29.16 The anionic sites on the polymer will abstract protons from the added water.

29.17

The monomer is 2,3-dimethyl-1,3-butadiene.

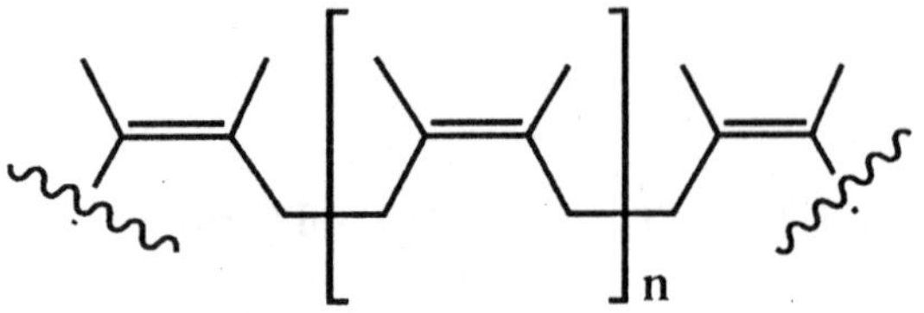

Solution of Study Guide Practice Problems

29.1

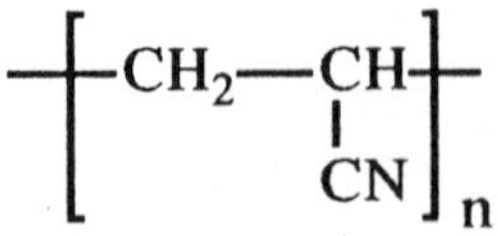

$$\left[-CH_2-\underset{CN}{CH}- \right]_n$$

29.2
We would use methyl methacrylate [CH$_2$=C(CH$_3$)CO$_2$CH$_3$] as the monomer for the synthesis of lucite by free radical polymerization.

29.3

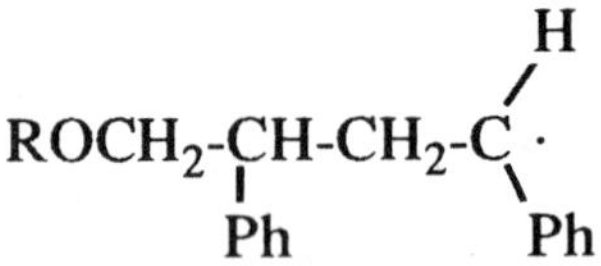

ROCH$_2$-CH(Ph)-CH$_2$-C·(H)(Ph)

29.4
Atactic and isotactic polymers differ in the order of stereogenic sites along the polymer chain. With atactic polymers the stereogenic centers are randomly arrayed, whereas with isotactic polymers the stereogenic sites are generated with the same stereochemistry throughout. Ziegler-Natta catalysts are particularly useful for producing isotactic polymers. Due to their regular stereochemistry, isotactic polymers are more crystalline than are atactic polymers.

29.5
Polyisobutylene can not be prepared in atactic, isotactic, and syndiotactic forms because it contains no stereogenic centers. Isobutylene, its monomer precursor, can not exist in geometrically isomeric forms.

29.6
The repeating unit is as shown below.

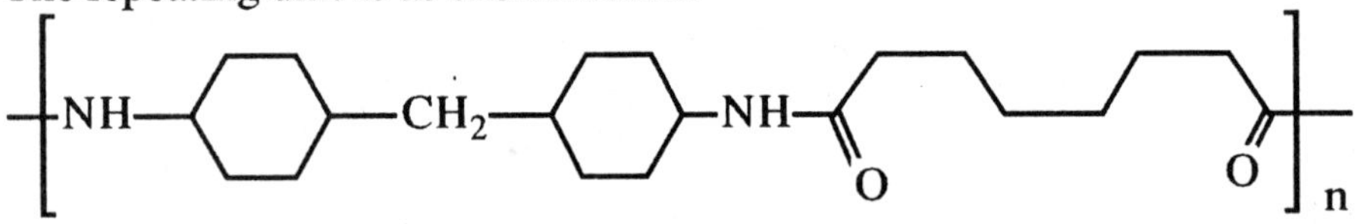

$$\left[-NH-\bigcirc-CH_2-\bigcirc-NH-\underset{O}{C}-(CH_2)_4-\underset{O}{C}- \right]_n$$

29.7
The cyclic structure forms by initial reaction of formaldehyde and the primary amine to generate a *N*-substitutedformaldimine. The intermediate is attacked at carbon by another molecule of the primary amine to generate a *N,N'*-disubstituted-diaminomethane which rapidly reacts with another equivalent of formaldehyde to generate an iminium ion, further reacting by attack of a third primary amine molecule at carbon. The final molecule of formaldehyde reacts with the two amino ends of the chain, eliminating water, to complete the six-membered ring. In order to generate the polymeric material shown, we would use urea [(H$_2$N)$_2$C=O] in place of the primary amine.

29.8

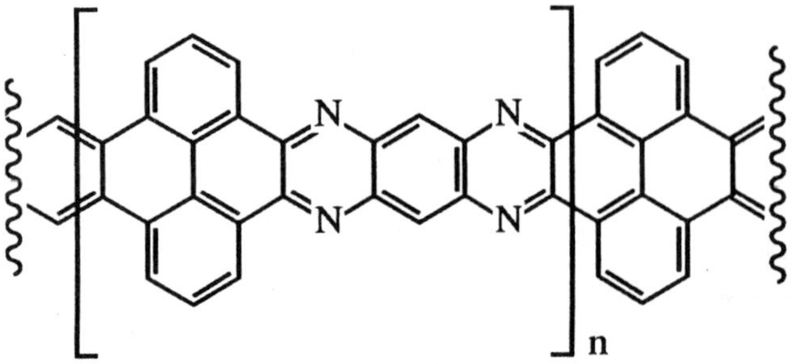

CHAPTER 30
MOLECULAR ORBITALS IN CONCERTED REACTIONS

Key Points

• Understand how to recognize simple cycloaddition and electrocyclic reactions. Know the $[\pi_x+\pi_y]$ classification system used to describe cycloaddition reactions.

Practice Problem 30.1
Predict the products of the thermally stimulated electrocyclic reactions of the following molecules.
   (a) 2,3-dimethyl-1,3-butadiene
   (b) *cis,cis*-1,3,5,7-octatetraene

Practice Problem 30.2
Give the appropriate $[\pi_x+\pi_y]$ designation for the cycloaddition reaction shown below.

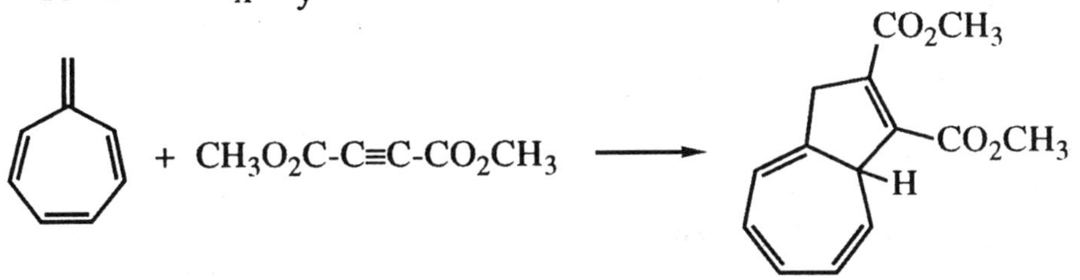

Practice Problem 30.3
The reaction shown below is a $[\pi_6+\pi_4]$ cycloaddition. What is the structure of the π_4 component?

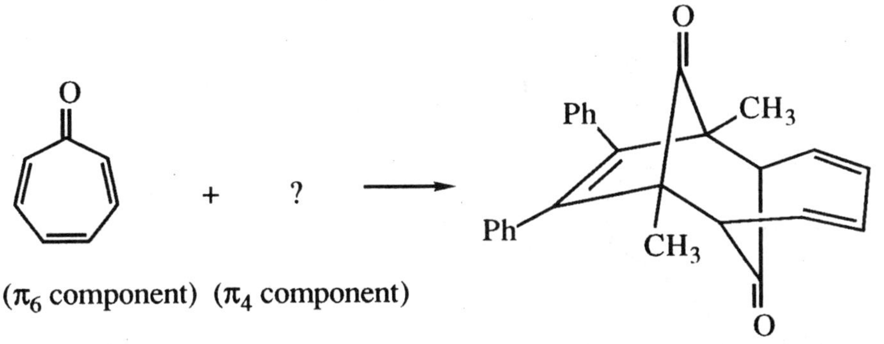

(π_6 component)   (π_4 component)

($[\pi_6+\pi_4]$ cycloaddition product)

• A satisfactory understanding of electrocyclic reactions, cycloadditions, and other pericyclic reactions depends on the application of concepts from fundamental molecular orbital theory. Therefore, you should review the basics of molecular orbital theory as presented in Chapter 16, and should be able to depict accurately the HOMO and LUMO for molecules and ions containing π bonds.

Practice Problem 30.4
Sketch the HOMO and LUMO for the electronic ground state of each of the following:
   (a) 1,3-butadiene
   (b) allyl cation
   (c) pentadienyl anion

• Be able to use the HOMOs and LUMOs of the reactants to predict if a proposed concerted cycloaddition reaction will be favorable under thermal or photochemical stimulation.

Practice Problem 30.5
Predict whether the following reactions are likely to be concerted under conditions of thermal stimulation. To make this determination, consider the interactions of the LUMO of component A with the HOMO of component B.

286 Study Guide and Solutions Manual

	A	B	Product
(a)	CH₃CH₂O₂C-C≡C-CO₂CH₂CH₃	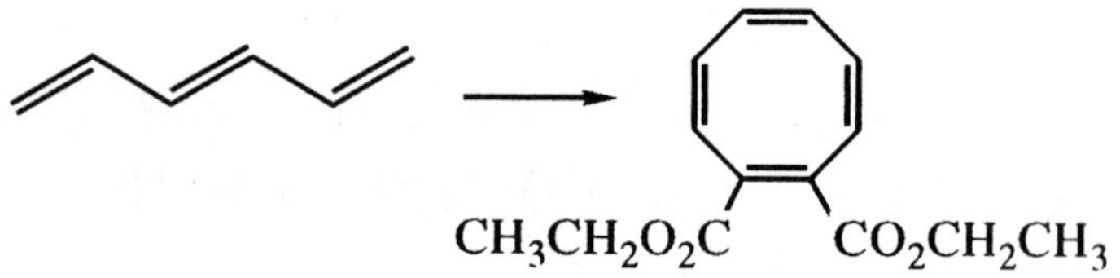 (hexadiene)	(cyclooctatriene with CH₃CH₂O₂C and CO₂CH₂CH₃ substituents)
(b)	H₂C=CH-CH₂⁺	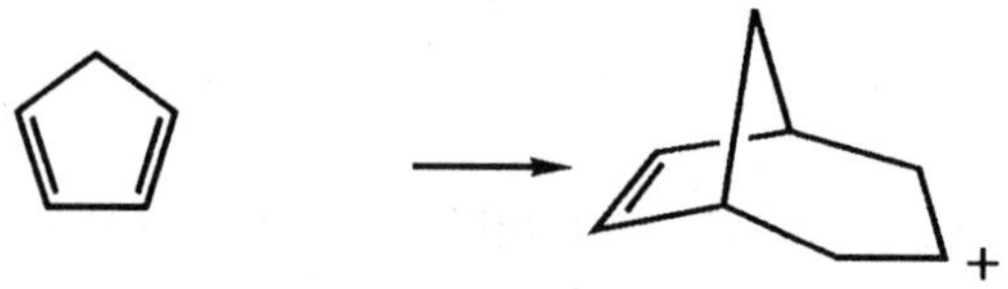 (cyclopentadiene)	(norbornene) +

Practice Problem 30.6
Consider the molecule A shown below. When A is heated, it is converted into a cyclooctatriene (B). In turn, B is converted by heating into C which has a substituted bicyclo[4.2.0]octane skeleton. Suggest structures for B and C and tell whether the conversions A → B and B → C are conrotatory or disrotatory.

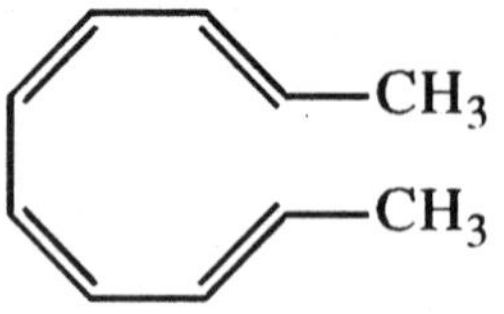

A (cyclic polyene with two -CH₃ substituents)

* Understand how to classify sigmatropic rearrangements using the [x,y] notation, and be able to analyze these reactions using orbital symmetry considerations.

Practice Problem 30.7
Give the [x,y] classification for each of the following sigmatropic rearrangements.

(a) [rearrangement of substituted allyl system with H₃C, H, Ph, Ph, CH₃ substituents → product with H₃C, Ph, Ph, H, CH₃]

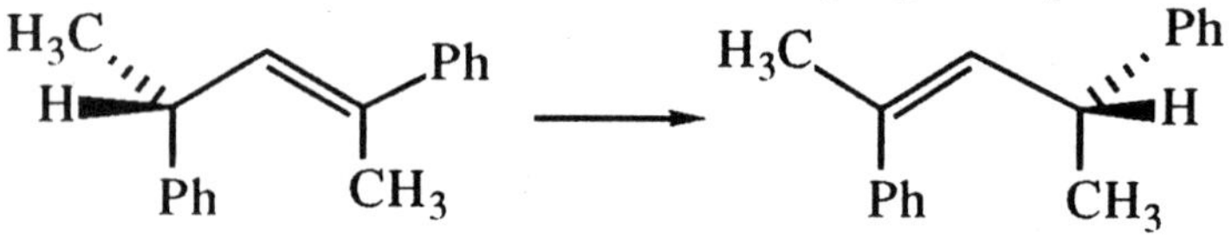

(b) [sulfonium ylide with allyl group → sulfoxide-like rearranged product]

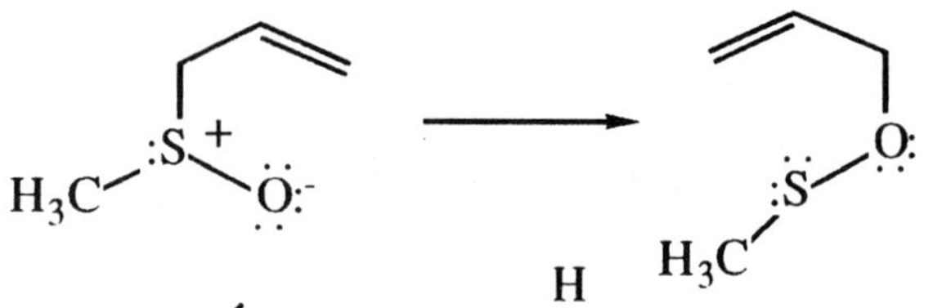

(c) [allyl vinyl ether type → rearranged carbonyl product]

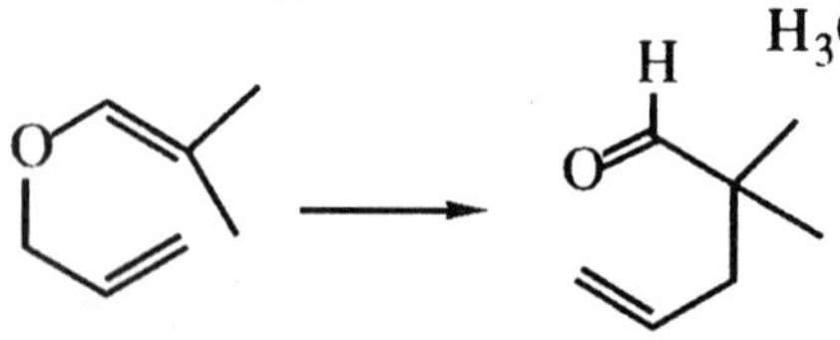

Practice Problem 30.8
A student has proposed the scheme shown below for the generation of ketone E from alcohol D, based on a sigmatropic rearrangement. Do you think that the sigmatropic rearrangement is likely to occur under thermal stimulation? Explain your conclusion.

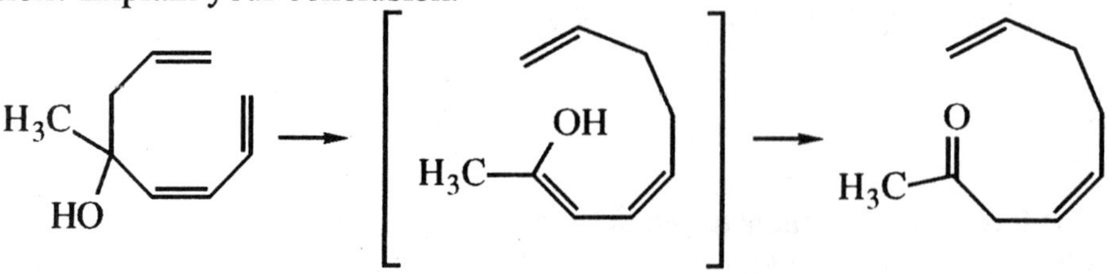

[D: cyclic compound with H₃C and HO substituents] → [intermediate with H₃C and OH on ring] → [E: cyclic ketone with H₃C and O]

Solution of Text Problems

30.1 (a) cycloaddition
(b) electrocyclic reaction
(c) electrocyclic reaction

30.2 (a) symmetric - π_1, π_3, and π_5^*
antisymmetric - π_2, π_4^*, and π_6^*

(b) The π_3 is the HOMO and π_4^* is the LUMO.

(c) The π_4 is the HOMO and π_5^* is the LUMO

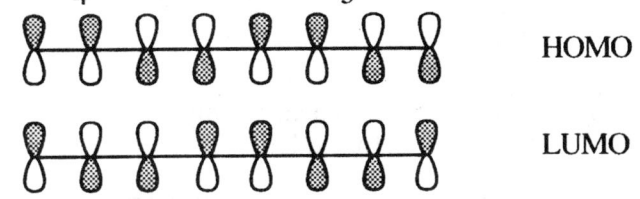

HOMO

LUMO

30.3 (a) thermally forbidden
(b) thermally allowed
(c) thermally forbidden
(d) thermally forbidden

30.4 (a) photochemically allowed
(b) photochemically forbidden
(c) photochemically allowed
(d) photochemically allowed

30.5

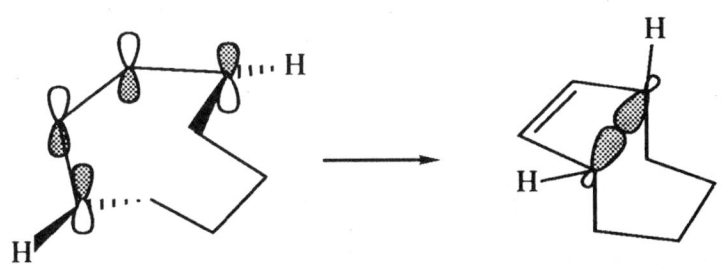

30.6 There is a *trans* double bond in an eight-membered ring.
30.7 B Thermal electrocyclic ring opening for this system is conrotatory.
30.8 The electrocyclic ring opening is conrotatory, which produces 1,3-cycloheptadiene having one of the double bonds *cis* and the other *trans*.
30.9 (a) allowed photochemically - disrotatory
(b) allowed photochemically - conrotatory
(c) forbidden photochemically - conrotatory motion would be required

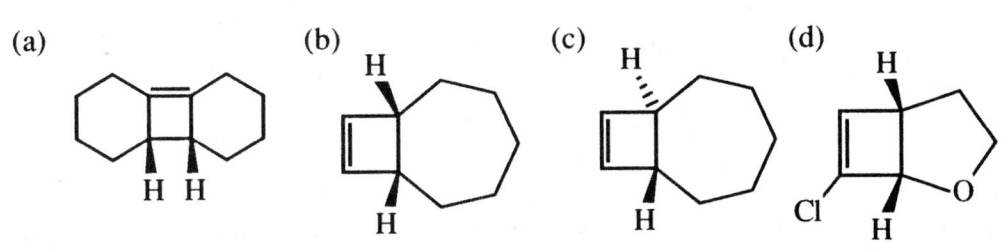

30.10
(a) (b) (c) (d)

30.11 The major product is (*E,E*)-1,4-dichloro-1,4-butadiene and the minor product is (*Z,Z*)-1,4-dichloro-1,3-butadiene.

30.12 (a) (b) (c)

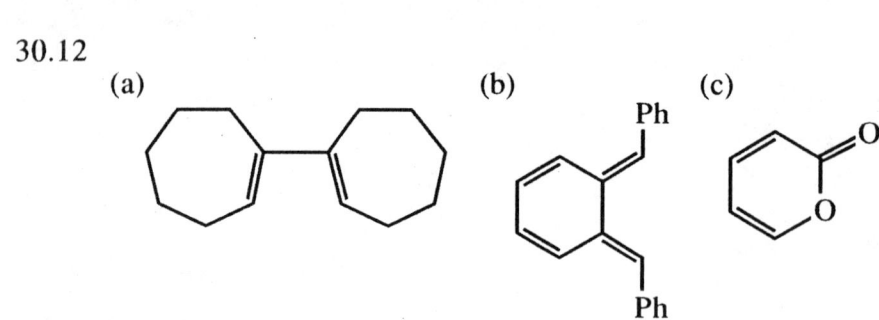

30.13

thermally allowed

photochemically forbidden

30.14 The observed product is 2,7-octanedione. This is expected by a [3,3] sigmatropic reaction of the starting compound which leads initially to a dienol form of the product, which tautomerizes immediately to the diketone product.

30.15

30.16

The carbon atoms are marked by asterisks as they become part of the cyclopropane ring with successive sigmatropic rearrangements. After five sigmatropic rearrangements, all ten carbon atoms have been part of the cyclopropane ring and thereby have occupied equivalent positions of the bullvalene molecule.

30.17 (a) photochemically allowed
(b) photochemically allowed
(c) photochemically forbidden
(d) photochemically forbidden
(e) photochemically forbidden
(f) photochemically allowed

30.18 (a) photochemical
(b) thermal
(c) photochemical
(d) thermal
(e) thermal

30.19 (a) (b) (c) (d)

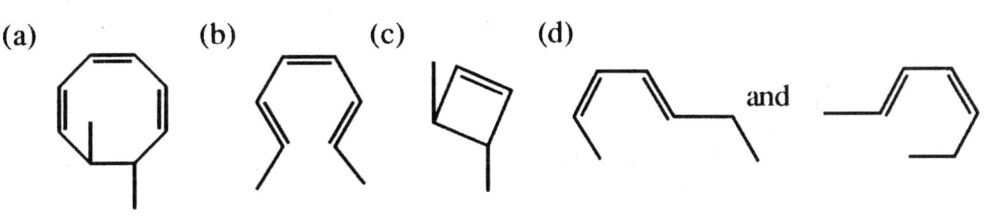

30.20 Bicyclo[2.1.0]pentane can be prepared photochemically from 1,4-pentadiene. The product is kinetically stable since the thermal ring opening is symmetry forbidden.

30.21 Conrotatory ring opening would generate a *trans* double bond in a seven-membered ring.

30.22 (a) [π6s+π6s] photochemical
(b) [π6s+π4s] thermal
(c) [π2s+π2s] photochemical
(d) [3,3] sigmatropic thermal

30.23 We would expect it to be able to add to an alkene. The reaction would be described as [π4s+π2s].

30.24 Compound V is formed.

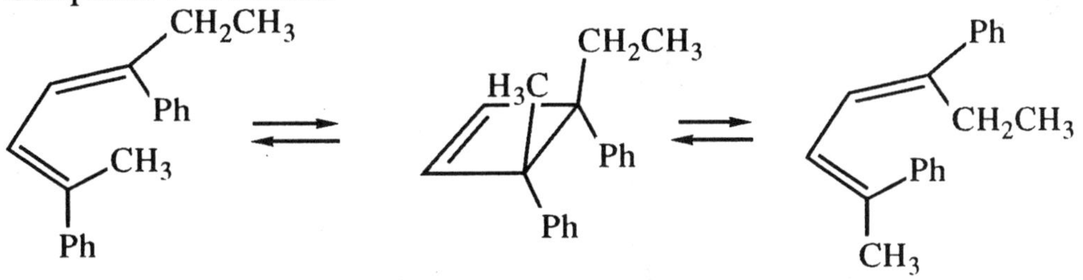

30.25 This is a [3,3] sigmatropic rearrangement proceeding through a chair conformation activated complex.

30.26 (a) No - thermal reaction proceeds in a conrotatory manner.
(b) Yes

30.27 (a) No - the [1,7] suprafacial sigmatropic rearrangement is thermally forbidden.
(b) Yes - the [1,9] suprafacial sigmatropic rearrangement is thermally allowed.

30.28 The *trans* compound undergoes thermal ring opening at 90°. The *cis* isomer would produce a *trans*-cyclohexene product by symmetry allowed conrotatory opening.

30.29

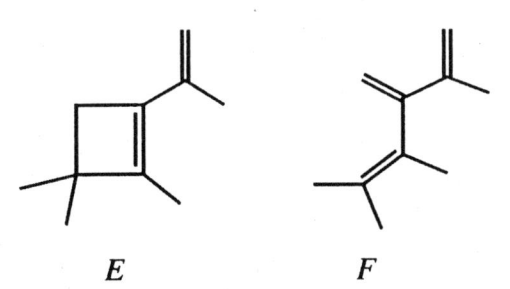

E   F

Solution of Study Guide Practice Problems

30.1 (a) 1,2-dimethylcyclobutene
(b) 1,3,5-cyclooctatriene

30.2 [π8+π2] cycloaddition

290 Study Guide and Solutions Manual

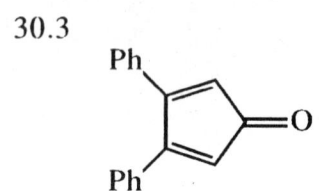

30.3

[Structure: cyclopentadienone with Ph groups at 3,4-positions]

30.4

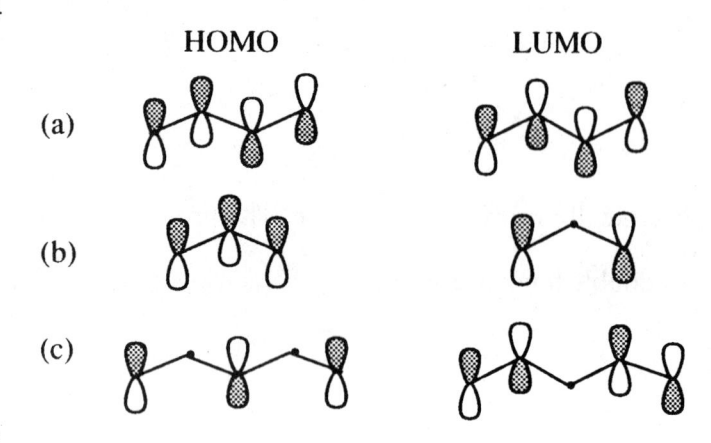

	HOMO	LUMO
(a)	[orbital diagram]	[orbital diagram]
(b)	[orbital diagram]	[orbital diagram]
(c)	[orbital diagram]	[orbital diagram]

30.5

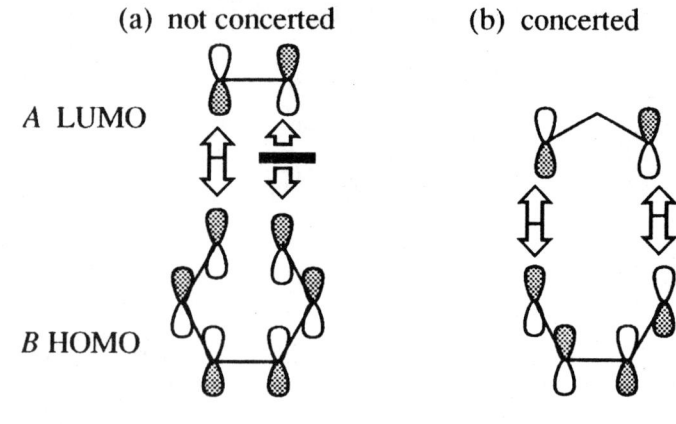

(a) not concerted   (b) concerted

A LUMO

B HOMO

30.6

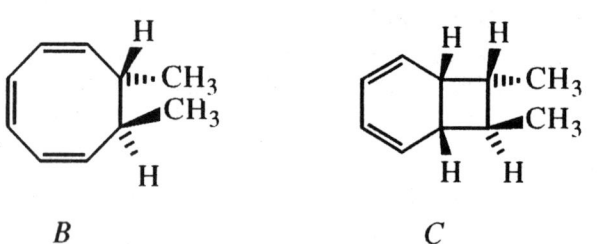

[Structures B and C shown]

B   C

A → B is conrotatory; B → C is disrotatory

30.7 (a) [1,3] (not concerted under conditions of thermal stimulation)
(b) [2,3]
(c) [3,3]

30.8 The first step in the proposed scheme is a [5,3] sigmatropic rearrangement. This type of reaction can not be concerted under thermal stimulation.

PRACTICE EXAMINATION SIX

A time limit of 90 minutes should be set for completion of this entire practice examination. Answers should be written out completely as they would be when presented for independent grading. No text or supplemental materials should be consulted during the testing period, and you should not check your answers until you have worked out the complete examination and the time limit has been reached.

1. Give structures matching the descriptions given below (3 points for each part).
    (a) *N*-acetylleucine
    (b) the compound that, on final hydrolysis in the Strecker synthesis, yields phenylalanine
    (c) deoxyadenosine
    (d) the compound formed by the reaction of alanine with *tert*-butylazidoformate
    (e) the repeating unit in the polymer prepared by the reaction of the following components:

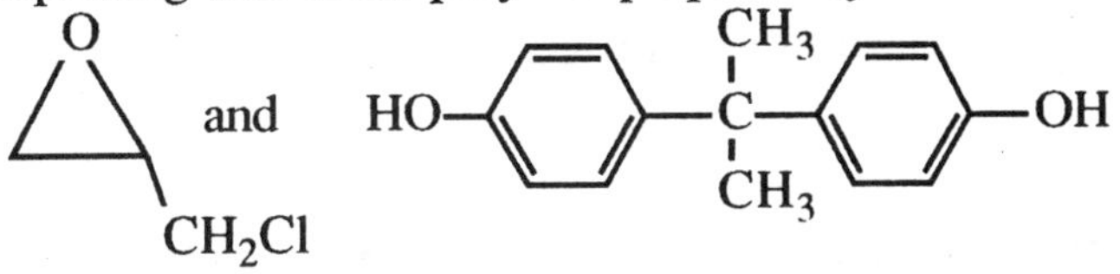

    [epoxide with CH₂Cl substituent] and [HO—C₆H₄—C(CH₃)₂—C₆H₄—OH]

    (f) the cation produced by the reaction of 2-methylpropene with the *tert*-butyl cation
    (g) the product obtained on concerted thermal ring-opening of *trans*-3,4-dimethylcyclobutene
    (h) the product obtained when 1,5-heptadiene undergoes a thermal [3,3] sigmatropic rearrangement
    (i) the monomer that on free radical initiated polymerization yields the polymer indicated below, used in "super-glue"

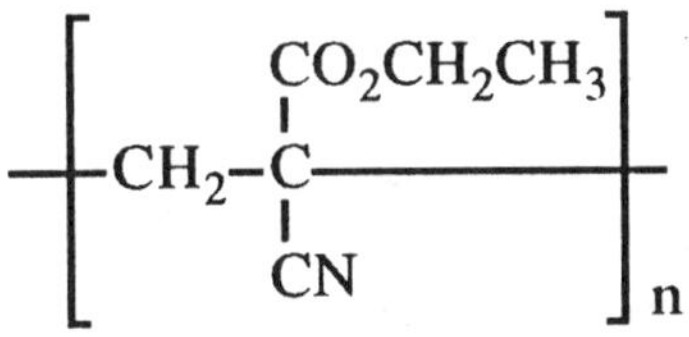

    $$\left[-CH_2-\underset{CN}{\overset{CO_2CH_2CH_3}{C}}-\right]_n$$

    (j) the dinucleoside phosphate deoxythymidylyl(3'→5')deoxyguanosine

2. A nonapeptide has the gross composition
    Ala₃, Phe₂, Glu, Gly, Lys, Val.
Tell what is revealed about the structure by each of the pieces of information in parts (a)-(f), and finally, in part (g), give the entire primary structure of the nonapeptide (3 points for each part).
    (a) When the nonapeptide is treated with carboxypeptidase, glycine is the initially detected amino acid in the reaction solution.
    (b) When the nonapeptide is treated with phenylisothiocyanate, followed by treatment with trifluoroacetic acid and then aqueous acid, compound *I* shown below is formed.

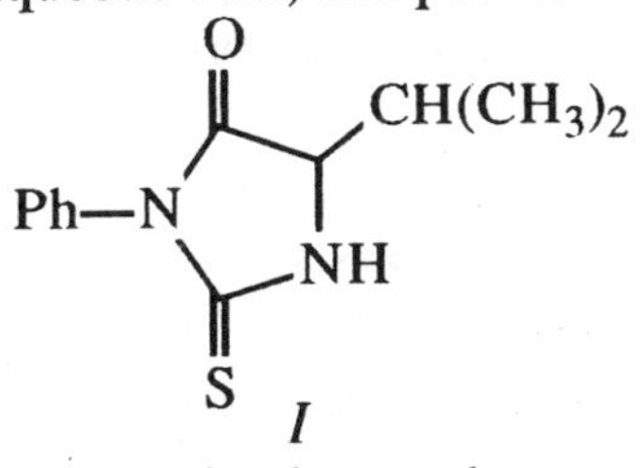

    [Structure of compound I: Ph—N, C=O, CH(CH₃)₂, NH, C=S ring]
    I

    (c) Chymotrypsin cleaves the nonapeptide into three tripeptides, *A*, *B*, and *C*.
    (d) On treatment with phenylisothiocyanate followed by trifluoroacetic acid and then aqueous acid, tripeptide *A* yields *I* (structure above), but tripeptides *B* and *C* both yield compound *II*, shown below.

(e) On treatment with carboxypeptidase, tripeptide *C* initially yields glycine, while under the same conditions both *A* and *B* yield phenylalanine.

(f) When subjected to electrophoresis at pH 7, tripeptide *B* migrates toward the anode while tripeptide *C* migrates toward the cathode.

(g) Give the complete primary structure of the nonapeptide.

3. Consider the following pericyclic reactions (7 points for each part).

(a) Predict whether the thermally stimulated ring opening of the cyclopropyl cation to form the allyl cation is disrotatory or conrotatory. Show your orbital analysis in arriving at your conclusion.

(b) 1,5-Dimethyl-1,3,5,7-cyclooctatetraene, *D*, on being heated is isomerized by a succession of pericyclic reactions as shown below. Give structures for compounds *E-H*. (NOTE: There are *two* correct answers for compounds *E-G*, depending on the particular ring closure route followed, but still only *one* correct answer for compound *H*. That is, both routes ultimately lead to the same product.

(c) The compound *J* shown below is also isomerized by a series of pericyclic reactions. Give structures for the compounds *K-M*.

4. Give the structure of the indicated organic product in each of the following reactions (2 points for each part).

(a) treatment of 2,3-dichloropyridine with KCN in ethanol to form a product of formula $C_6H_3N_2Cl$

(b) treatment of furan with acetic anhydride to form a product of formula $C_6H_6O_2$

(c) treatment of furfural (2-formylfuran) with acetone in the presence of base to form a product of formula $C_8H_8O_2$

(d) treatment of 2,5-dimethylfuran with dilute aqueous sulfuric acid to form a product of formula $C_6H_{10}O_2$

(e) treatment of 2,3-diaminotoluene with glycerol under Skraup reaction conditions to form a product of formula $C_{10}H_{10}N_2$

5. The "road-map" below depicts the synthesis of a heterocyclic compound, *V*. Based on this reaction scheme, and the additional information that *Q* is oxidized by hot aqueous basic potassium permanganate solution to a compound *W* which exhibits two singlets in its ^{1}H NMR spectrum (one at δ 7.9 of relative area 2, and the other at δ 11.4 of relative area 1), deduce the structures of compounds *N-V*.

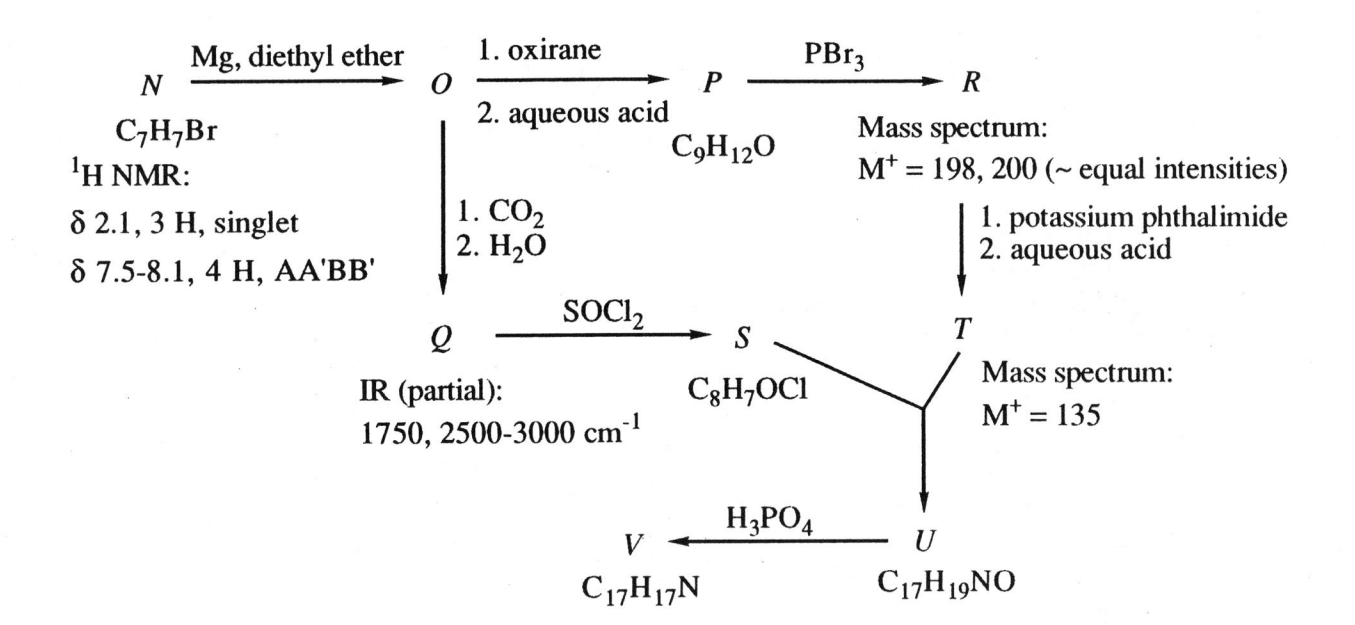

Examination Six 293

N $\xrightarrow{\text{Mg, diethyl ether}}$ O $\xrightarrow[\text{2. aqueous acid}]{\text{1. oxirane}}$ P $\xrightarrow{PBr_3}$ R

N: C_7H_7Br
^{1}H NMR:
δ 2.1, 3 H, singlet
δ 7.5-8.1, 4 H, AA'BB'

P: $C_9H_{12}O$

R: Mass spectrum: M$^+$ = 198, 200 (~ equal intensities)

O $\xrightarrow[\text{2. H}_2\text{O}]{\text{1. CO}_2}$ Q

R $\xrightarrow[\text{2. aqueous acid}]{\text{1. potassium phthalimide}}$ T

Q $\xrightarrow{SOCl_2}$ S

Q: IR (partial): 1750, 2500-3000 cm^{-1}

S: C_8H_7OCl

T: Mass spectrum: M$^+$ = 135

$S + T \longrightarrow U$

V $\xleftarrow{H_3PO_4}$ U

V: $C_{17}H_{17}N$
U: $C_{17}H_{19}NO$

ANSWERS TO PRACTICE EXAMINATION ONE

1. (a) 2,4-dimethylpentane
   (b) 5-isopropyl-6-methyl-2-nitrooctane
   (c) 1-bromo-2-ethyl-2,3-dimethylbutane
   (d) spiro[2.4]octane
   (e) cis-1,3-diisobutylcyclohexane

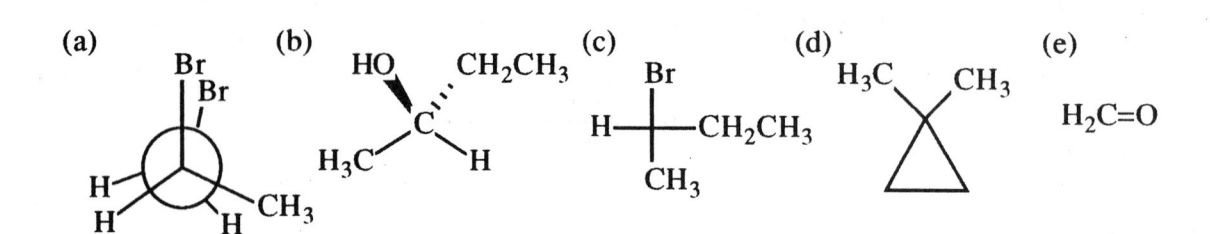

2. (a) [Newman projection with Br, Br, CH₃, H, H, H]
   (b) [structure with HO, CH₂CH₃, H₃C, H on central C]
   (c) [structure: H—C(Br)(CH₃)—CH₂CH₃]
   (d) [cyclopropane with H₃C, CH₃]
   (e) H₂C=O

3. (a) $\overset{..}{\underset{..}{S}}=\overset{+}{\underset{..}{S}}-\overset{..}{\underset{..}{O}}:^-$

   (b) $\left[\overset{..}{\underset{..}{S}}=\overset{+}{\underset{..}{S}}-\overset{..}{\underset{..}{O}}:^- \longleftrightarrow :\overset{..}{\underset{..}{S}}-\overset{+}{\underset{..}{S}}=\overset{..}{\underset{..}{O}}\right]$

   (c) bent - The central S atom has three associated groups of valence level electrons (one group being the unshared electron pair, a second group being the four electrons of a double bond, and a third group being the two electrons of a single bond). The VSEPR model thus predicts a planar trigonal structure.

4. ethanol (CH_3CH_2OH), acetylene (HC≡CH), and bromide ion

5. (a) isopropyl alcohol
   (b) H_2S
   (c) pentane
   (d) both are protic solvents
   (e) hydrogen sulfide
   (f) cyclopropane
   (g) acetic acid
   (h) both
   (i) methane and lithium hydroxide
   (j) hydrogen

6. (a) CH_3CH_2Cl - The nucleophile is chloride ion and the leaving group is a water molecule.
   (b)
   $CH_3CH_2-\overset{..}{\underset{..}{O}}H + H^+ \longrightarrow CH_3CH_2-\overset{+}{\underset{..}{O}}H_2$

   $:\overset{..}{\underset{..}{Cl}}:^- \quad CH_3CH_2-\overset{+}{\underset{..}{O}}H_2 \longrightarrow :\overset{..}{\underset{..}{Cl}}CH_2CH_3 + \overset{..}{\underset{..}{O}}H_2$

(c) Zinc chloride acts as a Lewis acid to convert the hydroxyl group of the ethanol into a better leaving group, in a manner completely analogous to the action of protic acids.

7. (a) The products are ammonia and potassium ethoxide.
   (b)

$$H_2\ddot{N}:^- \quad H-\ddot{O}-CH_2CH_3 \longrightarrow H_2\ddot{N}-H + {}^-:\ddot{O}-CH_2CH_3$$

(The K$^+$ is a spectator ion in this process.)

ANSWERS TO PRACTICE EXAMINATION TWO

1. (a) none
   (b) *A*
   (c) *B*
   (d) *A*
   (e) none

2. (a) sodium borohydride
   (b) CrO_3/pyridine
   (c) HBr (or phosphorus tribromide)
   (d) $(CH_3)_3CCH_2Br$
   (e)
   $$BrCH_2\text{-}\underset{\underset{H}{|}}{C}\text{-}CH_3$$

3. (a) (*R*)-2-methyl-3-buten-1-ol
   (b) (2*E*,4*E*)-3-ethyl-2,4-hexadiene
   (c) (*E*)-(5*S*,6*S*)-5,6-dibromo-3-methyl-3-heptene
   (d) (*E*)-(4*S*,5*S*)-5-bromo-4-ethyl-4,5-dimethyl-2-heptene
   (e) (*E*)-(*R*)-4-methoxy-5-methyl-2-hexene

4. (a) 2-methyl-2-butanol
   (b) 1-butene or 2-methylpropene
   (c) *trans*-2-methylcyclopentanol
   (d) *trans*-3-hexene
   (e) 2-bromo-2,3-dimethylbutane

5. [The answers listed below are given in a "shortened" form". That is, the specific reagents are listed in the order in which they are used. With an examination question of this type you would usually also be expected to show the structures of the intermediate substances produced. You should present your answers by drawing the starting structure on the left. Then, use an arrow with the reagents for the first step written over the arrow. Show the product of the first step to the right of this arrow, and continue to use a similar procedure for the further steps of the synthesis.

   (a) 1. borane; 2. hydrogen peroxide, aqueous KOH; 3. heat with aqueous basic potassium permanganate solution

   (b) 1. elimination using potassium *tert*-butoxide in *tert*-butyl alcohol; 2. aqueous acid

   (c) 1. bromine under light irradiation; 2. potassium *tert*-butoxide in *tert*-butyl alcohol; 3. ozone; 4. workup with zinc and acetic acid or hydrogen sulfide

   (d) 1. treatment with dilute aqueous basic potassium permanganate at room temperature; 2. chromic anhydride in pyridine

6.

(a), (b), (c) — reaction mechanism schemes.

ANSWERS TO PRACTICE EXAMINATION THREE

1.
(a) 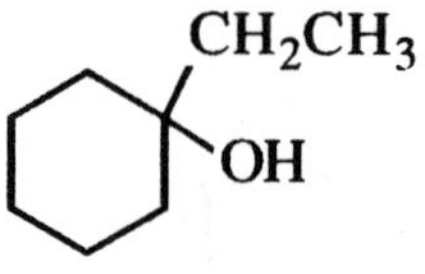 1-ethylcyclohexanol (structure: cyclohexane with CH$_2$CH$_3$ and OH on same carbon)

(b) 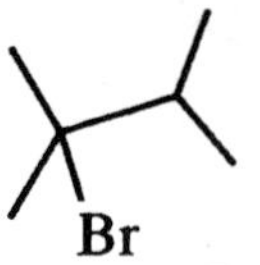 2-bromo-2,3-dimethylbutane

(c) 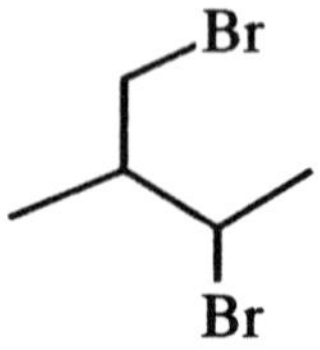 1,3-dibromo-2-methylbutane

(d) 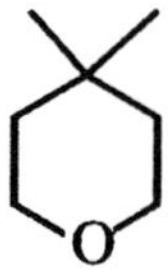 4,4-dimethyltetrahydropyran

(e)  (E,E)-2,4-hexadiene

2. (a) 109°28'
(b) 120°
(c) 3
(d) 0.9 kcal/mole
(e) 8:1

3. (a) 1-iodobutane - With acetone as the solvent, the S$_N$2 reaction is favored. The compound which is least hindered for backside attack will react faster.

(b) both are the same - The S$_N$1 reaction is favored with benzylic halide in a polar solvent. The concentration of the nucleophile has no influence on the rate of an S$_N$1 reaction.

(c) HI in water - The acid will protonate the hydroxyl group converting it into a good leaving group.

(d) KCN in DMSO - The cyanide ion is less solvated and thereby more reactive in DMSO than in aqueous acetone.

(e) cis-1-chloro-4-tert-butylcyclohexane - The tert-butyl group occupies an equatorial position, leaving the bromide group in an axial position in the cis isomer. This allows for the alignment of the involved bonds in an anti periplanar arrangement for efficient elimination reaction.

4. A 2-pentyne
B cis-2-pentene
C (2R,3S)- and (2S,3R)-pentane-2,3-diol (resolvable enantiomers)
D (2R,3R)- and (2S,3S)-pentane-2,3-diol (resolvable enantiomers)
E 3-pentanone

5. (a) S$_N$2
(b) F is 1-bromo-2,2-dimethylpropane (CH$_3$)$_3$CCH$_2$Br
G is 1-ethoxy-2,2-dimethylpropane (CH$_3$)$_3$CCH$_2$OCH$_2$CH$_3$
H is 2-ethoxy-2-methylbutane (CH$_3$)$_2$C(OCH$_2$CH$_3$)CH$_2$CH$_3$

(c) 1. slow loss of bromide ion accompanied by migration of a methyl group as (H$_3$C$^-$) to form the *tert*-pentyl cation; 2. combination of the carbocation with an ethoxide ion to generate *H*, or removal of a proton from the site adjacent to the carbocation site to generate 2-methyl-2-butene.

6. (a) 1. treatment of cyclohexene with *m*-chloroperoxybenzoic acid to form the epoxide; 2. treatment with sodium ethoxide to add to carbon and open the epoxide ring; 3. addition of iodomethane to generate the methyl ether linkage

(b) 1. treatment of *tert*-butyl alcohol with potassium metal to generate the *tert*-butoxide anion; 2. addition of iodoethane to form the ether linkage

(c) 1. treatment of PhCH=CH$_2$ with bromine in carbon tetrachloride solution; 2. treatment of the dibromide with excess potassium amide in liquid ammonia to form the anion of the terminal alkyne; 3. addition of iodoethane to form the internal alkyne

(d) 1. conversion of ethene to iodoethane by the addition of HI; 2. another portion of ethene is converted to oxirane by reaction with *m*-chloroperoxybenzoic acid; 3. the iodoethane is converted to the Grignard reagent by addition to magnesium metal in ether; 4. the oxirane is added to the Grignard reagent, and water added for workup; 5. treatment of the adduct with chromic anhydride in sulfuric acid to generate the carboxylic acid

(e) 1. treatment of 2-butyne with lithium in liquid ammonia; 2. addition of bromine in carbon tetrachloride solution

ANSWERS TO PRACTICE EXAMINATION FOUR

1. (a) cyclohexylamine - The basicity of arylamines is low as the nitrogen valence level unshared electron pair is delocalized into the aromatic ring.
   (b) benzyl chloride - Benzylic halides are highly reactive toward nucleophiles, but aryl halides are not.
   (c) imidazole - Protonation of pyrrole destroys the aromatic character and is therefore disfavored.
   (d) 1,3-cyclopentadiene - Loss of a proton leads to an aromatic anion.
   (e) 4,4-dimethylcyclohexanol - Tertiary alcohols cannot be prepared by reduction of carbonyl compounds.
   (f) [structure: 1,3-dioxolane] - The -OCH$_2$O- protons should produce a signal downfield of that from the -OCH$_2$CH$_2$O- protons, and indeed, the less intense signal *is* downfield. The *larger* signal would be downfield for the alkyne.
   (g) allyl anion - The HOMO of the allyl cation has no vertical nodes.
   (h) benzene HOMO - The benzene HOMO has an additional bonding interaction.
   (i) anisole - The methoxy group is electron donating and thereby activating.
   (j) *N*-methylaniline - Only secondary amines react with aldehydes to give enamines.

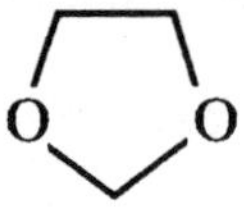

2. (a) Ethanal will be formed. The cyclic substrate contains three acetal linkages which will be hydrolyzed in aqueous acid.
   (b) [structure shown]
   (c) [structure shown]
   (d) *B* is 3-phenyl-1-butene and *C* is 2,3-diphenylbutane
   Protonation of 1,3-butadiene yields a secondary allylic-type cation (shown as *I* below) which then acts as an electrophile toward benzene to produce *B*. Protonation of *B* also leads to a cation (secondary, non-allylic, shown as *II* below) which also acts as an electrophile toward benzene yielding *C*.

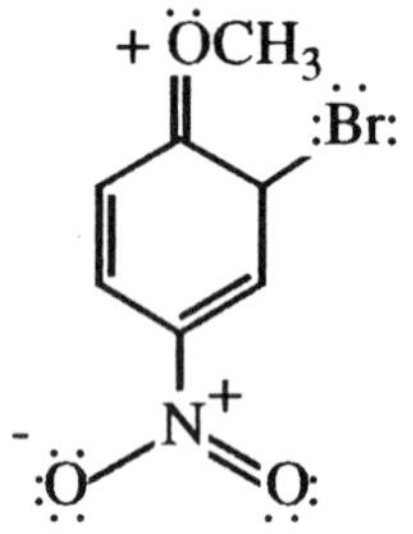

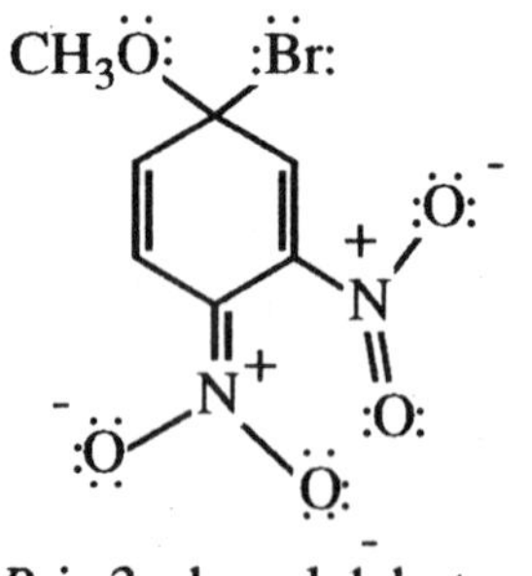

   $$\text{H}_2\text{C=CH-}\overset{+}{\text{C}}\text{H-CH}_3 \qquad \text{PhCH-}\overset{\overset{\text{CH}_3}{|}}{\overset{+}{\text{C}}}\text{H-CH}_3$$
   $$\quad I \qquad\qquad\qquad\quad II$$

   (e) *D* is (CH$_3$)$_2$CHC(O)Cl   *E* is isopropyl phenyl ketone   *F* is PhCH$_2$CH(CH$_3$)$_2$

3. (a) *G* is propanoic acid
   (b) *H* is acetone dimethyl acetal [(CH$_3$)$_2$C(OCH$_3$)$_2$]
   (c) *I* is dichloroethanal

(d) *J* is 2,3-dimethy-12,3-butanediol
(e) *K* is 4-ethoxyaniline

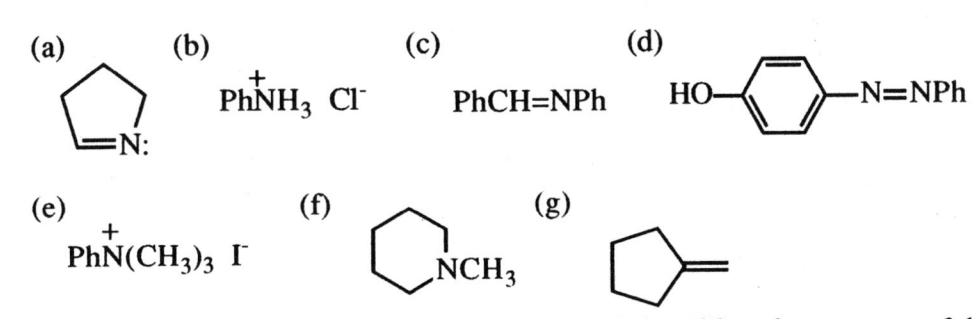

4.
(a) 3,4-dihydro-2H-pyrrole (cyclic C=N)
(b) PhNH₃⁺ Cl⁻
(c) PhCH=NPh
(d) HO–C₆H₄–N=NPh (para)
(e) PhN(CH₃)₃⁺ I⁻
(f) N-methylpiperidine
(g) methylenecyclopentane

5. (a) 1. treatment of 2-iodopropane with triphenylphosphine; 2. treatment of the resultant phosphonium salt with butyllithium to generate the ylide; 3. addition of benzaldehyde with heating to generate the target material
(b) 1. treatment of benzene with acetyl chloride in the presence of aluminum chloride; 2. treatment with nitric acid and sulfuric acid; 3. reduction of the nitro group to an amino group and the ketone to a methylene group with zinc and hydrochloric acid
(c) treatment of benzene with chloroethane and aluminum chloride; 2. treatment with nitric acid and sulfuric acid with isolation of the *para* isomer; 3. bromine and light irradiation; 4. potassium hydroxide in ethanol to give the target material
(d) 1. reduction of 4-nitrotoluene with tin and hydrochloric acid; 2. sodium nitrite and sulfuric acid; 3. cuprous cyanide; 4. lithium aluminum hydride reduction of the nitrile to the primary amine (water workup)

302 Study Guide and Solutions Manual

ANSWERS TO
PRACTICE EXAMINATION FIVE

1.

(a) N(CH₂CH₃)₂ on cyclohexene

(b) 4-bromobenzonitrile (Br and CN on benzene)

(c) 1-methyl-2-oxocyclohexane-1-carboxylic acid (cyclohexanone with CH₃ and CO₂H at α-carbon)

(d) 4-bromo-N-propionylaniline (Br and NHC(O)CH₂CH₃ on benzene)

(e) 5-methylhexanoic acid

(f) β-D-ribofuranose (furanose ring with CH₂OH, OH groups)

(g) O₂NCH₂CH₂CH₂C(O)CH₃

(h) CH₃CH₂O₂C(CH₂)₄CO₂CH₂CH₃

(i) HO₂CCH₂OCH₂CH₃

(j) CH₂=CHCH₂CH=CH₂ (1,4-pentadiene)

(k) CH₃CH₂CH₂C(O)Cl

(l) 2-(ethoxycarbonyl)-1,3-indandione

(m)
CHO
H——OH
H——OH
H——OH
H——OH
CH₂OH

(n) CH₃O₂C(CH₃)₂CH₂OH

(o) HOCH₂CH₂CH₂OH

2. The anion derived from dihydroxyacetone attacks the carbonyl group of D-glyceraldehyde. Four diastereoisomeric ketohexoses are possible, in principle, since the stereochemistry at the third and fourth carbon could be either R or S.

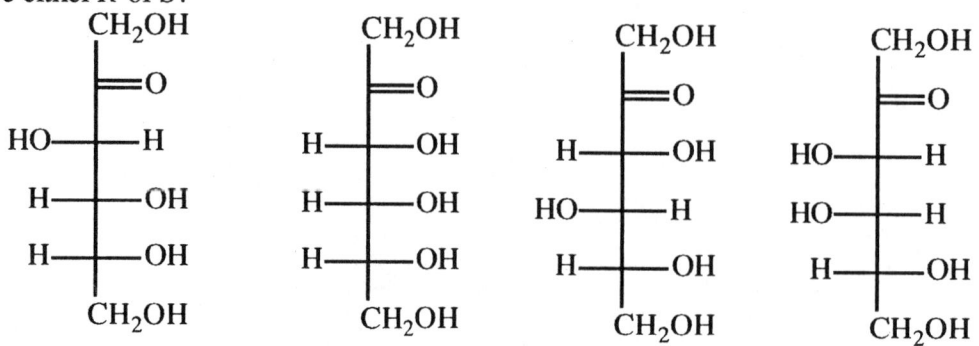

```
CH₂OH         CH₂OH         CH₂OH         CH₂OH
 =O            =O            =O            =O
HO—H          H—OH          H—OH          HO—H
H—OH          H—OH          HO—H          HO—H
H—OH          H—OH          H—OH          H—OH
CH₂OH         CH₂OH         CH₂OH         CH₂OH
```

We do not expect these diastereoisomers to form in equal amounts.

3. Compound A is PhCH₂C(O)CH₃ and compound B is PhCH(CN)C(O)CH₂Ph.

Formation of A: Methylmagnesium iodide acts as a nucleophile, attacking the nitrile group to form an intermediate which is no longer reactive with Grignard reagent and upon hydrolysis yields *A*.

Formation of *B*: Methylmagnesium iodide acts as a base pulling a proton from the α-carbon atom of phenylacetonitrile, which then attacks a second molecule of phenylacetonitrile in an aldol-type addition. The resultant intermediate is hydrolyzed to yield *B*.

4. (a) 1. D-glucose is treated with methanol in the presence of an acid catalyst; 2. treatment with dimethyl sulfate to form four methyl ether linkages

(b) 1. treatment of ethyl acetoacetate with sodium ethoxide to generate the anion; 2. addition of 1-bromo-3-methylbutane; 3. heat with aqueous acid to hydrolyze the ester and decarboxylate the resultant acid, forming the target material

(c) 1. treatment of acrylonitrile with dimethylamine (conjugate addition); 2. reduction of the nitrile linkage to a primary amine with lithium aluminum hydride; 3. amide formation by addition of acetyl chloride

(d) 1. treatment of benzaldehyde with acetone in the presence of sodium hydroxide to effect an aldol condensation and form an α,β-unsaturated ketone; 2. addition of the sodium salt of diethyl malonate to effect a Michael addition; 3. heat with aqueous acid to accomplish ester hydrolysis and decarboxylation; 4. ketone reduction using sodium borohydride; 5. treatment with acid to effect intramolecular ester formation

ANSWERS TO PRACTICE EXAMINATION SIX

1.

(a) CH₃-C(=O)-NH-CH(CH₂CH(CH₃)₂)-CO₂H

(b) PhCH₂CH(NH₂)-CN

(c) 2'-deoxyadenosine (adenine attached to deoxyribose with HO-CH₂ and OH shown)

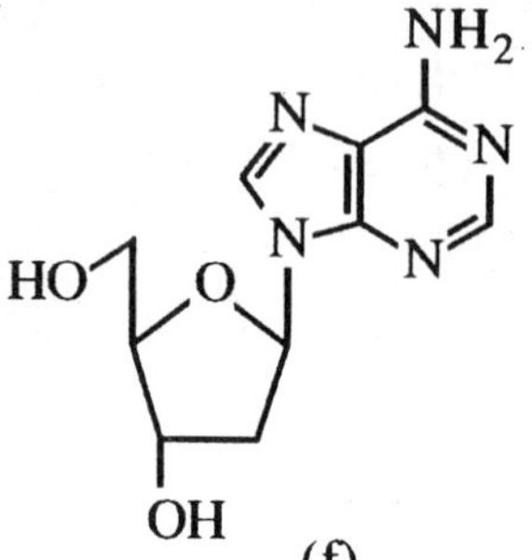

(d) (CH₃)₃COC(=O)NHCH(CH₃)CO₂H

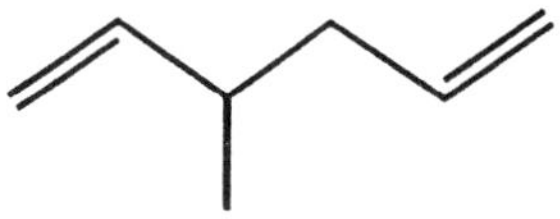

(e) [-O-C₆H₄-C(CH₃)₂-C₆H₄-OCH₂CH(OH)CH₂-]ₙ

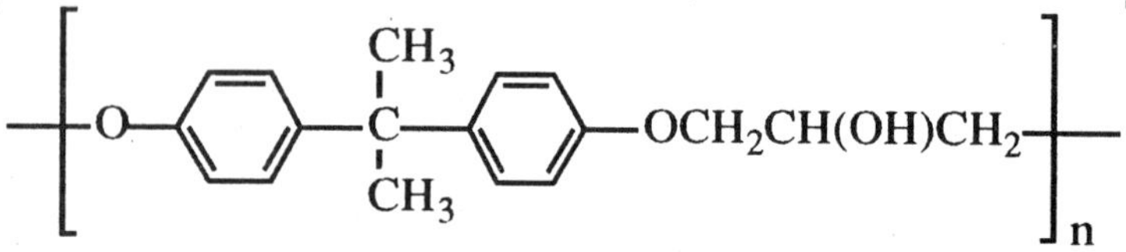

(f) (CH₃)₂C⁺CH₂C(CH₃)₃

(g) CH₂=CH-CH=CH-CH₂-CH₃ (hexa-1,3-diene skeletal)

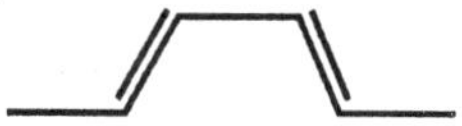

(h) CH₂=CH-CH₂-CH(CH₃)-CH=CH₂ (3-methyl-1,5-hexadiene)

(i) H₂C=C(CO₂CH₂CH₃)(CN)

(j) dinucleotide: thymidine 3'-phosphate linked to 5'-deoxyadenosine

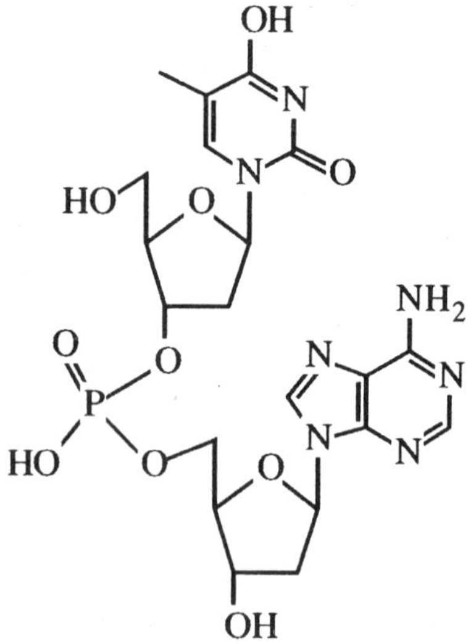

2. (a) Gly is the C-terminal amino acid.
    (b) Ser is the N-terminal amino acid.
    (c) Phe units must be the third and sixth amino acids of the peptide, counting from the N-terminus.
    (d) Compound A must be a tripeptide derived from the original N-terminus. The fourth and sixth amino acids (counting from the C-terminus) must be Ala.
    (e) Compound C must be derived from the C-terminus.
    (f) Compound B must contain an acidic amino acid (Glu is the only possibility) as it will exist as an anion at pH 7 and migrate toward the anode. Compound C must be a basic amino acid (Lys is the only possibility) as it will exist as a cation at pH 7 and migrate toward the cathode.
    (g) Ser-Ala-Phe-Ala-Glu-Phe-Ala-Lys-Gly

3.

(a)

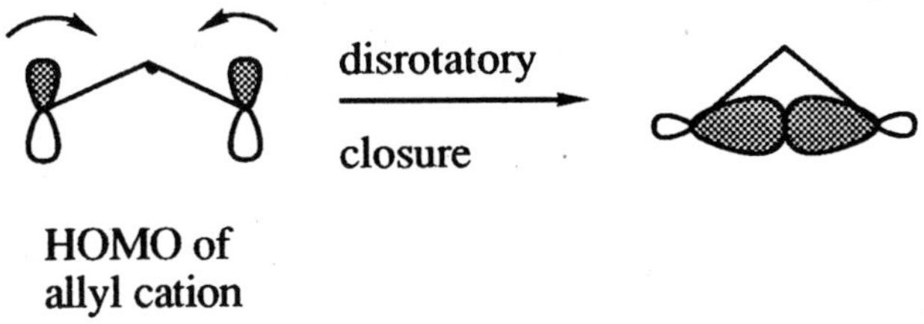

HOMO of allyl cation → disrotatory closure

The principle of microscopic reversibility allows us to predict that the reverse process is also disrotatory.

(b)

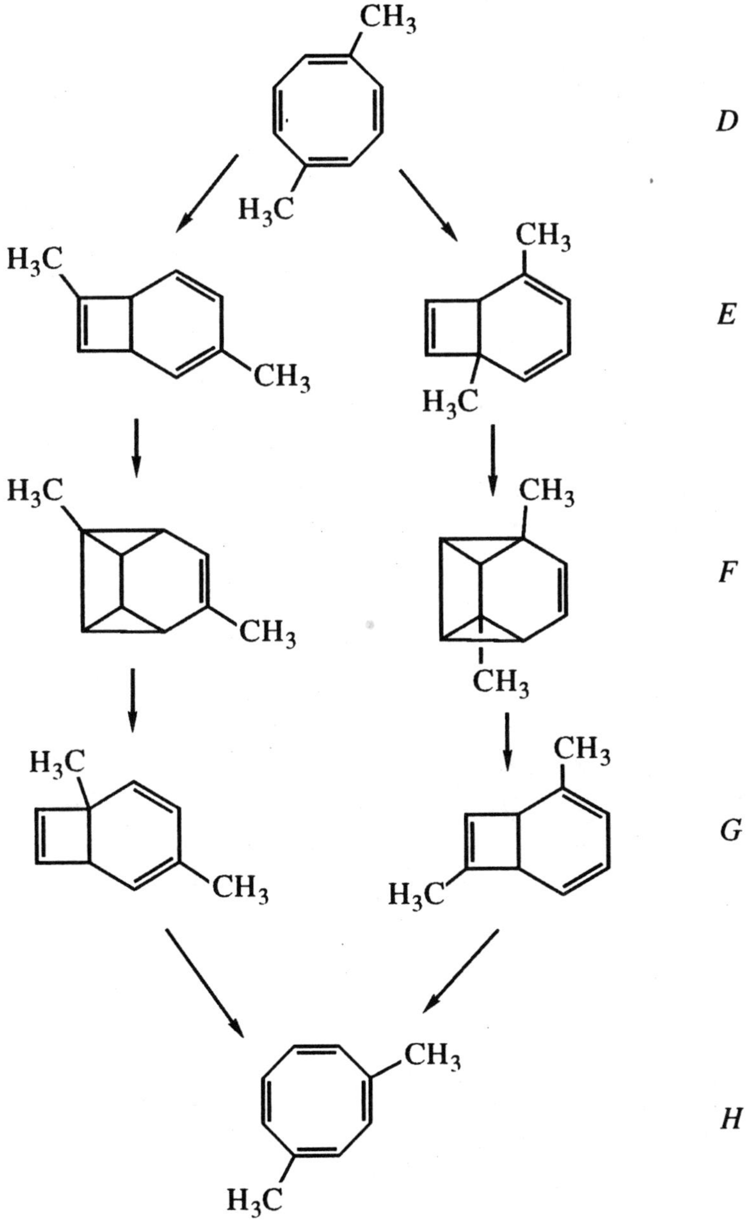

D, E, F, G, H

(c) K, 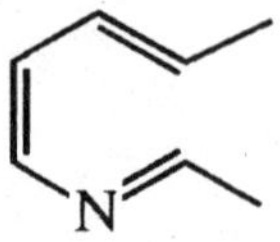 L, 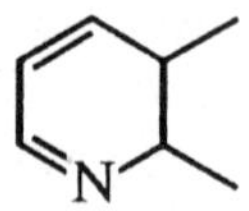 M (structures shown)

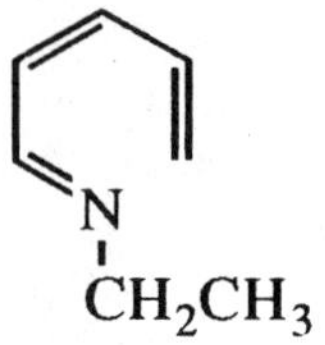

4. (a) 3-chloro-2-cyanopyridine
   (b) 2-acetylfuran
   (c) furan-CH=CH-C(=O)-CH₃
   (d) CH₃CCH₂CH₂CCH₃ (2,5-hexanedione, with two C=O)
   (e) 8-amino-7-methylquinoline

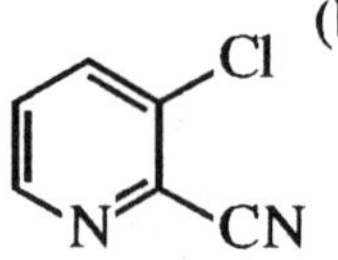

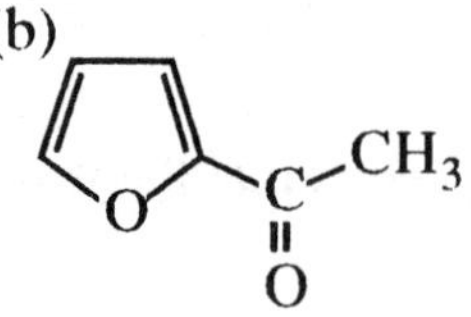

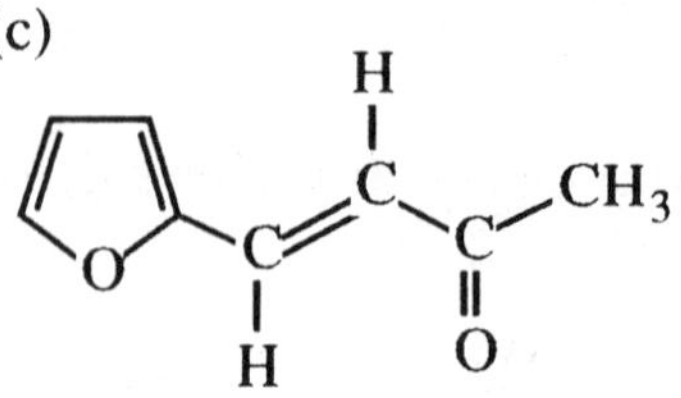

5. N: p-CH₃-C₆H₄-Br
   O: p-CH₃-C₆H₄-MgBr
   P: p-CH₃-C₆H₄-CH₂CH₂OH
   Q: p-CH₃-C₆H₄-CO₂H
   R: p-CH₃-C₆H₄-CH₂CH₂Br
   S: p-CH₃-C₆H₄-COCl
   T: p-CH₃-C₆H₄-CH₂CH₂NH₂
   U: p-CH₃-C₆H₄-CH₂CH₂NHC(=O)C₆H₅
   V: 7-methyl-1-(4-methylphenyl)-3,4-dihydroisoquinoline

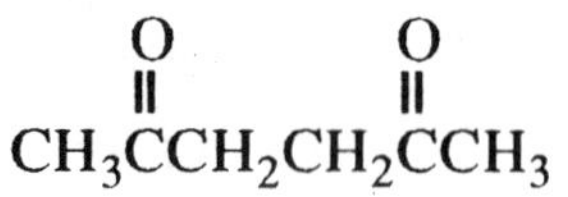

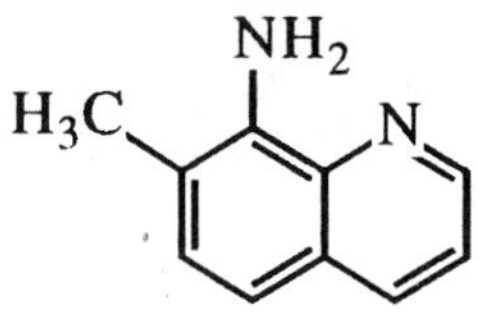

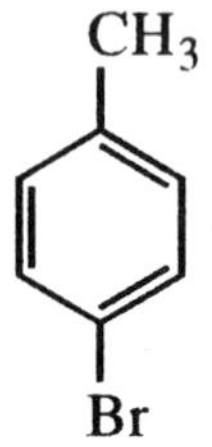

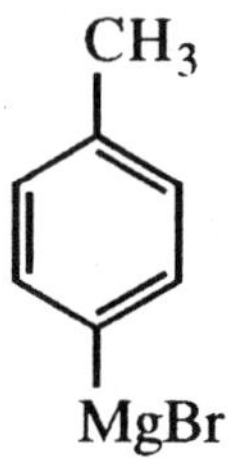

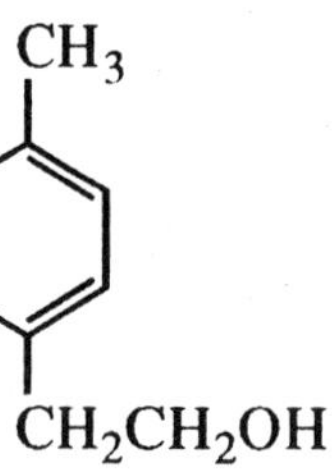

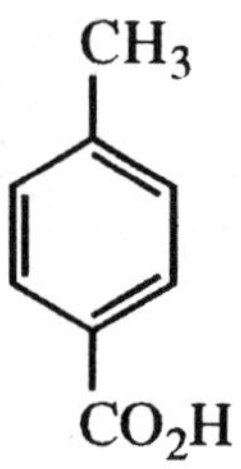

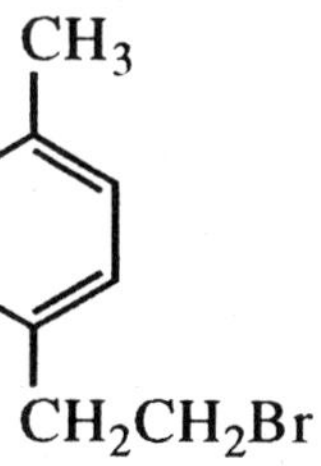

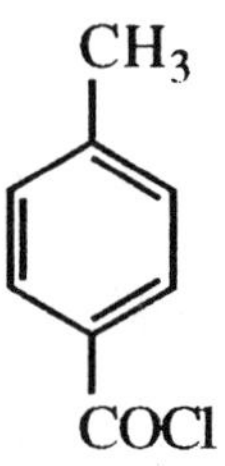

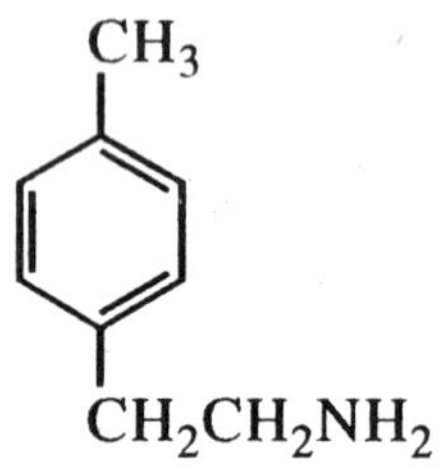

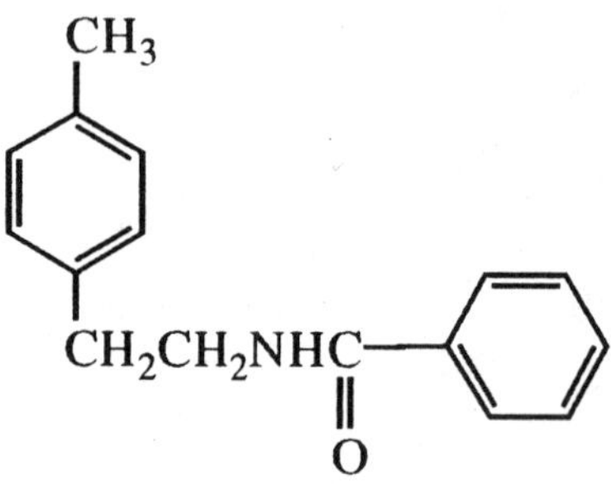

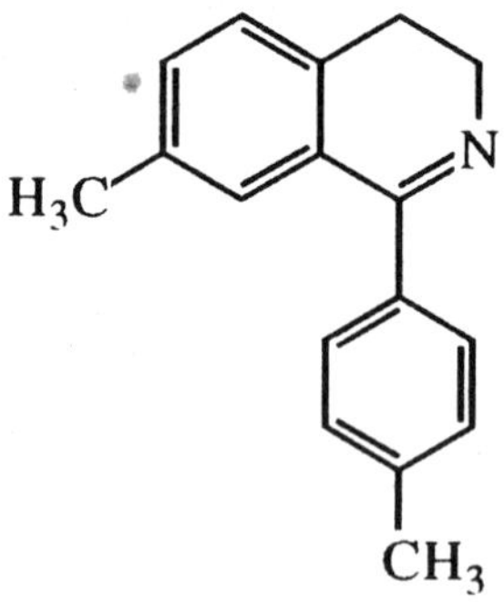